# FORMULA WEIGHTS

| Element | Formula Weight | Element | Formula Weight |
|---|---|---|---|
| AgBr | 187.78 | $HgCl_2$ | 271.50 |
| AgCl | 143.35 | HgO | 216.59 |
| AgCN | 133.89 | $Hg_2Cl_2$ | 472.08 |
| $Ag_2Cr_2O_7$ | 413.73 | KBr | 119.01 |
| $Ag_2CrO_4$ | 331.73 | KCl | 74.55 |
| AgI | 234.77 | $KHC_8H_4O_4$(KHP) | 204.23 |
| $AgNO_3$ | 169.87 | KI | 166.01 |
| $Ag_3PO_4$ | 418.58 | $KIO_3$ | 214.00 |
| AgSCN | 165.95 | $KIO_4$ | 230.00 |
| $Ag_2SO_4$ | 311.80 | $KMnO_4$ | 158.04 |
| $Al(OH)_3$ | 78.00 | KSCN | 97.18 |
| $Al_2O_3$ | 101.96 | $K_2Cr_2O_7$ | 294.19 |
| $AsCl_3$ | 181.27 | $MgCl_2$ | 95.22 |
| $As_2O_3$ | 197.84 | $Mg(NH_4)PO_4$ | 137.32 |
| $As_2O_5$ | 229.84 | $NH_2CO_2NH_2$ (urea) | 60.05 |
| $BaCl_2$ | 208.25 | $NH_2(C_4H_9O_3)$ (Tham) | 121.14 |
| $BaCl_2 \cdot 2H_2O$ | 244.27 | $NH_4Cl$ | 53.49 |
| $BaCO_3$ | 197.35 | $NH_4OH$ | 35.05 |
| $Ba(OH)_2$ | 171.35 | NaBr | 102.90 |
| $BaSO_4$ | 233.40 | $NaC_2H_3O_2$ (acetate) | 82.04 |
| $Bi_2O_3$ | 465.96 | $Na(C_6H_5CO_2)$ benzoate | 144.10 |
| $CaCO_3$ | 100.09 | NaCl | 58.44 |
| $CaC_2O_4$ | 128.10 | NaF | 41.99 |
| $Ca(OH)_2$ | 74.10 | $NaHCO_3$ | 84.01 |
| $CaSO_4$ | 136.14 | $NaH_2PO_4$ | 119.98 |
| $Ca_3(PO_4)_2$ | 310.18 | $Na_2HPO_4$ | 142.00 |
| CdS | 144.46 | NaOH | 40.00 |
| CuS | 95.60 | $Na_2CO_3$ | 105.99 |
| $CuSO_4 \cdot 5H_2O$ | 249.68 | $Na_2C_2O_4$ | 134.00 |
| $FeSO_4 \cdot 7H_2O$ | 182.02 | $Na_2H_2Y \cdot 2H_2O$ (EDTA) | 372.24 |
| $HC_2H_3O_2$ (acetic acid) | 60.05 | $NaNO_2$ | 69.00 |
| HCl | 36.46 | $Na_2O$ | 61.98 |
| $HNO_3$ | 63.02 | $Na_2S_2O_3$ | 158.10 |
| $HSO_3NH_2$ (sulfamic acid) | 97.09 | $Ni(C_4H_8N_2O_2)_2$ (DMG) | 288.94 |
| $H_2C_2O_4$ (oxalic acid) | 90.04 | $P_2O_5$ | 141.94 |
| $H_2C_2O_4 \cdot 2H_2O$ | 126.09 | $PbO_2$ | 239.19 |
| $H_2O_2$ | 34.02 | $PbI_2$ | 461.00 |
| $H_2SO_3$ | 82.08 | $SnCl_2$ | 189.61 |
| $H_2SO_4$ | 98.08 | TlCl | 239.82 |
| $H_3PO_4$ | 98.00 | $Zn_2P_2O_7$ | 304.68 |

# Analytical Chemistry

**SECOND EDITION**

# Analytical Chemistry

**SECOND EDITION**

**DONALD J. PIETRZYK**
**CLYDE W. FRANK**
University of Iowa

**ACADEMIC PRESS**

New York    San Francisco    London
A Subsidiary of Harcourt Brace Jovanovich, Publishers

*ACADEMIC PRESS, INC.*
111 Fifth Avenue, New York, New York 10003

*United Kingdom Edition published by*
*ACADEMIC PRESS, INC. (LONDON) LTD.*
24/28 Oval Road, London NW1

ISBN: 0–12–555160–6
Library of Congress Catalog Card Number: 77–80792

PRINTED IN THE UNITED STATES OF AMERICA

**To our parents**

# Contents

## CHAPTER THIRTY
**Experimental Techniques    614**

**EXPERIMENTS    638**

# Preface

This text is designed for introductory courses in analytical chemistry, especially those shorter courses servicing chemistry majors and life and health science majors.

Before undertaking this revision, we discussed the role of analytical chemistry with advisers from the various academic disciplines whose students traditionally require exposure to analytical courses. Two major objectives arose from these discussions: first, students should become acquainted with the fundamental principles encountered in modern chemical and instrumental methods of analysis, and second, students need to master the basic quantitative skills and techniques required to perform careful measurements in the laboratory. These are essentially the same goals one strives for in a course for chemistry majors.

Within the pages of this text, we attempt to present information that will accommodate both the chemistry major and students majoring in other disciplines without sacrificing the fundamentals of analytical chemistry. This was accomplished by using examples from related disciplines to illustrate the fundamental principles. In addition, most experiments involve procedures identical to those that would be used for real samples. We have not utilized real samples in the experiments, since students at this level need to know if they have obtained the correct answer. This is accomplished best by using well-characterized unknowns.

The text begins with a core of six chapters containing concepts basic to all of analytical chemistry. Five major areas are then emphasized. These include neutralization, potentiometry, spectroscopy, chromatography, and electrolysis methods. Each of these are subdivided into units. The first unit provides the fundamentals specific to that area, while the following units build by presenting additional concepts, applications, calculations, instrumentation, and chemical reactions specific to that particular area.

In neutralization, concepts relating to solutions of strong acids and bases, weak acids and bases, their salts, and buffer solutions are stressed. Final expressions are written when appropriate in a Henderson–Hasselbalch form. All approximations are clearly defined and discussed, while concepts necessary for exact calculations are introduced in chapters dealing with complexes in analytical chemistry.

Oxidation–reduction reactions and instrumental measurements of these reactions via cell potential determinations are discussed in several chapters. Accompanying this section is a broad development of the fundamental concepts and the utility of ion-selective electrodes.

In spectroscopy, atomic and molecular absorption, emission, and luminescence techniques are discussed. The first chapter of this unit provides the fundamental concepts, while subsequent chapters contain specific details for each technique including a discussion of the key components of the required instrumentation. Many practical examples are cited. For example, a comparison is made for the various ways of applying Beer's Law.

Separations are introduced in a series of six chapters with an emphasis on chromatography. The first of these chapters provides a brief survey of separation techniques and introduces the concepts common to all of chromatography. Subsequent chapters describe sheet and column methods, gas chromatography, ion-exchange chromatography, and solvent extraction. All the fundamentals are provided while stressing operational features for each technique.

Electrochemical techniques are introduced but not covered in depth. However, the principles discussed are sufficient to provide a background for understanding electrolysis, coulometry, and polarography.

Complementing these five areas are chapters devoted to the discussion of precipitation and complexes in analytical chemistry. Principles and applications and the relationship of these reactions to the other areas are stressed.

The remaining portion of the book is devoted to the laboratory. In one chapter, the basic laboratory operations are discussed with an emphasis on safety. This is followed by a series of experiments that are designed to reinforce the concepts developed in the chapters.

Many changes have been made in this second edition. Readily apparent ones are the omission of the chapter on radiochemistry and a restructuring of the chapter order. The first was done as a compromise due to the space needed for the many additions, while the second was based on user response. It should be emphasized that after the core of introductory chapters the presentation of the other areas is independent enough to allow instructors to assign chapters according to their own course outline.

Other changes in the second edition deal with improvements in the presentation, addition of new concepts or expansion of previously discussed concepts, and the inclusion of many new examples and problems. In the core of introductory chapters, Chapters 1 to 6, discussion of the problems of method development and of obtaining standards and suitable samples has been broadened. Particularly noteworthy is that the chapter on Statistical Handling of Analytical Data has been broadened and introduces the concept of propagation of error.

In the chapters dealing with precipitation, neutralization, and oxidation–reduction in analytical chemistry (Chapters 7 to 12), several new examples have been introduced which illustrate typical calculations. Precipitation titrations (Chapter 14) has been separated from the chapter on precipitation methods (Chapter 7) and follows the chapter on ion-selective electrodes. Thus, its development is tied with cell potential measurements. The ion-selective electrode chapter has been broadened to contain more applications, including a discussion of gas-permeable membranes.

Additions in the chapter on spectroscopy include more practical examples and discussion of the instrumental requirements for measurement of atomic emission and related techniques. In chromatography, emphasis has been increased on the discussion of the chromatographic peak and how it is used in qualitative and quantitative determinations. Also, chromatographic detection techniques have been broadened. More examples have been introduced in the chapters on electrochemistry.

Throughout the presentation we have illustrated principles by often referring to biological, clinical, pharmaceutical, environmental, and industrial problems. Not only do they illustrate practical analytical chemistry but they also illustrate the mathematical steps, approximations, etc. encountered in analytical chemistry. Many new practical problems are also included at the end of each chapter.

We would like to thank the many students and colleagues at The University of Iowa for their timely advice, help in proofreading, and constructive criticism. Particularly valuable to us have been the many comments from students and faculty who used the first edition.

**DONALD J. PIETRZYK**
**CLYDE W. FRANK**

# Chapter One
# Introduction to Analytical Chemistry

## TO THE STUDENT

For many of you, this course will be your first experience in analytical chemistry. The authors assume that you are here because you have an interest in chemistry, pharmacy, medical technology, medicine, the biological sciences, or a combination of these fields, and because you want to learn how analytical chemistry will be of use to you. For this reason, the presentation in this book is not designed just for the student majoring in chemistry. Instead, the emphasis is on the knowledge and experimental techniques of analytical chemistry that are most often encountered in the disciplines listed above.

The authors feel that there are four major areas of analytical chemistry that are of importance in their application to diverse scientific disciplines. These areas are spectroscopy, acid–base methods, potentiometry (ion selective electrodes, etc.), and chromatography. Therefore, much of this book is devoted to the background and techniques necessary to understand and carry out these analytical methods.

The goal of the authors in this book is to answer in a concise and logical way the following questions: What is analytical chemistry? What does an analytical chemist do? Where does analytical chemistry fit into the overall scheme of science?

## ANALYTICAL CHEMISTRY

Analytical chemistry deals with the solving of qualitative and quantitative problems. In qualitative analysis the goal is to determine *what* the constituents are in the sample while in quantitative analysis the goal is to determine *how*

**Table 1-1.  A Survey of Analytical Problems**

Problems in Analysis

1. Qualitative analysis for elements, ions, atomic groups, and functional groups in mixtures of substances
2. Qualitative analysis of a single chemical species or of each in mixtures of species
3. Quantitative analysis for elements, ions, atomic groups, and functional groups in mixtures of substances
4. Quantitative analysis of a single chemical species or mixtures of species throughout the sample or at its surface

*much* of each constituent is in the sample. Table 1-1 surveys in general terms the types of problems encountered in analytical chemistry.

Many paths are available to achieve the information sought, and many individual techniques may be incorporated into the route leading to a completed analysis. An important part of the analytical chemist's task is choosing the optimum pathway, a choice which is simplified only through the assimilation of knowledge and experience. Thus, in solving analytical problems an analytical chemist is often required to design or repair electronic systems, arrange optical systems, design instruments, interpret spectra and other instrumental data, perform classical analyses with simple chemicals and solutions, develop and evaluate new procedures or modify old ones, separate simple and complex mixtures, purify samples, and write computer programs. It is not likely that any one analytical chemist will be capable in all of these areas.

The nature of the chemical and physical systems which an analytical chemist encounters is a further complication. An analytical chemist must deal with inorganic and organic mixtures composed of metallurgical, biochemical, pharmaceutical, or medicinal compounds. For example, the dissolution and separation of a 12-component, high-temperature alloy, the separation of a mixture of polymeric materials which differ only in molecular weight, or possibly the separation of a mixture of all the amino acids are typical problems. The analytical chemist may be asked to analyze polluted air or the insecticides in fish, birds, plants, or animals; to measure the rate of a reaction; or to determine the number of electrons and intermediates involved in an electrochemical reaction.

## ANALYTICAL CHEMISTRY
## IN ALLIED AREAS

Historically, chemistry could be easily divided into five areas: analytical, biochemical, inorganic, organic, and physical. Today such divisions are very difficult to make because of the contributions of each to one another. Analytical chemistry, like other areas of chemistry and of all sciences, has gone through a period of rapid growth and change. At present, new areas such as chemical physics, biophysics, and molecular biology are rapidly developing. Many advances in these areas were made possible by analytical results.

The importance of analytical chemistry in related scientific areas can be illustrated by considering its impact on clinical analysis, in pharmaceutical research and quality control, and in environmental analysis.

**Clinical Analysis.**    In the past, the medical profession used clinical results qualitatively and many of their diagnoses were based on symptoms and/or X-ray examinations. This was true even though it was known that many physiological diseases were accompanied by chemical changes in the metabolic fluids. Eventually, sensitive chemical and instrumental tests were devised to detect both the abnormal and normal components of body fluids. Furthermore, the tests were improved to the point where it became possible to quantitatively determine these components. As accuracy improved and normal levels were established, it became clear that these results could be used for diagnostic purposes. A patient requesting a general physical examination or the diagnosis of a specific set of symptoms is now often required to have several quantitative tests on their body fluids, and in the future such tests will become more and more numerous. The test results will be available to the doctor at the first meeting with the patient and will play a major role in influencing the diagnosis.

Currently, over one billion laboratory tests are performed in clinical laboratories per year and the number is going to increase. The major portion of these tests deal with blood and urine samples and include the determination of glucose, urea nitrogen, protein nitrogen, sodium, potassium, calcium, $HCO_3^-/H_2CO_3$, uric acid, and pH.

Table 1–2 lists the components of the blood, all of which can be analyzed quantitatively, and gives the range for the amount of each normally present in the blood. Urine contains an even larger number of components, which have much broader normal ranges because of a greater dependency on such factors as diet, liquid intake, and urinary tract conditions.

Table 1–3 lists approximate ranges for electrolytes in urine. The accurate determination of urine electrolyte concentrations permits the diagnosis of the general disorders listed in Table 1–4.

Clinical analysis is not limited to such general determinations; it may also help to pinpoint specific metabolic diseases. For example, for 1 in 10,000 newborn babies the enzyme phenylalanine hydroxylase is absent. This causes a secondary metabolism to take place which, if not corrected at an early stage of life, leads to a brain which is not fully developed and severe mental retardation. This disease is known as phenylketonuria because one result of the

**Table 1–2.  Normal Range of Components of Blood and Cerebrospinal Fluid (CSF)**[a]

| Determination | Amount of whole blood (ml) | Normal range |
|---|---|---|
| Albumin | 6 | 3.8 to 5.0 g/100 ml |
| Amino acid nitrogen | 5 | 4 to 6 mg/100 ml |
| Ammonia | | 40 to 125 μg/100 ml |
| Amylase | 3 | Up to 150 units |
| Barbiturates | 10 | None |
| Bilirubin | 6 | Direct up to 0.4 mg/100 ml |
| | | Total up to 1.0 mg/100 ml |
| Bromide | 8 | None |
| Bromsulphalein (BSP) | 6 | Up to 5% in 45 min |
| Calcium | 3 | 4.5 to 5.5 mEq/liter |
| Carbon dioxide content | 5 | 25 to 32 mEq/liter |
| Carotene | 10 | 50 to 200 μg/100 ml |
| Cephalin–cholesterol flocculation | 3 | Up to 2+ in 48 hr |
| Chloride, serum | 3 | 100 to 108 mEq/liter |
| Chloride, CSF | 0.6 (CSF) | 120 to 130 mEq/liter |
| Cholesterol, total | 3 | 100 to 350 mg/100 ml |
| Cholesterol, esters | 6 | 70 to 75% of total |
| Colloidal gold (CSF) | 0.2 (CSF) | 0 |
| Creatinine | 10 | 0.3 to 1.0 mg/100 ml |
| Creatinine clearance | | 100 to 180 ml/min |
| Fatty acids | 10 | 200 to 400 mg/100 ml |
| Globulins, total | 5 | 2.5 to 3.5 g/100 ml |
|    alpha-1 | | 0.1 to 0.4 g/100 ml |
|    alpha-2 | | 0.3 to 0.7 g/100 ml |
|    beta | | 0.4 to 0.9 g/100 ml |
|    gamma | | 0.6 to 1.3 g/100 ml |
| Iron, serum | 10 | 50 to 180 μg/100 ml |
| Iron-binding capacity, total (TIBC) | 5 | 250 to 400 μg/100 ml |
| Lactic dehydrogenase (LDH) | 3 | Up to 180 units |
| Leucine aminopeptidase (LAP) | 3 | Up to 230 units |
| Lipase | 10 | Up to 1.5 units |
| Lipids, total | 10 | 350 to 800 mg/100 ml |
| Magnesium | 12 | 1.5 to 2.5 mEq/liter |
| Natural fat | | 0 to 150 mg/100 ml |
| Nonprotein nitrogen (NPN) | 4 | 25 to 40 mg/100 ml |
| Oxygen | | |
| pH (blood) | | 7.38 to 7.42 |
| Phosphatase, acid | 6 | Up to 4 Gutman units |
| Phosphatase, alkaline | 6 | Up to 4 Bodansky units |
| Phospholipids as phosphorus | 10 | 100 to 250 mg/100 ml |
| | | 4 to 10 mg/100 ml |
| Phosphorus, inorganic | 5 | 3 to 4.5 mg/100 ml |
| Potassium | 5 | 3.8 to 5.0 mEq/liter |

**Table 1–2.  Continued**

| Determination | Amount of whole blood (ml) | Normal range |
|---|---|---|
| Protein, total serum | 3 | 6.5 to 8 g/100 ml |
| Protein, CSF | 1.1 (CSF) | 20 to 45 mg/100 ml |
| Protein-bound iodine (PBI) | 10 | 3.5 to 8 $\mu$g/100 ml |
| Salicylate | 6 | None |
| Sodium | 5 | 138 to 146 mEq/liter |
| Sugar, blood | 3 | 65 to 90 mg/100 ml |
| Sugar, CSF | 1.1 (CSF) | 50 to 70 mg/100 ml |
| Sulfa | 3 | None |
| Thymol turbidity | 3 | Up to 2 units |
| Transaminase (SGO) | 3 | Up to 40 units |
| Transaminase (SGP) | 3 | Up to 30 units |
| Triglycerides | 10 | 1.5 to 5.5 mEq/liter |
| Urea nitrogen (BUN) | 3 | Up to 20 mg/100 ml |
| Uric acid | 6 | 3 to 6 mg/100 ml |
| Vitamin A | 12 | 20 to 60 $\mu$g/100 ml |
| Vitamin C (ascorbic acid) | 10 | 0.5 to 1.5 mg/100 ml |

[a] Taken from J. S. Annino, "Clinical Chemistry," 3rd ed.  Little, Brown and Co., Boston, 1964.

faulty metabolism is an excess of phenylpyruvate in the urine. Therefore, detection of phenylpyruvate in urine is an important clinical test, and it is done routinely on newborn babies by treating a urine sample with an $FeCl_3$ reagent. A green-blue coloration confirms the presence of the phenylpyruvate. For a quantitative measurement a more careful test must be made. Fortunately, once the disease is detected, it is readily taken care of through a controlled diet.

**Table 1–3.  Approximate Normal Values for Electrolytes in Urine[a]**

| Substance | Concentration in mEq/24 hours |
|---|---|
| Bicarbonate | None |
| Chloride | 75 to 200 |
| Sodium | 75 to 200 |
| Potassium | 40 to 80 |
| Calcium | 5 to 15 |

[a] Taken from J. S. Annino, "Clinical Chemistry," 3rd ed. Little, Brown and Co., Boston, 1964.

**Table 1-4. Disorders Indicated by Electrolyte Patterns in Urine**[a,b]

| $CO_2$ | Cl | Na | K | pH | General disorder |
|---|---|---|---|---|---|
| → | ↑ | ↑ | → | → | Dehydration |
| ↓ | ↓ | ↓ | → | → | Water intoxication |
| ↓ | ↓ | → | → ↑↓ | ↓ | Metabolic acidosis |
| ↑ | ↓ | → | ↓ | ↑ | Metabolic alkalosis |
| ↑ | ↓ | → | → ↑ | ↓ | Respiratory acidosis |
| ↓ | → | → | → ↓ | ↑ | Respiratory alkalosis |

[a] Taken from J. S. Annino, "Clinical Chemistry," 3rd ed. Little, Brown and Co., Boston, 1964.
[b] ↑ = increased concentration.
↓ = decreased concentration.
→ = unchanged.

**Pharmaceutical Problems.**    In the pharmaceutical industry the quality of the manufactured drugs in tablet, solution, and emulsion form must be carefully controlled. Slight changes in composition or in the purity of the drug itself can affect the therapeutic value. Hence, the student in pharmacy must be aware of and capable of developing new methods for this purpose in order to be in a viable position in the pharmaceutical industry and government testing laboratories.

In other pharmaceutical studies, it is necessary to establish the properties and therapeutic value of a drug before the drug is approved and made available to the public. Establishment of the permissible level of dosage of a drug requires the determination of its decomposition products, its toxicity, and its metabolite products at various stages. Accurate and sensitive tests must be developed since the levels that are being determined are often as little as 1 nanogram ($1 \times 10^{-9}$ gram) in a complicated biological sample.

Presently, the laws of the United States government on how a new drug must be tested are very strict and a considerable amount of time and money must be expended to establish the quality of the drug. Figure 1–1 shows the stages at which analysis must be done. If this very careful evaluation were not to be made, public disasters such as that of the drug thalidomide (approved in foreign countries but not in the United States) would be more common.

(Thalidomide was prescribed in Europe for thousands of pregnant women before it was discovered that it caused the birth of seriously deformed babies.)

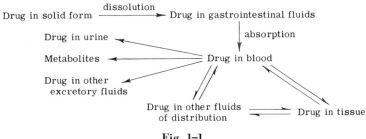

Fig. 1–1.

**Environmental Problems.** Environmental science is concerned with the chemical, physical, and biological changes in the environment through contamination or modification, with the physical nature and biological behavior of air, water, soil, food, and waste as they are affected by man, and with the application of science and technology toward the control and improvement of environmental quality. Analytical methods provide the basis for these studies.

Consider the problem of common air contaminants. Analytical methods have shown that approximately 15% of the total settleable dust and about 25% of the suspended particulate matter in air is pollution of natural origin. (These percent natural contaminants will vary according to the makeup of the region being sampled.) However, analyses have also shown the presence of many other contaminants.

Probably more time, effort, and money have been spent on the study of fuel combustion products as air contaminants than on any other source in industrial technology. But the automobile added only one new class of particulates. In fact, the development of analytical methods of separation, identification, and subsequent determination have provided information about the presence of such particle contaminants in air as lime, limestone, and cement dust from kilning operations, coke dust and polycyclic aromatics from coking operations, iron oxides from ore smelting procedures, and fluorides from metallurgical processes. Asphalts, solvents, synthetic monomers, butyl rubbers, and carbon black have been shown to be present. Other contaminants are fly ash from coal–fired electrical power plants and wind-blown slag and insulation from industrial operations. Added to this complex list of contaminants are gaseous air contaminants as well as particulates that are present due to local pollution.

Table 1–5 lists some of the organic pollutants that appear in typical industrial waste waters. Clearly, water is as complex a system as air when it comes to environmental monitoring. As in the study of air, analytical chemistry has played a major role in the environmental science of water.

Table 1–5.   Organic Components of Industrial Waste Effluents[a]

| Plant | Composition of Waste Effluent |
| --- | --- |
| Mines, ore, treatment plants | Humus, coal sludge, flotation agents |
| Foundries | Cyanides, phenol, tar components, coal sludge |
| Iron and steel processing | Wetting agents and lubricants, cyanides, inhibitors, hydrocarbons, solvent residues |
| Coal production, coking plants | Humus, coal particles, cyanides, rodanines, phenols, hydrocarbons, pyridine bases |
| Wood charcoal production | Fatty acids, alcohols, particularly methanol, phenols |
| Petroleum industry | Oil emulsions, naphthenic acids, phenols, sulfonates |
| Sulfite pulp | Methanol, cymol, furfurol, soluble carbohydrates, lignosulfonic acids |
| Rayon and cellulose | Xanthogenates, alkali hemicelluloses |
| Paper manufacture | Resinic acids, polysaccharides, mucins, cellulose fibers |
| Textile industry | Scouring and wetting agents, leveling agents, sizers, desizing agents, fatty acids, finishes, Trilon (nitrolotriacetic acid), dyes |
| Laundries | Detergents, carboxymethyl cellulose, enzymes, optical brighteners, colorants, soil, protein, blood, cocoa, coffee |
| Leather and tanning industry | Protein degradation products, soaps, tanning agents, emulsified lime soap, hair |
| Sugar refineries | Sugar, plant acids, betaine, pectin |
| Starch plants | Water-soluble components (protein compounds, pectins, soluble carbohydrates) |
| Dairies | Milk components (protein, lactose, lactic acid, fat emulsions), washing and rinsing agents |
| Grease and soap factories | Glycerine, fatty acids, fat emulsions |
| Canning factories | Soluble plant components |
| Beer breweries | Water-soluble plant components, beer residues, rinsing agents |
| Fermentation industry | Fatty and amino acids, alcohols, unfermented carbohydrates |
| Slaughter houses | Blood, water-soluble and emulsified meat components |

[a] W. Leithe, "The Analysis of Organic Pollutants in Water and Waste Water." Ann Arbor Science Publishers, Ann Arbor, Michigan, 1973.


# THE ANALYTICAL CHEMIST AND THE ANALYST

The concept of measurement is basic to analytical chemistry. A simple measurement may involve properties such as mass, current, voltage, volume, or time. The measurement of such properties as absorption or emission of energy, optical rotation, overvoltage, refractive index, equilibrium constant, rate constant, activation energy, or heat of reaction is more complex. Whether simple or complex, the reliability, utility, accuracy, interpretation, and specificity of these measurements are the responsibility of the analytical chemist. The analytical chemist is not only concerned with carrying out the analysis, but also with the how, why, and where of using such measurements for analysis, separation, or for the elucidation of the fundamental chemistry involved in a problem.

The responsibility of the analyst is to make the measurement, which is usually based on a tried and tested procedure, in a routine manner. Consequently, the analyst is primarily concerned with the final answer.

The development of analytical chemistry in this text provides both the experience of being an analytical chemist and the experience of being an analyst. The fundamental details presented in each chapter are the basis for analytical chemistry, while the experiments illustrate the practice of analytical chemistry.

Not every analytical principle, detail, application, and experiment can be covered in this book. For this reason it is important to relate the subject matter to your own scientific interests. This is important particularly with respect to applications.

# Chapter Two
# Development of an Analytical Method

## THE ANALYTICAL PROBLEM

The first step in solving analytical problems is the identification of the problem. Only by clearly identifying the problem can one devise a logical path toward its solution. Many questions should be asked. For example, what kind of sample is it, inorganic or organic? What information is being sought? What accuracy and precision is demanded? Is the sample large or small? Are the components of interest major or minor constituents? What are the interferences? Are standards available? How many samples have to be analyzed? What is the cost of the analysis? Are suitable equipment and personnel available?

It is an important task for the practicing analytical chemist to choose an analytical method which provides the best solution to the problem. In some cases freedom of choice is limited. For example, water analyses must be done by procedures approved by the APHA (American Public Health Association) or ASTM (American Society for Testing Materials). Methods for quantitative testing and quality control of pharmaceuticals are controlled by the FDA (Federal Drug Administration). Environmental analytical procedures are approved by the EPA (Environmental Protection Agency). Other governmental, state, and science agencies also provide guidance as well as determine the legal procedures. In some cases licensing is required before the analyst can carry out the procedures. An unfortunate complication is that many countries will have their own regulatory agencies and because of this, the approved procedures and the expectations of these procedures often differ from country to country.

## CHOOSING AN ANALYTICAL METHOD

Once the problem is defined the following important factors are considered in choosing the analytical method. These are: concentration range, required accuracy and sensitivity, selectively, time requirements, and cost of analysis.

**Concentration Range.**    Table 2–1 classifies samples according to the amount of the substance in the sample that is to be determined. Thus, for the macro-samples the substance to be determined is present as the major constituent on a percent basis. In contrast, for the micro- and smaller samples the substance is at a trace level at less than percent concentrations.

Certain analytical methods are best suited for macro-amounts, while others are useful for trace amounts only. Consequently, analytical methods can be divided according to their suitability to the size of the sample. In general, chemical methods tend to be best for the determination of macro-amounts and instrumental methods for trace amounts. The ability to match the method to the optimum sample size is usually gained through experience and aware-ness of the different methods.

**Sensitivity and Accuracy.**    Sensitivity, as it applies to an analytical method, corresponds to the minimum concentration or lowest concentration of a substance that is detectable with a specified reliability. It is often expressed numerically as a detection limit or sensitivity. Different analytical methods will provide different sensitivities and the one chosen will depend on the sensitivity that is required to solve a particular problem. The smaller the sample, for example a trace level, the more sensitive the method must be.

Accuracy refers to the correctness of the result achieved by the analytical method. As the sensitivity varies among the methods, so does the accuracy. In practice, the analytical chemist will choose methods that provide the required level of accuracy. In order to achieve the necessary accuracy, higher costs and excessive consumption of time, equipment, and personnel may be required.

**Table 2–1.   Classification of Analytical Methods According to the Amount of Substance to Be Determined in the Sample**

| Name | Approximate size |
|---|---|
| Macro methods | 100 mg |
| Semimicro methods | 10 mg |
| Micro methods | 1 mg |
| Ultramicro methods | 0.001 mg (1 $\mu$g) |
| Submicrogram methods | 0.010 $\mu$g |

**Selectivity.**     Selectivity is an indication of the preference that a particular method shows for one substance over another. The more complex the sample, the more selective the chosen analytical method must be if one component is to be determined in the presence of the rest. Another term that is often used is specificity. Where selectively shows a preference, specificity in an analytical method implies a specific response. In general, analytical methods are not completely specific toward an individual component.

**Time and Cost.**     Time and cost often go hand in hand and usually are a reflection of the equipment, personnel, and space required to complete a determination. The number of samples will also influence the choice of the method. If there are many similar samples, such as in quality control, provision for automation becomes feasible. Often a shortening of the analysis time is at the expense of accuracy. However, in emergency situations this sacrifice may be warranted.

## TYPES OF ANALYTICAL METHODS

Analytical methods can be grouped in several ways. For example, the classification can be based on the type or physical state of the sample, on the purpose of the analysis, on the size of the sample as in Table 2–1, or by the type of analytical method. Only the latter will be considered further.

The different analytical methods can be divided into chemical or instrumental methods. Chemical methods in this book are those that depend on chemical operations in combination with the manipulation of simple glassware and the simplest of instruments. In general, the measurement of mass or volume is the essential part in the method. An instrumental method encompasses the use of more complicated instrumentation based on electronic, optical, or thermal principles. In these cases, energy is measured and related to the composition of the sample.

Upon close examination of the possible solutions to a problem, it is often found that a combination of chemical and instrumental techniques is best employed. It is perhaps reasonable to state that in recent years many of the advances in analytical measurements have been in the general area of instrumentation.

Both instrumental and chemical methods offer advantages.

*Advantages of instrumental methods:*

1. The determination is fast (down to 1/100 second response).
2. Small samples can be used.
3. Complex samples can be handled.
4. High sensitivity is obtained.
5. Reliable measurements are obtained.

*Advantages of chemical methods*:

1. The procedures are inherently simple.
2. The procedures are accurate.
3. Generally, the methods are based on absolute measurements.
4. Specialized training is often not required.
5. The equipment needed is not expensive.

Although recent trends in research have been in instrumentation, it would be dangerous to conclude that instrumentation has replaced chemical methods. In practice, it is soon realized that preliminary chemical steps are often an intregal part of an instrumental method. Such steps could include sampling, dissolution, change in oxidation state, removal of excess reagent, adjustment of pH, addition of complexing agents, precipitation, concentration, and the removal of interferences. Perhaps the most complicated or difficult step is the last one, removal of interferences. Usually, this is accomplished by employing a separation technique.

Both chemical and instrumental methods suffer from limitations. Many of these will become more obvious as the two are discussed in more detail.

*Limitations of chemical methods*:

1. There is a lack of specificity.
2. Procedures tend to be time consuming.
3. Accuracy falls off with decreasing amounts.
4. There is a lack of versatility.
5. Chemical environment is critical.

*Limitations of instrumental methods*:

1. An initial (or continuous) calibration is needed.
2. The sensitivity and accuracy depend on the reference instrumental or wet chemical method used for calibration.
3. Final accuracy is often in the region of $\pm 5\%$.
4. Initial cost and upkeep of complex equipment is large.
5. Concentration range is limited.
6. Sizable space is usually required.
7. Specialized training is needed.

The most important criterion for any analysis or measurement is to choose the method or procedure, regardless of whether it is instrumental or chemical, that best solves the problem. It is essential to develop an intuitive feeling for selecting the correct approach. Therefore chemical and instrumental methods and procedures are introduced in this text in an order which leads in this direction. A critical comparison of the chemical method with the instrumental method from the standpoint of solving the same problem is very helpful. In addition, the student should attempt to see how the two can complement each other and, thus, produce a more powerful means of solving chemical problems.

## METHODS OF ANALYSIS

*Quantitative analysis* is based upon the measurement of a property which is related, directly or indirectly, to the amount of the desired constituent (substance being determined) present in the sample. The ideal situation would be the case in which no constituent other than the desired one would contribute to the quantity being measured. Unfortunately, such selectivity is seldom encountered. Table 2–2 summarizes the basic steps in an analytical procedure.

There are five basic types of techniques which are important to analytical chemistry*: (1) gravimetric; (2) volumetric; (3) optical; (4) electrical; and (5) separation. In general, the first two are chemical methods (mass and/or volume measurements) while the remaining ones are instrumental methods (energy–sample composition relationship). Often, two or more of these basic techniques are encountered in combination.

**Table 2-2.   Steps in Analysis**

1. Obtaining statistically meaningful sample
2. Preparation of the sample
3. Analytical Procedure
   A. Methods
      a. Chemical
      b. Physical with or without changes in the substance
   B. Conditions
      a. Determined by the analytical problem
      b. Determined by the substance being investigated
   C. Requirements
      a. Fast, accurate, economical
      b. Adaptable to automation
4. Critical evaluation of the result

**Gravimetry.**    Gravimetry is based on the ability to convert the desired constituent into a pure weighable form. This can be done by precipitation, electrodeposition, and volatilization.

---

* A sixth and seventh technique can be designated under the heading of resonance and thermal methods. Resonance methods are not included in this book while thermal methods are only briefly introduced in Chapter 7.

The operations in obtaining a pure sample by *precipitation* are (1) precipitation of the desired constituent, (2) filtration, (3) drying, and (4) weighing of the precipitate. An example is the quantitative precipitation of chloride ion by silver ion or vice versa.

$$Ag^+ + Cl^- \rightarrow AgCl_{(s)} \tag{2-1}$$

After allowing the reaction to take place for the appropriate period of time, the rest of the operations are completed. From the final weight of AgCl found, the amount of chloride ion (or silver ion) originally present in the sample can be calculated.

In *electrodeposition* the desired constituent is isolated at an electrode by passage of a current. If the weight of the electrode is taken before and after the plating process the difference in weight corresponds to the amount of the desired constituent.

*Volatilization* is similar to the electrodeposition technique in that a weight difference is recorded. In this case, the sample is decomposed by a known stoichiometric reaction in which one of the products is volatilized. The difference in weight, before and after volatilization, indicates the amount of vaporized constituent. In general, the weight loss is due to decomposition of a compound

$$CaCO_{3(s)} \xrightarrow{\Delta} CaO_{(s)} + CO_{2(g)} \tag{2-2}$$

or to the loss of volatiles associated with the sample.

$$KCl \cdot xH_2O_{(s)} \xrightarrow{\Delta} KCl_{(s)} + xH_2O_{(g)} \tag{2-3}$$

**Volumetric Methods.**    Volumetric methods are comprised of techniques in which the volume of a gas is measured or a volume of a titrant is measured. The latter technique is known as a *titration*.

In a *titration* one of the reactants is prepared as a solution of known concentration and is called the *titrant*. The titrant is then added to a solution of the sample from a *buret*. A stoichiometric reaction between the titrant and the sample is required and a technique must be available to detect the point at which this reaction is complete. By using a titrant of known concentration (a standard solution) and determining with the buret the volume of titrant needed for the complete reaction, the amount of the reactant in the sample can be calculated.

There are four types of volumetric titrations, each of which is based on a certain type of reaction: (1) acid–base titration; (2) oxidation–reduction titration; (3) precipitation titration; (4) chelometric titration.

An *acid–base titration* involves a neutralization reaction. Thus, acids are determined by titration with an appropriate standard base solution and bases by titration with an appropriate standard acid solution. In a typical example either HCl or NaOH is the titrant and the other is the sample.

$$H_3O^+ + Cl^- + Na^+ + OH^- \rightarrow Na^+ + Cl^- + 2H_2O \qquad (2\text{-}4)$$

It is possible to perform an acid–base titration in a solvent other than water and, frequently, the choice of the solvent can determine the success or failure of the analysis.

An *oxidation–reduction titration* involves a change in oxidation state for both the substance being determined and the titrant. An example of this is the titration of an iron(II) solution with a standard dichromate ion solution.

$$6Fe^{2+} + Cr_2O_7^{2-} + 14H^+ \rightarrow 6Fe^{3+} + 2Cr^{3+} + 7H_2O \qquad (2\text{-}5)$$

It is vital that the sample and titrant are at the correct oxidation state in stoichiometric quantities.

A *precipitation titration* is similar to the gravimetric method in that a precipitate is formed. The difference is that, rather than using an excess of precipitating agent, a stoichiometric amount of precipitating agent (the titrant) is added and the volume needed for the reaction is measured. Since the precipitating agent solution is a standard solution, the amount of the desired constituent can be calculated. In general, the precipitation method applies to only a small number of species and suffers from a lack of useful and easy ways of detecting the stoichiometric point of the titration.

In a *complexometric titration* the titrant is a complexing agent and the reaction that takes place results in the formation of a complex. An example of this is the titration of Cu(II) with ethylenediaminetetraacetic acid disodium salt (EDTA).

$$
\text{Cu}^{2+} +
\begin{array}{c}
\text{HO}_2\text{CCH}_2 \qquad\qquad \text{CH}_2\text{CO}_2\text{H} \\
\diagdown \qquad\qquad\qquad \diagup \\
\text{N}-\text{CH}_2-\text{CH}_2-\text{N} \\
\diagup \qquad\qquad\qquad \diagdown \\
\overset{+}{\text{Na}}\overset{-}{\text{O}}_2\text{CCH}_2 \qquad\qquad \text{CH}_2\text{CO}_2^-\text{Na}^+
\end{array}
\rightarrow \text{Cu}-\text{EDTA} + 2\text{Na}^+ + 2\text{H}^+
\qquad (2\text{-}6)
$$

Not all complex-formers can be used as titrants. In fact, most do not meet the requirements needed to be a successful titrant. This type of titration has been exploited primarily in the last 15 to 20 years.

For all volumetric techniques a method must be available for determining the stoichiometric point of the titration. This may be done by chemical action (color indicators) or by instrumental action. Regardless of which is employed, the detection system chosen should provide some visible change which coincides with the stoichiometric point of the reaction. The experimentally determined position is called the *end point*, while the actual *stoichiometric point* of the titration is the *equivalence point*. The difference between the two is the *titration error*.

**Optical Methods.**     Optical methods are based on how the sample acts toward electromagnetic radiation. The absorption or emission of radiant energy, the bending of radiant energy, the scattering of radiant energy, and the delayed emission of radiant energy are typical optical properties which can be correlated to concentration. Construction of the instruments for these measurements generally involves the use of lenses, mirrors, prisms, and gratings. There are some instruments which are included in this general classification that do not have optical parts. The more important techniques in this group are mass spectrometry, nuclear magnetic resonance, and electron spin resonance.

**Electrical Methods.**     Electrical methods involve electronic instruments that are used to measure or produce electrical phenomena. Current flow as a function of time, potential produced or required, ability to pass a current, and resistance are typical properties which are related to the reaction taking place or are causing a reaction to take place. The fundamental measurements then are resistance, voltage, time, and current.

**Separations.**     Often it is necessary to simplify the sample by removing the interferences prior to the final measurement. The techniques for doing this come under the general heading of separations. However, the idea of separation, which can be based on chemical or physical phenomena, should not be associated only with interference removal. Separating the components of a mixture may also be of qualitative or quantitative importance, useful for purification, or needed for concentrating one or all of the components. Many industrial processes for isolating metals, organic compounds, and other materials are based on a separation scheme. In essence, the concept of separation makes other analytical techniques more powerful because increased selectivity through separation becomes possible.

# LITERATURE IN ANALYTICAL CHEMISTRY

Books and journals dealing with all fundamental and practical facets of analytical chemistry are available in science libraries and through numerous scientific organizations. Some books are directed to a specific chemical or instrumental topic, while others are more general in that they contain many aspects of chemical or instrumental methods. Some books catalog analytical methods according to the element or class of substance being determined. In general, the scientific journals and periodicals devote most of their space to original research articles which outline the fundamentals and applications of newly developed methods and techniques or are critical evaluations of older methods and techniques. The remaining sections of journals are comprised of review articles and essays on scientific problems.

The literature in analytical chemistry, as in any other scientific area, is its life blood. The books provide the newcomer with the required background while the journals provide a vehicle for critical discussion about the old and the new. Several of the more general books and journals in analytical chemistry are listed below.

## Books

I. M. Kolthoff, P. J. Elving, and E. B. Sandell, Eds. (1959). "Treatise on Analytical Chemistry." Wiley (Interscience), New York. This series of books is divided into volumes and parts, and surveys analytical chemistry by topic and substance to be determined.

C. L. Wilson and D. W. Wilson. (1959). "Comprehensive Analytical Chemistry." Elsevier, New York. This is a series of books surveying analytical methods and techniques.

L. Meites, Ed. (1963). "Handbook of Analytical Chemistry." McGraw-Hill, New York. This book stresses the listing of physical and chemical constants and associated chemical and instrumental methods of analysis for the elements.

Several other series of books are available. These include numerous ones published by the American Society for Testing Materials and series titled *Advances in Analytical Chemistry* and *Methods of Biochemical Analysis*. "The National Formulary" and the "United States Pharmacopeia" list standard methods of analysis that are used routinely in pharmaceutical disciplines.

## Journals and Periodicals

*Published by the American Chemical Society*

*Chemical Abstracts.* This publication provides a short summary of scientific articles. American and foreign journals are abstracted. All chemistry and allied scientific areas are covered.

*Analytical Chemistry.* Original research articles in all facets of analytical chemistry appear in this journal.

*Environmental Science and Technology.* The application of analytical methods to the environment are included in this journal.

*From Other Societies and Private Publishers*

*Journal of the Association of Official Analytical Chemists.* This is an applied journal which is devoted to methodology for the analysis of foods, drugs, cosmetics, colors, pesticides, feeds, plants, fertilizers, beverages, additives, hazardous substances, and flavors

*Applied Spectroscopy.* This journal, which is characteristic of several, deals specifically with absorption and emission techniques.

*Journal of Chromatographic Science.* This journal, which is characteristic of several, deals specifically with the chromatographic techniques.

*Journal of the Electrochemical Society.* This journal, which is characteristic of several, deals specifically with the area of electrochemistry.

*Analytical Chimica Acta, Analyst,* and *Talanta* are other general analytical journals published in the English language.

A variety of general and specific analytical journals are published by foreign chemistry societies in their native languages.

# Chapter
# Three
# Stoichiometry

## INTRODUCTION

All analytical procedures are based on either stoichiometric or nonstoichiometric methods.

In a stoichiometric analytical procedure the constituent whose amount is being measured undergoes a reaction with another substance or is made to decompose in accordance with a well-defined equation which can be written in terms of reactants (R) and products (P).

$$R_A + R_B \rightarrow P_C + P_D \qquad (3\text{-}1)$$

By measuring the amount of any one of the products ($P_C$ or $P_D$) or of the reagent used ($R_A$ is the desired constituent and $R_B$ is the reacting reagent) the amount of the desired constituent can be calculated by applying the laws of definite and combining proportions.

Nonstoichiometric methods are just the opposite, in that exact, well-defined reactions cannot always be written. In most cases nonstoichiometric methods are based on a measurement of a physical property that changes in proportion to the concentration of the desired constituent. Since many of these physical properties are easily measured, often with great accuracy, it only becomes necessary to calibrate the procedure. The calibration empirically defines the relationship between the concentration of the desired constituent and the magnitude of the physical property for a given set of conditions.

In general, wet chemical methods, such as gravimetric and volumetric methods, and certain types of separation techniques are stoichiometric, while most instrumental methods, including optical and electrical techniques, are nonstoichiometric. Regardless of the nonstoichiometric relationship, instrumental methods after calibration very frequently offer speed, selectivity, sensitivity, and accuracy.

**Table 3–1.  Stoichiometric (S) and Nonstoichiometric (N) Analytical Methods of Measurement**

I. GRAVIMETRIC.  Isolation of weighable precipitate
   A. *Inorganic Precipitating Agents* (S)
   B. *Organic Precipitating Agents* (S)
   C. *Electrodeposition* (S)


II. TITRIMETRIC.  Reaction of unknown with standard solution
   A. *Acid–Base Titrations* (S)
   B. *Precipitation Titrations* (S)
   C. *Complexometric Titrations* (S)
   D. *Oxidation–Reduction Titrations* (S)


III. OPTICAL.
   A. *Absorption of Energy.*  Attenuation of radiation by an absorbing sample
      1. Colorimetry (N)
      2. Ultraviolet spectrophotometry (N)
      3. Infrared spectrophotometry (N)
      4. Reflectance measurement of reflected light off sample (N)
   B. *Emission of Energy.*  Addition of energy (heat, light, etc.) and subsequent observation of photon emission
      1. Arc emission, electric arc excitation (N)
      2. Flame photometry, flame excitation (N)
      3. Fluorescence, excitation by photons; emitted photons observed (N)
      4. Phosphorescence.  Excitation by photons; delayed emission of photons observed (N)
      5. Chemiluminescence.  Observation of photons released by chemical reaction (N)


IV. RESONANCE.  Interaction of radiowaves with atomic nuclei in a strong magnetic field (N)


V. GAS ANALYSIS.
   A. *Volumetric.*  Measurement of volume of a gas (S)
   B. *Manometric.*  Measurement of the gas pressure (S)


VI. ELECTRICAL.
   A. *Potentiometry.*  Measure of the potential of an electrochemical cell (N)
   B. *Conductivity.*  Measurement of the resistance of a solution (N)
   C. *Coulometry.*  Measurement of the quantity of electricity needed to cause a reaction quantitatively (S)
   D. *Polarography.*  Current voltage characteristics of a solution containing ions which can be oxidized or reduced (N)


VII. THERMAL.  Change in a physical property with temperature (N)

*(continued)*

Table 3–1.   Continued

---

VIII. Other Methods.
  A. *X-Ray Fluorescence*.   X-Ray excitation of sample; observation of emitted X-rays (N)
  B. *Mass Spectrometry*.   Measurement of the number ions of given masses (N)
  C. *Refractometry*.   Measurement of the refractive index of a sample (N)
  D. *Polarimetry*.   Measurement of the rotation of light by a solution (N)
  E. *Optical Rotatory Dispersion*.   Measurement of the rotation of light by a sample as a function of wavelength (N)
  F. *Light-Scattering Photometry*.   Measurement of the amount of light scattered by a suspension (N)
  G. *Activation Analysis*.   Formation of artificial radioactive materials; count particles (N)

---

In addition, there are several useful wet chemical methods which are non-stoichiometric. Not only is calibration required, but it is also necessary to carefully repeat each operation when the procedure is carried out. A non-stoichiometric wet chemical method is not desirable and an important decision must be made as to whether such a method or an alternate stoichiometric method is more suitable, even though the accuracy or sensitivity of the latter might be less.

Table 3–1 lists typical stoichiometric and nonstoichiometric types of measurements.

## UNITS

The basic metric units of interest in analytical chemistry, which are taken from the International System of Units, the SI units, are listed in Table 3–2. Many of these have been and are still routinely used. However, some are new and are just now being adopted. The approved prefixes are also listed in Table 3–2.

## CONCENTRATION

The three most useful ways of expressing concentration in stoichiometric reactions are molarity, normality, and formality. (Gas concentrations are usually expressed by their pressure or partial pressure.) Definitions for these and several others are listed in Table 3–3.

**Molarity and Formality.**   Molarity is a unit of concentration expressed as the number of moles of dissolved solute per liter of solution. If the number of moles and the volume are divided by 1000, then molarity is expressed as the

**Table 3–2.  International System of Units and Approved Prefixes**

| SI units | |
|---|---|
| Length | meter, m |
| Mass | kilogram, kg |
| Time | second, s |
| Current, electrical | ampere, A |
| Temperature | kelvin, K |
| Amount of substance | mole, mol |
| Luminous intensity | candela, cd |
| Pressure | pascal, Pa |

| Approved prefixes | | | | | |
|---|---|---|---|---|---|
| $10^{12}$ | tera, T | 10 | deka, da | $10^{-9}$ | nano, n |
| $10^{9}$ | giga, G | $10^{-1}$ | deca, d | $10^{-12}$ | pico, p |
| $10^{6}$ | mega, M | $10^{-2}$ | centi, c | $10^{-15}$ | femto, f |
| $10^{3}$ | kilo, k | $10^{-3}$ | milli, m | $10^{-18}$ | alto, a |
| $10^{2}$ | hecto, h | $10^{-6}$ | micro, $\mu$ | | |

number of millimoles per milliliter of solution. To make a 1.000 $M$ solution of KCl, exactly 1.000 mole of KCl (74.55 g) is dissolved in an amount of water or some other solvent to yield exactly 1.000 liter of solution.

Actually, this unit of concentration is often misused. For example, if water is used as the solvent the solution is exactly 1.000 $M$ in $K^+$ and 1.000 $M$ in $Cl^-$ since KCl is completely dissociated.

$$KCl \xrightarrow{\text{water}} K^+ + Cl^- \qquad (3\text{-}2)$$

Therefore, the molar concentration of KCl is zero. If the KCl were only partially dissociated, its molar concentration would be determined by the amount of dissociation that takes place.

Many compounds are only partially dissociated. For example, if 0.1000 mole or 5.905 g of acetic acid ($HC_2H_3O_2$) is dissolved in 1.000 liter of solution ($H_2O$), the solution will be 0.00134 mole/liter in $H_3O^+$, 0.00134 mole/liter in $C_2H_3O_2^-$, and 0.0987 mole/liter in un-ionized $HC_2H_3O_2$. Chemically, this partial dissociation can be represented by

$$HC_2H_3O_2 + H_2O \rightleftharpoons H_3O^+ + C_2H_3O_2^- \qquad (3\text{-}3)$$

To avoid the misuse of molarity, the concentration term formality ($F$) was introduced and is defined as the number of gram-formula weights of the

**Table 3–3.  Concentration Expressions**[a]

| Unit | Symbol | Definition | Relationship |
|------|--------|------------|--------------|
| Molarity | $M$ | Number of moles of solute per liter of solution | $M = \dfrac{\text{moles}}{\text{liters of solution}}$ |
| Formality | $F$ | Number of gram-formula weights (GFW) of solute per liter of solution | $F = \dfrac{\text{(GFWs)}}{\text{liters of solution}}$ |
| Normality | $N$ | Number of equivalents of solute per liter of solution | $N = \dfrac{\text{equivalents}}{\text{liters of solution}}$ |
| Molality | $m$ | Moles of solute per kilogram of solvent | $m = \dfrac{n_2}{\text{kg solvent}}$ |
| Mole fraction | $X$ | Ratio of the moles of solute to the total moles of solute plus solvent | $X_2 = \dfrac{n_2}{n_1 + n_2}$ |
| Percent by weight | wt % | Ratio of the weight of solute to the total weight of solute plus solvent | $\text{wt}_2\,\% = \dfrac{g_2}{g_1 + g_2} \times 100$ |
| Percent by volume | vol % | Ratio of volume of solute to volumes of solute plus solvent needed to reach total volume | $\text{vol}_2\,\% = \dfrac{V_2 \times 100}{V_2 + V_1 \text{ (needed to reach total)}}$ |

[a] Subscripts 1 and 2 refer to solvent and solute, respectively.

solute initially taken in the preparation of the solution. This unit of concentration is a summation of all the existing forms of the solute in the solution. For the KCl solution the concentration can be expressed as $1.000\,F$ KCl. Since the salt is completely dissociated the concentration can also be expressed as $1.000\,F$ $K^+$ and $1.000\,F$ $Cl^-$.

In the acetic acid solution there are three different species in solution, $H_3O^+$, $C_2H_3O_2^-$, and $HC_2H_3O_2$. However, dissolving 0.1000 mole of $HC_2H_3O_2$ in 1.000 liter of solution results in a 0.1000 $F$ $HC_2H_3O_2$ solution.

Molarity as a unit of concentration should be used to describe the actual concentration of the species in the solution. Frequently, this is called the equilibrium concentration. Formality, on the other hand, is a concentration which represents the total amount of a given species regardless of its state in the solution. Another term that is often used to describe this is the analytical (molecular) concentration. (Usually, if the concentration is clearly stated as analytical concentration, it will be stated in terms of molarity.) Throughout this book the two terms formality (or analytical concentration) and molarity are used according to their definitions.

**Normality.**    Normality is defined as the number of gram-equivalent weights of solute per liter of solution. If the number of gram-equivalents and volume are divided by 1000, normality becomes the number of milliequivalents per milliliter of solution. Unlike molarity and formality, normality varies according to the reaction in which the solute participates. This means that the exact calculation of the gram-equivalent weight can be made only after determining the specific changes in chemical identity of the solute in the course of the reaction.

In general, there are $n$ gram-equivalents per mole and since $n$ is always a small whole number, the gram-equivalent weight is always equal to the molecular weight or a fraction of it:

$$\text{gram-equivalent weight} = \frac{\text{molecular weight (g/mole)}}{n \text{ (equivalents/mole)}} \qquad (3\text{-}4)$$

Although the equation contains molecular weight, normality can also be used to express concentration of the ionic or atomic state. The number of gram-equivalents in any number of grams of a solute is given by

$$\text{gram-equivalents} = \frac{\text{grams of solute}}{\text{gram-equivalent weight of solute (g/equivalent)}} \qquad (3\text{-}5)$$

The equivalency of a reaction is defined by $n$. The value of $n$ is determined by an oxidation state, the number of replaceable $H_3O^+$ or $OH^-$ ions, or by a change in oxidation state. Several examples are shown in Table 3–4.

Several examples can be used to illustrate the need of knowing the specific reaction in which the solute is participating. Phosphoric acid, $H_3PO_4$, has three replaceable hydrogen ions and can consume one, two, or three $OH^-$ ions in an ordinary acid–base reaction. If 1 mole of $H_3PO_4$ is dissolved per liter of solution, the solution is 1 $N$ in $H_3PO_4$ if reaction (3-6) is followed, 2 $N$ if reaction (3-7) is followed, and 3 $N$ if reaction (3-8) is followed:

$$H_3PO_4 + OH^- \rightarrow H_2O + H_2PO_4^- \qquad (3\text{-}6)$$

$$H_3PO_4 + 2OH^- \rightarrow 2H_2O + HPO_4^{2-} \qquad (3\text{-}7)$$

$$H_3PO_4 + 3OH^- \rightarrow 3H_2O + PO_4^{3-} \qquad (3\text{-}8)$$

Potassium hydrogen oxalate, $KHC_2O_4$, is an interesting compound in that 1 mole of $KHC_2O_4$ per liter of solution is also 1 $N$ if it were to be used as an acid (one replaceable hydrogen ion) and 2 $N$ if it were to be used as a reducing agent where a change in oxidation state occurs in the transformation of $C_2O_4^{2-}$ to $CO_2$. (Carbon changes from $3+$ to $4+$ or a change of $+1$ occurs, but, two carbon atoms change, so the total change is 2.)

Nitric acid, $HNO_3$, has one replaceable hydrogen ion via the reaction

$$HNO_3 + H_2O \rightarrow H_3O^+ + NO_3^- \qquad (3\text{-}9)$$

**Table 3–4.  Examples of Equivalency**

| Solute | $n$ |
|--------|-----|
| | Oxidation State |
| NaCl | 1 (where $Na^+$ is of interest) |
| $BaCl_2$ | 2 (where $Ba^{2+}$ is of interest) |
| $AlCl_3$ | 3 (where $Al^{3+}$ is of interest) |
| $Na_2SO_4$ | 2 (where $SO_4^{2-}$ is of interest) |
| | Replaceable Hydronium and Hydroxide Ions |
| HCl | 1 |
| $HNO_3$ | 1 |
| $H_2SO_4$ | 1 (where product is $HSO_4^-$) |
| | 2 (where product is $SO_4^{2-}$) |
| NaOH | 1 |
| KOH | 1 |
| $Ba(OH)_2$ | 2 |
| | Change in Oxidation State |
| $FeCl_3$ | 1 (where product is $Fe^{2+}$) |
| $MnO_4^-$ | 5 (where product is $Mn^{2+}$) |
| $SnCl_4$ | 2 (where product is $Sn^{2+}$) |

but $NO_3^-$ can undergo a variety of changes in oxidation state. A solution of 1 mole of $HNO_3$ per liter of solution is therefore 1 $N$ if used as an acid, and can be 1 $N$, 3 $N$, or 8 $N$ if used as an oxidizing agent where the products are $NO_2$, NO, and $NH_4^+$, respectively. There are also other possible products and consequently different normalities would result.

Expressing a concentration of a solution in terms of normality must be done carefully. Since the normality label is associated with a specified chemical behavior, any departure from this behavior will require a recalculation of the normality. Although this is inconvenient and can possibly be misleading, the use of normality as an expression of concentration is universally applied to reaction stoichiometry of solutions. Since one equivalent of one reactant will always react with one equivalent of another reactant, it becomes possible to easily calculate the stoichiometric relationship for the reaction.

**Percent Solute.**    The percent of solute is an ambiguous designation which may refer to *percent by weight* or *percent by volume*. In the former case it is the fraction of the total solution weight that is contributed by the solute. For example, a 3% by weight $H_2O_2$ solution would be 3 g of $H_2O_2$ per 100 g of solution or 3 g of $H_2O_2$ mixed with 97 g of water. Note that the weights of solute and solvent are additive. This is not true when reference is made on a volume basis. If the solution was made to be 3% by volume $H_2O_2$, 3 ml of $H_2O_2$ would be diluted to a total volume of 100 ml. However, this does not mean that 3 ml of $H_2O_2$ and 97 ml of $H_2O$ are mixed. Since hydration and solvation effects cause volume alterations, the correct solution (percent by volume) would not be obtained by mixing 3 ml and 97 ml. In fact, expansion or contraction will occur and its extent is dependent on the nature of the solute and also its physical state (solid, liquid, or gas).

**Activity.** Three types of interactions are present in solutions of non-electrolytes. For example, in a solution of alcohol in water interactions occur between alcohol molecules, between water molecules, and between each other. Various properties of this solution, such as vapor pressure, freezing-point lowering, boiling-point elevation, solubility, and many other chemical and physical properties that are a function of concentration, will be influenced by these interactions. Thus, departures from the "true concentration" value are observed.

In solutions of electrolytes such as NaCl in water, the interactions are potentially of greater magnitude since electrostatic charges are involved. These interactions include attraction between oppositely charged ions, between ions and solvent molecules, between solvent molecules, and repulsion between like-charged ions.

As a result of these various types of interactions the number of solute ions or molecules that affect a particular solution property can be quite different from what was originally added. Thus, the ideal concentrations are the types previously mentioned ($M$, $N$, etc.) and correspond to what was originally added. In contrast, the true or effective concentration which takes into account the interactions is called the *activity*.

For most solutions of moderate concentration the activity is proportional to the concentration of the solute

$$a \propto c \tag{3-10}$$

where $a$ is the activity of the solute and $c$ is the concentration with both expressed in molar units. If a proportionality constant is included, the equation becomes

$$a = \gamma c \tag{3-11}$$

where $\gamma$ is the activity coefficient and carries no units.

It is important to realize that in the typical titrimetric and gravimetric methods, and in most other ordinary analytical determinations, it is the molar concentration or analytical concentration of a substance that is being measured and not the activity. Thus, a solution prepared by diluting 0.1000 mole of HCl to 1.000 liter will analyze to be 0.1000 $M$ $H_3O^+$ and 0.1000 $M$ $Cl^-$. For example, the hydronium ion could be titrated with standard base and the chloride titrated with standard $AgNO_3$ solution. The various types of interactions and repulsions, although present, are not strong enough to interfere with the titration reactions.

However, it must not be concluded that the activity concept is an insignificant one. It is always possible that in the measurement of some particular physical property of a solution the attractions or repulsions between the ions and molecules present might be significant enough to affect the measurement. In general, not only the concentration of the solutes, but also the nature of the chemical reactions involving them must be considered. Such interactions become less important as the solution becomes more dilute, because the solutes have a tendency to be farther apart and consequently have less of an effect on each other.

**pX.** Many numerically small values, such as equilibrium constants and ionic or molecular concentrations, are often encountered in analytical chemistry. A convenient way to express these small numbers is by using $pX$, which is defined as

$$pX = \log \frac{1}{X} = -\log X \tag{3-12}$$

where $X$ is a molar concentration or an equilibrium constant. Since $pX$ is for base 10 logarithms, Eq. (3-12) can also be written as

$$X = 10^{-pX} \tag{3-13}$$

For all numbers smaller than 1, $pX$ values have the advantage of being positive, while negative $pX$ values, which are representative of numbers larger than 1, are seldom encountered.

Probably the most widely used $pX$ is where $X = [H_3O^+]$ and thus takes the form

$$pH = \log \frac{1}{[H_3O^+]} = -\log[H_3O^+] \tag{3-14}$$

where the bracket means moles/liter. A more precise and fundamental definition involves hydrogen ion activity. However, the difference between expression (3-14) and one based on activity is very slight and for almost all practical and common applications expression (3-14) can be used.

Alternatively, pOH, which indicates the hydroxyl ion concentration, could be used:

$$pOH = \log \frac{1}{[OH^-]} = -\log[OH^-] \tag{3-15}$$

For any tenfold change in $H_3O^+$ (or $OH^-$) concentration there is a one unit change in pH (or in pOH), and the sum of pH and pOH for the same solution is equal to 14 or

$$pH + pOH = 14 \tag{3-16}$$

**Titer.** The titer of a solution is defined as the weight of a pure substance which is chemically equivalent to or reacts with a fixed volume of the solution (usually to 1 ml). An example of a calculation using titer is presented in Example 3-8 in this chapter.

**Parts Per Thousand and Parts Per Million.** Percentage concentration on a weight per weight basis can also be expressed as parts per thousand (ppt) or parts per million (ppm). Expressed in units of weight ppt is milligrams per gram (mg/g) while ppm is milligrams per kilogram (mg/kg) or micrograms per gram ($\mu$g/g). For example, 1 ppt Cu in an alloy means that the alloy contains 1 part by weight Cu in 1000 parts by weight of alloy. If it were 1 ppm, it would mean that there is 1 part Cu in 1 million parts by weight of alloy.

Solutions would be handled in the same way. Therefore, a solution of 1 ppm Cu would be 1 part Cu by weight in 1 million parts by weight of sample solution. Since parts per million is used to express concentration of trace quantities in dilute solution, a good approximation is that milligrams per kilogram is equal to milligrams per liter.

## CALCULATIONS

**Calculations Based on Molarity.** Calculations in analytical procedures can be handled by manipulating expressions based on moles and molar solutions. To do the calculations this way it is essential to know the stoichiometry of the reaction.

The useful equations are

$$M = \frac{\text{moles of solute}}{\text{liter}} \tag{3-17}$$

or

$$M = \frac{\text{mmoles of solute}}{\text{ml}} \tag{3-18}$$

A mole and millimole (mmole) are given by

$$\text{number of moles} = \frac{\text{wt of solute(g)}}{\text{formula wt(g/mole)}} \tag{3-19}$$

$$\text{number of mmoles} = \frac{\text{wt of solute(mg)}}{\text{formula wt(mg/mmole)}} \tag{3-20}$$

Thus,

$$\text{moles of solute} = M \times \text{liter} \tag{3-21}$$

$$\text{mmoles of solute} = M \times \text{ml} \tag{3-22}$$

For a stoichiometric reaction such as

$$A + B \rightarrow C + D \tag{3-23}$$

the calculation is straightforward since A and B are reacting on a mole to mole basis. Thus,

$$M_A \times \text{liter}_A = M_B \times \text{liter}_B \tag{3-24}$$

and

$$M_A \times \text{ml}_A = M_B \times \text{ml}_B \tag{3-25}$$

which state that at the stoichiometric point of the reaction the

$$\text{number of moles of A} = \text{number of moles of B} \tag{3-26}$$

and

$$\text{number of mmoles of A} = \text{number of mmoles of B} \tag{3-27}$$

Equations (3-20) and (3-22) can be combined to give

$$\frac{\text{wt of A (mg)}}{\text{formula wt of A (mg/mmole)}} = \text{ml}_B \times M_B \text{(mmole/ml)}$$

$$= \frac{\text{wt of B (mg)}}{\text{formula wt of B (mg/mmole)}} \tag{3-28}$$

and

$$\text{wt of A (mg)} = \text{ml}_B \times M_B \times \text{formula wt of A} \tag{3-29}$$

If A represents part of a sample, the % A in the sample can be calculated based on the weight of sample taken.

$$\% \text{ A} = \frac{\text{wt of A (mg)} \times 100}{\text{wt of sample (mg)}} \tag{3-30}$$

Alternatively, Eqs. (3-29) and (3-30) can be combined to give

$$\% \text{ A} = \frac{\text{ml}_B \times M_B \text{(mmole/ml)} \times \text{formula wt of A (mg/mmole)} \times 100}{\text{wt of sample (mg)}}$$

$$\tag{3-31}$$

Many reactions do not occur on a one-to-one basis. The preceding equations must, therefore, be modified to account for the ratio of moles of reactants involved in the stoichiometric reaction. Assume that reaction (3-23) is of the general form

$$aA + bB \rightarrow cC + dD \tag{3-32}$$

Since $a$ moles (or mmoles) of A react with $b$ moles (or mmoles) of B, expression (3-26) must be modified to take care of the mole ratio relating A and B in the balanced reaction. Hence,

$$\text{number of moles of A} = \text{number of moles of B} \times a/b \tag{3-33}$$

where $a/b$ is the mole ratio, A is the analyte or sample being analyzed, and B is the titrant or reactant of known concentration. Alternatively,

$$\text{number of mmoles of A} = \text{number of mmoles of B} \times a/b \qquad (3\text{-}34)$$

can be written. Substituting (3-20) and (3-22) into (3-34) gives

$$\frac{\text{wt of A (mg)}}{\text{formula wt of A}} = \text{ml}_B \times M_B \times \frac{a}{b} \qquad (3\text{-}35)$$

and

$$\text{wt of A (mg)} = \text{ml}_B \times M_B \times \frac{a}{b} \times \text{formula wt of A} \qquad (3\text{-}36)$$

For % A, the expression is

$$\% \text{ A} = \frac{\text{ml}_B \times M_B (\text{mmole/ml}) \times (a/b) \times \text{formula wt of A (mg/mmole)} \times 100}{\text{wt of sample (mg)}}$$

$$(3\text{-}37)$$

Several examples using these equations and illustrating typical calculations in quantitative analysis are given below. Since the examples deal with analytical concentrations the molar symbol will be replaced by $F$ (formality). This in no way alters the derived expressions.

*Example 3-1.* How many milliliters of 0.5100 $F$ HCl should be used to make a solution of 600 ml of 0.0100 $F$ HCl?

$$F_{\text{initial}} \times \text{ml}_{\text{initial}} = F_{\text{final}} \times \text{ml}_{\text{final}}$$

$$0.5100 \text{ mmole/ml} \times \text{ml}_{\text{initial}} = 0.0100 \text{ mmole/ml} \times 600 \text{ ml}$$

$$\text{ml}_{\text{initial}} = 11.76 \text{ ml}$$

Therefore, 11.76 ml of 0.5100 $F$ HCl should be diluted to a total volume of 600 ml.

*Example 3-2.* Calculate the formality of an HCl solution whose specific gravity (sp gr) is 1.1878 and which contains 37.50% HCl by weight.

$$\text{g of HCl/ml of solution} = \text{sp gr} \times \text{density} \times \% \text{HCl}$$

$$\text{g of HCL/ml of solution} = 1.188 \times 1.000 \text{ g/ml} \times 0.3750 = 0.4455 \text{ g/ml}$$

$$\frac{\text{wt of HCl (g)}}{\text{formula wt of HCl}} = \text{number of moles of HCl}$$

$$\frac{0.4455 \text{ g/ml}}{36.47 \text{ g/mole}} = 0.01222 \text{ mole/ml} \times 1000 \text{ ml/liter} = 12.22 \text{ } F$$

*Example 3-3.* Calculate the $Na^+$ concentration in g/liter for the solution made by mixing 100.0 ml of 0.1200 $F$ NaCl and 200.0 ml of 0.05000 $F$ NaOH. Assume that volumes are additive.

$$ml_{NaCl} \times F_{NaCl} = \text{mmoles } Na^+$$

$$100.0 \text{ ml} \times 0.1200 \text{ mmole/ml} = 12.00 \text{ mmoles } Na^+$$

$$ml_{NaOH} \times F_{NaOH} = \text{mmoles } Na^+$$

$$200.0 \text{ ml} \times 0.05000 \text{ mmole/ml} = 10.00 \text{ mmoles } Na^+$$

$$\text{total mmoles } Na^+ = 22.00$$

$$\text{total volume} = 300.0 \text{ ml}$$

$$\frac{22.00 \text{ mmoles}}{0.300 \text{ liter}} = 73.33 \text{ mmole } Na^+/\text{liter}$$

$$1 \text{ mmole } Na^+ = 0.02299 \text{ g}$$

$$\text{g } Na^+/\text{liter} = 73.33 \text{ mmoles/liter} \times 0.02299 \text{ g/mmoles} = 1.686 \text{ g/liter}$$

*Example 3-4.*    Calculate the volume of 0.2500 $F$ HCl required to make exactly 500 ml of 0.08000 $F$ HCl.

$$ml_{HCl} \times F_{HCl} = \text{mmoles HCl}$$

$$500.0 \text{ ml} \times 0.08000 \ F = 40.00 \text{ mmoles HCl}$$

$$ml_{HCl} = \frac{\text{mmoles HCl}}{F_{HCl}}$$

$$ml_{HCl} = \frac{40.00 \text{ mmoles}}{0.2500 \ F} = 160.0 \text{ ml}$$

160.0 ml of 0.2500 $F$ diluted to 500.0 ml will yield a 0.08000 $F$ HCl solution.

*Example 3-5.*    How many grams of NaOH should be weighed out to make 1.000 liter of approximately 0.1 $F$ NaOH solution.

$$1000 \text{ ml} \times 0.1 \text{ mmole/ml} = 100 \text{ mmole of NaOH in solution}$$

$$\frac{\text{wt of NaOH (mg)}}{\text{formula wt NaOH}} = \text{number of mmoles NaOH}$$

$$\frac{\text{wt of NaOH (mg)}}{40.00 \text{ mg/mmole}} = 100 \text{ mmole}$$

$$\text{wt of NaOH} = 4000 \text{ mg} \quad \text{or} \quad 4.0 \text{ g}$$

*Example 3-6.*    What is the formality of an NaOH solution if 26.45 ml of it is needed to exactly react with 0.5644 g of pure potassium acid phthalate, KHP (KHC$_6$H$_4$O$_4$, 204.2 molecular weight), in solution?

$$\text{number of mmoles of KHP} = \frac{\text{wt of KHP (mg)}}{\text{formula wt of KHP}}$$

$$\text{number of mmoles of KHP} = \frac{564.4 \text{ mg}}{204.2 \text{ mg/mmole}}$$

$$\text{number of nmoles of KHP} = 2.764 \text{ mmole}$$

Since the reaction takes place on a 1:1 mole basis, it can be stated that at the stoichiometric point

$$\text{number mmoles of KHP} = \text{number mmoles of NaOH}$$

$$\text{ml}_{\text{NaOH}} \times F_{\text{NaOH}} = \text{mmoles of NaOH} = \text{mmoles of KHP}$$

$$26.45 \text{ ml} \times F_{\text{NaOH}} = 2.764 \text{ mmole}$$

$$F_{\text{NaOH}} = 0.1045 \ F$$

*Example 3-7.* What is the percent purity of oxalic acid if 0.1683 g of the acid in solution is completely neutralized by 34.65 ml of 0.1045 $F$ NaOH?

$$H_2C_2O_4 + 2NaOH \rightarrow Na_2C_2O_4 + 2H_2O.$$

number of mmoles of H$_2$C$_2$O$_4$ = number of mmoles of NaOH $\times \frac{1}{2}$

number of mmoles of H$_2$C$_2$O$_4$ = ml$_{\text{NaOH}} \times F_{\text{NaOH}} \times \frac{1}{2}$

number of mmoles of H$_2$C$_2$O$_4$ = 34.65 ml $\times$ 0.1045 mmole/ml $\times \frac{1}{2}$ = 1.811 mmoles

$$\frac{\text{wt of H}_2\text{C}_2\text{O}_4 \text{ (mg)}}{\text{formula wt of H}_2\text{C}_2\text{O}_4 \text{ (mg/mmole)}} = \text{number of mmoles}$$

$$\frac{\text{wt of H}_2\text{C}_2\text{O}_4 \text{ (mg)}}{90.04 \text{ mg/mmole}} = 1.811 \text{ mmole}$$

$$\text{wt of H}_2\text{C}_2\text{O}_4 = 163.0 \text{ mg (0.1630 g)}$$

$$\% \text{ purity} = \frac{\text{wt of H}_2\text{C}_2\text{O}_4 \text{ (g) found} \times 100}{\text{wt of H}_2\text{C}_2\text{O}_4 \text{ (g) taken}}$$

$$\% \text{ purity} = \frac{0.1630 \text{ g} \times 100}{0.1683 \text{ g}} = 96.85\%$$

Alternatively,

$$\%H_2C_2O_4 = \frac{ml_{NaOH} \times F_{NaOH} \times \frac{1}{2} \times \text{formula wt } H_2C_2O_4 \times 100}{\text{wt of sample (mg)}}$$

$$\%H_2C_2O_4 = \frac{34.65 \text{ ml} \times 0.1045 \text{ mmole/ml} \times \frac{1}{2} \times 90.04 \text{ mg/mmole} \times 100}{168.3 \text{ mg}}$$

$$\%H_2C_2O_4 = 96.85\%$$

If the reaction was

$$H_2C_2O_4 + NaOH \rightarrow NaHC_2O_4 + H_2O$$

the calculation would be simplified since the acid and base react on a 1 to 1 basis. Thus, the reaction ratio is 1. A purity of 96.85% for the $H_2C_2O_4$ sample would still be calculated since only 17.33 ml of the base solution would be required according to this reaction.

Titer and parts per million are frequently encountered in routine analysis and trace analysis, respectively. Examples 3-8 to 3-12 illustrate the use of these concentration units in calculations.

*Example 3-8.*      Express the standard NaOH solution in Example 3-6 as a titer with respect to KHP.

$$F_{NaOH} = 0.1045 \, F \quad \text{and} \quad \text{mg KHP}/1.0 \text{ ml NaOH} = \text{titer}$$

At the stoichiometric point of the reaction

$$\text{mmoles of NaOH} = \text{mmoles of KHP}$$

$$F_{NaOH} \times ml_{NaOH} = \frac{\text{mg KHP}}{\text{formula wt KHP}}$$

$$\frac{\text{mg KHP}}{ml_{NaOH}} = \text{formula wt KHP} \times F_{NaOH}$$

$$\frac{\text{mg KHP}}{ml_{NaOH}} = 204.2 \text{ mg/mmole} \times 0.1045 \text{ mmole/ml} = 21.34 \text{ mg/ml NaOH}$$

*Example 3-9.*      An impure KHP sample (0.5000 g) was titrated to an end point with 10.01 ml of 0.1045 $F$ NaOH. With the data in Example 3-8, calculate the % KHP in the sample.

$$\text{mg KHP} = \text{titer} \times \text{ml}$$

$$\text{mg KHP} = 21.34 \text{ mg/ml}_{NaOH} \times 10.01 \text{ ml} = 213.6 \text{ mg}$$

$$\% \text{ KHP} = \frac{213.6 \text{ mg}}{500.0 \text{ mg}} \times 100 = 42.72\%$$

*Example 3-10.*    A 50.00-ml aliquot of 0.1200 $F$ HCl was added to 40.00 ml of a KOH solution. The excess acid was titrated with 23.50 ml of a 0.09910 $F$ NaOH solution. Calculate the formality of the KOH solution.

$$HCl + NaOH \text{ (or KOH)} \rightarrow NaCl \text{ (or KCl)} + H_2O$$

The reaction ratio for both reactions is 1, therefore

$$ml_{HCl} \times F_{HCl} = mmoles_{HCl} \text{ (added)}$$

$$50.00 \text{ ml} \times 0.1200 \text{ } F = 6.000 \text{ mmoles}$$

$$ml_{NaOH} \times F_{NaOH} = mmoles_{NaOH} \text{ (titration)}$$

$$23.50 \text{ ml} \times 0.09910 \text{ } F = 2.329 \text{ mmoles}$$

$$mmoles_{KOH} = mmoles_{HCl} - mmoles_{NaOH}$$

$$3.671 \text{ mmoles} = 6.000 \text{ mmoles} - 2.329 \text{ mmoles}$$

$$\frac{mmoles_{KOH}}{ml_{KOH}} = F_{KOH}$$

$$\frac{3.671 \text{ mmoles}}{40.0 \text{ ml}} = 0.09178 \text{ } F$$

*Example 3-11.*    Calculate the concentration of an 0.001000 $F$ NaOH solution in parts per million (wt/vol).

$$ppm = \frac{mg \text{ NaOH}}{1000 \text{ ml solution}} = \frac{mgNaOH}{liter}$$

$$\frac{mg \text{ NaOH}}{liter} = F_{NaOH} \times \text{formula wt} \times 1000 \text{ mg/g}$$

$$\frac{mg \text{ NaOH}}{liter} = 1.000 \times 10^{-3} \text{ mole/liter} \times 40.00 \text{ g/mole} \times 1000 \text{ mg/g} = 40.00 \text{ mg/liter}$$

$$ppm \text{ NaOH} = 40.00 \text{ ppm}$$

*Example 3-12.*    Calculate the mg/ml and ppm $Hg^{2+}$ and $NO_3^-$ in a $4.15 \times 10^{-5}$ $F$ $Hg(NO_3)_2$ solution.

$Hg^{2+}$:

$$4.15 \times 10^{-5} \text{ mole/liter} \times 201 \text{ g/mole} = 8.34 \times 10^{-3} \text{ g/liter}$$

$$8.34 \times 10^{-3} \text{ g/liter} \times 1000 \text{ mg/g} \times \frac{1 \text{ liter}}{1000 \text{ ml}} = 8.34 \times 10^{-3} \text{ mg/ml}$$

$$8.34 \times 10^{-3} \text{ mg/ml} \times 1000 \text{ ml/liter} = 8.34 \text{ mg/liter} = 8.34 \text{ ppm}$$

$NO_3^-$:

$$Hg(NO_3)_2 \rightarrow Hg^{2+} + 2NO_3^-$$

$4.15 \times 10^{-5}$ mole/liter $\times$ 62.0 g/mole $\times$ 2 = $5.15 \times 10^{-3}$ g/liter

$5.15 \times 10^{-3}$ g/liter $\times$ 1000 mg/g $\times \dfrac{1 \text{ liter}}{1000 \text{ ml}} = 5.15 \times 10^{-3}$ mg/ml

$5.15 \times 10^{-3}$ mg/ml $\times$ 1000 ml/liter = 5.15 mg/liter = 5.15 ppm

**Calculations Based on Normality.**    It is useful to define normality again, since data for stoichiometric reactions in solution can also be calculated this way. The useful equations are

$$N = \frac{\text{eq of solute}}{\text{liter}} \tag{3-38}$$

$$N = \frac{\text{mEq of solute}}{\text{ml}} \tag{3-39}$$

where eq and mEq are equivalents and milliequivalents, respectively. Equations (3-38) and (3-39) can be rewritten in the form

$$N(\text{eq/liter}) \times \text{liters} = \text{number of equivalents of solute} \tag{3-40}$$

$$N(\text{mEq/ml}) \times \text{ml} = \text{number of milliequivalents of solute} \tag{3-41}$$

In a stoichiometric reaction such as

$$A + B \rightarrow C + D \tag{3-42}$$

1 equivalent of reactant A will exactly react with 1 equivalent of reactant B, or 8 milliequivalents of reactant A will exactly react with 8 milliequivalents of reactant B, and so on. The reaction will always involve equal equivalents even if one of the reactants happens to be in excess. The number of equivalents or milliequivalents of a solute A in a solution of known concentration in units of normality can be calculated by Eq. (3-40) or (3-41), provided the volume of the solution is known. A similar situation would exist for B. If enough time is allowed, A and B will react on an equivalent per equivalent basis and therefore Eq. (3-43) can be written

$$\text{liter}_A \times N_A = \text{liter}_B \times N_B \tag{3-43}$$

which states

$$\text{number of eq A} = \text{number of eq B} \tag{3-44}$$

at the stoichiometric point of the reaction. Alternatively, the following can be written:

$$ml_A \times N_A = ml_B \times N_B \tag{3-45}$$

and

$$\text{number of mEq A} = \text{number of mEq B} \tag{3-46}$$

The number of milliequivalents of A (or B) can be calculated if its weight and milliequivalent weight are known. To determine the latter, the exact reaction between A and B must be known. The equation to use is then

$$mEq\ A = \frac{\text{wt of A(mg)}}{\text{mEq wt (mg/mEq)}} \tag{3-47}$$

or if equivalents are desired

$$eq\ A = \frac{\text{wt of A(g)}}{\text{eq wt(g/eq)}} \tag{3-48}$$

Equations (3-4) and (3-5) are used to calculate the milliequivalent weight or equivalent weight of A. Equations (3-45) and (3-47) can be combined to give

$$\frac{\text{wt of A(mg)}}{\text{mEq wt of A(mg/mEq)}} = ml_B \times N_B = \frac{\text{wt of B(mg)}}{\text{mEq wt of B(mg/mEq)}} \tag{3-49}$$

and

$$\text{wt of A(mg)} = ml_B \times N_B \times \text{mEq wt of A} \tag{3-50}$$

To calculate the percent of A in a sample a quantity of the sample is weighed and after carrying out the appropriate experimental steps with suitable reagents, in which careful measurements are made, the weight of A can be calculated. Therefore

$$\%\ A = \frac{\text{wt of A(mg)} \times 100}{\text{wt of sample(mg)}} \tag{3-51}$$

or by combining Eqs. (3-50) and (3-51), Eq. (3-52) is obtained.

$$\%\ A = \frac{ml_B \times N_B(\text{mEq/ml}) \times \text{mEq wt of A(mg/mEq)} \times 100}{\text{wt of sample(mg)}} \tag{3-52}$$

The most critical part of the calculation is the establishment of the correct equivalent weight. For this reason it is absolutely necessary to know the type of reaction that is taking place and its stoichiometry. Several examples illustrating the calculations are given below. These examples are the same as Examples 3–5 to 3–7 except that normality is used.

*Example 3-13.* How many grams of NaOH should be weighed out to make 1.000 liter of approximately 0.1 $N$ solution?

$$1000 \text{ ml} \times 0.1 \text{ N} = 100 \text{ mEq of NaOH in solution}$$

$$100 \text{ mEq} = \frac{\text{wt of NaOH (g)}}{\text{mEq wt of NaOH (g/mEq)}}$$

Assume NaOH is to be used in an acid–base reaction; therefore, one replaceable OH is present.

$$\text{mEq wt of NaOH} = \frac{40.00 \text{ g/mole}}{1 \text{ Eq/mole} \times 1000 \text{ mEq/eq}} = 0.0400 \text{ g/mEq}$$

$$\text{wt of NaOH} = 100 \text{ mEq} \times 0.0400 \text{ g/mEq} = 4.00 \text{ g}$$

Approximately 4.0 g of NaOH diluted to 1 liter of solution.

*Example 3-14.* What is the normality of a NaOH solution if 26.45 ml of it is needed to exactly react with 0.5644 g of pure potassium acid phthalate, KHP ($KHC_8H_4O_4$, 204.2 molecular weight), in solution?

$$\text{NaOH} + \underset{CO_2^- K^+}{\overset{CO_2^- H^+}{\bigcirc}} \longrightarrow \underset{CO_2^- K^+}{\overset{CO_2^- Na^+}{\bigcirc}} + H_2O$$

$$\text{number of mEq of KHP} = \frac{\text{wt of KHP}}{\text{mEq wt of KHP}}$$

$$\text{number of mEq of KHP} = \frac{0.5644 \text{ g} \times 1000 \text{ mg/g}}{204.2 \text{ mg/mEq}} = 2.764 \text{ mEq}$$

$$\text{mEq of KHP} = \text{mEq of NaOH (at the stoichiometric point)}$$

$$\text{ml}_{\text{NaOH}} \times N_{\text{NaOH}} = \text{mEq of NaOH} = \text{mEq of KHP}$$

$$26.45 \text{ ml} \times N_{\text{NaOH}} \text{ (mEq/ml)} = 2.764 \text{ mEq}$$

$$N_{\text{NaOH}} = 0.1045 \text{ N}$$

*Example 3-15.* What is the percent purity of oxalic acid ($H_2C_2O_4$, 90.04 molecular wt) if 0.1683 g of the acid in solution is completely neutralized by 34.65 ml of 0.1045 $N$ NaOH?

$$H_2C_2O_4 + 2NaOH \rightarrow Na_2C_2O_4 + 2H_2O$$

$$\text{mEq of NaOH} = \text{mEq of } H_2C_2O_4 = \text{ml}_{\text{NaOH}} \times N_{\text{NaOH}}$$

$$\text{mEq of } H_2C_2O_4 = 34.65 \text{ ml} \times 0.1045 \text{ N}$$

$$\text{mEq of } H_2C_2O_4 = 3.621 \text{ mEq}$$

$$\text{mEq wt of } H_2C_2O_4 = \frac{\text{molecular wt of } H_2C_2O_4 \text{ (g/mole)}}{n(\text{eq/mole}) \times 1000 \text{ mEq/eq}}$$

$$\text{mEq wt of } H_2C_2O_4 = \frac{90.04}{2 \times 1000} \text{ (2 replaceable H)}$$

$$\text{mEq wt of } H_2C_2O_4 = 0.04502 \text{ g/mEq} = 45.02 \text{ mg/mEq}$$

$$\text{number of mEq } H_2C_2O_4 = \frac{\text{wt of } H_2C_2O_4 \text{ (g)}}{\text{mEq wt of } H_2C_2O_4}$$

$$\frac{\text{wt of } H_2C_2O_4 \text{ (g)}}{0.04502 \text{ g/mEq}} = 3.621 \text{ mEq}$$

$$\text{wt of } H_2C_2O_4 = 0.1630 \text{ g}$$

$$\% \text{ purity} = \frac{\text{wt of } H_2C_2O_4 \text{ (g) found}}{\text{wt of } H_2C_2O_4 \text{ (g) taken}} \times 100$$

$$\% \text{ purity} = \frac{0.1630 \times 100}{0.1683} = 96.85\%$$

Alternatively,

$$\% \ H_2C_2O_4 = \frac{ml_{\text{NaOH}} \times N_{\text{NaOH}} \times \text{mEq wt } H_2C_2O_4 \times 100}{\text{wt of sample (mg)}}$$

$$\% \ H_2C_2O_4 = \frac{34.65 \text{ ml} \times 0.1045 \ N \times 0.04502 \text{ mg/mEq} \times 100}{0.1683 \text{ mg}} = 96.86\%$$

If the reaction was

$$H_2C_2O_4 + \text{NaOH} \rightarrow \text{NaHC}_2O_4 + H_2O$$

the milliequivalent weight of $H_2C_2O_4$ would be

$$\text{mEq wt of } H_2C_2O_4 = \frac{90.04 \text{ g/mole}}{1 \text{ Eq/mole} \times 1000 \text{ mEq/eq}}$$

$$\text{mEq wt of } H_2C_2O_4 = 90.04 \text{ mg/mEq}$$

since only one replaceable H is involved. The $H_2C_2O_4$ is still only 96.85% pure since only 17.33 ml of the NaOH would be needed according to this reaction.

*Example 3-16.*    A tablet weighing 2.212 g was dissolved in water and the acidic solution was titrated with 24.12 ml of 0.1109 $F$ NaOH. Calculate the mEq of acid per tablet and %$H^+$ in the tablet.

$$ml_{\text{NaOH}} \times F_{\text{NaOH}} = \text{mEq}_{\text{NaOH}}$$

$$24.12 \text{ ml} \times 0.1109 \ F = 2.675 \text{ mEq}$$

$$\text{mEq}_{\text{NaOH}} = \text{mEq}_{\text{acid}}$$

2.675 mEq acid per tablet

$$\%\text{H}^+ = \frac{\text{ml}_{\text{NaOH}} \times F_{\text{NaOH}} \times \text{mEq}_{\text{H}^+} \times 100}{\text{wt of sample (mg)}}$$

$$\%\text{H}^+ = \frac{24.12 \text{ ml} \times 0.1109 \, F \times 1.008 \text{ mg/mEq} \times 100}{2212 \text{ mg}} = 0.122\%$$

**Uses of Molarity and Normality Calculations.**    Throughout the remainder of this book the molarity concept will be used in all illustrations, examples, and discussion. Analytical chemistry is in a transition period during which "normality" and "equivalency" are gradually being dropped from the vocabulary. One important reason for this is that the fundamental relationships and the calculations in the study of equilibrium and kinetics are based on the use of moles, molarity, and formality.

It is important, of course, to understand normality and equivalency, because their use is still encountered. For example, the concept of equivalents is employed in resistance measurements and in Faraday's electrolysis law. In clinical laboratories and in industry it is still common to report many quantitative results, particularly in acid–base analyses, in terms of equivalents or milliequivalents.

## STANDARDS

It is very important to establish a standard, or reference point, for any kind of measurement. For measuring physical properties the primary standard, that is, the ultimate reference point, is a very precisely defined unit of measurement. In chemistry, a substance whose purity has been verified may be a primary standard. Table 3–5 gives examples of the descriptions of primary standards for units used in measuring mass, length, time, and wavelength.

Since primary standards are not always available, other references which have been compared to the original reference material or which closely satisfy the requirements, are used. These are called secondary standards.

Many international scientific groups, with the full support of the scientific community, have provided leadership in the establishment of universal standards. In the United States, the National Bureau of Standards has been an important contributer, particularly in the development of chemical standards.

The requirements for a standard vary with the physical or chemical property to be measured. For example, powders of known surface area or porosity, a polymer of known molecular weight or cross-linking, an alloy of a certain tensile strength, or a fiber of known elasticity could all be suitable standards for particular kinds of measurement problems. Obviously, the designation of a substance as a standard is a very important decision.

Table 3–5.  **Primary Standards for Physical Properties**

| Physical property | Unit | Definition of unit |
|---|---|---|
| Mass | Gram | 1/1000 the quantity of matter in the International Prototype Kilogram |
| Length | Centimeter | 1/100 the length of the International Prototype meter (Paris) at 0°C, or 1,553,164.13 times the wavelength of the red Cd line in air at 760 mm Hg at 15°C |
| Time | Ephemeris second | 1/31,556,925.9747 of the tropical year for 1900 January $0^d12^h$ ephemeris time. 1 ephemeris day equals 86,400 ephemeris seconds |
| Wavelength | Ångstrom | 6438.4696 Å, which is the wavelength for the red radiation from cadmium relative to the meter (adopted 1907) |

It should be pointed out that the word "standard" is often used in a somewhat different way than the one we have been talking about. "Standards" are set for the maximum allowable amount of pollutants in air, of impurities in food or drugs, or of pesticide residues in agricultural products. The problem of an analytical chemist may well be to determine whether a product has been manufactured in such a way as to meet standards of this kind.

**Chemical Standards.**    The success of all analytical techniques is based on the choice of the standard reference material used for the calibration of the technique. A primary standard should be sought. However, a secondary standard, or a substance which has been mutually designated as a standard and validated by the parties concerned may be the only one available.

The essential requirements for a primary chemical standard are that the substance must (a) be commercially available at reasonable cost, (b) have known purity of at least 99% or better (99.99% or better preferred), (c) be soluble in the solvent used, (d) be stable and nonhygroscopic, (e) undergo stoichiometric reactions, and (f) possess a high molecular weight. Unfortunately, the number of substances satisfying all of these requirements is limited.

Virtually all analytical methods require a chemical standard. This point is illustrated by considering one of the simplest of analytical techniques, the titrimetric (volumetric) determination. A typical procedure involves the addition of a measured volume of a reagent of known concentration into a solution of the substance to be determined. Addition of the reagent is continued until an amount is added which is chemically stoichiometric to the amount of the substance being determined (the *stoichiometric point*). To permit the calculation of the amount of substance in the solution, the concentration of the reagent solution must be exactly known. Such a solution is called a *standard solution.*

To prepare a primary standard solution an amount of a substance designated as a primary standard is accurately weighed with the analytical balance and diluted to a known volume in a volumetric flask. Thus, a solution of known concentration is obtained; the concentration can be expressed as molarity, normality, or whatever is required. If the primary standard is acidic, a standard solution of known acidity is obtained; if basic a solution of known basicity is obtained.

Assume that a standard solution of NaOH is required. It is not possible to prepare one by directly weighing NaOH pellets since they fail to meet the requirements of a primary standard. Consequently, a solution containing the approximate concentration is prepared. Then, an accurately measured amount of an acidic primary standard solution is taken or an amount of the acidic primary standard is accurately weighed and this is titrated to the stoichiometric point with the NaOH solution. Since the reaction stoichiometry is known, the concentration of the unknown solution is easily calculated from the amount of primary standard taken. This calculation was illustrated in Examples 3-6 and 3-14. The NaOH solution is now standardized and may be referred to as a secondary standard. Other acidic solutions can be standardized by comparison to this NaOH solution.

What about the $Na^+$ concentration? Does a solution prepared in this way also furnish a known concentration of $Na^+$? If the NaOH pellets are pure with respect to NaOH but impure with respect to water of hydration, the solution is standard with regard to $Na^+$. On the other hand, if the NaOH contains $Na_2O$, NaCl, or other basic material as impurities, the solution is not standardized with respect to the $Na^+$.

A better approach to the preparation of a solution with known $Na^+$ concentration would be to use a primary standard which contains $Na^+$ or to standardize the NaOH solution by a reaction which involves the $Na^+$. The former is a more satisfying approach since NaCl is an appropriate primary standard. The latter approach, although possible, is more difficult because there aren't many convenient quantitative reactions involving $Na^+$.

Primary standards are needed for all of the types of quantitative reactions used in analysis. Thus, a list of primary standards must include oxidizing agents, reducing agents, basic and acidic compounds, neutral salts, and salts that furnish specific cations or anions. Each chapter will include a discussion of the primary standards useful for calibration of the particular methods described in the chapter.

There is a definite need for the improvement or development of new standards. What was considered "pure" a few years ago may not be "pure" by today's standards. In the United States, the National Bureau of Standards, which has provided the pioneer work in the development of standards, continues to develop new Standard Reference Materials (SRMs). Information about the type and availability of SRMs can be easily obtained (see footnote to Fig. 3–1).

Other scientific groups and societies have recognized the need for the development and improvement of SRMs. For example, the Committee on Analytical Chemistry, Division of Chemistry and Chemical Technology, National Academy of Sciences—National Research Council in 1968 formed a Subcommittee on Reference Materials. This committee representing the field of analytical chemistry and other recognized committees will continue to evaluate the status of reference materials.

**Clinical–Environmental Standards.**      As analytical chemistry has grown, more and more complex samples (including samples containing trace substances) have been analyzed with a greater degree of accuracy. Two disciplines that have benefited from this growth are clinical and environmental chemistry. However, to meet these needs, there also had to be a parallel development in clinical and environmental standards.

In 1962, the Standards Committee of the College of American Pathologists surveyed clinical laboratories by submitting two serum samples each of which contained a known concentration of cholesterol. Over 1000 useful replies were received and the results are shown in Fig. 3–1. The wide scatter around the "true value," which includes errors as high as 35–50%, is the result of many sources of error. One major one is the difference in quality of the chole-

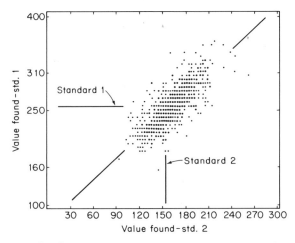

**Fig. 3–1.**    National cholesterol survey in 1962. Results of one standard are plotted against results of a second standard for each testing site. Standard 1 was 259 mg cholesterol/100 ml and standard 2 was 152 mg/100 ml. [Reprinted with permission from W. W. Meinke, *Anal. Chem.* **43(6),** 28A (1971). Copyright by the American Chemical Society.]

Table 3–6.   Standard Reference Materials for Clinical Measurements That Are Available from the National Bureau of Standards

| SRM No.[a] | Name | Purity, % | Property certified |
|---|---|---|---|
| 84h | Acid potassium phthalate | 99.993 | Acidimetric standard |
| 186Ic | Potassium dihydrogen phosphate | 99.9 | pH |
| 186IIc | Disodium hydrogen phosphate | 99.9 | pH |
| 911 | Cholesterol | 99.4 | Identity and purity |
| 912 | Urea | 99.7 | Identity and purity |
| 913 | Uric acid | 99.7 | Identity and purity |
| 914 | Creatinine | 99.8 | Identity and purity |
| 915 | Calcium carbonate | 99.9 | Identity and purity |
| 916 | Bilirubin | 99 | Identity and purity |
| 917 | D-Glucose | 99.9 | Identity and purity |
| 922 | Tris(hydroxymethyl)aminomethane | 99.9 | pH |
| 1571 | Orchard leaves | | Major and trace constituents |
| 2201 | NaCl | 99.9 | pNa |
| | | | pCl |
| 2202 | KCl | 99.9 | pK |
| | | | pCl |

[a] Orders and requests for information about these SRMs should be directed to the Office of Standard Reference Materials, National Bureau of Standards, Washington, D.C. 20234. See W. W. Meinke, *Anal. Chem.* **43(6)**, 28A (1971).

sterol that each of the laboratories used as a standard to calibrate their method of analysis. Several other similar evaluations of clinical as well as environmental testing laboratories also emphasized the need for the development of suitable standards.

Much of the development in analyses in these two areas included advances in automation of the analytical procedures and their computer control. Such procedures require continued, careful, accurate standardization to prevent the generation of meaningless data.

The National Bureau of Standards has provided leadership in the development of suitable reference standards for the clinical and environmental laboratory. Table 3–6 lists several SRMs for the clinical laboratory while Table 3–7 lists SRMs that contain trace components; most of which are useful in environmental testing laboratories.

Many other standards are at various stages of development. These include standards such as mercury in water, lead in gasoline, other kinds of leaves (similar to orchard leaves), different gas permeation tubes (tubes deliver a constant trace level of the gas at a known flow rate), river sediment standards, hydrocarbons in air, vinyl chloride in air, trace elements in oyster meat, in spinach, in grains, and in yeast, trace elements in high temperature alloys, in glasses, and in water, and many industrially related pollutant systems.

Table 3–7. **Standard Reference Materials for Trace and Environmental Analysis That Are Available from the National Bureau of Standards**[a]

| SRM No. | Name | Certified for: |
|---------|------|----------------|
| 1571 | Orchard leaves | 14 trace elements, 5 major and minor elements |
| 1577 | Bovine liver | 9 trace elements, 3 major and minor elements |
| 608–619 | Trace element glasses | 35 trace elements (not all certified) |
| 1632 | Trace elements in coal | 14 trace elements including Pb, Hg, Cd |
| 1633 | Trace elements in fly ash | 12 trace elements |
| 1630 | Mercury in coal | Hg 0.03 ppm |
| 1610–1613 | Hydrocarbons in air | Methane in air 0.0001 to 0.1 mol/percent |
| 1604–1608 | Oxygen in nitrogen | Oxygen 1.5 ppm to 978 ppm |
| 1601–1603 | Carbon dioxide in nitrogen | $CO_2$ 300–400 ppm |
| 1677–1681 | Carbon monoxide in nitrogen | Carbon monoxide 10–1000 ppm |
| 1665–1669 | Propane in air | Propane 2.8–475 ppm |
| 1625–1627 | $SO_2$ permeation tubes | Permeation rate of $SO_2$ @ .56–3 $\mu$g/min |
| 607 | Potassium feldspar | Rb, Sr content, Sr isotopic ratio |
| 685 | High purity gold | Cu, Fe, In |
| 680–681 | High purity platinum | 12 trace elements |
| 682–683 | High purity zinc | 6 trace elements |
| 726 | Intermediate purity selenium | 24 trace elements |

[a] See P. D. La Fleur, Ed., "Accuracy in Trace Analysis: Sampling, Sample Handling, Analysis," Vol. 1 and 2. *National Bureau Standards Publication 422*, Washington, D.C. 1974.

# *Questions*

1. What properties characterize a stoichiometric reaction; a nonstoichiometric reaction?
2. What is a calibration curve?
3. Explain why nonstoichiometric instrumental methods must be calibrated.
4. Explain the difference between molarity and formality.
5. What is a gram-formula weight; gram-equivalent weight?
6. Suggest where molality and mole fraction might be useful as expressions of concentration.
7. Explain the difference between percent by weight and percent by volume.
8. Prove that pH + pOH = 14.
9. Define ppm. Where is this unit of concentration useful?

10. Why is it necessary to have standards?
11. What is the difference between a primary and a secondary standard?
12. List the properties that a chemical primary standard must possess.
13. List some typical physical primary standards that are useful to the analytical chemist.
14. What is the difference between the equivalence point and end point in a volumetric titration procedure?
15. Distinguish between a primary and secondary standard solution.

## *Problems*

1. Calculate the formality of each of the following solutions.
  *a. 10.00 g of $H_2SO_4$ in 500 ml of solution.
   b. 4.00 g of NaOH in 250 ml of solution.
  *c. 6.00 g of $CuSO_4 \cdot 5H_2O$ in 1000 ml of solution.
   d. 1.50 g of $NH_3$ in 750 ml of solution.

2. Calculate the volume of water that must be added to the following solutions to give the desired formality. Assume that the volumes are additive.
  *a. 250 ml of 0.1511 $F$ to 0.1000 $F$.
  *b. 500 ml of 0.2000 $F$ to 0.1250 $F$.
   c. 100 ml of 1.000 $F$ to 0.1500 $F$.
   d. 20 ml of $2.00 \times 10^{-3}$ $F$ to $2.00 \times 10^{-4}$ $F$.

3. Calculate the weight of each substance needed to prepare the following solutions.
  *a. 500 ml of 0.2500 $F$ NaOH.
  *b. 1000 ml of 0.1000 $F$ $FeCl_3 \cdot 6H_2O$.
   c. 250 ml of 0.1250 $F$ $K_2Cr_2O_7$.
   d. 700 ml of 0.2111 $F$ $I_2$.

4. Calculate the weight of substance required to prepare the following solutions.
  *a. 500 ml of 1% by weight $NH_4Cl$ in water.
   b. 1000 ml of 10% by weight $NaC_2H_3O_2$ in water.
   c. 250 ml of 5% by weight NaCl in water.
   d. 550 ml of 10% by volume ethanol in water.
   e. 1000 ml of 2% by volume $HC_2H_3O_2$ in water.

5. Calculate the following for a mixture made from 100 ml of 0.100 $F$ $NaNO_3$, 50 ml 0.0100 $F$ NaOH, and 25.0 ml of 0.0500 $F$ KCl. Assume volumes are additive.
  *a. Calculate the mmoles of $Na^+$ in the mixture.
  *b. Calculate the formal concentration of $Na^+$ in the mixture.
   c. Calculate the mg of $Na^+$ in the mixture.
   d. Calculate the formal concentration of $Cl^-$ in the mixture.

---

* Answers are listed at the end of the book for problems marked with an asterisk.

e. Calculate the moles of $Cl^-$ in the mixture.

f. Calculate the grams of $K^+$ in the mixture.

6. Calculate the formality for each of the following:
   *a. 37% HCl; specific gravity 1.18.
   b. 70% $HClO_4$; specific gravity 1.67.
   c. 96% $H_2SO_4$; specific gravity 1.84.

7.* Calculate the volume of each of the acids in Question 6 required to prepare 500 ml of a 0.1000 $F$ acid solution.

8. Calculate the concentration of the metal ion in g/ml for the following solutions:
   *a. 1.00 g $CuSO_4 \cdot 5H_2O$ in 1000 ml of solution.
   *b. 20.0 ml of 0.1250 $F$ NaCl.
   c. 100.0 ml of 10 ppm NaCl solution.
   d. 2.00 g $Ni(NO_3)_2 \cdot 6H_2O$ in 500 ml of solution.

9.* If each of the solutions in question 8 were diluted to exactly 2.000 liters, calculate the new concentration in ppm.

10. Calculate the volume of water required to prepare the following solutions:
    *a. 100 ml of 0.20 $F$ $H_2SO_4$ from 6.0 $F$ $H_2SO_4$.
    b. 1 liter of 0.10 $F$ HCl from 37% HCl that has a density of 1.18 g/ml.
    c. 500 ml of 0.10 $F$ NaOH from 6.0 $F$ NaOH by weight.
    *d. 250 ml of 0.010 $F$ $ZnCl_2$ from 0.25 $F$ $ZnCl_2$.
    e. 500 ml of 0.10 $F$ NaOH from 0.55 $F$ NaOH.

11.* Calculate the mmoles of each solute present in the solutions listed in Question 1.

12.* Calculate the grams of each solute present in the solutions listed in Question 10.

13. Fifty (50.0) ml of 0.0100 $F$ NaOH, 40.0 ml of 0.00500 $F$ $NaNO_3$, and 40.0 ml of 0.0100 $F$ HCl are mixed. Calculate the following for the mixture considering that the reaction

$$HCl + NaOH \rightarrow NaCl + H_2O$$

takes place and that volumes are additive:
    a. mmoles of $Na^+$ present.
    b. mmoles of $Cl^-$ present.
    c. mmoles of $NO_3^-$ present.
    d. mg of $NO_3^-$ present.
    e. formality for $NO_3^-$.
    f. mmoles of $OH^-$ present.

# Chapter
# Four
# Statistical Treatment
# of Analytical Data

## INTRODUCTION

Assume that a particular experimental procedure has been selected to solve an analytical problem. The procedure should be examined to determine where errors are most likely to occur, or where to be especially concerned about the procedure. Awareness of the limitations of the glassware and instruments, how to use them, and what to expect from them is vital. Once the final answer is obtained the question that should be asked is: Is the answer acceptable and if it is, how reliable is it?

Every physical measurement is subject to a degree of uncertainity. Modification of the procedure and instruments employed in making the measurement can improve the uncertainity; it cannot be eliminated.

Many factors employing human judgment are involved in the determination of any "accepted, reliable" value.

1. In the use of measuring instruments, the last figure in the measurement must be obtained by estimating the smallest division in the measuring device.

2. A number of measurements are usually performed and an "average" is taken. Often certain data are excluded as being unreliable in the opinion of the analyst.

3. Experiments of different types have different uncertainties. In calculating the final value, a weighted average of the separate values are taken with the more accurate experiments being given greater importance.

4. The error of human judgment enters into all of the above items. Although certain rules are available for guidance, they themselves are based on arbitrary assumptions and are influenced by opinion.

In this chapter, the typical types of errors encountered in quantitative analysis are briefly described. Following this, mathematical procedures for estimating and reporting the magnitudes of the errors are evaluated.

## DEFINITIONS

*Accuracy* is a measure of the departure of the measurement from the true value. This implies that the exact or true value is known, which, of course, is normally not the case. What is known, however, is that value which has been compared to an acceptable standard. This is used as the exact or true value.

*Precision* is a measure of the reproducibility of a measurement. As the difference between the repetitive measurements increases, the precision decreases. However, good precision does not mean good accuracy since it is possible to make the same mistake over and over again. Usually, for an experienced scientist, good precision can be used as a guideline towards suggesting that good accuracy has been achieved. However, the scientist who is both wise and experienced will not be lulled into rigidly applying this guideline.

The goal in any measurement should be to obtain both acceptable precision and accuracy.

*Uncertainty* is an expression of the extent of an error in a measurement. It is determined statistically or by propagation of estimated errors. A "true value" of an experimentally measured quantity is never known. The value reported which is often the average of a series of measurements is a "best value." In reporting a "best value" it is useful to estimate the errors and express them as an uncertainty by a $\pm$ sign. This gives the range within which the true value is believed to lie. Consider a $0.1106 \pm 0.0007$ $M$ HCl solution. The solution contains no more than 0.1113 and no less than 0.1099 moles of HCl per liter. The best value is 0.1106 $M$ and the uncertainty is $\pm 0.0007$ $M$. An equivalent expression is 0.1106 $M \pm 0.63\%$, where 0.0007 is 0.63% of 0.1106. The *absolute uncertainty* or *absolute error* is $\pm 0.0007$ $M$ and the *relative uncertainty* or *relative error* is $\pm 0.63\%$ or $\pm 0.0063$. It is also customary to express relative uncertainty or error as parts per thousand.

*Significant figures* are the number of figures necessary to express a measurement to the precision with which it is made. If the figure is assigned some degree of reliability, it is a significant figure. The figure "zero" can be a significant figure or it can be used to locate the decimal point.

In general, data are reported in such a way that only the last figure is an uncertain one. Consider the following illustration: An object is weighed on an ordinary triple-beam balance and found to weigh 3.45 g. With this type of balance it is not possible to weigh the object to any greater degree of accuracy. If the reproducibility of the balance is $\pm 0.02$ g, the second figure past the decimal is uncertain. Any effort to determine the number in the third decimal place, such as 3.457 g, by weighing on this balance is wasted effort and is an incorrect use of the reproducibility of the balance. Thus, the weight for the object will be in the range 3.43 to 3.47 g and the weight would be recorded as 3.45 ± 0.02 g. If the same object is weighed on an analytical balance that has a reproducibility of ±0.0002 g, the weight could be determined as 3.4557 g. Now, the figure 7 is the uncertain figure and the weight would be recorded as 3.4557 ± 0.0002 g.

The use of zero as a significant figure can be illustrated by using a different object and the same two balances as in the previous example. On the triple-beam balance the object was found to weigh 4.52 g and by chance on the analytical balance it was found to weigh 4.5200 g.

If these numbers were recorded in milligrams, the number in the first case would be 4520 mg while the second would be 4520.0 mg. At first sight, there does not appear to be much difference between the two. However, even though the numbers are listed in milligrams the 2 is still the uncertain figure for the triple-beam balance weight while the last zero is the uncertain figure for the analytical balance weight. Considering the reproducibility of the balances the weights would be written as 4520 ± 20 mg and 4520.0 ± 0.2 mg. In the former case the zero is being used to locate the decimal point. To avoid confusion this number can be written as $4.52 \times 10^3$ mg. It would not be correct to write 4520.0 mg as $4.52 \times 10^3$ mg since in doing this the number of significant figures is reduced from five to three.

Powers of ten are very useful in identifying the significant numbers. However, since there is still the possibility of ambiguity in the use of zeros, it is very important to be aware of the accuracy of the different devices used in the measurements and the capabilities of the reaction if a chemical step is involved.

Using significant figures in measurements requires personal judgments. In many cases, arbitrarily applied rigid rules can often lead to poor use of the figures. As a guideline, in multiplication and division the number of significant figures in the final answer is determined by that number having the least number of significant figures. In addition and subtraction the number of significant figures is determined by the location of the decimal point and explicit information about whether zeros are significant or are present only to indicate the location of the decimal point. The following example illustrates significant figures by using an 8-place calculator to carry out the mathematics.

*Example 4-1.*    Calculate the following and give the final answer in significant figures.

| A | B | C | D |
|---|---|---|---|
| | | 14.1 | |
| 14.6 | | 1.2 | 14,050 |
| × 3.12 | 14.106 ÷ 4.204 | 112.14 | − 121 |
| 45.552 | = 3.3553758 | 127.44 | 13,929 |
| ans. 45.6 | ans. 3.355 | ans. 127.4 | ans. 13,929 |

An *error* by definition is (1) the difference between a measured value and the true value or (2) the estimated uncertainty in the experiment expressed in terms of statistical quantities. Although the term error is used to describe the difference between two measured values (for example, the difference between results obtained by two students), this technically is incorrect. This difference should be called a *discrepancy*. Errors are designated as being random errors, systematic errors, or illegitimate errors.

*Random errors* are indicated when a measurement is repeated and the resulting values do not exactly agree. The reasons for the disagreement between individual measurements must be the same reasons for their disagreement with the true value. Random errors are experimental or accidental errors and include errors of judgment, fluctuating experimental conditions, instrumental fluctuations, and imperfections in the measuring process.

*Systematic errors* are indicated if all the individual values are in error by the same amount. They are the result of errors in calibration of the instrument or procedure, in personal errors which are caused by personal habits, in experimental conditions which change relative to those during calibration, and errors due to known imperfections in the technique.

Random and systematic errors are present in most experiments. Sometimes both may arise from the same source. If a particular experimental procedure suffers only from small random errors, it will provide a high degree of precision. If the systematic errors are small, it is said to have high accuracy.

The terms determinate and indeterminate errors are also often used. The former are errors that can be evaluated in a logical way by a theoretical or experimental procedure. These are random errors. The latter cannot be logically evaluated. These are systematic errors.

*Illegitimate errors* are clearly avoidable errors which have no place in an experiment. These include outright mistakes in reading instruments or balances, failing to adjust experimental conditions, or failing to do the calculations properly.

The *mean* is the numerical value obtained by dividing the sum of a set of replicate measurements by the number of individual results in the set. Other terms having the same meaning are "arithmetic mean" and "average."

The mean, $m$, is calculated from the sum of the individual measurements

$$m = \frac{\sum M_n}{n} \tag{4-1}$$

where $M$ is the individual measurement and $n$ is the total number of measurements.

The *median* is a value about which all the others are equally distributed. Half of the values are larger and the other half are smaller than the median value. The mean and the median may or may not be the same.

As an example of absolute and relative error, consider the two numbers, 20.0 and 40.0, to have an absolute uncertainty of $\pm 0.1$. The range of the acceptable values would be 19.9 to 20.1 and 39.9 to 40.1, respectively. On the other hand, if the relative uncertainty of the same numbers is $\pm 0.1$ (or 1 part per ten), the acceptable values would now be in the range 18.0 to 22.0 and 36.0 to 44.0. It is clear then that the manner of expressing uncertainty must be carefully designated.

*Average deviation* is the average of the deviations from the mean computed without regard to sign. The average deviation, $\bar{d}$, is calculated by

$$\bar{d} = \frac{\sum |M_n - m|}{n} \tag{4-2}$$

where $|M_n - m|$ is the absolute value of the deviation of the $M_n$th number from the mean.

As an example, assume that an object is weighed five times and the values listed in the following tabulation are obtained. In the value for $\bar{d}$ the 2 is set below the line. This indicates that the 2 is not a significant figure.

| Weight (g) | Deviation ($|M_n - m|$) |
|---|---|
| 0.1010 | 0.0006 |
| 0.1020 | 0.0004 |
| 0.1005 | 0.0011 |
| 0.1030 | 0.0014 |
| 0.1015 | 0.0001 |
| $\sum M_n = 0.5080$ | $\sum |M_n - m| = 0.0036$ |
| $m = 0.1016$ | $\bar{d} = 0.0007_2$ |

*Standard deviation,*[*] $s$, can be obtained from the deviations from the mean by the equation

$$s = \left[ \frac{1}{n-1} \sum (M_n - m)^2 \right]^{1/2} \tag{4-3}$$

The standard deviation is preferred to the average deviation because it is statistically interpretable.

---

[*] The standard deviation, $\sigma$, in a more precise statistical form is given by the equation

$$\sigma = \left[ \frac{1}{n} \sum (M_n - m)^2 \right]^{1/2}$$

However, this equation implies that a large number of measurements have been made. In practice, a number considerably fewer than this are made and thus, the standard deviation expressed as $s$ is used.

The calculation of the standard deviation is illustrated by considering the previous set of data.

| Weight (g) | Deviation | $(M_n - m)^2$ |
|---|---|---|
| 0.1010 | 0.0006 | $3.6 \times 10^{-7}$ |
| 0.1020 | 0.0004 | $1.6 \times 10^{-7}$ |
| 0.1005 | 0.0011 | $12.1 \times 10^{-7}$ |
| 0.1030 | 0.0014 | $19.6 \times 10^{-7}$ |
| 0.1015 | 0.0001 | $0.1 \times 10^{-7}$ |

$$\sum | M_n - m | = 0.0034 \qquad\qquad 37.0 \times 10^{-7}$$

$$\bar{d} = 0.0006_8 \qquad s = \left(\frac{3.70 \times 10^{-6}}{4}\right)^{1/2} = 0.0009_6$$

# ELIMINATION OF DATA

Normally, it can be assumed that the larger the number of measurements on a given sample, the greater the confidence that the average of these numbers is the true answer. Therefore, it is desirable to obtain as many measurements as possible. However, there is a practical limit set by either the method of measurement itself or by the amount of sample available. The question still remains as to which of the measurements should be accepted and which should be rejected. There are a number of tests available which can be used for this purpose. The test or tests chosen are dependent upon the degree of accuracy expected in the measurements.

AVERAGE DEVIATION.   After the average deviation for a series of measurements has been obtained, the data may be tested as follows. If a measurement deviates from the mean by more than $4\bar{d}$, it may be excluded from the original average. This test, however, is only used when the number of measurements exceeds three.

STANDARD DEVIATION.   If many measurements are taken on a single sample, the observations will approach a normal distribution curve (see Fig. 4–1*), where the frequency of occurrence of a measurement is plotted versus the value of the measurement. The curve indicates that upon taking another measurement, there would be a 68.26% chance it would fall between $\pm 1\sigma$ of the mean. If $\pm 2\sigma$ were taken, the chance would be 95.46%, while a $\pm 3\sigma$ would give a 99.7% chance that the answer would fall within this range. Normally, when using the standard deviation as a criterion for rejecting a measurement the value of $3\sigma$ is used. If the value of a particular measurement falls outside this range, it may be rejected.

---

* See footnote on page 52.

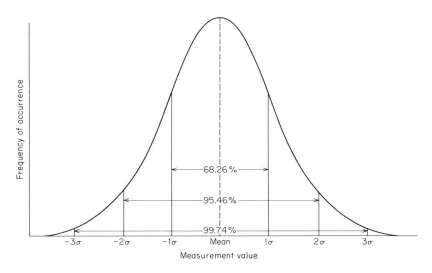

**Fig. 4–1.** Ideal sampling distribution.

The use of the $4\bar{d}$ and $3s$ rejection tests are both dependent on the total number of measurements of a quantity that are available. If enough data are available it can be shown that the two rejection tests are essentially equal.

THE $Q$ TEST.    The $Q$ test for rejection of a measurement may be made on a population of 3 to 10 measurements of a quantity. The $Q$ test may be applied at any confidence level; however, under normal circumstances the 90% level is used. This means that if the test indicates that a measurement should be rejected, it may be eliminated with 90% confidence. In order to apply this test, the values of the measurements are arranged in increasing order. This places the most divergent values at the top or bottom of the table. The range $(M_n - M_1)$ is calculated and the quotients, $Q$, for the upper and lower values are calculated:

$$Q_1 = \frac{M_2 - M_1}{M_n - M_1} \tag{4-4}$$

$$Q_n = \frac{M_n - M_{n-1}}{M_n - M_1} \tag{4-5}$$

Since the values which would be eliminated are either $M_1$ or $M_n$, the $Q$ test is applied to these two values. The quotients, $Q_1$ and $Q_n$, are obtained by dividing the difference in $M_1$ and $M_n$ and their neighboring values by the range. These values are then compared with the $Q$ table, given in Table 4–1. If either $Q_1$ or $Q_n$ is larger than the value given in the table, the measurement may be rejected at a certain confidence limit. For example, if $Q_1$ has a value

**Table 4-1. Values of $Q$ for Rejection of Data at the 90% Confidence Limit**

| Number of observations | $Q_{90\%}$ |
|---|---|
| 3 | 0.94 |
| 4 | 0.76 |
| 5 | 0.64 |
| 6 | 0.56 |
| 7 | 0.51 |
| 8 | 0.47 |
| 9 | 0.44 |
| 10 | 0.41 |

of 0.60 and the number of measurements, $n$, is 7, the value may be eliminated at the 90% confidence limit.

THE $t$ DISTRIBUTION. If enough measurements of a quantity are taken, a distribution similar to that shown in Fig. 4–1 will be obtained. Unfortunately, in most determinations it is not possible to make a large number of measurements. The result is that the calculated average may be different than the distribution mean. The $t$ distribution predicts the limits within which an average should agree with the distribution mean.

In order to determine the confidence limits, the standard deviation of the measurements is determined and a value of $t$ is obtained from Table 4–2 for a given probability level. The limits for the given confidence level are then calculated by

$$\text{average} \pm \frac{ts}{\sqrt{n}} \qquad (4\text{-}6)$$

**Table 4-2. Values of $t$ for Different Levels of Probability**

| Number of measurements | Confidence level | | | |
|---|---|---|---|---|
| | 80% | 90% | 95% | 99% |
| 3 | 1.89 | 2.92 | 4.30 | 9.92 |
| 4 | 1.64 | 2.35 | 3.18 | 5.84 |
| 5 | 1.53 | 2.13 | 2.78 | 4.60 |
| 6 | 1.48 | 2.02 | 2.57 | 4.03 |
| 7 | 1.44 | 1.94 | 2.45 | 3.71 |
| 8 | 1.42 | 1.90 | 2.36 | 3.50 |
| 9 | 1.40 | 1.86 | 2.31 | 3.36 |
| 10 | 1.38 | 1.83 | 2.26 | 3.25 |

where $t$ is obtained from the $t$ tables, $s$ is the standard deviation, and $n$ is the number of measurements on a particular sample.

As an example, assume that six measurements are made on a particular sample, giving an average of 9.46% with a standard deviation of 0.17%. There is a 90% confidence that if a large number of measurements were made, the mean would be within the limits of 9.46% $\pm$ 0.14%.

*Example 4-2.*     The following data are the results for a series of iron determinations which were made on the same sample. Evaluate statistically which of the data should be eliminated.

(a) Calculation of the average deviation and standard deviation.

| % Fe | $(\,|\,M_n - m\,|\,)$ | $|\,M_n - m\,|^2$ |
|------|------|------|
| 14.28 | 0.30 | 0.090 |
| 14.20 | 0.22 | 0.048 |
| 13.99 | 0.01 | 0.0001 |
| 13.18 | 0.80 | 0.64 |
| 13.92 | 0.06 | 0.0036 |
| 14.30 | 0.32 | 0.102 |
| Average 13.98 | $\bar{d} = 0.285$ | $s = 0.420$ |

If no data are eliminated, the answer is

$$\% \text{ Fe} = 13.98\% \pm \text{ average deviation of } 0.285$$

$$\% \text{ Fe} = 13.98\% \pm \text{ standard deviation of } 0.420$$

(b) Application of average deviation and standard deviation. In practice, it is not necessary to do both since if enough measurements of a quantity are available, both will give the same result. Each calculation is done here primarily as an illustration of the procedure.

(1) Average deviation

The average deviation $\pm 0.285$ was calculated by considering all the data. If the poorest measurement 13.18% is discarded the new average and average deviation is

$$\text{Average} = 14.14\% \qquad \bar{d} = 0.14_6$$

and

$$4 \times \bar{d} = 4 \times 0.15 = 0.60$$

The limit for acceptance now becomes

$$\% \text{ Fe} = 14.14 \pm 0.60\%$$

and clearly the value 13.18% is not within these limits and can be discarded. This procedure is repeated for the next poorest value, 13.92%, and the new $d$ calculated. If this is done one would find that this result is within the limits and the measurement is retained.

(2) Standard deviation

The standard deviation $\pm 0.420$ was calculated by considering all the data. If the poorest measurement, 13.18% is discarded the new average and standard deviation are

$$\text{Average} = 14.14\% \qquad s = 0.17_3$$

and

$$3 \times s = 3 \times 0.17 = 0.51$$

The limit for acceptance now becomes

$$\%\text{Fe} = 14.14 \pm 0.51\%$$

and clearly the value 13.18% is not within these limits and can be discarded. This procedure is repeated for the next poorest value, 13.92%, and the new $s$ calculated. If this is done one would find that this result is within the limits and the measurement is retained.

(c) The $Q$ test

13.18%
13.92%
13.99%
14.20%
14.28%
14.30% Range $(M_6-M_1) = (14.30\text{--}13.18) = 1.12$

By Eqs. (4-4) and (4-5)

$$Q_1 = \frac{M_2-M_1}{M_6-M_1} = \frac{13.92 - 13.18}{1.12} = 0.66$$

$$Q_2 = \frac{M_6-M_5}{M_6-M_1} = \frac{14.30 - 14.28}{1.12} = 0.01_8$$

For a 90% confidence limit $Q_{0.90} = 0.56$ (Table 4–1). The first value is not within this limit, thus the value 13.18 is eliminated. Repeating the calculation would show that all remaining data are acceptable and the average would be reported as

$$\%\text{Fe} = 14.14\%$$

(d) The $t$ distribution

By Eq. (4-6) and a probability level of 95% (see Table 4–2), the confidence limit is

$$\text{mean} \pm \frac{ts}{\sqrt{n}}$$

$$13.98 \pm \frac{2.57 \times 0.42_3}{\sqrt{6}}$$

$$\%\text{Fe} = 13.98 \pm 0.44_0$$

The value 13.18% is out of this range and this value should be eliminated.

Example 4–2 illustrates the basic statistical techniques for eliminating data. How the final result is reported will depend on which statistical test procedure is used. Most often either the average deviation $(4\bar{d})$ or standard deviation $(3s)$ test is used. Consequently, in Example 4-2 the %Fe is reported as 14.14% Fe with an average deviation of $\pm 0.60$ or 14.14% Fe with a standard deviation of $\pm 0.051$.

In the beginning analytical laboratory the student is faced with making decisions with a small set of measurements of a quantity, for example, three or four determinations on a single sample. In eliminating data the first step is to consider the possibility of systematic and illegitimate errors. Clearly, data containing the second type of error should be eliminated. The remaining data (3 or more) can be evaluated by the average or standard deviation tests or by the $Q$ test. However, it is important to realize that applying statistical tests which are based on a large set of data to small sets of data ($<5$) can be misleading. In many cases, with small sets of data, it is preferred to arrive at the uncertainty by evaluating the procedural errors in arriving at the measurement. This technique is called propagation of errors.

## PROPAGATION OF ERRORS

The technique of propagation of errors is one whereby the estimated uncertainties in measured quantities (independent variables) are combined to give an uncertainty in a quantity calculated from them (the dependent variable). This estimation of the uncertainty can be compared with the average deviation to find out whether all the sources contributing significantly to the total uncertainty is recognized or not.

If only one or a few experiments are done, the estimated uncertainty from the propagation of errors can be used as a substitute for the average deviation. First, the uncertainty in each measured value, such as in each weight, volume, length, etc., must be estimated. Table 4–3 lists typical uncertainties in the measurements encountered in the beginning analytical laboratory. Second, the uncertainties are combined according to two rules:

1. If the measured values are added or subtracted to get the result, the *absolute uncertainties* are added to get the *absolute uncertainty* in that result. If the measured values are $a$ and $b$ and their uncertainties are $\pm \Delta a$ and $\pm \Delta b$, then

$$a + b = c, \qquad \Delta c = \pm (\Delta a + \Delta b)$$

$$a - b = c, \qquad \Delta c = \pm (\Delta a + \Delta b)$$

*Example 4-3.* In measuring volume with a buret, the volume delivered is the difference between the initial and final levels in the buret.

$$V = V_{\text{final}} - V_{\text{initial}}$$

Table 4–3.  **Typical Error Values of Some Common Laboratory Equipment**

| Weighing | | $2 \times 10^{-4}$ g |
|---|---|---|
| Pipets | 1 ml | 0.006 ml |
| | 2 ml | 0.006 ml |
| | 5 ml | 0.01 ml |
| | 10 ml | 0.02 ml |
| | 25 ml | 0.03 ml |
| | 50 ml | 0.05 ml |
| Volumetric flasks | 10 ml | 0.02 ml |
| | 25 ml | 0.03 ml |
| | 50 ml | 0.05 ml |
| | 100 ml | 0.08 ml |
| | 200 ml | 0.10 ml |
| | 500 ml | 0.15 ml |
| | 1000 ml | 0.30 ml |
| Buret reading | | 0.02 ml |
| Chart paper | | ½ division |

$V_{final}$ and $V_{initial}$ will have uncertainties of $\pm 0.02$ ml because the buret is marked only to the nearest 0.1 ml and it is necessary to estimate the nearest 0.01 ml. Then

$$\pm \Delta V = \pm (\Delta V_{final} + \Delta V_{initial})$$

$$\pm \Delta V = \pm (0.02 + 0.02) = \pm 0.04 \text{ ml}$$

2. If the measured values are multiplied or divided to get the final result, their *relative* uncertainties are added to get the *relative or fractional* uncertainty in the final result. Thus, if $a \times b = c$ or $a/b = c$

$$\pm \frac{\Delta c}{c} = \pm \left( \frac{\Delta a}{a} + \frac{\Delta b}{b} \right)$$

The absolute uncertainty in $c$ is then obtained by multiplying its relative uncertainty by the value of $c$,

$$\pm \Delta c = \pm \frac{\Delta c}{c} \times c$$

*Example 4-4.*    In the titration of $10.00 \pm 0.04$ ml of $0.1018 \pm 0.0002$ $F$ HCl, $23.02 \pm 0.04$ ml of NaOH was required for neutralization. Calculate the formality of NaOH and the uncertainty in this value. All volumes were measured by a buret.

(The uncertainties in the volumes were determined in Example 4-3 while the uncertainty in $F_{HCl}$ was supplied by the person that did the standardization.

$$F_{NaOH} = \frac{ml_{HCl} \times F_{HCl}}{ml_{NaOH}}$$

$$F_{NaOH} = \frac{10.00 \text{ ml} \times 0.1018 \text{ } F}{23.02 \text{ ml}} = 0.04422 \text{ } F$$

The total estimated uncertainty in $F_{HCl}$ is

$$\pm \frac{\Delta F_{NaOH}}{F_{NaOH}} = \frac{\Delta ml_{HCl}}{ml_{HCl}} + \frac{\Delta F_{HCl}}{F_{HCl}} + \frac{\Delta ml_{NaOH}}{ml_{NaOH}}$$

$$\pm \frac{\Delta F_{NaOH}}{F_{NaOH}} = \frac{0.04 \text{ ml}}{10.00 \text{ ml}} + \frac{0.0002 \text{ } F}{0.1018 \text{ } F} + \frac{0.04 \text{ ml}}{23.02 \text{ ml}} = \pm 0.0077$$

The absolute uncertainty in $F_{NaOH}$ is

$$\frac{\Delta F_{NaOH}}{F_{NaOH}} \times F_{NaOH} = \pm 0.0077 \text{ ml} \times 0.04422 \text{ } F = \pm 0.00034 \text{ } F$$

Thus, the formality of the NaOH in four significant figures is

$$F_{NaOH} = 0.04222 \pm 0.0003 \text{ } F$$

It should be noted that the uncertainty in $F_{HCl}$ was included in the propagation of errors, and so $\pm 0.0003$ is the total estimated uncertainty in $F_{NaOH}$.

Suppose successive samples of NaOH were titrated and an average and average deviation were obtained. This average deviation is a measure of the *random error* of measurement and the same identical value for $F_{HCl}$ was used each time in calculating $F_{NaOH}$. Thus, the uncertainty in $F_{HCl}$ does not enter into this *random error*.

To estimate the *random error* by a propagation of estimated errors the uncertainty of $F_{HCl}$ should not be included for the reasons just given. Thus, the estimated *fractional random error* or *uncertainty* in $F_{NaOH}$ is

$$\pm \frac{\Delta F_{NaOH}}{F_{NaOH}} = \frac{\Delta ml_{HCl}}{ml_{HCl}} + \frac{\Delta ml_{NaOH}}{ml_{NaOH}} = \frac{0.04 \text{ ml}}{10.00 \text{ ml}} + \frac{0.04 \text{ ml}}{23.02 \text{ ml}} = \pm 0.0057 \text{ ml}$$

The *absolute random uncertainty* is

$$\Delta F_{NaOH} = \frac{\Delta F}{F} \times F = \pm 0.00025 \text{ F}$$

which is slightly smaller than the *total* estimated uncertainty or $\pm 0.00034$. (In this case, of course, both should be rounded to $\pm 0.0003$, which shows that the uncertainty in the calculated $F_{NaOH}$ is controlled by the uncertainty in the volumes, not by that in $F_{HCl}$.

Suppose you had performed several titrations of this NaOH and had obtained the following average formality and average deviation thereof, $0.0442 \pm 0.0002 \, F$. The average deviation and estimated uncertainty are approximately the same size. We conclude that the uncertainties in volume used in the propagation of errors are reasonable estimates, and there are no other significant sources of error. On the other hand, if the average deviation had turned out to be $\pm 0.0008$, there is either an additional source of error to look for, or else you have estimated the errors in volume too conservatively.

## Questions

1. Discuss the need for statistical evaluation of experimental data.
2. What is the difference between accuracy and precision?
3. Criticize the statement, "Good precision means good accuracy."
4. When can a zero be used as a significant figure?
5. Describe the types of errors that are often encountered in weighing a sample; in carrying out an experimental procedure.
6. Explain why standards are often the limitation in establishing uncertainties.
7. What is the difference between mean and median?
8. What is the difference between average deviation and standard deviation?

## Problems

1. Express the result for the following arithmetic operations with the proper significant figures. Assume that each value is uncertain to $\pm 1$ in the last figure.

*a. $6.78 + 44.67 =$  
b. $56.9 + 4.89 + 22.67 - 14.1 =$  
c. $4.50 \times 5.67 \times 23.45 =$  
d. $77.9 \times 0.00891 + 44.56 =$  

e. $(444.6 - 34.56)/0.776 =$  
f. $1.78 \times 10^4 - 2.444 \times 10^5 =$  
g. $(3.456 \times 10^{-4})(2.4 \times 10^{-6}) =$  
h. $4.562 \times 10^{-7}/6.7 \times 10^{-3} =$  

2.* A sample is weighed 10 times on a balance. The data are given below:

*Weight in grams*

| | | | |
|---|---|---|---|
| 3.6052 | 3.6053 | 3.6060 | 3.6051 |
| 3.6049 | 3.6052 | 3.6081 | 3.6021 |
| 3.6063 | 3.6047 | 3.6048 | 3.6098 |

---

* Answers are listed at the end of the book for problems marked with an asterisk.

Determine the mean, the median, average deviation, and standard deviation. By applying the tests for elimination of data, should any of the weighings be eliminated?

3. The thickness of a board is measured as a function of distance:

| Distance (cm) | Thickness (cm) | Distance (cm) | Thickness (cm) |
|---|---|---|---|
| 1 | 23.52 | 7 | 20.67 |
| 2 | 24.67 | 8 | 19.34 |
| 3 | 22.47 | 9 | 27.23 |
| 4 | 23.74 | 10 | 26.24 |
| 5 | 22.67 | 11 | 27.32 |
| 6 | 21.70 | 12 | 23.54 |

Determine the average thickness of the board, the average deviation, and the standard deviation.

4.* On a triplicate gravimetric determination the precipitates are found to weigh 0.3417 g, 0.3426 g, and 0.3342 g. Calculate the mean and the average deviation, and determine whether any data should be eliminated.

5. Should any of the following data for a triplicate titrimetric determination be eliminated? % Composition: 22.64%, 22.64%, 23.01%.

6.* A triplicate determination of the amount of glucose in blood gave the following results: 67.1 mg/100 ml, 72.5 mg/100 ml, and 94.4 mg/100 ml glucose. Should any of the measurements be eliminated?

7.* A 25-ml pipet (see Table 4–3) was used to transfer an aliquot of $0.2081 \pm 0.0008$ F HCl to a beaker. The acid required $41.51 \pm 0.05$ ml of NaOH for neutralization. Calculate the formality of the NaOH and the uncertainty in this value.

8. A 50-ml pipet was used to transfer an HCl solution to a 250-ml volumetric flask (see Table 4–3) and diluted to volume. A 25-ml aliquot of the diluted HCl was taken by pipet, transferred to a beaker, and required $21.12 \pm 0.04$ ml of a $0.1050 \pm 0.0002$ F NaOH solution for neutralization. Calculate the formality of the HCl and the uncertainty in this value.

# Chapter Five Preliminary Operations in Quantitative Analysis

## INTRODUCTION

Several basic laboratory operations are common to all quantitative analysis procedures. These are sampling, drying, weighing, and dissolving. Dissolving is the one operation not always necessary, since there are several instrumental methods in which the measurements are made directly on the sample.

For the experienced analyst these operations are rather obvious and often are performed so routinely that little thought is given to them. In certain respects this is unfortunate since proper preparation for the measurement is as important as the measurement itself. In other words, the final measurement is no better than the preparation for the measurement.

## SAMPLING

The objective in obtaining a suitable sample is to take a portion that is representative of all the components and their amounts that are contained in the bulk sample. If the bulk sample is homogeneous, no problem is encountered in obtaining a laboratory size sample, regardless of whether it is solid, liquid, or gas. It is when the bulk sample is heterogeneous that special precautions must be taken in order to obtain a representative sample. For example, the iron content of a rod of iron alloy may vary widely through the length of the rod, on the surface, and in the center. If the rod was homogeneous, it wouldn't matter what section or part of the rod was selected as the sample, since the composition would be uniform throughout the rod. Unfortunately, blunders in sampling are made and much time and effort is wasted in trying to get good determinations using improper samples.

A bulk sample may be reduced to a laboratory size by a random choice or it may be reduced according to a statistically based plan, which theoretically gives every particle or portion of the substance an equal chance of appearing in the sample. In general, statistical sampling requires removal of portions from every section of the sample. These are subsequently combined, mixed, and resampled until a suitable laboratory size sample is obtained. The details of this general technique will differ according to the physical state of the substance being sampled.

Sampling by random choice is difficult, as personal prejudices often exert an influence. Some deliberate thoughts about how to carry out the random choice are necessary before actually doing it. Nothing should be done to the sample which might inadvertently change its composition. Thus, crushing or grinding, labeling, and storage of the sample must be employed cautiously.

Different techniques are employed for sampling gases, liquids, and solids. Heterogeneous mixtures such as emulsions, powders, suspensions, or aerosols require statistical handling. Only in homogeneous samples are the techniques simple and generally straightforward.

**Gases.** There are three basic methods of collecting gases. These are expansion into an evacuated container, flushing, and displacement of a liquid. In all cases the collection vessels are of known volume, and the temperature and/or pressure within the vessel must be known. Usually, the collection vessels are made from glass (other inert material can be used), and they must be fitted with an inlet and an outlet both of which can be conveniently opened and closed.

Contamination from previous samples can be a serious problem, and its elimination is affected by extensive flushing of the container with the gas to be sampled. The design of the sample device should easily permit this procedure.

**Air Sampling.** Air is a complex mixture containing gases as well as particulate matter. Its actual composition is very dependent on the location and environment from which it is taken. Presently, because of pollution studies, much effort is directed toward monitoring the quality of air.

In collecting an air sample and assuming that the determination of dissolved or suspended pollutants is what is being sought, two goals must be satisfied.

1. An accurate flow device must be used to measure the volume of air sample. Such a device must be calibrated as one would calibrate a piece of volumetric glassware, since these devices represent a measurement by volume.

2. A sample collector, such as a filter or an absorbing solution, must be used to trap the contaminants in the air. The actual efficiency of the collector

must be determined experimentally with standards since few collectors of this type operate with 100% collection efficiency.

Atmospheric sampling is much more difficult. Factors such as wind, temperature, or rain are variables which are difficult to overcome or control. The type of sampling method that is selected depends on the chemical and physical properties of the substances in the atmosphere that are being determined. In general, the atmospheric sample is passed through a series of fine filters or through a trapping solution. In the first case, which is excellent for isolating particulate matter, the filtering action is controlled by the porosity of the filtering device. In the latter, as the atmospheric sample is percolated through a column of solution, a chemical reaction traps the sought-for components. Figure 5–1 illustrates several designs for collecting air samples.

**Liquids.**    Sampling pure or homogeneous liquids is straightforward and usually any sampling device which doesn't destroy the purity or homogeneity is an appropriate one. It is, however, a good idea to statistically sample even "pure" liquids.

Heterogeneous liquid mixtures present a more difficult problem and the technique employed depends on whether the mixture is a suspension, an

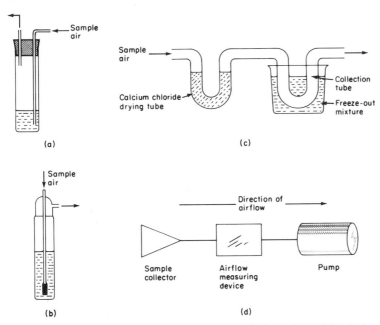

**Fig. 5–1.**  Collection of air samples (a) by a trapping solution using a fritted absorber, (b) and (c) by a cold trap, and (d) by a general arrangement of absorber, flow-measuring device, and pump.

emulsion, a mixture of immiscible liquid phases, or a liquid containing solid residues. Additional complications arise if the liquid mixture is unstable (for example, an emulsion), contains volatile components, or contains dissolved gases.

In general, aliquots are randomly withdrawn from various depths and from all locations in the liquid sample. These can be analyzed separately or combined to provide a composite sample statistically representative of the original sample.

**Solids.**     If a solid is homogeneous, any portion of it can be selected as being representatitive. For a heterogeneous solid a careful plan must be prepared which will statistically allow selection of all sections of the bulk solid. The sampling can be done by hand or by a mechanical sampling machine. The latter technique is useful if the bulk sample is a large mass.

It is not always possible to statistically obtain a representatitive sample. Consider the difficult task of determining the composition of the moon's crust. Obviously, a statistical selection of the moon's surface was not possible. From the limited quantity of moon rocks and dust, the sampling was based partly on size and partly on physical state.

The effect of particle size is introduced if a solid substance is being sampled, since composition of particles of different sizes can vary. In general, the conversion of a large sample to one suitable for analysis requires, first, reducing the sample to a uniform particle size and, second, reducing the mass of the sample. A uniform particle size is obtained by putting the sample through either crushers, pulverizers, mills, or mortars. Sieving can also be used. Whatever the procedure, it is necessary to ensure that contamination is not introduced into the sample by these operations.

A very important part of the analysis of the moon rocks or of any material from beyond the earth is to establish whether or not there is organic matter present. A recent report on the analysis of a meteorite demonstrates the need for very careful handling of the samples to prevent organic contamination on earth. This is illustrated in the following example.

Figure 5–2 shows a flow sheet describing the sampling technique employed on the meteorite (February 8, 1969, date of falling; February 15, 1969, date of sample collection; and March 1 and 10, 1969 date of duplicate analyses). The sample was egg shaped and weighed about 2.5 kg.

A 250-g fragment was taken and portions of its surface and its interior used for analysis. The surface chips represented a depth of about 0.25 inch and included all of a fusion crust and a fresh surface break. This entire operation and all succeeding operations were carried out in a clean cabinet through which a filtered air stream was continuously passed. The flow sheet describes some of the details of the procedure. The actual data can be found

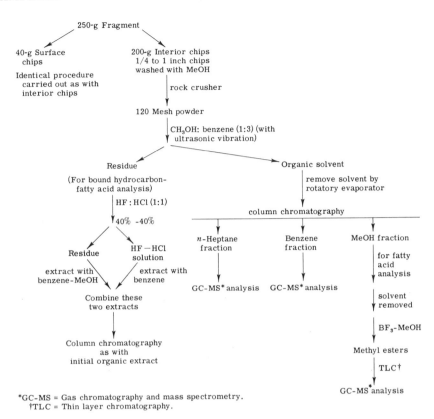

**Fig. 5–2.** Flow sheet for analysis of meteorite sample. [From J. Han, B. R. Simoneit, A. L. Burlingame, and M. Calvin, *Nature* **222**, 364 (1969).]

in the reference. From these data it was concluded that organic material found in the surface layer of this meteorite was of biological origin and could not be other than terrestrial contamination acquired during the short period of time before collection (February 8 to 15). Since the sample was contaminated it is doubtful that any conclusions can be drawn with respect to organic matter being originally present in the sample.

Only correct sampling procedures will yield useful information. Regardless of how precise or accurate the chemical operations or measurements are, the collected data will not be better than the type of sample taken.

Solid sampling can involve large masses such as a train-car load of coal or large amounts of small items such as pharmaceutical tablets. In quality control in the manufacture of pharmaceutical tablets a large number of tablets are randomly selected, weighed, and then pulverized into a powder. A weighed amount of the powder is taken for analysis and the results reported on a per tablet basis.

**Clinical Samples.**     Clinical samples are most often whole blood, serum or plasma, cerebrospinal fluid, gastric juice, ascitic fluid in the abdominal cavity, pleural fluid in the chest cavity, synovial fluid in joints, draining wounds, urine, and tissue. Each of these samples presents its own unique problems with regard to the collection and handling of the sample. It is beyond the scope of this book to document these details.

A factor that the analytical chemist or analyst encounters in clinical analysis more so than in industrial or enviromental analysis is that someone else is responsible for collecting the sample. The physician, nurse, or medical technician obtains the sample and the analyst must rely on their procedural ability to properly collect and handle the sample. For example, in collecting blood or other physiological samples the individual responsible for the collection must be aware of the following precautions.

1. The immobilization of the fluid flow in the individual should be minimized, since prolonged immobilization may alter the chemical values.

2. Samples of the fluid, in most cases, should not be taken while intravenous solutions are being administered.

3. Syringes must be scrupulously clean.

4. The container used for transporting the sample must be scrupulously clean. If anticoagulants are added the sample should be thoroughly mixed and this information provided to the analyst.

5. The amount of sample collected must be sufficient to allow the analyst to perform the determination. Different tests will require different amounts of sample, and consequently, the technician collecting the sample must be aware of this.

6. The analyst has no control over the diet, environment, etc. of the patient. The levels of the components of most clinical samples are an average, and the aforementioned factors have a significant effect on these levels. It is the responsibility of the individual collecting the sample to control these factors.

7. The analyst should be made aware of any special steps taken to preserve the sample, particularly if materials are added to the sample.


**Summary.**     Some analytical methods allow measurements to be made without destroying or changing the sample, while others consume the entire sample or require only trace amounts of sample material. It is also possible to analyze a surface by instrumental techniques. In this way the randomness of the entire surface of the sample can be determined.

In summary, it is not possible to describe a set of general methods for sampling all substances under all conditions. However, the aim in all cases is statistical sampling, the use of chemical common sense, and cooperation between the analyst and the individual providing the sample. Many sources are available to aid the analyst in selecting the sampling procedure. These include general procedures,* clinical procedures,** and environmental procedures for air† and water.‡

## DRYING

Once a suitable sample is obtained, a decision must be reached on whether the analysis is to be made on the sample as received or after it has been dried. Most samples contain varying amounts of moisture, either because the sample is hygroscopic or because water is absorbed in the surface. Analysis on an "as-received basis" treats water as part of the sample's composition and in handling the sample it is necessary that no gain or loss of water occur prior to analysis. Analysis on a dried basis means that the water has been removed. This is usually done by heating in an oven, a muffle furnace, or by Bunsen or Meeker burners. Techniques for this are shown in the chapter on laboratory techniques.

Since heat is used for drying, it is possible to decompose the sample or to remove volatile components from the sample. In some cases a dried basis implies that this is true, while in others, perhaps the majority, only water removal is implied.

Frequently, drying procedures are such that only a fixed amount of moisture is removed. An example of this is the removal of water to form a stable stoichiometric hydrate by controlled heating. Continued heating at an elevated temperature would be needed to remove the remaining water of hydration. Drying to a known hydrated state is an example of producing reproducible dryness, while the complete removal of water is the production of a state of absolute dryness.

Regardless of which state is used, the handling of the dried sample must be rapid and its storage must be in the absence of water. In addition, the heated sample must be cooled, usually in a desiccator, before weighing.

---

* P. D. La Fleur, Ed., "Accuracy in Trace Analysis: Sampling, Sample Handling, Analysis," Vols. 1–2, NBS Publication 422, U.S. Government Printing Office, Washington, D.C. (1976).

I. M. Kolthoff, P. J. Elving, E. B. Sandell, "Treatise on Analytical Chemistry," Parts I to III, Wiley–Interscience, New York, 1963.

** J. S. Annino, "Clinical Chemistry: Principles and Procedures," third ed., Little Brown and Co., Boston, 1964.

M. M. Wintrobe, "Clinical Hematology," Lea and Febiger, Philadelphia, 1961.

† P. O. Warner, "Analysis of Air Pollutants," J. Wiley & Sons, New York, 1976.

A. C. Stern, Ed., "Air Pollution," Academic Press, New York, 1968.

‡ "Standard Methods for the Examination of Water and Waste Water," 13th ed., American Public Health Association, New York, 1970.

W. Leithe, "The Analysis of Organic Pollutants in Water and Waste Water," Ann Arbor Science Publishers, Inc., Michigan, 1973.

## WEIGHING

Weighing of the sample is described in detail in the chapter on laboratory techniques. In ordinary analysis this measurement is one of the most accurate ones that can be made in the laboratory.

Samples are usually weighed in triplicate and each of these is independently carried through the experimental procedure. An alternate technique is to weigh one sample and dissolve it in a known volume in a volumetric flask. Aliquots of this solution can then be taken for the quantitative determination. The weight of sample that is actually taken will depend on the concentration level of the substances in the sample that is being determined.

## DISSOLVING

Usually, after weighing the sample the next step is dissolution. If the sample is soluble in water the problem is solved, although occasionally the sample slowly hydrolyzes in water to produce insoluble hydrous oxides. Organic materials are usually dissolved in organic solvents or mixtures of organic solvents and water, if water alone is not a suitable solvent. Although the most direct and easily used solvent is water, there are a variety of chemical and instrumental procedures which require the use of a certain solvent composition. In other cases, the dissolving process is not even required. For example, in atomic emission, where the sample is excited by an arc or spark and the resulting radiant energy emitted is instrumentally analyzed, a solid or liquid sample can be used directly.

There are general guidelines concerning the solubility of salts in water. These are briefly summarized in Table 5–1.

Modern technology has resulted in a vast number of alloys and inorganic and organic mixtures. With such samples, the dissolution step is not a simple case of following the solubilities of salts. If the organic part of a mixture is to be analyzed, the solvents and techniques of organic chemistry must be used. For inorganic analysis, if water is not suitable, the sample must be dissolved in an acid or by fusion with a flux.

**Acid Treatment.**     In using acids several questions must be asked. Is a nonoxidizing or oxidizing acid needed? What kind of anion is desirable or most appropriate in solution? What are the chemical properties of the sample? Is it necessary to get rid of the excess acid?

Knowledge of inorganic reactions and chemical common sense are needed to answer these questions. For example, $H_2SO_4$ would not be used to dissolve a sample containing barium metal or ion, or HCl would not be used to dissolve a silver metal or silver salt sample. In other situations acid treatment should

Table 5–1.   General Guidelines for the Solubility of Inorganic Salts

---

Acetates are soluble, except a few sparingly soluble ones like $AgC_2H_3O_2$ and $Hg_2(C_2H_3O_2)_2$.

Arsenates, borates, and carbonates are insoluble, except those of $NH_4^+$, $Na^+$, and $K^+$. The bicarbonates of these three and of $Ba^{2+}$, $Ca^{2+}$, $Fe^{2+}$, $Mg^{2+}$, $Mn^{2+}$, and $Sr^{2+}$ are soluble. Many borates are soluble in $NH_4Cl$ solution, and all are soluble in hot dilute acids. Any salts of these ions with tervalent or quadrivalent metals are so completely hydrolyzed that they may be considered as not existing.

Fluorides are insoluble, except those of $Na^+$, $K^+$, $NH_4^+$, $Ag^+$, and $Sn^{2+}$. $FeF_3$ is slightly soluble, but like many other metals, iron forms a very soluble complex ion with excess $F^-$.

Oxalates are insoluble, except those of $Na^+$, $K^+$, $NH_4^+$, and $Fe^{3+}$. A number of oxalate complexes of other metals are soluble.

Chromates are insoluble, except those of $Na^+$, $K^+$, $NH_4^+$, $Ca^{2+}$, $Mg^{2+}$, $Zn^{2+}$, and $Fe^{3+}$.

Sulfides are insoluble, except $Na_2S$, $K_2S$, and $(NH_4)_2S$. Slightly soluble sulfides are BaS, SrS, CaS, and MgS.

Phosphates are insoluble, except those of $NH_4^+$, $Na^+$, and $K^+$. Phosphates of alkaline earth metals are soluble in acid but insoluble in neutral or alkaline solution.

Chlorides and bromides of $Ag^+$ and $Hg_2^{2+}$ and iodides of $Ag^+$, $Hg_2^{2+}$, $Hg^{2+}$, and $Pb^{2+}$ are insoluble. $PbCl_2$, $PbBr_2$, and $HgBr_2$ are slightly soluble. Some basic halides are insoluble; for example, SbOCl.

Thiocyanates are soluble, except $Ag^+$, $Pb^{2+}$, $Hg_2^{2+}$, and $Hg^{2+}$.

Sulfates are soluble, except $BaSO_4$, $Hg_2SO_4$, $PbSO_4$, and $SrSO_4$. $Ag_2SO_4$ and $CaSO_4$ are slightly soluble. Some basic sulfates are insoluble; for example, $(BiO)_2SO_4$.

Sulfites are generally similar to carbonates. $MgSO_3$ is slightly soluble.

Nitrates are soluble, except a few basic salts like $BiONO_3$. Nitrates do not exist for arsenic or tin (IV).

Nitrites are soluble, except $AgNO_2$, which is slightly soluble. Nitrites are very unstable to oxidation or reduction.

---

not be used because the sample may be volatilized. For example, acid treatment of carbonate or sulfide samples can result in the loss of $CO_2$ or $H_2S$, unless arrangements are made to trap the gases. But if the analysis is not concerned with the sulfide or carbonate content of the sample, then there is no difficulty. In some cases acid cannot be used because it causes parts of the sample to become passive and not take part in the reaction, usually because an oxide coating forms.

It is difficult to describe acid conditions for dissolving inorganic samples since not all samples are simple pure metals, metal oxides, or alloys. For example, a high-temperature alloy might contain 13 different metals, including such elements as Nb, Ta, W, Zr, and rare earths. Sometimes, one technique is used to dissolve one part of the sample and another to dissolve the residue.

As a general guide it is useful to classify the more common acid conditions according to whether they are oxidizing or nonoxidizing. The nonoxidizing acids are HCl, dilute $H_2SO_4$, and dilute $HClO_4$, while the oxidizing acids are $HNO_3$, hot concentrated $H_2SO_4$, and hot concentrated $HClO_4$.

Dissolution of metals by the nonoxidizing acids is a process of hydrogen replacement. Therefore, any scheme of relating the ability of metals to displace hydrogen can also be used to qualitatively predict solubility. The Table of Standard Reduction Potentials (see Appendix IV and Chapter 10) is such a scheme, and metals below hydrogen in the series will dissolve in a nonoxidizing acid. There are exceptions and these are usually due to a presence of a passive condition, oxide film formation, or insoluble salt formation.

Briefly, HCl will dissolve the metals above hydrogen, salts of weak acids, and many oxides. Dilute $H_2SO_4$ and $HClO_4$ are useful for metals above hydrogen, the difference being the solubility of the salts that are formed. Heat and concentrated $H_2SO_4$ will often dissolve metals below hydrogen. However, the problem of solubility of sulfate salts should be considered. Nitric acid will dissolve metals below and above hydrogen since its oxidizing power will vary according to whether the acid is dilute or concentrated. In general, the metal is oxidized to its highest oxidation state. The principle limitations are that some metals (Al and Cr) become passive while others (Sn, Sb, and W) form insoluble acids. Sulfide salts and salts of oxidizable anions are also dissolved by $HNO_3$. Perhaps the most powerful oxidizing conditions are obtained by using hot, concentrated $HClO_4$ which will dissolve all the common metals.

It is often an advantage to use acids in combination. The most familiar mixture is aqua regia (1:3 $HNO_3$–HCl). In essence, the features of both acids are included since the $HNO_3$ furnishes oxidizing power, while HCl furnishes complexing properties and strong acidity. Solubility of many metal ions is maintained only in the presence of complexing agents. In fact, this is a useful technique for solubilizing an otherwise insoluble ionic salt. Sometimes the addition of bromine or hydrogen peroxide to mineral acids increases their solvent action. An added advantage of such combinations is the hastened oxidation (and destruction) of organic materials in the sample.

Hydrofluoric acid is special in that it is a weak acid as well as a nonoxidizing acid, but yet it is still useful for solubilizing certain samples. Silicate samples, where silica is not to be analyzed are readily decomposed by HF, and the silica is volatilized as $SiF_4$. Hydrofluoric acid is better than HCl in that it furnishes a good complexing anion, $F^-$. In certain cases, the complex is very difficult to decompose and it can, therefore, affect future chemical steps. Since HF can cause serious injury upon contact with the skin, a convenient way of creating HF conditions is to add NaF to an HCl solution of the sample.

Perchloric acid when hot and concentrated is a potent oxidizer. If a dilute solution is boiled and the water is slowly evaporated, the oxidizing power increases, gradually reaching a maximum at 72% acid; at this point an azeotropic mixture is distilled. Other advantages of perchloric acid are that very soluble perchlorate salts are formed, it acts as a dehydrating agent, and the hot concentrated acid readily oxidizes organic materials. The latter technique, which is generally referred to as wet ashing and employs a $HNO_3$–

$HClO_4$ mixture, requires careful handling since an improper procedure can result in a violent explosion. Nitric acid is added first, heated, cooled, and then the $HNO_3$–$HClO_4$ mixture is added. Thus, the $HNO_3$ acts as a moderating force by oxidizing the more reactive compounds at lower temperatures.

Minerals and rocks, such as silicates, sulfides, phosphates, carbonates, sulfates, and very refractory minerals and oxides often require special treatment. The difficulties are enhanced if analysis of the anion is also desired as some of the more easily used acids can result in volatilization losses of this portion of the sample.

**Flux Treatment.**     The second general dissolving technique, fusion with a flux, is more potent than acid treatment for two reasons. First, since a flux is a fused salt media, the temperatures required for creating the condition (300° to 1000°C) are much higher than possible by the acid treatment. Second, there is a greater concentration of reagent in contact with the sample. The advantage of oxidizing or nonoxidizing conditions is not lost since there are fluxes which yield both of these conditions.

The difficulties in using the fluxes are several. Special containers that will withstand the temperature as well as the reaction conditions are needed. Pt, Ag, Ni, Au, and Fe are some of the more common crucible materials. However, before choosing the appropriate crucible the reactivity of the flux must be known. Upon completion of the fusion and after cooling, the solid material is usually dissolved in water or dilute acid. The solution that is obtained has a very high salt content which often must be considered in the following chemical or instrumental steps. Other possible difficulties are impurities introduced by the flux, volatilization due to the high temperature, and spattering losses as a result of the reaction.

Table 5–2 lists the common fluxes according to their oxidizing or nonoxidizing properties and also some of their applications. The fluxes can also be divided according to whether they yield acidic or basic conditions. Carbonates, hydroxides, and peroxides provide basic conditions, while boric oxide and pyrosulfates provide acidic conditions.

**Decomposition of Organic Matter.**     Decomposition of organic material can be done in several ways. The method that is finally chosen will depend on whether analysis of the organic matter is desired or whether only organic removal is desired.

The wet ashing procedure, already mentioned, in which $HClO_4$–$HNO_3$ is used, completely destroys organic matter. Some other chemical agents that can be used are metallic sodium in liquid ammonia, fuming nitric acid, and dry-ashing. Dry-ashing is the simplest to carry out. Usually, the sample is heated in an open crucible or dish in the presence of air or oxygen to red heat until all the carbonaceous material has been oxidized. The nature of

**Table 5–2.  Common Fluxes**

| Flux | Crucible | Application |
|---|---|---|
| *Nonoxidizing* | | |
| $Na_2CO_3$ | Pt | Silicates, phosphates, sulfates |
| NaOH | Au, Ni, Ag | Silicates, silicon carbides |
| KOH | Au, Ni, Ag | Silicates, silicon, carbides |
| $B_2O_3$ | Pt | Silicates, oxides |
| $CaCO_3$-$NH_4Cl$ | Ni | Silicates (known as J. Lawrence Smith method for analysis of alkali metals) |
| *Oxidizing* | | |
| $Na_2CO_3$ mixed with $KNO_3$, $KClO_3$ or $Na_2O_2$ | Pt, Ni | Common oxidizing conditions for readily oxidized samples such as Sb, S, Cr, Fe |
| $Na_2O_2$ | Fe, Ni (not Pt) | Sulfides, acid-insoluble alloys such as ferrochromium, ferrotungsten, Ni, Mo, W, and Pt alloys |
| $K_2S_2O_7$ | Pt porcelain | Insoluble oxides and oxide-containing samples, iron ore, and Zr, Hf, and Th phosphates |

the method requires some control to prevent loss of the desired portion of sample by spattering or volatilization.

# Questions

1. Explain why heterogeneous samples require a statistical sampling procedure while homogeneous samples do not.
2. Suggest a procedure that is suitable for obtaining a representative sample of milk of magnesia (a suspension of MgO in water).
3. What type of problems are encountered when obtaining representative samples of gases? of liquids? of aerosols? of suspensions?
4. Lead occurs in the environment as particulate matter. What kind of procedure should be used to obtain a representative atmospheric sample for a lead determination?
5. In sampling air with the devices shown in Fig. 5–1 why would the pump not be placed in front of the sample collector?
6. Why is it necessary to carefully monitor the flow rate in sampling air?
7. If air contains 1 ppm Pb per liter of air, how many liters of air would have to be passed through a lead trapping solution in order to obtain 1 g of lead?
8. Explain why it is preferred to reduce a solid heterogeneous sample to a uniform particle size.

9. What is the significance of an analysis reported on an "as-received" basis?
10. Criticize the statement, "Reproducible dryness is the same as absolute dryness."
11. Suggest several reasons why desiccants are useful to the analytical chemist.
12. Suggest a procedure for dissolving each of the following:

  a. $Fe(NO_3)_3$    e. Sodium silicate    i. Dried blood
  b. $Na_2CO_3$    f. Iron ore    j. Wool
  c. $CaCO_3$    g. Zn
  d. $BaSO_4$    h. Brass ($Cu-Zn-Pb-Sn$)

# Chapter Six
# Introduction to Chemical Reactions in Analytical Chemistry

## SCOPE

Throughout the development of analytical chemistry certain methods stand above others in terms of practical applications. Several of these methods are old and are referred to as classical methods. A characteristic of most classical methods is that they are based on chemical reactions rather than physical properties.

The chemistry of these reactions is fundamental to the development of an understanding of modern chemical and instrumental methods of analyses, measurement, or separation. Consequently, considerable attention will be directed toward a discussion of these reactions. As instrumental techniques are introduced, the relationships among the instrumental response, the chemistry of the solution or the sample, and how the data can be used for analysis should become apparent. For convenience the chemical reactions of analytical interest can be divided into four areas:

1. Acid–base reactions
2. Precipitation; gravimetry and titration
3. Oxidation–reduction reactions
4. Complex formation

## COMPLETENESS OF REACTION

*on test*

Following is a list of requirements for the *suitability of a reaction for use in chemical analysis*:

1. The reaction should be stoichiometric. If this is not true, there is no basis for calculating the amounts of the substances that are involved in the reaction.

2. The chemical reaction should take place rapidly. If it is not rapid, excessive time would be needed to complete the analysis. This property is particularly important in titration procedures where a complete reaction is desired after each addition of titrant.

3. The reaction should be quantitative. Generally, this means that the reaction should be at least 99.9% complete.

4. A convenient method to follow the progress of the reaction or to determine when it is complete is necessary. *indicator*

In general a reaction can be represented by

$$aA + bB \rightarrow products$$

It is not necessary to know what the products are for the reaction to be useful as long as the reaction ratio $a/b$ remains fixed. For example, consider the analysis of 1,3-butadiene by bromination.

$$H_2C=CHCH=CH_2 + Br_2 \rightarrow$$

$$H_2BrCCH=CHCH_2Br + H_2BrCCBrHCH=CH_2$$

The reaction produces two different brominated products. The ratio of the amounts of the two products can change with the experimental environment but the 1 : 1 ratio of 1,3-butadiene to $Br_2$ used in the reaction remains constant. Therefore, the amount of butadiene in the sample (the calculation is usually for percent double bond) can be determined.

A reaction which takes place with a fixed and reproducible reaction ratio, but with a mixed or nonstoichiometric product formation, is not desirable for analysis. Although some of the empirical reactions are successfully used in quantitative analysis, generally, this is possible only by careful control of the experimental conditions. For this reason stoichiometric reactions are preferred.

In general, completeness of a reaction will tend to occur when one of the following takes place:

1. Formation of un-ionized molecules
2. Precipitate formation
3. Chelate formation
4. Gas formation

Therefore, in choosing a reaction for use in chemical analysis these are desirable properties.

## EQUILIBRIUM CONCEPT

Does a reaction which goes to completion go only in one direction, that is to form the products? Actually, this is not the case. Consider the reaction between reactants A and B, to form the products C and D:

$$A + B \rightleftharpoons C + D \tag{6-1}$$

If the reaction goes only in one direction and to completion, only C and D would be present. However, if all four substances, A, B, C, and D are present, several possible conclusions could be drawn.

Perhaps not enough time was allowed for the reaction to take place, or possibly only a portion of A and B are able to react. The second conclusion should not be given much consideration, however, since it implies that only certain parts of A and B possess the required energetics to cause the reaction. Nevertheless, it should be realized that certain physical properties such as particle size and contact can have a bearing on the rate at which a reaction takes place. Hence, the second possible conclusion is really related to the first one.

A third possibility is that C and D, the products, can react to form A and B, the starting materials. Proof that ionic reactions can indeed proceed in both directions can be obtained if starting materials can be formed by products alone. As an example, consider the reaction of stoichiometric amounts of barium ion and sulfate ion to form a precipitate of barium sulfate. After precipitation is complete, it is still possible to detect barium and sulfate ions with sensitive devices. If solid barium sulfate alone is put into water, it is also possible to detect barium and sulfate ions in the solution. Therefore, the reaction must proceed in both directions.

$$Ba^{2+} + SO_4^{2-} \rightleftharpoons BaSO_{4(s)}$$

Similar arguments can be extended to other reactions. The reaction of barium sulfate in water is an ionization reaction: solvated ions (ions in solution) are formed from an insoluble precipitate. Many ionic reactions which proceed in two directions are possible.

The forward and reverse reactions do not stop at some moment of time or concentration, although eventually a point is reached where the concentrations do not appear to change. A simple experiment will prove that the forward–reverse reaction is a continuous one. A saturated $BaSO_4$ solution is prepared, that is, the barium and sulfate ions in solution are at maximum concentration levels and are in contact with solid barium sulfate. If solid barium sulfate containing radioactive Ba and S (or O) is added to the system no radioactive barium ion or sulfate ion should be found in the solution unless some of the salt dissolves. However, since the solution is already saturated, the salt should not dissolve unless the forward and reverse reactions take place. With time the tagged barium and sulfate ions are found in the solution. Thus, as the dissolving process takes place, simultaneously, the precipitation process also takes place.

When the reverse and forward processes occur continuously there is no change in the concentration of reactants and products. This means that the rate of formation of C and D is equivalent to the rate of formation of A and B.

In describing the equilibrium process it was assumed that the reaction products of the two directions were not lost. A system of this type is said to be in a state of dynamic equilibrium. If a product is continuously removed, more reactants will undergo reaction in order to replenish the supply of product. The tendency, then, is for the reaction to always strive for the equilibrium concentration. This is an example of Le Chatelier's principle, which states that when a stress is applied to a system at equilibrium, the position of equilibrium tends to shift in a direction which diminishes or relieves the stress.

An illustration of a system not at dynamic equilibrium is the decomposition of $CaCO_3$ in an open container. As the decomposition takes place the $CO_2$ gas is lost and thus, it is not possible for the reverse reaction to take place.

$$CaCO_{3(s)} \xrightarrow{\Delta} CaO_{(s)} + CO_{2(g)} \qquad (6\text{-}2)$$

The theory of equilibrium is very important to the analytical chemist. With this theory it is possible to describe quantitatively whether reactions approach completion and whether reactions are feasible for analysis.

## REACTION RATES

Not all reactions reach the point of equilibrium at the same time. Some undergo reaction in what appears to be an instantaneous fashion, while others are hopelessly slow. The study of the speed at which these reactions take place is called kinetics. If a chemical reaction is involved, the speed is described as a reaction rate.

For the reaction

$$A \rightarrow X \qquad (6\text{-}3)$$

the reaction rate is defined as the change in concentration of either the product X or reactant A with time, $t$. Mathematically, this is given by

$$\text{Rate} = \frac{\Delta X}{\Delta t} = -\frac{\Delta A}{\Delta t} \qquad (6\text{-}4)$$

where the negative sign states that A is disappearing. Hence, it can be concluded that the reaction is concentration dependent, since the reaction rate decreases as the concentration decreases. In addition, the rate of reaction is the amount of material formed or depleted per unit period of time.

For all reactions, the rate of reaction is equal to a constant called the rate constant, $k$, times the product of the concentration of reactants (or products) raised to powers of small integers. For the reaction

$$m_1A + m_2B \rightarrow m_3C \tag{6-5}$$

the rate expression is

$$\text{Rate} = k[A]^{n_1}[B]^{n_2} \tag{6-6}$$

The sum of the integers $(n_1 + n_2)$ is defined as the order of the reaction. In elementary reaction kinetics, these may be zero, first, second, etc., order. Establishing the order of the reaction defines the dependence of the rate of reaction on the individual components in the reaction. If $n_1 = 1$ and $n_2 = 2$, the rate of reaction will have a first order dependence on A and a second order dependence on B. The total order will be third order. The order of the reaction, which can only be determined experimentally, is not synonymous with molecularity. Molecularity is the stoichiometry of the reactants entering into the reaction and is represented by $m_1$, $m_2$, and $m_3$ in Eq. (6-5).

The rate constant is independent of concentration, and dependent on solvent, temperature, and presence of catalysts. Its units are determined by the sum of the integers and are expressed in concentration, (moles/liter) per unit time (second, minute, etc.).

A detailed kinetic investigation will lead to conclusions about how reactions occur. For example, proton transfer in acid–base reactions, electron transfer in oxidation–reduction reactions, and coordination and bond formation in complexometric reactions can be described quantitatively as a result of the kinetic studies.


## EQUILIBRIUM CONSTANT

When a reaction, represented by the general reaction

$$A + B \rightleftharpoons C + D \tag{6-7}$$

reaches equilibrium, the concentration of A, B, C, and D remain constant. This equilibrium state can be represented mathematically by

$$K = \frac{[C][D]}{[A][B]} \tag{6-8}$$

where $K$ is the equilibrium constant.

The concentrations in the expression are the equilibrium concentrations of A, B, C, and D. They should not be confused with initial concentrations or concentrations at any other time prior to reaching equilibrium. Arbitrarily, the convention of product concentrations divided by reactant concentrations has been established. Thus, by glancing at the magnitude of $K$ it is possible to predict whether a reaction favors the products or the reactants; the larger the equilibrium constant the further the reaction has proceeded toward the products.

It is essential to note that the magnitude of the equilibrium constant is not determined by the rate of reaction or vice versa. There are many reactions whose equilibrium states are far toward the products, but whose rates are negligible. Similarly, there are very fast reactions with very small equilibrium constants. For example, $H_2$ and $O_2$ can be mixed at room temperature and no evidence of a reaction is found even though the equilibrium constant is a favorable one. However, if finely divided Ag, Au, or Pt is added as a catalyst, the reaction will take place at a controllable rate at room temperature.

For the general reaction

$$aA + bB \rightleftharpoons cC + dD \qquad (6\text{-}9)$$

the equilibrium expression is

$$K = \frac{[C]^c[D]^d}{[A]^a[B]^b} \qquad (6\text{-}10)$$

If any concentration is changed, spontaneously the others adjust themselves to new values so that the same constant value, $K$, is obtained. Ideally, $K$ is independent of concentration.

It was stated previously that the value of the equilibrium constant is independent of concentration and dependent on temperature and pressure. Using Le Chatelier's principle it is possible to qualitatively predict how these experimental factors affect chemical reactions and the equilibrium position.

**Temperature.** Increasing the temperature will increase the thermal energy available to the system. If the reaction itself involves the loss of heat (exothermic reaction), raising the temperature will act as a stress on the product side of the reaction. To relieve the stress, the direction of the equilibrium will shift toward the reactants and the numerical value of the equilibrium constant will be less. If the reaction consumes heat (endothermic reaction) and an increase in temperature is applied to the reaction, the stress will be on the side of the reactants. This will lead to a shift in the equilibrium toward the products and an increase in the value for the equilibrium constant. In addition to affecting the equilibrium constant, a temperature change will also have a large influence on the rate of reaction.

**Pressure.**    For reactions in which gaseous reactants or products are involved a change in pressure will have a significant effect on the equilibrium position. However, if the entire reaction including reactants and products takes place in solution, a large pressure change must take place before any change in equilibrium is noted. From a practical viewpoint, it can be concluded that the equilibrium constant is independent of pressure.

The reaction

$$2H_{2(g)} + O_{2(g)} \rightleftharpoons 2H_2O_{(g)} \tag{6-11}$$

is affected by a pressure change since the volume requirements of the reactants are 1.5 times that for the products. A restatement of Le Chatelier principle as it applies to this type of situation is that a decrease in pressure causes a shift in the equilibrium position in the direction which increases the volume of the system. Thus, for reaction (6-11) a decrease in pressure decreases its equilibrium constant. If the number of volumes of reactants equals the number of volumes of products no significant effect in equilibrium will be observed.

**Concentration.**    If hydrogen, oxygen, and water in the gaseous state are in equilibrium [see reaction (6-11)] addition or removal of any one of these components will cause the equilibrium to reestablish itself. The direction of change is readily predicted, qualitatively, from the Le Chatelier principle. To illustrate, assume that hydrogen gas is added to the equilibrium mixture. The stress in this case is an increase in the concentration of one of the reactants and to relieve the stress the reaction shifts in the direction of the products. If water is added, the stress is relieved by shifting towards the reactants, while removing water, causes a shift toward the products.

Eventually, equilibrium is reestablished. If there is no change in pressure or temperature, the equilibrium position will be defined by the same equilibrium constant. Thus, at constant pressure and temperature there are an infinite number of combinations of concentrations of reactants and products for a given reaction which yields the same equilibrium constant.

**Catalysts.**    Catalysts do not have any effect on the value of the equilibrium constant. They do, however, alter the rate at which equilibrium is reached by affecting both the forward and reverse reaction rates. From the viewpoint of the analytical chemist this is very important since many reactions which normally are slow can be accelerated by the use of catalysts. Frequently, this increase in rate is sufficient to make the reaction useful in analysis.

# APPLICATION OF THE EQUILIBRIUM CONCEPT

In attempting to understand the different types of reactions that are of interest to the analytical chemist it is necessary to consider equilibrium theory. In many instances it is possible to qualitatively predict an experimental environment that would provide favorable equilibrium conditions. If a quantitative description is required, equilibrium constants must be utilized in calculations based on the appropriate formulas to describe the system. Consequently, the analytical chemist is also interested in measuring equilibrium constants by existing methods and in the development of new techniques for measuring the constants. The next section briefly identifies equilibrium constants according to the different types of reactions; calculations based on these expressions are discussed in later chapters.

**Ionization.** Strong electrolytes, when dissolved in water, exist predominately in the ionized form, while nonelectrolytes are not ionized. Weak electrolytes exist in solution between these two extremes. It should be emphasized that these concepts can be quantitatively extended to solvents other than water.

Typical examples of electrolytes are the following:

### Strong Electrolytes

$$NaCl_{(s)} \xrightarrow{H_2O} Na^+ + Cl^- \text{ (solvated ions)}$$

$$NaOH_{(s)} \xrightarrow{H_2O} Na^+ + OH^- \text{ (solvated ions)}$$

$$HCl_{(g)} \xrightarrow{H_2O} H_3O^+ + Cl^- \text{ (solvated ions)}$$

### Nonelectrolytes

$$C_6H_6 \xrightarrow{H_2O} \text{No reaction}$$

$$CCl_4 \xrightarrow{H_2O} \text{No reaction}$$

### Weak Electrolytes

$$2H_2O \rightleftharpoons H_3O^+ + OH^-$$

$$HC_2H_3O_2 + H_2O \rightleftharpoons H_3O^+ + C_2H_3O_2^-$$

$$NH_3 + H_2O \rightleftharpoons NH_4^+ + OH^-$$

The reactions involving weak electrolytes are typical equilibrium reactions and can be expressed in terms of equilibrium constants:

$$K = \frac{[H_3O^+][OH^-]}{[H_2O]^2} \tag{6-12}$$

$$K_a = \frac{[H_3O^+][C_2H_3O_2^-]}{[HC_2H_3O_2]} \tag{6-13}$$

$$K_b = \frac{[NH_4^+][OH^-]}{[NH_3][H_2O]} \tag{6-14}$$

All of these $K$'s are called "ionization constants" or "dissociation constants." If a weak electrolyte is acidic or basic, the ionization constant is identified as $K_a$, weak acid ionization constant, and $K_b$, weak base ionization constant, respectively. The magnitude of the $K_a(K_b)$ is a measure of the weak electrolyte's strength as an acid (base). The larger the $K_a$ or $K_b$ value is, the stronger its acidic or basic strength. For soluble salts which behave as weak electrolytes the symbol for the ionization constant is $K$. As the value for $K$ increases, the salt becomes more like a strong electrolyte.

The equilibrium concept is not usually applied to strong electrolytes and nonelectrolytes. In the former case, the amount of undissociated substance approaches a negligible level; hence, the $K$ approaches infinity. For nonelectrolytes, where the amount of ionized substances approaches zero, the $K$ will also approach zero.

In dilute aqueous solutions, water concentration, since it is the solvent, can be treated as a constant when it appears in the equilibrium expression. Consequently, for the weak electrolyte water, expression (6-12) becomes

$$K_w = [H_3^+O][OH^-] = 1.0 \times 10^{-14} \quad (25°C) \tag{6-15}$$

where $K_w$ is the ion product of water and represents the product of the two constants $K$ and $[H_2O]^2$. A similar situation exists for reaction (6-14) and is correctly written as

$$K_b = \frac{[NH_4^+][OH^-]}{[NH_3]} \tag{6-16}$$

where $[H_2O]$ is part of the constant. Ionization constants, such as $K_a$, $K_b$, and $K_w$, are essential to the understanding of acid–base equilibria.

**Oxidation–Reduction.**    Reactions in which changes in oxidation state occur can also be represented by an equilibrium constant. Under the appropriate conditions $Ce^{4+}$ will undergo a reaction with $Fe^{2+}$ according to

$$Ce^{4+} + Fe^{2+} \rightleftharpoons Ce^{3+} + Fe^{3+} \qquad (6\text{-}17)$$

and the equilibrium constant is given by

$$K = \frac{[Ce^{3+}][Fe^{3+}]}{[Ce^{4+}][Fe^{2+}]} \qquad (6\text{-}18)$$

Reaction (6-17), which is typical of an oxidation–reduction titration method, is a useful reaction for the analysis of Fe(II). In the procedure Fe(II) is titrated with a standard solution of Ce(IV).

A common characteristic of oxidation–reduction reactions is that they have large $K$ values but slow rates of reaction. Fortunately, the rates of reaction can be significantly increased by the use of catalysts.

**Solubility Product.** If an insoluble electrolyte is added to water, some of it will dissolve according to the general reaction

$$M_mX_{x(s)} \rightleftharpoons mM^+ + xX^- \qquad (6\text{-}19)$$

For the simplest case where the electrolyte is a strong electrolyte, the equilibrium constant expression is

$$K = \frac{[M^+]^m[X^-]^x}{[M_mX_{x(s)}]} \qquad (6\text{-}20)$$

Since the solid $M_mX_x$ is in excess, its concentration is at its standard state and the equilibrium expression can be rewritten as

$$K_{sp} = [M^+]^m[X^-]^x \qquad (6\text{-}21)$$

where $K_{sp}$ is the product of the constants $K$ and $[M_mX_{x(s)}]$ and is called the solubility product.

The solubility product in this form is applicable only to insoluble strong electrolytes. As the solubility of the electrolyte increases, the deviation from what is predicted by the solubility product constant expression will become larger.

**Complex Formation.** Formation of complexes involves equilibrium steps and can be represented by equilibrium constant expressions. For example, silver ions will react with ammonia to form a series of complexes

$$Ag^+ + NH_3 \rightleftharpoons Ag(NH_3)^+$$

$$Ag(NH_3)^+ + NH_3 \rightleftharpoons Ag(NH_3)_2^+$$

where each equilibrium step has its own constant:

$$K_{S_1} = \frac{[\text{Ag}(\text{NH}_3)^+]}{[\text{Ag}^+][\text{NH}_3]} \qquad (6\text{-}22)$$

$$K_{S_2} = \frac{[\text{Ag}(\text{NH}_3)_2{}^+]}{[\text{Ag}(\text{NH}_3)^+][\text{NH}_3]} \qquad (6\text{-}23)$$

These constants are called stability constants or formation constants. Frequently, complex reactions are written in a reverse fashion or as dissociation reactions. Hence,

$$\text{Ag}(\text{NH}_3)^+ \rightleftharpoons \text{Ag}^+ + \text{NH}_3$$

$$\frac{1}{K_{S_1}} = K_{I_1} = \frac{[\text{Ag}^+][\text{NH}_3]}{[\text{Ag}(\text{NH}_3)^+]} \qquad (6\text{-}24)$$

where $K_I$ is the reciprocal of $K_S$ and is called an instability or dissociation constant.

It does not matter which constants are used providing they are used with the correct reaction and expression. When consulting tables it is important to recognize whether the constants are listed as stability or instability constants.

## EQUILIBRIUM CONDITION: FUNDAMENTAL DEFINITION

The equilibrium condition is based on a fundamental thermodynamic definition. At constant temperature and pressure equilibrium occurs when the free energy, $G$, of a system is at a minimum value. Therefore, the change in free energy, $\Delta G$, on going from the reactants to the products is given by the difference in free energy for the products and reactants. For the reaction

$$a\text{A} + b\text{B} \rightleftharpoons c\text{C} + d\text{D} \qquad (6\text{-}25)$$

the change in free energy is

$$\Delta G = cG_\text{C} + dG_\text{D} - aG_\text{A} - bG_\text{B} \qquad (6\text{-}26)$$

and is related to the activities of the reactants and products rather than concentrations by

$$\Delta G = \Delta G^\circ + RT \ln \frac{a_\text{C}^c a_\text{D}^d}{a_\text{A}^a a_\text{B}^b} \qquad (6\text{-}27)$$

where $R$ is 1.987 cal mole$^{-1}$ deg$^{-1}$, $T$ is 298°K (25°C), $\Delta G^\circ$ is the standard free energy, and $a$ is the activity of each species.

The standard free energy change is given by

$$\Delta G^\circ = cG_C^\circ + dG_D^\circ - aG_A^\circ - bG_B^\circ \qquad (6\text{-}28)$$

If $\Delta G$ for the reaction is negative, the reaction occurs spontaneously as written (reaction is possible); if it is positive the reaction occurs in the opposite direction as written. The sign, or magnitude, of $\Delta G$, however, does not suggest whether the reaction is fast or slow. Since

$$K = \frac{a_C^c a_D^d}{a_A^a a_B^b} \qquad (6\text{-}29)$$

$$\Delta G = \Delta G^\circ + RT \ln K \qquad (6\text{-}30)$$

When $\Delta G = 0$, the system is at equilibrium and

$$\Delta G^\circ = -RT \ln \frac{a_C^c a_D^d}{a_A^a a_B^b} = -RT \ln K \qquad (6\text{-}31)$$

Therefore, equilibrium constants can be calculated from standard free energies. Values for $G^\circ$ are obtained by arbitrarily assigning pure elements and the formation of aqueous hydrogen ions zero and measuring free energy changes of chemical reactions involving these species. Absolute standard free energies for a substance can not be measured directly.

Equation (6-29) can be rearranged by replacing activity with activity coefficients and concentration [see Eq. (3-11)] to

$$K = \frac{[C]^c \gamma^c [D]^d \gamma^d}{[A]^a \gamma^a [B]^b \gamma^b} \qquad (6\text{-}32)$$

If this is applied to the previous sections in which the application of the equilibrium concept to ionization, redox, solubility, and complex formation was described, it would be necessary to rewrite all the equilibrium constant expressions.

For example, for (6-19)

$$K_{sp} = [M^+]^m [X^-]^x \qquad (6\text{-}33)$$

the expressions based on activity would be

$$K_{sp} = (a_{M^+}^m)(a_{X^-}^x) \qquad (6\text{-}34)$$

and

$$K_{sp} = [M^+]^m (\gamma_{M^+}^m)[X^-]^x (\gamma_{X^-}^x) \qquad (6\text{-}35)$$

However, in dilute solutions activity approaches the concentration and the activity coefficients become unity which results in expressions (6-34) and (6-35) simplifying to (6-33) or for the general case, expression (6-32) simplifies to (6-10).

For most practical applications, using the equilibrium constant expression as shown in (6-10) is perfectly acceptable. However, in fundamental discussions or to explain the observed behavior in concentrated and sometimes dilute solutions it is necessary to use the activity concept.

## SUMMARY

What has been done is to briefly cite the different types of equilibrium constants that are of most interest to the analytical chemist. Details, limitations, and calculations in their application in the general areas of acid–base, oxidation–reduction, gravimetric, and complexometric methods are considered in later sections.

It is very important to emphasize that the equilibrium expressions cited in the previous paragraphs are not exact. Previously, the statement was made that the equilibrium constant is independent of concentration. This is only approximately true. What does happen is that the equilibrium constant approaches a constant value over a certain concentration range (usually small). However, for most practical applications in analytical chemistry the approximation that the equilibrium constant is independent of concentration can be made.

## *Questions*

1. What constitutes a complete reaction in analytical chemistry?
2. What are the characteristics of a stoichiometric reaction?
3. Is it necessary for a reaction to be stoichiometric to be useful in quantitative measurements? Explain your answer.
4. Define dynamic equilibrium.
5. Describe the Law of Mass Action.
6. Describe Le Chatelier's principle.
7. Iron(II) reacts with $Cr_2O_7^{2-}$ according to the reaction

$$6Fe^{2+} + Cr_2O_7^{2-} + 14H^+ \rightarrow 2Cr^{3+} + 6Fe^{3+} + 7H_2O$$

   Criticize the statement, "The kinetics for the reaction between $Fe^{2+}$ and $Cr_2O_7^{2-}$ in acid solution is given by rate $= k[Fe^{2+}]^6[Cr_2O_7^{2-}][H^+]^{14}$."
8. What factors influence the rate constant?
9. What factors influence the equilibrium constant value?
10. Differentiate between electrolytes and nonelectrolytes.
11. Differentiate between strong and weak electrolytes. Make a list of common acids and bases which fit each category.

12. Write the equilibrium constant expression for the following (assume all reactions are in water).

   a. $AgCl_{(s)} \rightleftharpoons Ag^+ + Cl^-$
   b. $HC_2H_3O_2 + H_2O \rightleftharpoons H_3O^+ + C_2H_3O_2^-$
   c. $H_3PO_4 + H_2O \rightleftharpoons H_3O^+ + H_2PO_4^-$
   d. $H_2PO_4^- + H_2O \rightleftharpoons H_3O^+ + HPO_4^-$
   e. $SrF_{2(s)} \rightleftharpoons Sr^{2+} + 2F^-$
   f. $5Fe^{2+} + MnO_4^- + 8H^+ \rightleftharpoons 5Fe^{3+} + Mn^{2+} + 4H_2O$
   g. $Cu(NH_5)_4^{2+} \rightleftharpoons Cu(NH_3)_3^{2+} + NH_3$
   h. $Cu(NH_3)_3^{2+} \rightleftharpoons Cu(NH_3)_2^{2+} + NH_3$
   i. $Ca_3(PO_4)_{2(s)} \rightleftharpoons 3Ca^{2+} + 2PO_4^{3-}$
   j. $Cl_{2(g)} + Zn_{(s)} \rightleftharpoons Zn^{2+} + 2Cl^-$

13. Using equilibrium theory explain why the $H_3O^+$ decreases (pH increases) when $NaC_2H_3O_2$ is added to a solution of $HC_2H_3O_2$.

$$HC_2H_3O_2 + H_2O \rightleftharpoons H_3O^+ + C_2H_3O_2^-$$

14. Explain the effect of adding ammonia to a saturated AgCl solution using equilibrium theory.

$$AgCl_{(s)} + 2NH_3 \rightleftharpoons Ag(NH_3)_2^+ + Cl^-$$

15. Explain using equilibrium theory why the solubility of $BaF_2$ increases upon the addition of HCl to the solution.

16. For the equilibrium

$$Ag^+ + 2CN^- \rightleftharpoons Ag(CN)_2^-$$

predict the effect of each of the following on the concentration of $Ag(CN)_2^-$.

   a. Addition of NaCN
   b. Addition of $NaNO_3$
   c. Addition of $AgNO_3$
   d. Addition of NaI, consider $K_{sp}$ for AgI
   e. Addition of $HNO_3$
   f. Addition of $NH_3$, consider $K_{sp}$ for $Ag(NH_3)_2^+$

# Chapter
# Seven
# Precipitation
# Methods

## INTRODUCTION

The process of precipitation has a long history as a useful separation technique. In this technique, differences in solubility are the main goal. Thus, Ag(I) can be separated from a Ag(I)–Fe(III) mixture by the addition of an acidified chloride solution. A precipitate of AgCl is formed, while the Fe(III) remains in solution. Filtering of the solution completes the separation process.

Not all precipitation reactions are quantitative. For example, Pb(II) can be precipitated as $PbCl_2$. However, depending on the amount of Pb(II) originally present and the volume of solution, not all the Pb(II) is precipitated. Also, if the temperature is increased, the solubility of $PbCl_2$ is sharply increased. In fact, Ag(I) and Pb(II) can be separated by the addition of chloride ion at elevated temperature. The AgCl precipitate is filtered and after cooling the solution, the $PbCl_2$ precipitate that forms is filtered from the solution. The completeness of this separation, however, is very dependent on the initial Ag(I) and Pb(II) concentrations.

Often groups of metal ions are precipitated, filtered, and thus separated. A classical example of this is the separation of metal ions into simple groups according to the old so-called "hydrogen sulfide" separation scheme. A flow diagram for this is shown in Fig. 7–1.

In other cases, the main goal in precipitation is purification. Thus, the experimental steps are designed to yield a precipitate of highest purity. These steps may not provide the optimum conditions for complete precipitation. Because of the similarity between the techniques of gravimetry and separation by precipitation, it is not necessary to treat each individually.

## GRAVIMETRY

If the main goal is a quantitative precipitation, the entire process is called gravimetry and the basic laboratory measurement of weight or weight change

| Test for $NH_4^+$ on a small portion of the original solution. To the remainder of the solution, add dil. HCl and centrifuge. | | | | |
|---|---|---|---|---|
| Residue 1. AgCl, $Hg_2Cl_2$, $PbCl_2$. Wash and treat ppt. | Solution 1 contains cations of Groups 2-5. Adjust acidity and saturate with $H_2S$. Centrifuge. | | | |
| | Residue 2. HgS, PbS, $Bi_2S_3$, CuS, CdS, $As_2S_3$, $Sb_2S_3$, $SnS_2$. Treat with KOH and centrifuge. | Solution 2 contains arsenate and the cations of Groups 3, 4 and 5, HCl and $H_2S$. Precipitate the arsenic and centrifuge. Remove excess acid and $H_2S$ from the solution, add $NH_4Cl$ and $NH_3$, and treat with $H_2S$. Centrifuge. | | |
| | Residue 2A. HgS, PbS, $Bi_2S_3$, CuS, CdS. | Solution 2B. $HgS_2^-$, $AsO_2^-$, $AsS_2^-$, $Sb(OH)_4^-$, $SbS_2^-$, $Sn(OH)_6^-$, $SnS_3^-$. | | |
| | | | Residue 3. NiS, CoS, $Al(OH)_3$, $Cr(OH)_3$, $Fe_2S_3$, MnS, ZnS. | Solution 3 contains the cations of Groups 4 and 5. Acidify with acetic acid, boil out $H_2S$ and centrifuge. Discard residue. Evaporate solution and destroy $NH_4$. Add $H_2O$, $NH_4Cl$ and $(NH_4)_2CO_3$ and centrifuge. |
| | | | Residue 4. $BaCO_3$, $SrCO_3$, $CaCO_3$, $MgCO_3$. | Solution 4 contains $Mg^{2+}$ $K^+$ and $Na^+$. |

**Fig. 7–1.** A qualitative scheme based on hydrogen sulfide precipitation. Other schemes are presented in "Handbook of Analytical Chemistry" (L. Meites, ed.), 1st ed., McGraw-Hill, New York, 1963.

is made with the analytical balance. From these measurements the percent of a substance in a sample can be calculated.

Most gravimetric procedures include a precipitation step brought about by a chemical reaction. However, this is only one of many experimental steps that are part of the procedure. The entire procedure is outlined in the following:

1. An accurate weight for the sample being analyzed is recorded.
2. The weighed sample is dissolved.
3. Species which may interfere in the selected method are often removed by some suitable separation procedure.
4. The experimental environment is adjusted. This may include pH control by the addition of buffers, change of oxidation state, concentrating or diluting the sample, or addition of masking agents.
5. An appropriate inorganic or organic precipitation reagent is added.
6. Generally, the precipitation is done in dilute, hot solution.
7. Separation of the precipitate from the solution is usually achieved through filtration.

8. The precipitate is washed.

9. The dried precipitate or some suitable precipitate product formed through ignition is brought to constant weight.

10. The determination of the desired constituent in the sample is calculated from the recorded weight of the sample and the precipitate by accounting for the stoichiometry in the transformation of the constituent in the sample to the precipitate.

**Related Gravimetric Procedures.**   Other common gravimetric procedures are based on electrodeposition (electrogravimetry) and gas evolution. In the former, an appropriate adjustment of the current (or the potential) in an electrolysis cell containing a solution of the sample forces an electrochemical reaction to take place. In this reaction a product characteristic of the sample is deposited through reduction on the cathode (or oxidation at the anode). Since the species being determined is quantitatively plated on the electrode the weight of the electrode before and after plating is recorded and the amount of the species originally present in the solution is found by difference. Electrogravimetry is discussed in more detail in Chapter 28.

Evolution methods are based on weight loss of a sample due to volatilization of part or all of the sample. Thus, it is necessary to record the sample weight before and after the volatilization process. In some cases, the substance itself is volatilized by heating, while in others, a decomposition to volatile products produces the weight-loss step. An alternative procedure is to collect the volatile products in a trapping medium and record the weight increase of the trap.

**Gravimetric Factor.**   Assume that the sulfate content of an impure $K_2SO_4$ sample is being determined. A typical gravimetric procedure is to precipitate the sulfate ion as $BaSO_4$.

$$K_2SO_{4(aq)} + BaCl_{2(aq)} \rightarrow BaSO_{4(s)} + 2KCl_{(aq)} \tag{7-1}$$

If the correct procedure is followed, a weight of the sample containing $K_2SO_4$ and the weight of the $BaSO_4$ precipitate are recorded. From these data, the percent $K_2SO_4$ in the sample is calculated.

From Eq. (7-1) the following proportion can be written:

$$\frac{\text{wt } K_2SO_4 \text{ g}}{\text{formula wt } K_2SO_4 \text{ g/mole}} = \frac{\text{wt } BaSO_4 \text{ g}}{\text{formula wt } BaSO_4 \text{ g/mole}}$$

and rearranged to

$$\text{wt } K_2SO_4 \text{ g} = \text{wt } BaSO_4 \text{ g} \left( \frac{\text{formula wt } K_2SO_4 \text{ g/mole}}{\text{formula wt } BaSO_4 \text{ g/mole}} \right) \tag{7-2}$$

The term in brackets, which is the ratio of the molecular weights, is called the gravimetric factor and is used to convert weight of one chemical formula to another based on the stoichiometric relationship between the two. If the percent $K_2SO_4$ in the sample is desired, Eq. (7-2) becomes

$$\%K_2SO_4 = \frac{\text{wt BaSO}_4 \left(\dfrac{\text{formula wt } K_2SO_4 \text{ g/mole}}{\text{formula wt BaSO}_4 \text{ g/mole}}\right) \times 100}{\text{wt of sample}} \qquad (7\text{-}3)$$

Alternatively, the percent S in the sample could be reported by

$$\%S = \frac{\text{wt BaSO}_4 \left(\dfrac{\text{formula wt S g/mole}}{\text{formula wt BaSO}_4 \text{ g/mole}}\right)100}{\text{wt of sample}}$$

A general form of Eq. (7-3) would be written as

$$\%A = \frac{\text{wt B}\left(\dfrac{x \cdot \text{formula wt A g/mole}}{y \cdot \text{formula wt B g/mole}}\right)100}{\text{wt sample}} \qquad (7\text{-}4)$$

In this expression, $x$ and $y$ are chosen so that the numerator and denominator contain the same number of the atoms common to A and B. Or stated in another way, $x$ and $y$ are the coefficients of A and B, respectively, in the chemical equation showing the stoichiometric conversion of substance B into substance A. Several more examples that illustrate gravimetric calculations are also listed at the end of this chapter.

*Example 7-1.* Iron in a sample can be determined gravimetrically by precipitating it in basic solution. After the iron sample (0.2010 g) is dissolved, the hydroxide ion concentration is increased to precipitate iron(III) hydroxide. The precipitate is filtered and ignited (1000°C) to $Fe_2O_3$ and weighed (0.1106 g). Calculate the %Fe.
 The reactions are

$$Fe^{3+} + 3OH^- \xrightarrow{\text{H}_2\text{O}} Fe(OH)_3 \cdot xH_2O$$

$$2Fe(OH)_3 \cdot xH_2O \xrightarrow{\Delta} Fe_2O_3 + (x+3)H_2O$$

$$\%Fe = \frac{\text{wt Fe}_2O_3(2Fe/Fe_2O_3)100}{\text{wt of sample}}$$

$$\%Fe = \frac{0.1106 \text{ g}\left(\dfrac{2 \times 55.85 \text{ g/mole}}{159.7 \text{ g/mole}}\right)(100)}{0.2010 \text{ g}} = 38.49\%$$

*Example 7–2.*     Saccharin, **I**, and its salts are often used as noncalorie sweetening agents. A procedure for the gravimetric determination of saccharin in cider approved by the Food and Drug Administration is to convert the saccharin via oxidation to sulfate ion and subsequent precipitation as $BaSO_4$. [Details of the testing and evaluation of this procedure are described by J. R. Markus, *J. Assoc. Off. Anal. Chem.* **56,** 162 (1973).]

I

A 100.0-g sample of cider was reduced in volume, converted to sulfate, and the $BaSO_4$ (233.4) precipitate collected and weighed. If the precipitate weighed 0.02227 g, calculate (a) the % saccharin (184.2) in the cider sample and (b) the mg of saccharin per 100 g of cider which is the official reporting procedure.

(a)
$$\% \text{ saccharin} = \frac{\text{wt } BaSO_4 \ (C_7H_6NO_3S/BaSO_4) \times 100}{\text{wt of sample}}$$

$$\% \text{ saccharin} = \frac{0.02227 \text{ g } (184.2/233.4)100}{100.0 \text{ g}} = 0.0176\%$$

(b)
$$\text{mg saccharin} = \text{wt } BaSO_4 \text{ mg } (C_7H_6NO_3S/BaSO_4)$$

$$\text{mg saccharin} = 22.27 \text{ mg } (184.2/233.4) = 17.58 \text{ mg}$$

$$17.58 \text{ mg saccharin}/100 \text{ g cider}$$

In some cases there is no atom common to substance A and B. For this situation, it is imperative that all the chemical reactions for the transformation of A to B are known.

Several examples of gravimetric factors are listed on the inside front cover. These are taken from actual procedures and, therefore, the column listed as substances weighed also illustrate suitable weighing froms. The column listed "reported as" was chosen purely for illustration purposes since the sample can be reported in any desired form.

## FORMATION OF A PRECIPITATE

A reaction involving the precipitation of an electrolyte can be represented by

$$M^+ + X^- \rightleftharpoons MX_{(s)}$$

where the solid is in equilibrium with its ions. A typical example is the precipitation of AgCl by the addition of a silver ion solution to a chloride ion solution or vice versa. Formation of a precipitate of this kind has been shown to take place in a series of stages including supersaturation, saturation, nucleation, and finally, crystal growth.

**Saturation–Supersaturation.**    Saturation of a solution occurs when that solution contains the maximum amount of the salt permitted by its solubility at specified conditions. Supersaturation is a nonequilibrium condition that occurs when a solution phase contains more of the dissolved salt than described by the equilibrium condition.

Supersaturation is the first step in precipitation. Since it is a transient condition, the system will strive to relieve itself toward the saturation condition. This period of time will vary from system to system and the relief will be promoted by the availability of growth sites in the solution. The initial step from supersaturation to saturation is called nucleation.

**Nucleation.**    Nucleation is a process which leads to the smallest particles that are capable of spontaneous growth. For this to start, a minimum number of $M^+$ and $X^-$ ions must collect together, thus producing the smallest or beginning nuclei (microscopic or less in size) for the solid phase. The rate at which these nuclei form, in general, will increase with an increase in supersaturation.

Theoretically, the clustering of the $M^+$ and $X^-$ ions into nuclei from a supersaturated solution should occur in a spontaneous fashion. However, for most situations nucleation is induced. For example, introduction of seed particles of the precipitate into the solution will serve as sites for further precipitation. Extraneous particles, such as dust colloidal particles, impurities, and even scratches on the glass surface, can also serve as nucleation sites.

**Crystal Growth.**    Once a nucleus forms, it will continue to grow by a continued deposition of the precipitate particles on the nuclei. In addition, the $M^+$ and $X^-$ ions are deposited at specified sites in an orderly and uniform geometric pattern.

In general, greater supersaturation produces a much faster crystal growth rate. Detailed investigations suggest that either of two processes is the rate-determining factor. The first is the diffusion of the ions or molecules to the surface of the growing crystal, while the second is the actual deposition of the ions or molecules on the surface of the crystal. The experimental environment will influence these differently. For example, diffusion will be affected by temperature, stirring, concentration, and the actual properties of the ions or molecules involved in the crystal growth. Deposition rates, on the other hand, are affected by concentration, surface properties, and by the type of geometric arrangement that occurs as the crystal grows. Therefore, these are the experimental factors that must be controlled in gravimetry.

Experimentally, crystal growth is much easier to quantitatively study than nucleation. Consequently, much more is known about the properties of crystal growth. For example, by using X-ray diffraction and an appropriately formed crystal it is possible to completely elucidate the internal structure of the crystal including the spatial arrangement and distances between the atoms, ions, or molecules which make up the crystal. These kinds of investigations are at present the most basic type of measurement that the chemist can make. Crystal structures of all types of simple as well as complicated kinds of inorganic and organic compounds have been determined. Therefore, even if gravimetry was unknown as an analytical method, knowledge of the properties of precipitates would still be essential because of the tremendous impact of the crystallographic technique.

## PROPERTIES OF PRECIPITATES

There are several characteristics of precipitates which are important in separations and gravimetry. The significance of these is briefly described in the remainder of this section.

**Particle Size.**    The influence of supersaturation on the size of particles was recognized many years ago. During this period it was observed, experimentally, that at some level of supersaturation, a maximum particle size is obtained.

Figure 7–2 illustrates how particle size of lead sulfate varies with the initial concentration of lead perchlorate and various sulfate salts. In these studies, the lead salt and the sulfate salt were mixed at the stated concentration. After precipitation from a stirred solution the length of the crystals were carefully measured. From these and related data, it can be concluded that for a given level of supersaturation a maximum particle size is obtained. Second, as supersaturation increases, a decrease in particle size is observed. Several other investigations support these observations.

Since the precipitate is going to be separated from the reaction mixture, usually by filtration, large particles are required so that this step is quantitative. If they are too small, the particles will pass through the filtering device. There are several other reasons for desiring large precipitate particles and these will become apparent in later paragraphs.

**Colloidal State.**    In the early stages, the system is in the colloidal state with the particles having diameters of $1 \times 10^{-7}$ to $1 \times 10^{-5}$ cm. Such a condition is not desirable in precipitation since they are small enough to pass through ordinary filtering devices. Therefore, the experimental procedure must provide conditions for continued crystal growth in order to produce larger particles. Thus, after nucleation and crystal growth, one other step can

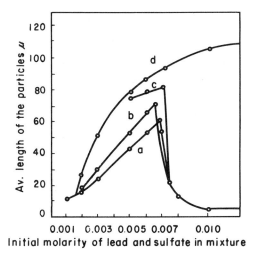

**Fig. 7–2.** Relation between particle size and concentration in mixtures of solutions of lead perchlorate and of various sulfates. (a) $H_2SO_4$; (b) $Na_2SO_4$; (c) $NaHSO_4$; (d) $K_2SO_4$. [Reprinted with permission from I. M. Kolthoff and B. Van't Riet, *J. Amer. Chem. Soc.* **63,** 817 (1959). Copyright by the American Chemical Society.]

be distinguished as part of the overall precipitation process. This is the *digestion* or *aging* (recrystallization) process.

In practice, the precipitate and mother liquor are heated at an elevated temperature for a fixed period of time. Since most substances have increased solubility at these conditions, the precipitation occurs at a slower rate. However, of greater importance is that the rate forward and reverse for the equilibrium

$$\text{dissolution} \rightleftharpoons \text{crystallization}$$

is significantly increased. The net result is that a large number of crystallizations take place.

Small crystals will have a slightly greater solubility than larger crystals. During digestion the smaller crystals will tend to dissolve and deposit on the remaining large crystals producing even larger crystals.

In passing from the colloidal state to a larger particle state two very important properties of colloids must be considered. These are (1) the presence of electrical charge on the colloid surface, and (2) the large surface area provided by the colloid.

## COPRECIPITATION

In general, as precipitates form and settle in solution, they become contaminated with the various ions and solvent present in the solution. This contamination is called coprecipitation and under normal conditions these impurities, which are carried along with the precipitate, are soluble.

Coprecipitation can be divided into two separate classes. These are adsorption at the precipitate surface and occlusion of impurities. Two other types of coprecipitation, which will not be discussed since they are minor problems in analytical precipitation procedures, are postprecipitation and isomorphous replacement.

In postprecipitation, partially soluble impurities are carried by the precipitate, the amount of which is determined by the rate of adsorption. Isomorphous replacement is a very specific type of contamination and involves replacement of ions in the crystal lattice by other ions of similar size and shape. The unwanted ions become a part of the crystal lattice and are not removed by washing.

When crystals grow, the ions orient themselves in a fixed pattern. Thus, at any surface on the crystal there are localized centers of charge. For example, on the surface of a crystal of AgCl, there are positive centers wherever silver ions are located and negative centers where chloride ions are located. Since there are equal numbers of cations ($Ag^+$) and anions ($Cl^-$), the entire surface has a net neutral charge.

The localized charge centers take on greater importance when considering the large surface area provided by a colloidal condition. Since colloidal particles are very small, the surface to mass ratio is enormous and the surface has a great tendency to adsorb cations or anions.

If a silver nitrate solution is slowly added to a sodium chloride solution, the silver chloride precipitates in the presence of excess silver ion and will adsorb silver ions. Consequently, the surface attains a positive charge and will repel other particles of the same type or other cations. But, anions will be attracted toward this surface. Thus, a second layer of ions, nitrate ions in this particular example, will be attracted to the particle. Since the number of anions in the second layer is equal to the number of cations in the first layer, the overall effect is an apparent neutral particle. However, it should be realized that the first layer is strongly attracted and is fixed on the particle surface, while the second layer resides at some small distance away from the surface to maintain electrical neutrality. Thus, the particle is said to have a positive charge, if the first layer is adsorbed cations, or a negative charge, if the first layer is adsorbed anions.

**Surface Adsorption.**     Every precipitate will have the tendency to adsorb cations or anions on to its surface. If cations are adsorbed, the surface acquires a positive charge while, if anions are adsorbed, a negative charge occurs on the surface. This type of condition promotes the colloidal state.

The first layer of adsorbed ions is usually called the primary layer. The second layer is called the counterlayer and is made up of counterions. Figure

| Ag | | Ag | $Cl^- \cdots Na^+$ |
|----|---|----|----|
| Cl | $Ag^+ \cdots NO_3^-$ | Cl | |
| Ag | | Ag | $Cl^- \cdots Na^+$ |
| Cl | $Ag^+ \cdots NO_3^-$ | Cl | |
| Ag | | Ag | $Cl^- \cdots Na^+$ |
| Cl | $Ag^+ \cdots NO_3^-$ | Cl | |

**Fig. 7–3.**   Adsorption on the surface of a silver chloride crystal.

7-3 illustrates the surface behavior for AgCl. Although not shown, molecules of water are also dispersed through the layers.

If the counterions completely neutralize the primary layer and are very close to the primary layer, there is little tendency for the colloidal particles to repel each other. Thus, they tend to collect together to form larger sized particles. This process is referred to as *coagulation*.

Coagulation is one direction of a reversible process. The breakdown of coagulated particles back to the colloidal state is the other direction. This process is called *peptization* and can be represented by

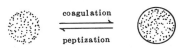

Washing a filtered precipitate that has been coagulated through charge neutralization with water can lead to peptization since the counter-primary ion interaction is disturbed. Therefore, precipitates are often washed with a solution containing a volatile or easily decomposed electrolyte. These electrolytes replace the initial adsorbed electrolyte impurities and are removed during the drying process.

Four main factors influence ion adsorption. These are briefly described in the following paragraphs.

PRECIPITATION EFFECT.   If several different ions are present in the solution and available for adsorption, the ion that is adsorbed will be the one that forms the most insoluble compound with one of the lattice ions.

CONCENTRATION EFFECT.   In general, the tendency is for the ion of greater concentration to be the one that is preferentially adsorbed, if all other factors are equal. A limited amount is adsorbed and it decreases with increased temperature.

CHARGE EFFECT.   The tendency is for the ion of greatest charge to be the one that is preferentially adsorbed if all other factors are equal.

ION SIZE EFFECT.   Ions that are of the same or nearly the same size as those in the crystal lattice are preferentially adsorbed over those of dissimilar size if all other factors are equal. Essentially, the adsorbed ion takes the place of those missing at the surface.

The experimental procedure in gravimetry is designed to minimize surface adsorption by the following steps:

1. Form the precipitate in dilute solution with respect to the ions being adsorbed.
2. Form the precipitate as large crystals.
3. Form the precipitate in hot solution (digestion).
4. Remove ions which are strongly adsorbed.
5. Replace adsorbing ions in the solution by other ions that are more easily eliminated.
6. Remove the precipitate, and dissolve in another solution only to re-precipitate it again.

**Occlusion.** In occlusion, foreign ions and solvent are heterogeneously trapped within the crystal. The available experimental evidence suggests that adsorption on the surface contributes to occlusion. However, it is also apparent that not all occluded impurities are present because of adsorption. During rapid crystal growth, solution can be physically trapped in small pockets within the crystal. Hence, solvent, as well as foreign ions, are occluded.

Washing the precipitate will not be effective in removing occluded substances even though these materials are soluble. Fortunately, through careful control of several experimental steps, occlusion is usually minor and can be minimized by precipitation from dilute solution and employing a digestion period.

**Types of Precipitates.** In general, precipitates are designated as being crystalline, curdy, or gelatinous. Crystalline precipitates ($BaSO_4$) are well-defined crystals whose shapes are a function of the composition of the salt, while curdy precipitates (AgCl) are aggregates of small, porous particles. Gelatinous precipitates [$Fe(OH)_3$], which are the most difficult to work with, are jellylike, hydrous masses. If particle sizes are of reasonable size, crystalline and curdy precipitates are filtered rapidly, easily, and completely.

# TYPES OF PRECIPITATING REAGENTS

The types of characteristics that a particular precipitating reagent, and the precipitate itself, must have to be useful will depend on the application. The more important characteristics are (1) solubility; (2) ease in filtering; (3) stoichiometry; (4) high formula weight; (5) nonhygroscopicity; and (6) selectivity.

The stoichiometry of the precipitate or at least that which is eventually weighed must be known and reproducible. Often the precipitate is ignited to a known stoichiometry. For example, $Mg^{2+}$ is precipitated as $MgNH_4PO_4$ $\cdot 6H_2O$ by the reaction

$$Mg^{2+} + NH_4^+ + PO_4^{3-} + 6H_2O \rightarrow MgNH_4PO_4 \cdot 6H_2O$$

However, the composition of the precipitate is not exact, and after filtration it is ignited (1100°C) and weighed as magnesium pyrophosphate, $Mg_2P_2O_7$. In general, coprecipitation errors, which also affect stoichiometry, are minimized by the experimental procedure.

Ideally, the precipitate should have a large molecular weight so that the precipitate represents a small amount of the cation or anion being determined. It it is also stable and nonhydroscopic, and special handling techniques are not required. Hence, errors in weighing are greatly reduced.

The one factor important to both gravimetry and separation by precipitation is selectivity. Ideally, one precipitating agent should precipitate only one cation or one anion. Similarly, at least one reagent should be available for each cation and for each anion. Such specificity is rarely encountered. More often the typical specificity displayed by a reagent is the precipitation of small groups of species.

**Inorganic Reagents.** A wide variety of inorganic precipitating agents are available. Table 7-1 lists some of the more common ones. Briefly, the reagents used, the precipitate formed, and the weighing form are listed.

Perhaps the two most general inorganic reactions used for separation of metal ions by precipitation are the formation of hydroxides or sulfides. For some metal ions these reactions can also be used for gravimetric purposes. However, hydroxide and sulfide precipitates often fail in meeting the gravimetric requirements, such as stoichiometry, readily filtered crystals, low coprecipitation, and free of precipitated interferences. Improvement in the quality of hydroxide and sulfide precipitants is possible through a technique known as homogeneous precipitation.

**Organic Reagents.** A wide variety of organic compounds are used as precipitating reagents and are often more versatile than inorganic precipitating reagents. The organic precipitating agents form products with cations or anions which can be classified as complexes or undissociated or partially dissociated salts. In contrast, inorganic precipitating agents produce predominately ionic products. Because of this difference in bonding, precipitates produced from organic precipitating agents possess different kinds of physical and chemical properties (see Chapter 15).

A typical example is the precipitation of $Ni^{2+}$ with dimethylglyoxime

Table 7-1. A Partial List of Inorganic Precipitating Conditions

| Element | Precipitant | Precipitated form | Wash | Ignition temperature (°C) | Weighing form[a] |
|---|---|---|---|---|---|
| Ag | HCl | AgCl | $HNO_3$ | 150 | AgCl |
| Al | $NH_3$ | $Al(OH)_3$ | $NH_4Cl$ | 1200 | $Al_2O_3$ |
| Ba | $H_2SO_4$ | $BaSO_4$ | $H_2O$ | 800 | $BaSO_4$ |
| Bi | KCl | BiOCl | $H_2O$ | 110 | BiOCl |
| Br, Cl, I | $AgNO_3$ | AgBr, Cl, I | $HNO_3$ | 110 | AgBr, Cl, I |
| Ca | $(NH_4)_2C_2O_4$ | $CaC_2O_4 \cdot H_2O$ | $H_2O$ | 950 | CaO |
| Cs | $H_2PtCl_6$ | $Cs_2PtCl_6$ | Alcohol | 100 | $Cs_2PtCl_6$ |
| F | $Pb(NO_3)_2$-HCl | PbClF | | 130 | PbClF |
| Fe | $NH_3$ | $Fe(OH)_3$ | $NH_3$ | 1000 | $Fe_2O_3$ |
| Hg | $H_2S$ | HgS | $H_2O$ | <100 | HgS |
| K | $H_2PtCl_6$ | $K_2PtCl_6$ | Alcohol | <270 | $K_2PtCl_6$ |
| K | $HClO_4$ | $KClO_4$ | Ethyl acetate | <650 | $KClO_4$ |
| Mg | $NH_4HPO_4$ | $MgNH_4PO_4 \cdot 6H_2O$ | $NH_4NO_3$ | 1050 | $Mg_2P_2O_7$ |
| Na | $KZn[UO_2(C_2H_3O_2)_3]_3$ | $NaZn[UO_2(C_2H_3O_2)_2]_3$ | Alcohol | 120 | $NaZn[UO_2(C_2H_3O_2)_2]$ |
| P | $MgSO_4$-$(NH_4)_2SO_4$ | $MgNH_4PO_4$ | $NH_4NO_3$ | 1050 | $MgP_2O_7$ |
| Pb | $H_2SO_4$ | $PbSO_4$ | $H_2O$ | 600 | $PbSO_4$ |
| Rb | $H_2PtCl_6$ | $Rb_2PtCl_6$ | Alcohol | 100 | $Rb_2PtCl_6$ |
| S | $BaCl_2$ | $BaSO_4$ | $H_2O$ | 800 | $BaSO_4$ |
| Sn | $HNO_3$ | $SnO_2 \cdot XH_2O$ | $H_2O$ | 1100 | $SnO_2$ |
| Transition metals | $(NH_4)_2HPO_4$ | $MHPO_4$ | $NH_4NO_3$ | 1000 | $M_2P_2O_7$ |
| Rare earths, Zr, Hf, Th, Sc, others | $H_2C_2O_4$ | $M(C_2O_4)_2$ or $M_2(C_2O_4)_3$ | $H_2O$ | 1000 | $MO_2$ or $M_2O_3$ |

[a] In some cases other weighing forms are possible if other ignition conditions are used. For further details and references see "Handbook of Analytical Chemistry," L. Meites, ed., 1st ed., McGraw-Hill, New York, 1963.

to form a brilliant carmen red Ni–DMG precipitate. This compound, like

$$Ni^{2+} + 2 \quad \begin{array}{c} OH \\ CH_3C=N \\ | \\ CH_3C=N \\ OH \end{array} \quad \longrightarrow \quad \begin{array}{c} O\cdots H-O \\ CH_3C=N \qquad N=CCH_3 \\ | \qquad Ni \qquad | \\ CH_3C=N \qquad N=CCH_3 \\ O-H\cdots O \end{array} \quad + \; 2\,H^+ \qquad (7\text{-}5)$$

DMG

most precipitants derived from organic precipitating agents, is covalent and has neither the obvious physical or chemical properties of $Ni^{2+}$ or DMG. In general, properties are typical of those found for organic compounds. For example, the Ni–DMG is soluble in organic solvents. This solubility property is also used advantageously in other instrumental (spectrophotometric) and chemical (extraction) procedures.

If an organic precipitating agent is used for gravimetric purposes, stoichiometry, suitable weighing and filtering form, high molecular weight, a precipitate free of coprecipitation and nonhygroscopic properties are still required. Since the products are covalent, coprecipitation problems like those in inorganic precipitates are not usually encountered.

Frequently, the organic reagent itself may have low solubility, thus, the precipitate can be contaminated by excess reagent. Most organic reagents will also contain acidic or basic sites which are only partially dissociated. This property can be used advantageously since it offers a route to selectivity because different species will precipitate in different pH ranges.

A procedural difficulty is often encountered due to the floating and creeping tendency of the organic-like precipitate. Frequently, it tends to stick to the glassware and exhibit other undesirable surface-related properties since the precipitates are not readily water-wetted.

Often the product of the reaction between the organic precipitating reagent and a cation or anion is not in a suitable weighing form. Thus, the precipitate is ignited to a known reproducible stoichiometric product. If metal ions are being precipitated, frequently, the ignited product is the oxide.

Table 7-2 lists some of the common, more selective organic precipitating agents. Also included are some brief comments about the applications of these reagents.

**Thermobalance.**    Heating the precipitate before weighing serves three purposes: (1) excess water is removed from the precipitate; (2) the electrolyte used in the wash solution is vaporized; and (3) the precipitate attains a constant weight.

Since heating temperatures (see Tables 7-1 and 7-2) vary from precipitate to precipitate, the need to establish correct ignition temperatures is important. The instrument which can be used for this purpose is a thermobalance. In its simplest form, the instrument is a balance with the pan sitting in an oven

**Table 7-2.  Some Organic Precipitants**

| Element | Conditions | Precipitant (L) | Precipitated form[a] | Ignition temperature (°C)[b] | Weighing form |
|---|---|---|---|---|---|
| Al, Bi, Cd, Co, Cu, Ga, Hf, Fe, In, Hg, Mo, Ni, Nb, Pd, Pa, Ag, Ta, Ti, Th, W, U, Zn, Zr | pH = 4.5 | 8-Hydroxyquinoline | $ML_4$ $ML_3$ $ML_2$ | 130 >1000 | $ML_{4,3,2}$ Metal oxide |
| Al, Be, Bi, Cd, Cu, Ga$^+$, Hf, Fe, In, Mg, Mn, Hg, Nb, Pd, Sc, Ta, Ti, Th, U, Zn, Zr, rare earths | $NH_3$ | 8-Hydroxyquinoline | $ML$ $ML_3$ $ML_2$ | 130 >1000 | $ML_{4,3,2}$ Metal oxide |
| Ni | $NH_3$ | Dimethylglyoxime | $ML_2$ | 150 | $ML_2$ |
| Pd | Acidic | Dimethylglyoxime | $ML_2$ | 150 | $ML_2$ |
| Co | Acidic | 1-Nitroso-2-naphthol | $ML_2$ | 900 | Metal oxide |
| Fe, Hg, Nb, Ta, W, Zr | Strong acid | Cupferron | $ML_3$ | >1000 | Metal oxide |
| Sb, Bi, Ga, Fe, Mo, Pd, Sn, Ta, Ti, V, W, Zr rare earths | Dilute acid | Cupferron | $ML_3$ | >1000 | Metal oxide |
| Al | Neutral | Cupferron | $ML_3$ | >1000 | Metal oxide |
| Cu, Cd, Ni | $HC_2H_3O_2$ | Anthranilic acid | $ML_2$ | <225 | $ML_2$ |
| Zn, Co, Pb | Neutral | Anthranilic acid | $ML_2$ | >1000 | Metal oxide |
| Cu | $NH_3$ | Benzoinoxime | $ML$ | >1000[b] | Metal oxide |
| Cs, K, Rb, Ag, Tl | Acidic–basic | Sodium tetraphenylborate | $ML$ | <250 | $ML$ |
| $Cr_2O_7^{-2}$, $MnO_4^-$, $ReO_4^-$, $ClO_4^-$ | Acidic | Tetraphenylammonium chloride | $ML$ | <225 | $ML$ |

[a] Stoichiometry of the precipitated complex will depend on the charge of the cation and its coordination number.

[b] In some cases it is not necessary to ignite to the metal oxide. A lower temperature can be used in which a metal–ligand (L) complex of known stoichiometry is the weighing form.

whose temperature can be carefully and slowly increased (for example, to 1000°C at a rate of 5°C/min). Consequently, weight change of a sample (8–10 mg) is recorded as a function of temperature. From this weight–temperature relationship, it is possible to see when decomposition takes place.

The graph obtained with the thermobalance is called a thermogram or pyrolysis curve. Several thermograms are shown in Fig. 7–4.

Consider curves C and D in Fig. 7–4. Four different ignition temperatures are possible for the calcium salt while three are possible for the magnesium salt. For both, the first weight-loss step in the curves is due to loss of water of hydration. Continued heating of the $MgC_2O_4$ to above 400°C leads to further decomposition and the loss of CO and $CO_2$ to produce MgO. In contrast, $CaC_2O_4$ decomposes stepwise, first forming $CaCO_3$ and then CaO as the temperature is raised.

The thermogram for $MgNH_4PO_4$, illustrated in Fig. 7–4E, shows a weight loss up to 477°C at which point $Mg_2P_2O_7$ is formed. This illustrates why $MgNH_4PO_4$ (the formation of $MgNH_4PO_4$ is routinely used for the gravimetric determination of Mg) is converted to the weighing form $Mg_2P_2O_7$ rather

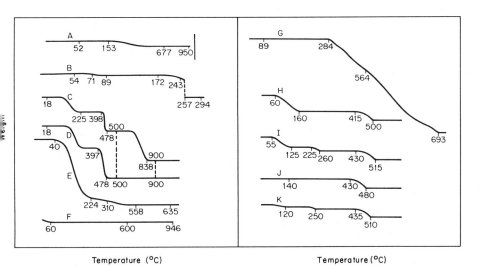

**Fig. 7–4.** Thermograms for several precipitates of analytical interest. (A) Barium sulfate; (B) bis(dimethylglyoximato)nickel(II); (C) calcium oxalate; (D) magnesium oxalate; (E) magnesium ammonium phosphate ($MgNH_4PO_4$); (F) silver chloride; (G) tris(8-hydroxyquinolinato)iron(III)($Fe(C_9H_6ON)_3$); (H) ammonium 12-molybdophosphate-$HNO_3$, washed, air-dried; (I) ammonium 12-molybdophosphate-$NH_4NO_3$, washed, air-dried; (J) ammonium 12-molybdophosphate-$HNO_3$, washed, oven-dried; (K) ammonium 12-molybdophosphate-$NH_4NO_3$, washed, oven-dried. (A–G from C. Duval, "Inorganic Thermogravimetric Analysis," Elsevier, Amsterdam, 1953; H–K from W. W. Wendlandt, "Thermal Method of Analysis," Wiley-Interscience, New York, 1964.)

than weighed as $MgNH_4PO_4$. It should be noted that the thermogram fails to indicate a well-defined loss of $NH_3$ or $H_2O$.

The thermogram for AgCl, Fig. 7–4F, is linear over a wide temperature range after first losing absorbed water. Iron, which can be precipitated as the 8-hydroxyquinoline complex where one iron combines with three 8-hydroxyquinoline molecules is stable as $Fe(C_9H_6OH)_3$ up to 284°C (see Fig. 7–4G). Continued heating causes decomposition and if the iron is to be weighed as $Fe_2O_3$, an ignition temperature of greater than 893°C must be used.

Washing and handling of the precipitate can result in variations in the drying procedures as evidenced by the thermograms H–K in Fig. 7–4. Formation of ammonium molybdophosphate is routinely used for the gravimetric determination of phosphate. The air-dried $HNO_3$-washed precipitate, H, loses water initially yielding the plateau region with the composition now being $(NH_4)_2HP(Mo_3O_{10}) \cdot H_2O$. Above 415°C additional weight loss occurs eventually reaching another flat region corresponding to $P_2O_5 \cdot 24MoO_3$, which is a suitable weighing form. For the oven-dried precipitate, J, initial water loss is not present. For the $NH_4NO_3$-washed precipitate, no initial water loss is observed as shown in curves I and K. More important, however, is that $NH_4NO_3$ is being lost in the 225–260°C region yielding $(NH_4)_3[P(Mo_3O_{10})_4]$, which upon heating above 510°C, yields the compound having the stoichiometry $P_2O_5 \cdot 24MoOH_3$.

## SOLUBILITY

Solubility in gravimetry is important since this is a measure of the quantitative nature of the precipitation process. Increased solubility can mean the difference between the process being classified as quantitative or not quantitative.

The idea of classifying substances as being soluble, partially soluble, or insoluble is a relative concept. No electrolyte, whether weak or strong, is completely insoluble. For example, the solubility of AgCl is $1.3 \times 10^{-5}$ $M$ or 1.43 mg/liter for a specified set of experimental conditions. If a AgCl precipitate weighs about 1 g and the volume of the precipitating solution is about 200 ml, the amount of AgCl remaining in solution is about 0.3 mg. This amount is barely detected by the ordinary analytical balance. As the weight of the AgCl decreases the amount that remains in solution becomes more significant. Therefore, attempts to isolate 10 mg from the precipitating solution by the usual gravimetric techniques must be reconsidered if quantitative results are expected.

The solubility of $KClO_4$ in water is 2.04 g/100 ml and would be classified as partially soluble. Hence, it is not possible to determine potassium ion quantitatively by precipitating it as $KClO_4$ in a water system. Even by using large amounts of a perchlorate ion solution as precipitating agent, the solubility is still too large. However, the addition of large quantities of ethanol decreases the solubility to the point where the procedure is quantitative.

**Solubility of Ionic Salts.**    In general, ionic compounds exhibit the greatest solubility in polar solvents such as water. The dissolution process takes place because the polar solvent molecules are sufficiently attracted to the ions and pull them away from their positions in the crystal lattice. During this process the ions become solvated.

$$MX_{(s)} \xrightarrow{x+yH_2O} M(H_2O)_x{}^+ + X(H_2O)_y{}^-$$

Even though the cation and anion are represented as solvated ions, the attraction of the water molecule to the metal ion is weak. Since the water molecules and ions in the solution are in rapid and constant motion, the number of water molecules about a metal ion tends to be variable. The average number is referred to as the *hydration number*. In crystals that are hydrated the number of water molecules is fixed.

**Solubility of Covalent Compounds.**    The solubility of covalent compounds will be dependent upon the structural features of the molecule. Some compounds will be ionic in character, while others will be almost completely covalent in character. An old chemical adage is that "like dissolves like." Essentially, this states that forces between similar molecules are comparable to those between identical molecules. Thus, new molecules can replace others if they are similar. Forces holding molecules together can be classified as London forces (induced dipole attractions), dipole–dipole forces, ion dipole forces, and hydrogen bonding (in order of increasing strength).

If a liquid is held together through hydrogen bonding, another solid or liquid should be soluble in it providing this solid or liquid is also a capable hydrogen bonder. Similarly, two different members of each of the other classes should form solutions providing one of them serves as the solvent.

In gravimetry it is very important to consider the solubility of the precipitating agent, particularly when using organic precipitating agents. If the precipitating agent is of limited solubility, the precipitate will be contaminated with excess reagent. Many useful organic precipitating agents contain acidic of basic functional groups such that solubilities of these compounds are increased in water and other polar solvents. Frequently, the solubility of an analytically useful reagent is increased by introduction of a sulfonic acid group into the molecule.

By realizing what factors influence solubility, a suitable solvent or solvent mixture can be selected or the precipitating agent can be structurally modified to increase solubility. Fortunately, solubility data are readily available in a variety of reference books that list physical properties of compounds.

## SOLUBILITY PRODUCT

The solubility product ($K_{sp}$) can be used for predicting optimum conditions for the formation and dissolving of precipitates. This includes variables, such as temperature, pH, precipitating reagent concentrations, inert salt concentration, and solvent composition.

For silver chloride, the equilibrium and $K_{sp}$ expressions are

$$AgCl_{(s)} \rightleftharpoons Ag^+ + Cl^-$$

$$K_{sp} = 1.8 \times 10^{-10} \ (M)^2 = [Ag^+][Cl^-]$$

At the point of equilibrium, opposing reactions in the saturated solution are taking place. In one direction ions escape from the surface of the cyrstal into the solution, while simultaneously, ions deposit on the crystal surfaces. These processes, although described in a very simplified fashion, actually involve complicated kinetics.

Frequently, a simple relationship exists between solubility product and solubility. For these cases it is possible to calculate the solubility of the salt from the solubility product or the solubility product from the solubility. The following are examples of the calculation.

*Example 7-3.*    Calculate the molar solubility of AgCl ($K_{sp} = 1.8 \times 10^{-10}$).

$$AgCl_{(s)} \rightleftharpoons Ag^+ + Cl^-$$

$$K_{sp} = [Ag^+][Cl^-] = 1.8 \times 10^{-10} \ (M)^2$$

It follows that

$$[Ag^+] = [Cl^-] = S \text{ (solubility)}$$

and therefore

$$1.8 \times 10^{-10} = S \cdot S$$

$$S = 1.34 \times 10^{-5} \ M$$

*Example 7-4.*    The solubility of $BaSO_4$ has been found to be $1.14 \times 10^{-5}$ F. Calculate the solubility product.

$$BaSO_{4(s)} \rightleftharpoons Ba^{2+} + SO_4^{2-}$$

$$K_{sp} = [Ba^{2+}][SO_4^{2-}]$$

$$[Ba^{2+}] = 1.14 \times 10^{-5} \ M$$

$$[SO_4^{2-}] = 1.14 \times 10^{-5} \ M$$

$$K_{sp} = [1.14 \times 10^{-5}][1.14 \times 10^{-5}]$$

$$K_{sp} = 1.30 \times 10^{-10} \ (M)^2$$

From the solubility product concept, it can be concluded that if the concentration of $Ag^+$ and $Cl^-$ are less than $1.34 \times 10^{-5}\ M$ a precipitate will not form. Once this concentration is reached additional AgCl that is formed or added will be present as a precipitate. If the ionic concentrations exceed $1.34 \times 10^{-5}\ M$ (and the temperature is not changed), the solution is supersaturated.

It is not always possible to suggest which of a series of salts is the most soluble by looking at the value for the solubility product. This is illustrated by considering the following example.

*Example 7–5.*    Calculate the molar solubility of $Ag_2CrO_4$ ($K_{sp} = 1.1 \times 10^{-12}$).

$$Ag_2CrO_{4(s)} \rightleftharpoons 2Ag^+ + CrO_4^{2-}$$

$$K_{sp} \rightleftharpoons [Ag^+]^2[CrO_4^-] = 1.1 \times 10^{-12}\ (M)^3$$

$$\left.\begin{array}{l} [Ag^+] = 2S \\ [CrO_4^-] = S \end{array}\right\} \text{based on reaction}$$

$$1.1 \times 10^{-12} = [2S]^2[S]$$

$$S = 6.5 \times 10^{-5}\ M$$

Silver chromate is more soluble than AgCl even though the solubility product of $Ag_2CrO_4$ is smaller than for AgCl. Only for electrolytes of the same valence type is it possible to qualitatively predict an order of solubility based on comparison of the solubility product constants.

*Example 7–6.*    A saturated solution of $PbF_2$ in equilibrium with solid $PbF_2$ was found to have a solubility of 0.514 mg $PbF_2$/ml. Calculate the $K_{sp}$ for $PbF_2$.

$$PbF_{2(s)} \rightleftharpoons Pb^{2+} + 2F^-$$

$$K_{sp} = [Pb^{2+}][F^-]^2$$

$$\text{g } PbF_2/\text{liter} \div \text{g } PbF_2/\text{mole} = \text{moles } PbF_2/\text{liter}$$

$$0.514\ \text{g/liter} \div 245\ \text{g/mole} = 2.10 \times 10^{-3}\ \text{moles/liter}$$

$$[Pb^{2+}] = 2.10 \times 10^{-3}M$$

$$[F^-] = 2 \times 2.10 \times 10^{-3}M$$

$$K_{sp} = [2.10 \times 10^{-3}][2 \times 2.10 \times 10^{-3}]^2$$

$$K_{sp} = 3.70 \times 10^{-8}(M)^3$$

**Factors That Affect Solubility Product.**    There are three main factors which affect $K_{sp}$. These are temperature, solvent, and particle size.

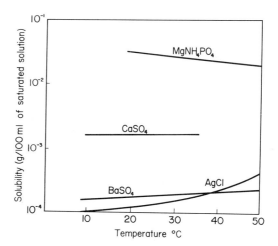

**Fig. 7-5.** Solubility of several precipitates of analytical interest as a function of temperature. [From H. Stephen and T. Stephen, "Solubilities of Inorganic and Organic Compounds," Vols. 1-2, Pergamon Press, Oxford, 1964.]

Most often, an increase in temperature increases the solubility and solubility product. For example, $PbCl_2$ ($K_{sp} = 1.6 \times 10^{-5}$) is partially soluble at room temperature but will readily dissolve at elevated temperature. However, there are exceptions and some salts dissolve with loss of energy, and their solubility and solubility product decreases or changes only slightly with increased temperature. The effect of temperature on the solubility of several salts of analytical interest is illustrated in Fig. 7–5.

The direction of solubility change as a function of temperature can be predicted by applying Le Chatelier's principle. Since increasing the temperature is a stress, the equilibrium between a precipitate and its ions in solution will shift according to whether the heat of solution is endothermic (increased solubility) or exothermic (decreased solubility).

Using a solvent of lower dielectric constant usually results in a lower solubility. Often the solubility in water of a moderately insoluble substance is reduced further by the addition of alcohol or some other water-miscible solvent. The effect of mixed solvent is shown in Fig. 7–6. The increased solubility for Ni(II)–dimethylglyoxime with increased alcohol concentration is the result of (1) the compound being covalent and (2) the presence of the organic structure in the compound.

As particle size decreases, solubility appears to increase. This is an initial rate effect and does not represent the equilibrium condition. If the solution is allowed to digest, the small particles gradually dissolve and saturation is maintained by depositing an equivalent amount on the larger crystals. Thus, the larger crystals grow at the expense of the smaller ones.

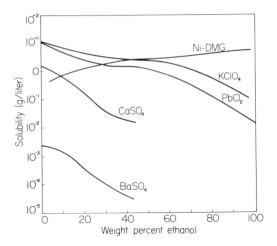

**Fig. 7–6.**   Solubility of several salts in water–alcohol mixtures. (See references in Fig. 7–5.)

**Common Ion Effect.**   If chloride ion in the form of potassium chloride is added to a saturated AgCl solution, a stress is placed on the equilibrium established for the system. The common ion, chloride ion, causes the equilibrium to shift toward the formation of solid AgCl. The net result is a reduction in the concentration of the silver ion concentration in the solution. If more KCl is added, the silver ion concentration decreases further. This is an example of the common ion effect and the magnitude of the effect can be calculated by the solubility product principle.

*Example 7-7.*      Calculate the solubility of AgCl in the presence of (a) 0.0001 *F* and (b) 0.1 *F* KCl.

$$AgCl_{(s)} \rightleftharpoons Ag^+ + Cl^-$$

$$K_{sp} = [Ag^+][Cl^-] = 1.8 \times 10^{-10} \, (M)^2$$

(a)

$$[Ag^+] = S(\text{solubility})$$

$$[Cl^-] = 0.0001 + S$$

$$[S][0.0001 + S] = 1.8 \times 10^{-10}$$

$$S = 1.75 \times 10^{-6} \, M$$

(b)

$$[Ag^+] = S$$

$$[Cl^-] = 0.1 + S \cong 0.1$$

$$[S][0.1] = 1.8 \times 10^{-10}$$

$$S = 1.8 \times 10^{-9} \, M$$

The effect of the common ion is apparent when this solubility is compared to the solubility in the absence of the common ion (see Example 7–2). It should also be noted that the approximation for $[Cl^-]$ is reasonable in (b) since $S$ is very small. The approximation, which should always be checked, should not be used in (a).

A comparison of calculated AgCl solubility vs experimentally determined solubilities as a function of KCl concentration is shown in Fig. 7–7. As the KCl concentration increases, a point is reached where the actual solubility increases, rather than decreases as predicted by the common ion effect. For example, the experimentally observed solubility for AgCl in 0.1 $F$ KCl is about 3000 times larger than the calculated value. Therefore, the common ion effect should not be carelessly applied in gravimetric procedures. A useful guideline, when exact data is not available, is to use the common ion in not greater than 0.1 $F$ in excess.

The fact that the calculated solubility curve fails to follow the experimental curve in Fig. 7–7 establishes that the system is not a simple one in which only insoluble AgCl is formed. Four additional equilibrium steps must be considered.

$$Ag^+ + Cl^- \rightleftharpoons AgCl_{(aq)} \qquad K_1 = \frac{[AgCl_{(aq)}]}{[Ag^+][Cl^-]}$$

$$AgCl_{(aq)} + Cl^- \rightleftharpoons AgCl_2^- \qquad K_2 = \frac{[AgCl_2^-]}{[AgCl_{(aq)}][Cl^-]}$$

$$AgCl_2^- + Cl^- \rightleftharpoons AgCl_3^{2-} \qquad K_3 = \frac{[AgCl_3^{2-}]}{[AgCl_2^-][Cl^-]}$$

$$AgCl_3^{2-} + Cl^- \rightleftharpoons AgCl_4^{3-} \qquad K_4 = \frac{[AgCl_4^{3-}]}{[AgCl_3^{2-}][Cl^-]}$$

At increasing levels of KCl each of these species takes on increased significance and the net result is increased solubility. Since the solubility product requires equilibrium concentrations for silver and chloride ion, an exact interpretation must account for all other equilibria, such as those shown for the silver–chloride system.

Ion-pair formation and/or hydrolysis may take place in other simple systems. If the solute is a salt of a weak acid or weak base, the solubility will be affected by pH. An example of the pH effect is the following:

$$CaC_2O_{4(s)} \rightleftharpoons Ca^{2+} + C_2O_4^{2-}$$

$$K_{sp} = 4 \times 10^{-9} \; (M)^2 = [Ca^{2+}][C_2O_4^{2-}]$$

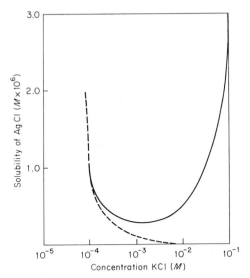

**Fig. 7–7.**   Solubility of AgCl in the presence of a common ion (KCl). (- - -) Calculated and (—) experimental [L. Lieser, *Z. Anorg. Allgem. Chem.* **292**, 97 (1957)]. See also A. Pinkus and A. M. Timmermans, *Bull. Soc. Chim. Belg.* **46**, 46 (1937).

With decreasing pH, association is going to take place according to the reactions

$$H_3O^+ + C_2O_4^{2-} \rightleftharpoons HC_2O_4^-$$

$$H_3O^+ + HC_2O_4^- \rightleftharpoons H_2C_2O_4$$

Thus, decreasing the pH would increase the solubility of $CaC_2O_4$ through the formation of $HC_2O_4^-$ and $H_2C_2O_4$.

**Diverse Salt Effect.**   A precise definition of solubility product is in terms of activities. Therefore, for MX

$$MX_{(s)} \rightleftharpoons M^+ + X^-$$

the solubility product expression is

$$K_{sp}^\circ = a_{M^+} a_{X^-}$$

where $K_{sp}^\circ$ is the activity solubility product constant.
  Substitution for activity gives

$$K_{sp}^\circ = \gamma_M [M^+] \gamma_X [X^-]$$

and

$$K_{sp} = \frac{K_{sp}^\circ}{\gamma_M \gamma_X} = [M^+][X^-] \tag{7-6}$$

In Eq. (7–6), $K_{sp}$ is the concentration solubility product and is a function of the ionic strength of the solution where ionic strength is a measure of the ionic content of the solution.

Increasing the concentration of a diverse ion (a diverse salt is one whose ions are not common to the ions of the insoluble salt under investigation) increases the solubility of the insoluble precipitate which, therefore, leads to an increase in the solubility product. If activity coefficients and $K_{sp}^{\circ}$ are known, solubility at a fixed concentration of a diverse ion can be calculated. Generally, however, this effect is observed experimentally rather than by calculation.

The charge of the electrolyte in the solution will affect the solubility. Experimentally, it is observed that greater solubility is obtained as the charge of the diverse electrolyte increases. Two typical examples are shown in Fig. 7–8A and B.

The charge of the ions within the precipitate also influences the solubility of the precipitate in the presence of a diverse electrolyte. For example, the solubility of $BaSO_4$ increases more rapidly than that found for $AgCl$ as the $KNO_3$ concentration increases. This is illustrated in Fig. 7–9. In general,

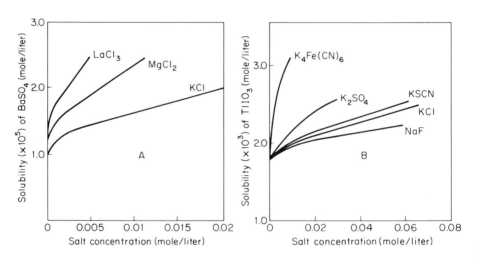

**Fig. 7–8.** Solubility of (A) barium sulfate and (B) thallous iodate in the presence of different diverse ions. (a) [From E. W. Neuman, *J. Amer. Chem. Soc.* **55**, 879 (1933). (b) From R. P. Bell and J. H. B. Geroge, *Trans. Farad. Soc.* **49**, 619 (1953).] See also W. F. Linke, "Solubilities of Inorganic and Metal-Organic Compounds," Vols. 1–2, fourth ed. Van Nostrand, New York, 1958.

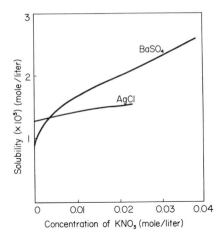

**Fig. 7–9.** Effect of valency of the precipitate on its solubility in the presence of a diverse ion. [From E. W. Neuman, *J. Amer. Chem. Soc.* **55**, 879 (1933). S. Popoff and E. W. Neuman, *J. Phys. Chem.* **34**, 1853 (1930).] See also W. F. Linke, "Solubilities of Inorganic and Metal-Organic Compounds," Vols. 1–2, fourth ed., Van Nostrand, New York, 1958.

the greater the charge of the precipitate ions the greater the effect the diverse ion has in increasing the solubility of the precipitate.

Of all equilibrium calculations, those involving solubility products are the most uncertain, particularly when used in their simplest form. Often many different equilibrium steps may be involved (complexation, hydrolysis, ion-pair formation, and association), while in other systems ionic strength is a determining factor. The different equilibrium constants for these competing reactions may not be available. Another complication is discovering that equilibrium steps not previously detected are also involved.

Values of $K_{sp}$ at known ionic strength are not available for all systems. Often those that are available do not cover the range of ionic strength encountered in gravimetric methods. For this reason the chemist is faced with carrying out the calculation knowing that an error is involved. Fortunately, in routine and in most applied quantitative or qualitative situations in the laboratory, the error is small. On the other hand, there are many instances where a more precise interpretation of the chemical system is required. Therefore, the formulism must include expressions describing the different equilibria and the effect of ionic strength. For these situations the chemist should anticipate the need to determine the solubility product at the ionic strength of interest and perhaps even to determine the different equilibrium constants for the other competing equilibria that are present.

## TYPICAL GRAVIMETRIC EXAMPLES

A wide variety of useful gravimetric procedures are known. In general, the vast majority of these are for the determination of inorganic species rather than organic species.

The gravimetric procedure tends to be time-consuming and cannot be automated. For these reasons other methods have been developed in areas where repetitive analyses are being done. Consequently, in clinical analysis and quality control, where organic species are being determined, the gravimetric procedure is not often encountered on a routine basis.

Several examples illustrate this. Total protein can be determined in serum by gravimetry. In this procedure, acetone is added to the serum sample, the total protein is precipitated, filtered, washed, dried, and weighed. All other clinical methods depend on the measurement of some component of the protein. Cholesterol, which is frequently determined in the clinical laboratory, can be precipitated with digitonin (a complex sugar which forms an equimolar species with cholesterol), filtered, washed, and weighed. Biological sodium and potassium can be precipitated as sodium uranyl zinc acetate and potassium cobalt nitrite, respectively.

Although these methods provide good accuracy and precision, they are not used because of the time required for each determination. Their value, however, is that these procedures are used to calibrate the standards needed for less time-consuming methods and to check the reliability of newly developed methods. For these reasons, these gravimetric procedures are still very valuable.

**Determination of Sulfur.**      Sulfur is readily determined by precipitation as $BaSO_4$

$$SO_4^{2-} + Ba^{2+} \rightarrow BaSO_{4(s)}$$

and is applicable to a wide sulfur concentration range. Although the reaction can be used for the determination of barium ion, its principal application is in sulfur analysis. The experimental procedure for the determination of sulfur is given at the end of this chapter.

From the solubility product ($K_{sp} = 1.3 \times 10^{-10}$), the solubility is calculated to be $1.14 \times 10^{-5}\,M$. In the experimental procedure the solubility is decreased because of the common ion effect (barium ion is added in about a 10% excess) and increased because of the presence of acid. For example, solubility of $BaSO_4$ was experimentally found to be 0.4 mg/100 ml in the absence of HCl and 8.1 mg/100 ml in the presence of 1.0 $F$ HCl. Increasing the temperature to close to 100°C increases the solubility about 1.5 times over the observed solubility at room temperature. These solubility studies in the presence of HCl and elevated temperature do not reflect the reverse contribution of the common ion effect.

A number of other anions form insoluble barium salts and can interfere in the determination of sulfate. These include $AsO_3^{3-}$, $AsO_4^{3-}$, $CO_3^{2-}$, $C_2O_4^{2-}$, $F^-$, $PO_4^{3-}$, and several less common anions. Most of these are anions of weak acids. Thus, if the solution is acidified, association will take place reducing the influence of the anion. For example, in the presence of $CO_3^{2-}$, the $CO_3^{2-}$ decreases in concentration with decreasing pH because

$$CO_3^{2-} + H_3O^+ \rightleftharpoons HCO_3^- + H_2O$$

$$HCO_3^- + H_3O^+ \rightleftharpoons H_2CO_3 + H_2O$$

Thus, the association that takes place will depend on the hydronium ion concentration and the $K_a$'s of the weak acids that are formed.

If barium ion is being determined by sulfate precipitation, several metal ions must be absent since they also form insoluble sulfates. The common interferences are $Ag^+$, $Ca^{2+}$, $Hg_2^{2+}$, $Pb^{2+}$, and $Sr^{2+}$. These interferences, as well as any other cation or anion that interferes, can be removed by some suitable separation method.

Many cations and anions are coprecipitated with $BaSO_4$. In general, coprecipitation of multivalent cations is the largest and often lowering the pH of the solution reduces the coprecipitation. The extent of coprecipitation of $NO_3^-$, $NO_2^-$, and $ClO_3^-$ is extensive enough to warrant removal before precipitation of the sulfate.

Several general techniques can be used for ion removal. For example $NO_3^-$, $NO_2^-$, and $ClO_3^-$ are removed by heating in HCl media. Several of the cations can be removed by prior precipitation as carbonates or hydroxides. However, extreme care must be exercised so that sulfate is not carried down with these precipitates. In some cases oxidation state is adjusted. For example, Fe coprecipitation is greatly reduced by changing its oxidation state from $3+$ to $2+$. A technique that is very useful is to convert the cation interference into a stable complex by the addition of a complexing agent (masking; see Chapter 15).

Study of the microcrystalline $BaSO_4$ has indicated that coprecipitation is by occlusion of the foreign ions rather than by an external surface coverage. In general, the coprecipitation of anions as barium salts or cations as metal sulfates follows the predictions suggested in the section on coprecipitation. For complex samples and for ultra accuracy, experimental factors which affect the extent of coprecipitation, such as concentration of foreign ions, temperature, order in which reagents are mixed, relative order of adsorption for the different ions, and treatment of the precipitate after formation, must be carefully controlled.

The precipitate is usually filtered through a fine ashless filter paper; however, filtration can also be done with a Gooch crucible or with a porcelain filtering crucible. If filter paper is used, it is transferred to a tared porcelain crucible and the paper is gradually dried and eventually charred and expelled as $CO_2$ and $H_2O$ vapor.

If the ashing process is done in an oxygen-deficient atmosphere carbon from the paper will reduce the $BaSO_4$ to $BaS$.

$$BaSO_4 + 4C \rightarrow BaS + 4CO_{(g)}$$

As the carbon is eventually expelled as $CO_2$, the oxygen in the atmosphere will reoxidize the $BaS$ to $BaSO_4$.

$$BaS + 2O_2 \rightarrow BaSO_4$$

However, if little oxygen is available or if extensive reduction takes place, it is simple and faster to treat the cooled precipitate with a drop of concentrated $H_2SO_4$ and carefully reheat it. The $H_2SO_4$ oxidizes the $BaS$ and the elevated temperature volatizes the excess $H_2SO_4$.

$$BaS + H_2SO_4 \rightarrow BaSO_4 + H_2S$$

The $BaSO_4$ is dried to constant weight in a muffle furnace or with a Meeker burner. Classical procedures suggest that heating of the crucible should involve 1 hour at a temperature which imparts redness to the crucible. From the thermogram (see Fig. 7–4), it can be seen that a slight weight loss occurs with increasing temperature. The weight loss is suggested to be due to the loss of adsorbed water. The ignition of the $BaSO_4$ to a constant weight should be done at a temperature of at least 800°C, preferably over 950°C.

In the discussion it has been assumed that the S in the sample was present as sulfate anion. Many compounds contain S in other oxidation states. The application of this method to the determination of S in these samples require the oxidation of sulfur to the 6+ oxidation state ($SO_4^{2-}$). After doing this the procedure is the same as if the sample contained the S originally as sulfate.

**Determination of Nickel as Ni–DMG.**     Of all the different organic precipitating agents dimethylglyoxime (DMG) is one that shows a high degree of selectivity. In acid solution only $Pd^{2+}$ is precipitated, while in weakly basic solution $Ni^{2+}$ is precipitated. Other metal ions such as $Co^{2+}$, $Cu^{2+}$, and $Zn^{2+}$ form soluble complexes. The insolubility of the $Pd^{2+}$ and $Ni^{2+}$ complexes in comparison to the $Co^{2+}$, $Cu^{2+}$, and $Zn^{2+}$ complexes is due to the structure of the complex. X-ray diffraction studies have shown that the $Ni^{2+}$ complex is planar and that the nickel atoms of neighboring molecules are stacked one above another (3.25 Å) in lines at right angles to the plane of the molecules. Thus, a bond between Ni atoms, although weak, exists in the molecule. In contrast, in the copper structure the copper atoms are lined up over an oxygen atom (Cu—O, 2.43 Å) and the organic part of the complex are stacked in a parallel manner. The $Pd^{2+}$ complex with DMG is similar to the $Ni^{2+}$ complex

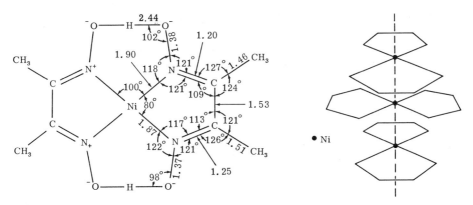

**Fig. 7–10.** Crystal structure of the nickel–dimethylglyoxime complex. [From L. E. Godycki and R. E. Rundle *Acta Cryst.* **6**, 487 (1953).]

structure in that Pd–Pd weak bonds are present. This crystal configuration is illustrated in Fig. 7–10. Additional properties of the Ni–DMG complex are cited in Chapter 15.

The procedure for the gravimetric determination of $Ni^{2+}$ is straightforward. A 1% by weight DMG solution in alcohol is added to an acidified solution of $Ni^{2+}$ which is then made basic with $NH_3$. The precipitate, which is generally free of coprecipitation, is digested for about an hour at 60°C to increase the crystal size. After cooling, the precipitate is filtered through a tared fritted glass crucible (Gooch crucibles can also be used), dried at 100–120°C, and weighed as Ni–DMG [$Ni(C_8H_{14}O_4N_4)$]. The precipitate is a brilliant carmine red color and has a bulky, creeping appearance. For this latter reason only small quantities of $Ni^{2+}$ are easily handled. A typical thermogram was shown in Fig. 7–4. From the thermogram it is concluded that the stoichiometry of the precipitate corresponding to $Ni(C_8H_{14}O_4N_4)$ is stable between 80 and 170°C. After this, decomposition of the organic portion of the molecule takes place.

**Determination of Carbon and Hydrogen.**    The total number of C and H determinations performed may exceed all others combined. Briefly, from C and H analyses it is possible to establish compound stoichiometry and empirical ratio of atoms in the compound. These data, in combination with other physical and chemical measurements, permit the chemist to characterize the substance. In general, this information is obtained from a sample weight of about 10 mg.

The micro determination is based on the Pregl combustion system and is an example of a gravimetric method in which volatilized products are weighed. This method involves the decomposition of the substance in a stream of oxygen in the presence of catalysts.

$$2C_xH_y + 2xyO_2 \rightarrow yH_2O_{(g)} + 2xCO_{2(g)} \qquad (7\text{-}7)$$

Pregl's apparatus can be described by subdividing it into three sections. These are (1) oxidizing catalysts, (2) collecting and weighing of the water and carbon dioxide, and (3) control and passage of oxygen. A typical diagram describing the Pregl apparatus is shown in Fig. 7–11 where the apparatus is separated.

The sample is placed in a small boat, inserted into the combustion tube, vaporized by heating, and decomposed in a slow stream of oxygen in the combustion tube. A mixture of copper oxide–lead chromate serves as the catalyst. Addition of metallic silver removes oxidation products of sulfur and halogens, while granular lead dioxide removes oxidation products of nitrogen. Part D serves the purpose of drying and removing interfering gases from the oxygen stream.

Water and carbon dioxide are swept out of the combustion tube by the oxygen into the weighing tubes J and K in Fig. 7–11. In tube J, water is adsorbed on a desiccant (usually $Mg(ClO_4)_2$, also known as anhydrone), while in tube K carbon dioxide is adsorbed on a mixture of NaOH–asbestos which is known as "Ascarite." If the reverse order were used the $CO_2$ tube would adsorb both the $CO_2$ and $H_2O$. The weight of each of the tubes containing the packing is of the order of 6–10 g. To determine the amount of water and carbon dioxide from the sample each tube is weighed before and after adsorption.

To obtain precise and accurate results it is necessary to handle and weigh the $H_2O$ and $CO_2$ tubes very carefully since the weight increase will be in the 1–20-mg range depending on sample size and C and H contents. The precision will depend on the weight of the sample and the type of microbalance; better precision is possible with specialty type microanalytical balances. Even with suitable balances it is necessary to establish a routine that is followed very scrupulously. For example, the glass tubes are uniformly wiped with a chamois for a known number of times always in the same direction. Also, the time between the wiping and weighing of the tube must be controlled. This entire procedure must be exactly duplicated before and after weighing. An oxygen tank is the source of oxygen and parts A, B, C, and M have the important function of maintaining a precise oxygen flow.

In recent years, gas chromatography (Chapter 24) has been used for the determination of $CO_2$ and $H_2O$. This method is not as accurate as the gravimetric method but provides the analyses much quicker.

## HOMOGENEOUS PRECIPITATION

Precipitation from a supersaturated condition will always occur when a precipitating and sample solution are combined. Even if rapid stirring is

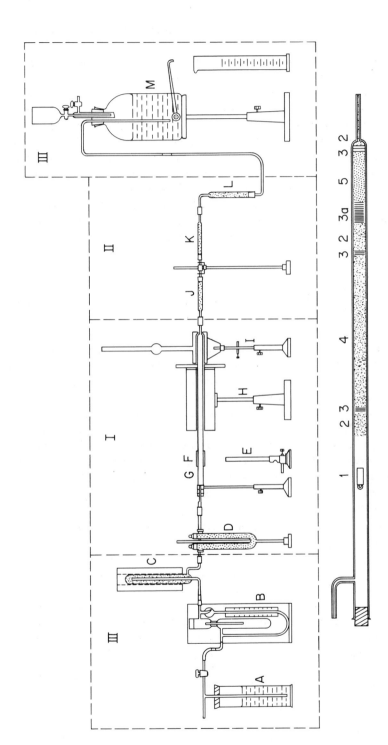

**Fig. 7-11.** Pregl-type combustion apparatus and combustion tube packing. A, pressure regulator; B, flowmeter; C, preheater; D, U-tube; E, bunsen burner; F, nickel sheath; G, combustion tube; H, electric furnace; I, heating mortar; J, water absorption tube; K, carbon dioxide absorption tube; L, drying tube; M, Mariotte bottle. Combustion tube packing, 1, sample boat; 2, silver wire; 3, asbestos plug; 3a, asbestos choking plug; 4, copper oxide–lead chromate mixture; 5, lead dioxide granules. (From G. Ingram, "Methods of Organic Elemental Microanalysis," © by G. Ingram. Reprinted by permission of Van Nostrand Reinhold Company.)

121

coupled with slow addition of very dilute solutions, the formation of irregular crystals and coprecipitation will not be completely eliminated.

One way of keeping and maintaining a very low degree of supersaturation is to generate the precipitating conditions *in situ*. This technique, which is called homogeneous precipitation, will not provide a localized area of supersaturation and, hence, supersaturation is negligible. The net result is the formation of large, near-perfect crystals which are also almost free of coprecipitation.

Two different techniques are used for producing the precipitating conditions. In the first, reagents are added which will increase or decrease the pH of the solution *in situ*. The second case is where the precipitating reagent is slowly generated through a hydrolysis reaction or through a synthesis reaction.

Several different reactions can be used to change the pH of the solution. One of the best reactions to increase the pH is by the hydrolysis of urea, $H_2NCONH_2$.

$$H_2N\overset{\overset{\displaystyle O}{\|}}{C}NH_2 + H_2O \xrightarrow{\Delta} 2\,NH_3 + CO_2 \qquad\qquad (7\text{-}8)$$

Using this reaction many different metal ions can be precipitated as the hydroxide. For example, to precipitate $Al^{3+}$ as the hydroxide, urea is added to the acidified $Al^{3+}$ solution. As the solution is warmed the urea hydrolyzes yielding $NH_3$ which neutralizes the hydronium ion in the solution and $Al^{3+}$ precipitates as the hydroxide when a pH of about 4.1 is reached. Other metal ions are precipitated in a similar fashion. An added advantage is that the precipitate is denser, larger, and more easily filtered. Often the best precipitating conditions will include the presence of sulfate or succinate ion with the product of the precipitation being a basic sulfate or succinate, respectively. Other reactions used to increase the pH are the hydrolysis of acetamide, potassium cyanate, or hexamethylenetetramine.

The improvement in the crystalline form for the precipitation of $Al^{3+}$ homogeneously over conventional precipitation as the hydroxide is illustrated in the thermograms shown in Fig. 7–12. Weight loss corresponds to the loss of water from $Al(OH)_3 \cdot xH_2O$ to $Al_2O_3$. For precipitation by ammonia solution, a gradual loss is observed for the gelatinous-hydrated aluminum precipitate. When the aluminum is precipitated homogeneously water loss is more uniform and stoichiometry is more clearly defined for the more well-defined crystalline precipitate. The large drop in curve (D) is due to the decomposition of the succinate. The thermograms also show that a suitable ignition temperature is more easily attained for the homogeneously precipitated aluminum.

The hydrolysis of sulfamic acid or potassium persulfate can be used to decrease the pH or to generate sulfate ion for the precipitation of barium ion.

$$NH_2SO_3H + 2H_2O \xrightarrow{\Delta} NH_4^+ + H_3O^+ + SO_4^{2-}$$

$$S_2O_7^{2-} + 3H_2O \xrightarrow{\Delta} 2SO_4^{2-} + 2H_3O^+$$

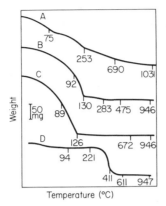

**Fig. 7–12.** Thermograms for aluminium hydroxide precipitated by different conditions. (A) Aqueous ammonia; (B) gaseous ammonia; (C) urea; (D) urea/succinic acid. (From W. W. Wendlandt, "Thermal Method of Analysis," Interscience–Wiley, New York, 1964.)

**Table 7–3.    Precipitation from Homogeneous Solution[a]**

| Precipitant | Reagent | Element precipitated |
|---|---|---|
| Hydroxide | Urea | Al, Ga, Th, Fe(III), Sn, Zr |
| | Acetamide | Ti |
| | Metal chelate and $H_2O_2$ | Fe(III) |
| Phosphate | Triethyl phosphate | Zr, Hf |
| | Urea | Mg |
| Oxalate | Dimethyl or diethyl oxalate | Th, Ca, Am, Ac, rare earths, Mg, Zn, Ca |
| | Urea and an oxalate | Ca |
| Sulfate | Dimethyl sulfate | Ba, Ca, Sr, Pb |
| | Sulfamic acid | Ba, Pb, Ra |
| | Ammonium persulfate | Ba |
| Sulfide | Thioacetamide | Pb, Sb, Bi, Mo, Cu, As, Cd, Sn, Hg, Mn |
| Iodate | Iodine and chlorate | Th, Zr |
| | Periodate and ethylene diacetate | Th, Fe(III) |
| | Ce(III) and bromate | Ce(IV) |
| Carbonate | Trichloroacetate | Rare earths, Ba, Ra |
| Chromate | Urea and dichromate | Ba, Ra |
| | Cr(III) and bromate | Pb |
| Chloride | Ag ammonia complex and β-hydroxyethyl acetate | Ag |
| Dimethylglyoxime | Urea and metal chelate | Ni |
| 8-Hydroxyquinoline | Urea and metal chelate | Al |
| Fluoride | Fluoroboric acid | La |

[a] Taken from M. L. Salutsky, "Treatise on Analytical Chemistry," Part I, Vol. 1, Interscience, New York, 1959, page 741. See also L. Gordon, M. L. Salutsky, and H. H. Willard "Precipitation from Homogeneous Solution," Wiley, New York, 1959.

Hydrolysis of esters is a very useful homogeneous route to the production of anions that form precipitates with metal ions. In addition ester hydrolysis is also a technique to lower the pH of the solution. Typical examples are the homogeneous production of sulfate, phosphate, and oxalate from the hydrolysis of dimethyl sulfate, trimethyl phosphate, and dimethyl oxalate, respectively. Sulfide precipitates are homogeneously precipitated from thioacetamide.

$$CH_3\overset{\overset{\text{S}}{\|}}{C}{-}NH_2 \ + \ H_2O \ \xrightarrow{\ \Delta\ } \ CH_3\overset{\overset{\text{O}}{\|}}{C}{-}NH_2 \ + \ H_2S$$

This reaction is frequently used in qualitative schemes thus eliminating the need of storing $H_2S$ gas.

A typical example in which the precipitating agent is produced through synthesis is the *in situ* preparation of dimethylgloxime.

$$\begin{matrix} CH_3{-}C{=}O \\ | \\ CH_3{-}C{=}O \end{matrix} \ + \ 2\,H_2NOH \ \xrightarrow{\text{slow}} \ \begin{matrix} CH_3{-}C{=}NOH \\ | \\ CH_3{-}C{=}NOH \end{matrix} \ + \ 2\,H_2O$$

As DMG is produced homogeneously, it forms a well-defined, crystalline, easily filtered precipitate with nickel ion. Table 7–3 lists several other reagents that are useful in homogeneous precipitation.

## CALCULATIONS

*Example 7–8.*    A sample of alum, $K_2SO_4 \cdot Al_2(SO_4)_3 \cdot 24H_2O$, contains only inert impurities and weighs 0.9237 g. Upon dissolution the aluminum is precipitated as the $Al^{3+}$-8-hydroxyquinoline complex $[Al(C_9H_6NO)_3]$. The precipitate is filtered, washed, and ignited to $Al_2O_3$ which weighed 0.09170 g. Calculate the %Al and %S in the sample.

(a)         $$\%Al = \dfrac{\text{wt } Al_2O_3 \times \dfrac{2Al}{Al_2O_3} \times 100}{\text{wt sample}}$$

$$\%Al = \dfrac{0.09170 \text{ g} \times \dfrac{2 \times 26.98}{102.0} \times 100}{0.9237 \text{ g}} = 5.252\% = 5.25\%$$

(b) From the stoichiometry of alum, 2 aluminums are stoichiometric to 4 sulfurs; therefore

$$\%S = \frac{wt\ Al_2O_3 \times \dfrac{4S}{Al_2O_3} \times 100}{wt\ sample}$$

$$\%S = \frac{0.09170\ g \times \dfrac{4 \times 32.06}{102.0} \times 100}{0.9237\ g} = 12.48\%$$

*Example 7–9.* An organic compound that had been purified through repeated recrystallizations and had a sharp melting point of 121–122°C was oxidized in the combustion train shown in Fig. 7–11. The sample weight, the weight of the $MgClO_4$ packed tube, and the weight of the Ascarite tube were determined with a micro balance* to be 4.432 mg, 30.148 mg, and 104.710 mg, respectively. If the $Mg(ClO_4)_2$ and Ascarite tube weighed 32.616 mg and 115.307 mg after combustion calculate the %H and %C in the sample.

$$C_xH_y + (x + y/2)O_2 \xrightarrow{\Delta} (y/2)H_2O + xCO_2$$

Mg($ClO_4$)$_2$ tube

wt of tube + $H_2O$ = 32.616 mg
wt of tube = 30.148 mg
wt of $H_2O$ = 2.468 mg

$$\%H = \frac{wt\ H_2O \times \dfrac{2H}{H_2O} \times 100}{wt\ sample}$$

$$\%H = \frac{2.468\ mg \times \dfrac{2 \times 1.008}{18.02} \times 100}{4.432\ mg}$$

$$\%H = 6.23_0\%$$

Ascarite tube

wt of tube + $CO_2$ = 115.307 mg
wt of tube = 104.710 mg
wt of $CO_2$ = 10.597 mg

$$\%C = \frac{wt\ CO_2 \times \dfrac{C}{CO_2} \times 100}{wt\ sample}$$

$$\%C = \frac{10.60\ mg \times \dfrac{12.01}{44.01} \times 100}{4.432\ mg}$$

$$\%C = 65.27\%$$

* Weights can be recorded to 0.001 mg on a free swing micro balance using the sensitivity value for the balance.

## Questions

1. Outline the essential steps in a typical gravimetric procedure.
2. Which steps in Question 1 are essential if purification of the sample is prime goal?
3. Define the term gravimetric factor and describe how it is used.
4. Should a gravimetric factor be large or small? Why?
5. List the gravimetric factors for each of the following using chemical formulas for formula weights.

| Substance weighed | Substance sought | Substance weighed | Substance sought |
|---|---|---|---|
| a. $AlPO_4$ | Al | e. $Fe_2O_3$ | $Fe_3O_4$ |
| b. $AlPO_4$ | $Al_2O_3$ | f. $KClO_4$ | $Cl_2$ |
| c. $Mg_2P_2O_7$ | $P_2O_5$ | g. BiOCl | $Bi_2O_3$ |
| d. $Fe_2O_3$ | Fe | h. $CaCO_3$ | $CO_2$ |

6. What is the difference between precipitation and weighing form?
7. What is the difference between "constant weight" and "absolute weight?"
8. Describe the stages that take place in the formation of a precipitate.
9. What is the difference between a colloid and a nuclei?
10. Why is it valuable to be able to determine the crystal structure?
11. What kinds of information about a substance can be concluded from its crystal structure?
12. What is digestion and what role does it serve?
13. List the types of coprecipitation.
14. Suggest how coprecipitation must be used advantageously.
15. What is surface adsorption? How can surface adsorption be minimized? Be increased?
16. Why is a gelatinous precipitate difficult to filter?
17. What are the advantages and disadvantages of organic precipitating agents compared to inorganic precipitating agents?
18. Compare the solubility of inorganic to organic compounds in water and in ethanol. Why is there this difference?
19. Write the solubility product expression for the following:
    a. $CaF_2$         c. $AlPO_4$         e. $Ag_2CrO_4$         g. $Cu_2S$
    b. $Mg_2P_2O_7$    d. $Mg(OH)_2$       f. $Ca_3(PO_4)_2$      h. HgS
20. What factors influence the solubility product constant?
21. What factors influence the solubility of an "insoluble" salt?
22. Differentiate between concentration and ionic strength.
23. Write equations that illustrate how $PO_4^{3-}$ can be eliminated as an interference in the gravimetric determination of sulfate.

24. What chemical step would have to be included if sulfur in a sulfide ore were to be determined by the gravimetric barium sulfate procedure?
25. Why is $Ba(NO_3)_2$ not used as a precipitating solution for sulfate?
26. What would happen to the Ni–DMG complex if it were dried at 1000°C? Write a reaction for this.
27. Why is alcohol used for the solvent in preparing the DMG reagent?
28. Explain why the solubility of Ni–DMG in alcohol–water mixtures increases with increased alcohol concentration while $BaSO_4$ decreases?
29. List the possible sources of error in determining C and H by the Pregl method.
30. In the combustion train in Fig. 7–11 why is the $Mg(ClO_4)_2$ tube placed before the Ascarite tube?
31. Explain the following statement. The structure of a compound containing C and H cannot be deduced from the determination of %C and %H.
32. What is a homogeneous precipitation and why is it useful?
33. Write the reaction for the hydrolysis of each of the following:
    a. $(CH_3CH_2O)_3PO$     b. $(CH_3O)_2SO_2$     c. $(CH_3O)_2C_2O_2$.

# Problems

1. Calculate the solubility from the solubility product for each of the following (see Appendix II for values of $K_{sp}$).
    *a. AgI          c. $Pb(IO_3)_2$          e. $Ca_3(PO_4)_2$          g. $Pb_3(PO_4)_2$
    *b. $MgF_2$       d. $Ag_2CrO_4$          f. CdS                    h. $Hg_2Cl_2$

2. Calculate the solubility product from the solubility for each of the following.
    *a. 3.205 g TlCl per 1000 ml          d. 26.05 mg $Ag_2SO_4$ per 50 ml
    b. 8.510 mg $Mg(OH)_2$ per 1000 ml     *e. 60.6 mg $Ag_3PO_4$ per 100 ml
    c. $6.41 \times 10^{-4}$ moles/liter $PbI_2$     f. 0.167 mg AgSCN per 1000 ml

3. Calculate the concentration of $Ca^{2+}$
    a. in a 1.000-liter saturated solution of $CaF_2$ and
    *b. after the addition of 0.1 mole of NaF.

4. Calculate the concentration of $SO_4^{2-}$
    *a. in 500 ml of a saturated solution of $Ag_2SO_4$ and
    *b. after the addition of 0.1 mole of $AgNO_3$ to the solution in part a.

5. Calculate the ppm of $Bi^{3+}$
    *a. in 1.000 liter of a saturated solution of $Bi_2S_3$ and
    b. after the addition of 0.1 mole of $Na_2S$ to the solution in part a.

6. If 25.0 ml of 0.1230 $F$ $AgNO_3$ and 75.0 ml of 0.04100 $F$ NaCl are mixed, calculate the concentration of $Ag^+$ in the solution as mg $Ag^+$/ml.

---

* Answers are listed at the end of the book for problems marked with an asterisk.

7.* If 50.0 ml of 0.0250 $F$ AgNO$_3$ and 150 ml of 0.0250 $F$ Na$_3$PO$_4$ are mixed, calculate the concentration of Ag$^+$ and PO$_4^{3-}$ in the solution in mg/ml.

8. Which contains more silver ion, a saturated solution of AgCl or of Ag$_2$Cr$_2$O$_7$?

9. Which ion is precipitated first when a AgNO$_3$ solution is added to a solution that is 0.1 $F$ in chloride ion and 0.1 $F$ in bromide ion?

10.* A 1.0000-g sample of a zinc ore was dissolved and the zinc precipitated as the phosphate and weighed as Zn$_2$P$_2$O$_7$. If the Zn$_2$P$_2$O$_7$ weighed 0.6611 g, calculate the percent Zn in the sample.

11. A 1.1000-g sample of limestone was dissolved and the calcium precipitated as the oxalate and weighed as CaO. If the CaO weighed 0.5110 g, calculate the percent CaCO$_3$ in the sample.

12. Sodium tetraphenylborate, NaB(C$_6$H$_5$)$_4$, is a useful precipitant for potassium. What is the percent K$_2$O in a fertilizer when a 0.4315-g sample gives 0.1880 g of KB(C$_6$H$_5$)$_4$.

13.* Histamine (**I**) in certain pharmaceutical preparations can be determined by precipitation with nitranilic acid (**II**). If one tablet weighed 0.9711 g and the precipitate (**III**)

$$C_5H_9N_3 + C_6H_2N_2O_8 \rightarrow C_5H_9N_3 \cdot C_6H_2N_2O_8$$
$$\textbf{I} \qquad\qquad \textbf{II} \qquad\qquad\qquad \textbf{III}$$

weighed 0.0899 g, calculate the mg of histamine per tablet.

14.* Serum sodium can be determined in the clinical laboratory by precipitation as NaZn(UO$_2$)$_3$(CH$_3$CO$_2$)$_9$·9H$_2$O. If 1.00 ml of serum was taken and after appropriate treatment, 0.2153 g of NaZn(UO$_2$)$_3$(CH$_3$CO$_2$)$_9$·9H$_2$O (1592) was obtained, calculate the mg of Na per ml of serum.

15. What is the weight of U in the precipitate in problem 14?

16.* A 0.6159-g sample of impure barium chloride dihydrate weighed 0.5401 g after it was dried at 250°C. Calculate the %H$_2$O in the sample.

17.* A sample of FeSO$_4$· (NH$_4$)$_2$SO$_4$·6H$_2$O containing only inert impurities weighs 1.1610 g. The iron after dissolution is oxidized, precipitated as the Fe$^{3+}$-8-hydroxyquinoline complex, [Fe(C$_9$H$_6$NO)$_3$], filtered, and ignited to Fe$_2$O$_3$. If the Fe$_2$O$_3$ weighed 0.2120 g calculate the %S in the sample.

18. Using the data in problem 17 calculate the iron content of the sample as %Fe$_3$O$_4$.

19. A 1.2510-g sample of a fertilizer sample was dissolved and the phosphorus was precipitated as MgNH$_4$PO$_4$. The precipitate was collected, ignited to Mg$_2$P$_2$O$_7$, and found to weigh 0.2612 g. Calculate the phosphorus content as %P$_2$O$_5$.

20.* Atropine sulfate, **IV**, an alkaloid which is used as an anticholinergic, can be determined gravimetrically by precipitation with ammonium tetrathiocyanodiammonochromate, known as Reinecke's salt, **V**,

$$(C_{17}H_{23}O_3NH^+) + NH_4[Cr(NH_3)_2(SCN)_4] \longrightarrow (C_{17}H_{23}O_3NH^+)[Cr(NH_3)_2(SCN)_4^-]_{(s)} + NH_4^+$$

IV                        V                                    VI

If the impure atropine sulfate weighed 0.8186 g and the precipitate, **VI**, weighed 1.2040 g, calculate the % atropine ($C_{17}H_{23}O_3N$), **IV**A, in the sample.

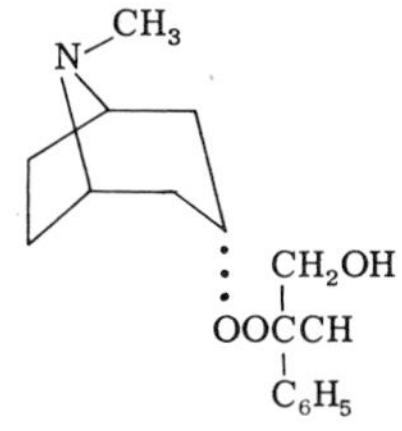

IVA

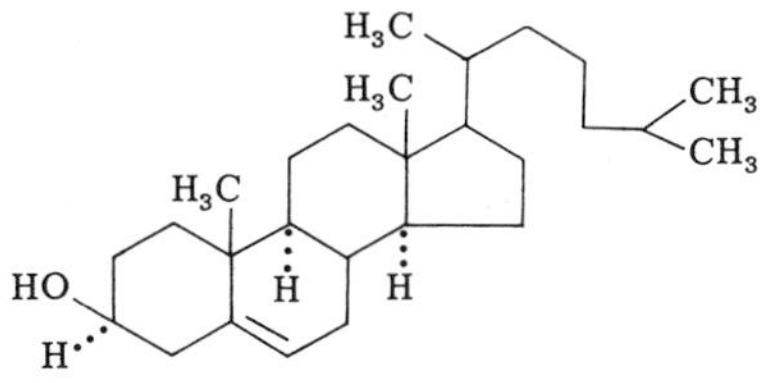

VII

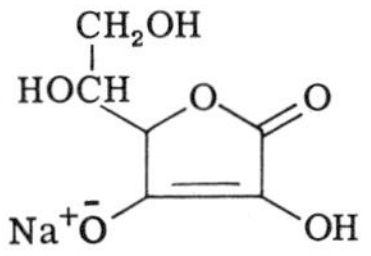

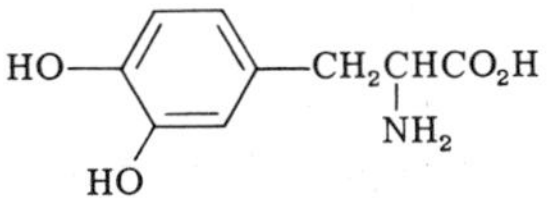

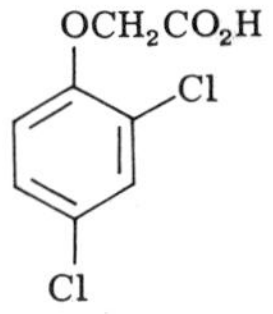

VIII

IX

X

21. A series of physical and chemical tests were used to confirm the purity of a sample of cholesterol ($C_{27}H_{46}O$), which was to be used as a standard for the determination of cholesterol (**VII**) in blood. Part of these measurements included the determination of C, H, and O (by difference) using the combustion train in Fig. 7-11. The sample weighed 6.415 mg. The $Mg(ClO_4)_2$ tube weighed 28.415 mg before and 35.330 mg after combustion, while the Ascarite tube weighed 112.162 mg before and 131.804 mg after combustion. Compare the %C, %H, and %O found in the laboratory to the actual %C, %H, and %O.

22. The sodium content of sodium ascorbate, **VIII** (ascorbic acid is Vitamin C and its salt as well as the free acid is used in Vitamin C preparations), can be determined on the micro scale by heating the sample with $H_2SO_4$ on a small

porcelain boat in a continuous supply of air. At dryness the sodium is converted to $Na_2SO_4$ and the organic matter is lost as $CO_2$ and $H_2O$. If the boat weighed 514.428 mg, the boat plus the sample weighed 524.622 mg, and the boat plus residue weighed 518.084 mg, calculate the %Na in the sodium ascorbate sample.

23.* The compound, 3-(3,4-dihydroxyphenyl)alanine, **IX**, known as DL-Dopa, is an important compound in the chemistry of the brain and in the study of Parkinson disease. If a 10.125-mg sample were taken for C and H analysis, calculate the mg of $CO_2$ and $H_2O$ that would be formed.

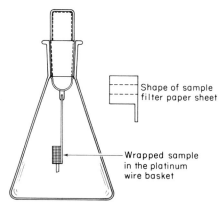

Shape of sample filter paper sheet

Wrapped sample in the platinum wire basket

Micro oxygen combustion flask

24.* The chlorine content of the herbicide, (2,4-dichlorophenoxy) acetic acid, **X**, was determined by combustion in a micro oxygen combustion flask. A 12.194-mg sample was wrapped in the filter paper, placed in the platinum basket, and the paper + sample ignited in an oxygen atmosphere in the flask which contained several mls of a $Na_2O_2$ solution. After combustion the solution is acidified with $HNO_3$, $AgNO_3$ solution added, and the silver chloride digested and filtered by a sintered glass crucible using micro techniques. If the empty crucible weighted 42.318 mg and the crucible plus silver chloride weighed 54.438 mg, calculate the %Cl in the sample.

25. A phosphorus containing organic sample weighing 8.966 mg was combusted by the micro oxygen flask method using a dilute $HNO_3$ trapping solution. The phosphorus which is converted to $PO_4^{3-}$ was precipitated by the addition of an ammonium molybdate solution. The precipitate was isolated in a filter stick. If the filter stick weighed 61.462 mg and the stick plus dried precipitate of $(NH_4)_3(PMo_{12}O_{40})$ weighed 68.393 mg, calculate the phosphorus content of the sample as $\%P_2O_5$.

# Chapter Eight
# Neutralization in Analytical Chemistry

## INTRODUCTION

Analytical procedures based on neutralization between an acid and a base have been used extensively in volumetric analysis. By this procedure, many inorganic and organic acids and bases can be determined with a high degree of precision and accuracy. In general, the procedure entails the dissolution of the acidic or basic sample and subsequent titration of the solution with a standard basic or acidic titrant, respectively.

Although the emphasis here is on the applications to analysis, it should be realized that the principles of neutralization are also very important in other scientific disciplines.

This chapter attempts to emphasize the significance of five main topics: (1) strong and weak acids and bases; (2) conjugate acid–base systems; (3) buffer systems; (4) expression and calculation of the hydronium ion concentration (pH) for a system; and (5) applications in quantitative analysis.

## ACID–BASE THEORIES

Although many contributed to the development of acids and bases, the main accomplishment from a modern viewpoint was the proposal developed by Arrhenius in 1884 as part of a general theory on electrolytic dissociation. In this proposal, he described acids and bases as those species which yield hydrogen ion and hydroxide ion, respectively, when dissolved in water.* Although this concept was, and still is, extremely useful, it possessed several limitations. For example, the definitions apply only to aqueous solutions.

---

* A proton will exist as the solvated species in aqueous solution. This species is called the hydronium ion and is designated as $H_3O^+$.

The next major step was in 1905 when Franklin attempted to include the solvent in an acid–base theory. In his definition an acid is a solute that yields a cation characteristic of the solvent and a base is a solute that yields an anion characteristic of the solvent. As an illustration, liquid ammonia ionizes according to the equation

$$NH_3 + NH_3 \rightleftharpoons NH_4^+ + NH_2^-$$

Thus, $NH_4Cl$ and $NaNH_2$ are acidic and basic salts in liquid ammonia, respectively, since they provide the $NH_4^+$ and $NH_2^-$ species.

In 1923 Brønsted and Lowry independently defined an acid as a substance capable of donating a proton to another substance and a base as a substance that accepts a proton. The two main advances of this theory are that the

**Table 8–1.    Conjugate Acid–Base Pairs Arranged According to Strength[a]**

| Conjugate acid[b] | | Conjugate base[c] | |
|---|---|---|---|
| Name | Formula | Formula | Name |
| Perchloric acid | $HClO_4$ | $ClO_4^-$ | Perchlorate ion |
| Sulfuric acid | $H_2SO_4$ | $HSO_4^-$ | Hydrogen sulfate ion |
| Hydrogen chloride | $HCl$ | $Cl^-$ | Chloride ion |
| Nitric acid | $HNO_3$ | $NO_3^-$ | Nitrate ion |
| Hydronium ion | $H_3O^+$ | $H_2O$ | Water |
| Hydrogen sulfate ion | $HSO_4^-$ | $SO_4^{2-}$ | Sulfate ion |
| Phosphoric acid | $H_3PO_4$ | $H_2PO_4^-$ | Dihydrogen phosphate ion |
| Acetic acid | $CH_3COOH$ | $CH_3COO^-$ | Acetate ion |
| Carbonic acid[d] | $H_2CO_3$ | $HCO_3^-$ | Hydrogen carbonate ion |
| Hydrogen sulfide | $H_2S$ | $HS^-$ | Hydrosulfide ion |
| Ammonium ion | $NH_4^+$ | $NH_3$ | Ammonia |
| Hydrogen cyanide | $HCN$ | $CN^-$ | Cyanide ion |
| Hydrogen carbonate ion | $HCO_3^-$ | $CO_3^{2-}$ | Carbonate ion |
| Phenol | $C_6H_5OH$ | $C_6H_5O^-$ | Phenoxide ion |
| Water | $H_2O$ | $OH^-$ | Hydroxide ion |
| Ethyl alcohol | $C_2H_5OH$ | $C_2H_5O^-$ | Ethoxide ion |
| Ammonia | $NH_3$ | $NH_2^-$ | Amide ion |
| Methylamine | $CH_3NH_2$ | $CH_3NH^-$ | Methylamide ion |
| Hydrogen | $H_2$ | $H^-$ | Hydride ion |
| Methane | $CH_4$ | $CH_3^-$ | Methide ion |

[a] Source: C. A. VanderWerf, "Acids, Bases, and the Chemistry of the Covalent Bond," D. Van Nostrand Company, New York, 1961. By permission.

[b] Listed in decreasing strength.

[c] Listed in increasing strength.

[d] This is the position for carbonic acid based on its apparent acidity; it appears to be a rather weak acid because only a small fraction of dissolved carbon dioxide is in the form $H_2CO_3$.

acid–base concept is not limited to water as solvent, and that the reaction is not just one that involves hydrogen ions and hydroxide ions.

Lewis in 1923 described the electronic theory of acids and bases by stating that an acid is a substance that can accept an electron pair and a base is one that can donate an electron pair. The significance of this theory is that the acid–base concept can be extended to many organic and inorganic reactions in which a proton is not involved. Since the interest in this chapter is with aqueous solutions and primarily with inorganic acids and bases, the Brønsted–Lowry and Arrhenius concepts are the most useful.

A loss of a proton by a Brønsted–Lowry acid results in the formation of a corresponding Brønsted–Lowry base and is called a conjugate base. Similarly, a Brønsted–Lowry base upon gaining a proton produces a Brønsted–Lowry acid and is called a conjugate acid. The equilibrium describing this inter-relationship can be expressed as

$$\text{acid} \rightleftharpoons \text{base} + \text{proton} \tag{8-1}$$

and typical examples would be

$$H_2O \rightleftharpoons OH^- + H^+ \tag{8-2}$$

$$NH_4^+ \rightleftharpoons NH_3 + H^+ \tag{8-3}$$

Hydroxide ion in reaction (8-2) is the conjugate base of the acid water, or, water is the conjugate acid of the base hydroxide ion. Similarly, ammonia is the conjugate base of the acid ammonium ion and ammonium ion is the conjugate acid of the base ammonia. Other conjugate acid–base pairs are listed in Table 8–1.

## ACID AND BASE STRENGTH

The strength of an acid or base, according to the Brønsted–Lowry concept, is determined by the ability to donate or accept a proton, respectively. Thus, the easier the proton is lost or gained, the stronger the acid or base, respectively.

The reaction that takes place when a Brønsted–Lowry acid and base are brought together can be represented as

$$\text{acid}_1 + \text{base}_2 \rightleftharpoons \text{base}_1 + \text{acid}_2 \tag{8-4}$$

$$HCl + H_2O \rightleftharpoons Cl^- + H_3O^+ \tag{8-5}$$

$$HC_2H_3O_2 + H_2O \rightleftharpoons C_2H_3O_2^- + H_3O^+ \tag{8-6}$$

$$H_2O + NH_3 \rightleftharpoons OH^- + NH_4^+ \tag{8-7}$$

$$H_3O^+ + OH^- \rightleftharpoons HOH + HOH \tag{8-8}$$

In these reactions $acid_2$ is the conjugate acid of $base_2$, and $base_1$ is the conjugate base of $acid_1$.

The position of equilibrium is different for reactions (8-5) to (8-6). The extent of this difference is a measure of the strengths of the acids and bases participating in the reactions.

The direction of the reaction will be toward the production of the weaker acid and base. For example, in reaction (8-5) HCl is a stronger acid than $H_3O^+$ and $H_2O$ is a stronger base than chloride ion. Similarly, in reaction (8-8) $OH^-$ is a stronger base than $H_2O$ and $H_2O^+$ is a stronger acid than $H_2O$. Consequently, both of these reactions lie far to the right.

Reactions (8-6) and (8-7) are not in the direction of the products since the acids produced, $H_3O^+$ and $NH_4^+$, are stronger than the original acids, $HC_2H_3O_2$ and $H_2O$, respectively. The bases produced, $C_2H_3O_2^-$ and $OH^-$, are also stronger than the original bases, $H_2O$ and $NH_3$. Table 8-1 correlates the strength of some of the more common conjugate acids and bases and can be used to predict the direction of acid–base reactions.

Many other acids act like HCl and are called strong acids. Similarly, many other bases act like $OH^-$ (the source being NaOH) and are called strong bases. These acids and bases are also part of a broad group of substances classified as strong electrolytes.

There are also a host of acids that act like $HC_2H_3O_2$ and bases that act like $NH_3$. These acids and bases are called weak acids and bases, respectively, and are part of the larger group of substances classified as weak electrolytes.

## LEVELING EFFECT

It should be noted in Table 8-1 that water can act as an acid or a base (amphoteric). Thus, if an acid is placed in water, hydronium ion is produced. Regardless of how strong the acid solute might be, the strongest acid that can exist in water is the hydronium ion. Similarly, the strongest base is the hydroxide ion. Even though there is a difference in acid strength between $HClO_4$, $H_2SO_4$, and HCl, they appear equal in strength when they are placed in water. An equal basic strength would be observed for LiOH, NaOH, and KOH in water. The effect of the solvent becomes very important, and the basic and/or acidic properties of the solvent will play an important role in determining the maximum acidic and basic limits in the solution. This property of the solvent is termed the _leveling effect_. Thus, water levels all strong acids to the strength of hydronium ion and all strong bases to the strength of hydroxide ion.

**Autoprotolysis.**     Pure water is slightly ionized, and because of its amphoteric nature, the ionization can be expressed as

$$H_2O + H_2O \rightleftharpoons H_3O^+ + OH^-$$

where one water molecule acts as a base and another water molecule acts as an acid. This type of equilibrium is called autoprotolysis. Examples of other solvents which behave similarly are

$$NH_3 + NH_3 \rightleftharpoons NH_4^+ + NH_2^-$$

$$HC_2H_3O_2 + HC_2H_3O_2 \rightleftharpoons H_2C_2H_3O_2^+ + C_2H_3O_2^-$$

$$H_2SO_4 + H_2SO_4 \rightleftharpoons H_3SO_4^+ + HSO_4^-$$

$$CH_3OH + CH_3OH \rightleftharpoons CH_3OH_2^+ + CH_3O^-$$

The equilibrium expression describing the autoprotolysis of water is

$$K = \frac{a_{H_3O^+} a_{OH^-}}{a^2_{H_2O}} \qquad (8\text{-}9)$$

Since water is the solvent, its activity is one because it is at its standard state. Furthermore, if activity equals concentration, Eq. (8-9) becomes

$$K_w = [H_3O^+][OH^-] \qquad (8\text{-}10)$$

where $K_w$ is the autoprotolysis constant.

Values of $K_w$ have been accurately measured as a function of temperature, and at 25°C $K_w$ is $1.01 \times 10^{-14}$. (The value $1.00 \times 10^{-14}$ will be used in this book.) Other values of $K_w$ at different temperatures are listed in Table 8-2. Consequently, for pure water it follows that $[H_3O^+]$ must equal $[OH^-]$ or be equal to $1.00 \times 10^{-7} M$. If the concentration of the hydronium ion is increased, hydroxide ion concentration must decrease since their product must equal $1.00 \times 10^{-14}$. Similarly, if hydronium ion concentration is decreased, the hydroxide ion concentration must increase.

## pH

A simple and convenient method of expressing hydronium ion concentration is by the following definition:

$$pH = \frac{1}{\log [H_3O^+]} = \log \frac{1}{[H_3O^+]} \qquad (8\text{-}11)$$

Also,

$$pOH = \frac{1}{\log [OH^-]} = \log \frac{1}{[OH^-]} \qquad (8\text{-}12)$$

**Table 8–2.  Autoprotolysis  Constants  for  Water  as  a Function of Temperature**[a]

| t (°C) | pKw | Kw |
|--------|--------|--------|
| 5 | 14.734 | $0.184 \times 10^{-14}$ |
| 10 | 14.535 | $0.292 \times 10^{-14}$ |
| 15 | 14.346 | $0.451 \times 10^{-14}$ |
| 20 | 14.167 | $0.681 \times 10^{-14}$ |
| 25 | 13.996 | $1.01 \times 10^{-14}$ |
| 30 | 13.833 | $1.47 \times 10^{-14}$ |
| 40 | 13.535 | $2.92 \times 10^{-14}$ |
| 50 | 13.262 | $5.47 \times 10^{-14}$ |

[a] Source: H. S. Harned and B. B. Owen, "The Physical Chemistry of Electrolytic Solutions,"   ACS Monograph No. 137, 3rd ed., © 1958.

and $K_w$ can be expressed as

$$pK_w = \frac{1}{\log K_w} \equiv \log \frac{1}{K_w} \tag{8-13}$$

Several problems prevent a theoretically precise establishment of a simple scale based on base-ten logarithms. In general, these have to do with the nature of the devices that are used for measuring hydrogen ion concentration. (Generally, activity is measured rather than concentration.) Much of the problem is solved, if concentrations are desired, by prior calibration of the measuring devices.

For the vast majority of routine applications, it can be assumed that activity coefficients approach unity, and concentrations and activities are equal. Thus, the expressions for pH, pOH, and $K_w$ can be defined in molar concentrations. Substitution into Eq. (8-10) for $K_w$ and taking −log of both sides of the equation leads to

$$pH + pOH = 14 \tag{8-14}$$

Thus, if the pH is known, the pOH is readily calculated, and, conversely, if the pOH is known, the pH is readily calculated.


## pH  SCALE

A pH scale covering 14 units is established by Eq. (8-14) and is illustrated in Table 8–3. In this scale acidic solutions are indicated by pH values <7.00 and basic solutions by pH values >7.00. The reverse is true if pOH values

**Table 8-3. pH–pOH Chart**

| | Dilute sulfuric acid | | Lemon juice | Vinegar | | | Tap water | Pure water | Albumin | | Milk of magnesia | Household ammonia | Lime water | Dilute sodium hydroxide | |
|---|---|---|---|---|---|---|---|---|---|---|---|---|---|---|---|
| pH | 0 | 1 | 2 | 3 | 4 | 5 | 6 | 7 | 8 | 9 | 10 | 11 | 12 | 13 | 14 |
| $H_3O^+$, $M$ | 1 | $10^{-1}$ | $10^{-2}$ | $10^{-3}$ | $10^{-4}$ | $10^{-5}$ | $10^{-6}$ | $10^{-7}$ | $10^{-8}$ | $10^{-9}$ | $10^{-10}$ | $10^{-11}$ | $10^{-12}$ | $10^{-13}$ | $10^{-14}$ |
| pOH | 14 | 13 | 12 | 11 | 10 | 9 | 8 | 7 | 6 | 5 | 4 | 3 | 2 | 1 | 0 |
| $OH^-$, $M$ | $10^{-14}$ | $10^{-13}$ | $10^{-12}$ | $10^{-11}$ | $10^{-10}$ | $10^{-9}$ | $10^{-8}$ | $10^{-7}$ | $10^{-6}$ | $10^{-5}$ | $10^{-4}$ | $10^{-3}$ | $10^{-2}$ | $10^{-1}$ | 1 |

←——— Acidic ———— Neutral ———— Basic ———→

are used. If $[H_3O^+]$ equals $[OH^-]$, pH = 7.00 and pOH = 7.00 and the solution is neutral. Thus, it is possible to differentiate between neutral, basic, and acidic solutions on the basis of pH as well as on the concentrations of $H_3O^+$ and $OH^-$ ions. Although acidity and basicity of a solution can be expressed in terms of pOH it is customary to use the pH scale.

It is possible to have negative pH values which means that the $H_3O^+$ concentration is greater than 1 $M$. However, this is not common since very concentrated solutions of strong acids are not fully dissociated. Second, the approximation that the activity of $H_3O^+$ and concentration of $H_3O^+$ are equal, that is the activity coefficient is 1 (see Chapter 3, page 27), is no longer valid in concentrated solutions.

## pH  OF  ACID  AND  BASE  SOLUTIONS

**Strong Acids and Bases.**    Strong acids and bases are classified as strong electrolytes, which means that they are virtually 100% dissociated in water. Thus, for a strong acid, such as hydrochloric acid, the concentration of hydronium ion in water is equal to the original analytical concentration of the acid. A strong base solution can be described in the same way. Therefore, the pH for solutions of strong acids and bases are easily calculated since the hydronium or hydroxide ion concentrations are arrived at directly from the analytical concentrations of the strong acid or base.

*Example 8-1.*    Calculate the $H_3O^+$, $OH^-$, pH, and pOH for 100 ml of 0.0250 $F$ HCl solution.

$$[H_3O^+] = 2.50 \times 10^{-2} \, M \ \text{(HCl is completely ionized)}$$

$$K_w = [H_3O^+][OH^-]$$

$$1.00 \times 10^{-14} = [2.50 \times 10^{-2}][OH^-]$$

$$[OH^-] = 4.00 \times 10^{-13} \, M$$

$$pH = -\log [H_3O^+]$$

$$pH = -\log 2.50 \times 10^{-2} = 1.60$$

$$pOH = -\log [OH^-]$$

$$pOH = -\log 4.00 \times 10^{-13} = 12.4$$

$$14.00 = pH + pOH$$

$$pOH = 14.00 - 1.60 = 12.4$$

*Example 8-2.*    Calculate the pH and pOH for the solution made by mixing 400 ml of water and 200 ml of 0.0500 $F$ NaOH. Assume that the volumes are additive.

$$ml_{NaOH} \times F_{NaOH} = mmoles_{NaOH}$$

$$200 \text{ ml} \times 0.0500 \, F = 10.0 \text{ mmoles}$$

$$\frac{\text{mmoles}_{\text{NaOH}}}{\text{ml}_{\text{NaOH}} + \text{ml}_{\text{H}_2\text{O}}} = F_{\text{NaOH}}$$

$$\frac{100 \text{ mmoles}}{200 \text{ ml} + 400 \text{ ml}} = 1.67 \times 10^{-2} \, F$$

$[\text{OH}^-] = 1.67 \times 10^{-2} \, M$ (NaOH is completely ionized)

$\text{pOH} = -\log [\text{OH}^-]$

$\text{pOH} = -\log 1.67 \times 10^{-2} = 1.78$

$14.00 = \text{pH} + \text{pOH}$

$\text{pH} = 14.00 - 1.78 = 12.22$

For very dilute solutions, $<10^{-6} \, M$, of strong acids or bases the ionization of water and its contribution to the equilibrium concentration of $[\text{H}_3\text{O}^+]$ and $[\text{OH}^-]$ must be considered. At higher acid or base concentrations the ionization of water is suppressed by the common ion effect (LeChatelier's principle) and the contribution from water becomes negligible.

*Example 8–3.* Calculate the $\text{H}_3\text{O}^+$, $\text{OH}^-$, pH, and pOH for a $5.00 \times 10^{-8} \, M$ HCl solution. [The pH of the solution cannot be 7.30 (pH $= -\log 5.00 \times 10^{-8}$) or basic since it was prepared with HCl.]

$$\text{HCl} \xrightarrow{\text{H}_2\text{O}} \text{H}_3\text{O}^+ + \text{Cl}^-$$

$$2\text{H}_2\text{O} \rightleftarrows \text{H}_3\text{O}^+ + \text{OH}^-$$

Based on the above dissociations

$[\text{H}_3\text{O}^+] = [\text{Cl}^-] + [\text{OH}^-]$

$[\text{H}_3\text{O}^+] = 5.00 \times 10^{-8} + K_w/[\text{H}_3\text{O}^+]$  $\qquad (K_w = [\text{H}_3\text{O}^+][\text{OH}^-])$

$[\text{H}_3\text{O}^+]^2 - 5.00 \times 10^{-8} [\text{H}_3\text{O}^+] - 1.00 \times 10^{-14} = 0$

This equation can be solved by the quadratic equation.*

$$[\text{H}_3\text{O}^+] = \frac{-(-5.00 \times 10^{-8}) \pm \sqrt{(-5.00 \times 10^{-8})^2 + 4(1)(1.00 \times 10^{-14})}}{2}$$

---

\* For the general equation

$$ax^2 + bx + c = 0$$

the quadratic equation is given by

$$x = \frac{-b \pm \sqrt{b^2 - 4\,ac}}{2a}$$

$$[H_3O^+] = \frac{5.00 \times 10^{-8} \pm 2.06 \times 10^{-7}}{2}$$

$$[H_3O^+] = \frac{5.00 \times 10^{-8} - 2.06 \times 10^{-7}}{2} \quad \text{(unreal solution)}$$

$$[H_3O^+] = \frac{5.00 \times 10^{-8} + 2.06 \times 10^{-7}}{2}$$

$$[H_3O^+] = 1.28 \times 10^{-7} \, M$$

$$pH = -\log [H_3O^+]$$

$$pH = -\log [1.28 \times 10^{-7}] = 6.89$$

$$K_w = [H_3O^+][OH^-]$$

$$1.00 \times 10^{-14} = [1.28 \times 10^{-7}][OH^-]$$

$$[OH^-] = 7.81 \times 10^{-8}$$

$$pOH = -\log [OH^-]$$

$$pOH = -\log [7.81 \times 10^{-8}] = 7.11$$

**Weak Acids and Bases.**     This is not the case for weak acids and weak bases since they are partially ionized in aqueous solutions. Therefore, the hydronium ion concentration of an acid, or hydroxide ion concentration for a solution of a base, is always less than the original analytical concentration. The extent of the difference is determined by the degree of ionization of the weak acid or base and is represented by the respective ionization constant. Hence, the equilibrium constant expression must be used in any calculations dealing with weak acids and bases.

**Ionization Constants.**     There are thousands of weak acids and bases with the majority of these being organic in nature. Appendix III lists the ionization constants for several common inorganic and organic weak acids and bases. In general, the reasons for the differences in ionization constants are due to inductive, resonance, steric, and other structural effects.

Consider an aqueous solution of $HC_2H_3O_2$ where the ionization step is given by

$$HC_2H_3O_2 + H_2O \rightleftharpoons H_3O^+ + C_2H_3O_2^- \tag{8-15}$$

If activities are assumed to be equal to concentrations, the ionization equilibrium expression is

$$K_a = \frac{[H_3O^+][C_2H_3O_2^-]}{[HC_2H_3O_2]} = 1.76 \times 10^{-5} \tag{8-16}$$

With this expression it is possible to calculate the hydronium ion concentration for an $HC_2H_3O_2$ solution. As the ionization constant for a series of acids decreases, the order of acidity decreases and the concentration of hydronium ion in solution also decreases.

*Example 8-4.*      Calculate the hydronium ion concentration and pH of a 0.100 $F$ $HC_2H_3O_2$ solution.

For each hydronium ion formed in reaction (8-15) an equivalent amount of acetate ion is formed. Thus,

$$[H_3O^+] = [C_3H_3O_2^-]$$

and

$$C_{HC_2H_3O_2} = [HC_2H_3O_2] + [C_2H_3O_2^-]$$

If it is assumed that very little $HC_2H_3O_2$ ionizes, $C_2H_3O_2^-$ must be negligible and

$$C_{HC_2H_3O_2} = 0.100 \cong [HC_2H_3O_2]$$

This approximation appears to be justified since the $K_a$ for $HC_2H_3O_2$ is small. Substitution into equilibrium constant expression (8-16) leads to

$$[H_3O^+] = \sqrt{K_a C_{HC_2H_3O_2}} \qquad (8\text{-}17)$$

$$[H_3O^+] = \sqrt{1.76 \times 10^{-5}(0.100)} = 1.32 \times 10^{-3} \, M$$

$$pH = -\log[H_3O^+] = 2.88$$

It is important to realize that the solution to the problem in Example 8–4 is approximate. The assumption, based on the $K_a$ value, was that the amount of weak acid that is ionized is negligible. For stronger acids (larger $K_a$ values) this approximation will not hold. A second assumption was that $H_3O^+$ comes only from the ionization of the acid. This will hold for all systems with the exception of very dilute solutions ($10^{-6} \, M$ and lower) and for very weak acids ($K_a = 10^{-12}$ and lower).

The approximation can always be checked by insertion of the calculated values back into the more exact expressions. For example, since $[C_2H_3O_2^-] = [H_3O^+]$ in Example 8–4, $[C_2H_3O_2^-] = 1.32 \times 10^{-3}$ and is therefore negligible with respect to the number 0.100 in the expression defining $C_{HC_2H_3O_2}$.

*Example 8-5.*      Calculate the pH and concentration for all the species in a 0.100 $F$ solution of chloroacetic acid, $ClCH_2CO_2H$, $K_a = 1.36 \times 10^{-3}$.

$$ClCH_2CO_2H + H_2O \rightleftharpoons H_3O^+ + ClCH_2CO_2^-$$

$$K_a = 1.36 \times 10^{-3} = \frac{[H_3O^+][ClCH_2CO_2^-]}{[ClCH_2CO_2H]}$$

$$[H_3O^+] = [ClCH_2CO_2^-]$$

(a)   $C_{ClCH_2CO_2H} = 0.100\ M \cong [ClCH_2CO_2H]$

$$\frac{[H_3O^+]^2}{0.100} = 1.36 \times 10^{-3}$$

$$[H_3O^+] = 1.17 \times 10^{-2}\ M; \quad pH = 1.93$$

$$[ClCH_2CO_2^-] = 1.17 \times 10^{-2}\ M$$

$$[ClCH_2CO_2H] = 0.100\ M$$

It is apparent that the approximation in (a) is in error since $[ClCH_2CO_2H]$ is less than 10 times as large as $[ClCH_2CO_2^-]$. Stated differently, more than 10% of the weak acid is ionized. Therefore, the following must be written

$$[H_3O^+] = [ClCH_2CO_2^-]$$

$$C_{ClCH_2CO_2H} = 0.100\ F = [ClCH_2CO_2H] + [ClCH_2CO_2^-]$$

and

$$[ClCH_2CO_2H] = 0.100 - [H_3O^+]$$

This equation states that the sum of all the chloroacetate-containing species are equal to the analytical concentration. Substituting into the equilibrium constant expression gives

$$1.36 \times 10^{-3} = \frac{[H_3O^+]^2}{0.100 - [H_3O^+]}$$

$$[H_3O^+]^2 + 1.36 \times 10^{-3}[H_3O^+] - 1.36 \times 10^{-4} = 0$$

which can be solved by the quadratic equation (see page 139).

$$[H_3O^+] = \frac{-1.36 \times 10^{-3} \pm \sqrt{(1.36 \times 10^{-3})^2 - 4(1)(-1.36 \times 10^{-4})}}{2(1)}$$

$$[H_3O^+] = \frac{-1.36 \times 10^{-3} + 2.34 \times 10^{-2}}{2}$$

and

$$[H_3O^+] = \frac{-1.36 \times 10^{-3} - 2.34 \times 10^{-2}}{2} \quad \text{(unreal solution)}$$

$$[H_3O^+] = 1.10 \times 10^{-2}\ M; pH = 1.96$$

$$[ClCH_2CO_2^-] = 1.10 \times 10^{-2}\ M$$

$$[ClCH_2CO_2H] = 8.90 \times 10^{-2}\ M$$

**Weak Base Solutions.**     The calculation for the hydronium ion concentration and pH of a weak base solution is analogous to a weak acid system except that a $K_b$ is involved. Thus, the analytical concentration of the weak base and the value for the $K_b$ will determine the type of approximations that can be made.

*Example 8-6.*     What is the $[OH^-]$, $[H_3O^+]$, and pH of a solution of 0.100 $F$ $NH_3$.

$$NH_3 + H_2O \rightleftharpoons NH_4^+ + OH^- \tag{8-18}$$

$$K_b = 1.79 \times 10^{-5} = \frac{[NH_4^+][OH^-]}{[NH_3]} \tag{8-19}$$

Since an equivalent amount of $NH_4^+$ and $OH^-$ are formed,

$$[NH_4^+] = [OH^-]$$

and

$$C_{NH_3} = [NH_3] + [NH_4^+]$$

If the amount of ammonia that ionizes is small, $[NH_4^+]$ is negligible and

$$C_{NH_3} = 0.100 \cong [NH_3]$$

Substitution into (8-19) leads to

$$[OH^-] = \sqrt{K_b C_{NH_3}} \tag{8-20}$$

$$[OH^-] = \sqrt{1.79 \times 10^{-5} \times 0.100} = 1.34 \times 10^{-3}\ M$$

$$[H_3O^+] = \frac{K_w}{[OH^-]} = \frac{1.0 \times 10^{-14}}{1.34 \times 10^{-3}} = 7.47 \times 10^{-12}\ M$$

$$pH = -\log[H_3O^+] = -\log 7.47 \times 10^{-12} = 11.1$$

Alternatively,

$$K_b = \frac{[OH^-]^2}{[NH_4OH]} = \frac{[K_w]^2}{[H_3O^+]^2[NH_4OH]}$$

$$[H_3O^+] = \sqrt{\frac{K_w^2}{K_b[NH_4OH]}} \tag{8-21}$$

Substitution for the $K$'s and $[NH_4OH]$ yields the $[H_3O^+]$ directly.

It should be apparent that as the $K_b$ for a base decreases, its strength as a base decreases. This means that the lower the $K_b$ the less hydroxide ion produced in a solution for a given concentration.

**Conjugate Acids and Bases.**     Expressions describing the ionization of conjugate acids or bases in water can be written as if they were weak acids or bases. For example, $C_2H_3O_2^-$ is a conjugate weak base of the acid $HC_2H_3O_2$, and its ionization and equilibrium expression are given by

$$C_2H_3O_2^- + H_2O \rightleftharpoons HC_2H_3O_2 + OH^- \tag{8-22}$$

$$K_b^{C_2H_3O_2^-} = \frac{[HC_2H_3O_2][OH^-]}{[C_2H_3O_2^-]} \tag{8-23}$$

Since $[OH^-] = K_w/[H_3O^+]$, substitution into Eq. (8-23) gives

$$K_b^{C_2H_3O_2^-} = \frac{[HC_2H_3O_2]\, K_w/[H_3O^+]}{[C_2H_3O_2^-]}$$

or

$$\frac{K_b^{C_2H_3O_2^-}}{K_w} = \frac{[HC_2H_3O_2]}{[H_3O^+][C_2H_3O_2^-]} = \frac{1}{K_a^{HC_2H_3O_2}}$$

Therefore, for any weak acid and its conjugate base

$$K_a^{HC_2H_3O_2}K_b^{C_2H_3O_2^-} = K_w \tag{8-24}$$

A similar relationship can be shown to hold for any weak base and its conjugate acid:

$$K_b^{NH_3}K_a^{NH_4^+} = K_w \tag{8-25}$$

The value of these relationships are that all the calculations of hydronium ion concentrations for solutions of weak acids, weak bases, and their salts can be based on either a $K_a$ or a $K_b$ value. If the former is chosen, all equilibria must be written as ionization of the acid or conjugate acid to the conjugate base or base, respectively (see Appendix III for $K_a$ and $K_b$ values). The advantage of this simple relationship is fully realized when hydronium ion concentrations for solutions of salts of weak acids or bases are calculated.

**Hydrolysis: pH of the Solution.**     If a salt of a weak acid and a strong base, such as $NaC_2H_3O_2$, is dissolved in water, the solution is basic because of the presence of the conjugate base $C_2H_3O^-$. The equilibrium, which is often called hydrolysis, is given by reaction (8-22) and the equilibrium constant expression by Eq. (8-23). The equilibrium constant expression can also be expressed in terms of $K_a$ rather than in the conjugate base form.

$$\frac{K_w}{K_a} = \frac{[HC_2H_3O_2][OH^-]}{[C_2H_3O_2^-]} \tag{8-26}$$

*Example 8-7.*     Calculate the $[OH^-]$, $[H_3O^+]$, and pH for a solution of $NaC_2H_3O_2$ whose analytical concentration is 0.100 $F$.

From reaction (8-22) an equivalent amount of $HC_2H_3O_2$ and $OH^-$ are formed. Thus,

$$[HC_2H_3O_2] = [OH^-]$$

Also,

$$C_{C_2H_3O_2^-} = [C_2H_3O_2^-] + [HC_2H_3O_2]$$

If very little of the acetate ion participates in hydrolysis, the amount of acetic acid formed must be very small. Therefore,

$$C_{C_2H_3O_2^-} = 0.100 \ M \cong [C_2H_3O_2^-]$$

Substitution into Eq. (8-23)

$$5.68 \times 10^{-10} = \frac{[OH^-]^2}{[0.100]}$$

or Eq. (8-26)

$$\frac{1.0 \times 10^{-14}}{1.76 \times 10^{-5}} = \frac{[OH^-]^2}{[0.100]}$$

where $K_a = 1.76 \times 10^{-5}$, $K_b{}^{C_2H_3O_2} = 5.68 \times 10^{-10}$ and $K_w = 1.0 \times 10^{-14}$, gives

$$[OH^-] = 7.53 \times 10^{-6} \ M$$

$$[H_3O^+] = 1.33 \times 10^{-9} \ M; \qquad \left([H_3O^+] = \frac{K_w}{[OH^-]}\right)$$

$$pH = 8.88$$

Equations (8-23) and (8-26) can be expressed in terms of $[H_3O^+]$ rather than $[OH^-]$, since $[OH^-] = K_w/[H_3O^+]$. Hence,

$$[H_3O^+] = \sqrt{\frac{K_w^2}{K_b{}^{C_2H_3O_2^-}C_{C_2H_3O_2^-}}} \quad \text{and} \quad [H_3O^+] = \sqrt{\frac{K_wK_a}{C_{C_2H_3O_2^-}}} \qquad (8\text{-}27)$$

where $[C_2H_3O_2^-] \cong C_{C_2H_3O_2^-}$.

Two approximations are part of the calculation in Example 8-7. First, it is assumed that the only source of $OH^-$ is from the hydrolysis of $NaC_2H_3O_2$. This is reasonable except at extreme dilution. Second, it is assumed that the amount of acetate converted to the undissociated form is negligible. This is reasonable, even though a poorly dissociated weak acid ($HC_2H_3O_2$) is formed in the hydrolysis, because, at the same time, a strong base ($OH^-$) is also formed. Neutralization (reverse of hydrolysis) is a much more favorable reaction.

Treatment of a solution of a salt of a weak base–strong acid is analogous to the weak acid–strong base case except that the solution must be acidic since a conjugate acid is involved. Therefore, the hydrolysis step can be expressed as a function of $K_b$ or in terms of the $K_a$ for its conjugate acid.

*Example 8–8.* Calculate the $[H^+]$ and pH of a solution in which the analytical concentration of $NH_4Cl$ is 0.100 $F$.

$$K_b^{NH_3} = 1.79 \times 10^{-5}, \quad K_a^{NH_4^+} = 5.59 \times 10^{-10}, \quad \text{and } K_w = 1.0 \times 10^{-14}$$

$$NH_4^+ + H_2O \rightleftharpoons NH_3 + H_3O^+ \tag{8-28}$$

$$\frac{[NH_3][H_3O^+]}{[NH_4^+]} = \frac{K_w}{K_b} \quad \text{or} \quad K_a^{NH_4^+} = \frac{[NH_3][H_3O^+]}{[NH_4^+]} \tag{8-29}$$

From (8-28) it can be stated that

$$[NH_3] = [H_3O^+]$$

If hydrolysis is small, $[NH_4^+] >>> [NH_4OH]$ and

$$C_{NH_4^+} = 0.100 \cong [NH_4^+]$$

Substitution into Eq. (8-29) yields

$$[H_3O^+] = \sqrt{\frac{K_w}{K_b} C_{NH_4}} \quad \text{or} \quad [H_3O^+] = \sqrt{K_a^{NH_4^+} C_{NH_4^+}} \tag{8-30}$$

$$[H_3O^+] = \sqrt{\frac{1.0 \times 10^{-14}}{1.79 \times 10^{-5}} (0.100)} \quad \text{or} \quad [H_3O^+] = \sqrt{5.59 \times 10^{-10} \times 0.100}$$

$$[H_3O^+] = 7.48 \times 10^{-6}; \quad \text{pH} = 5.13$$

The calculation is simplified because of two approximations. It is assumed that the only source of $H_3O^+$ is the hydrolysis step and that the amount of $NH_4^+$ converted to undissociated $NH_3$ is negligible. The reasoning behind these assumptions is analogous to the salt of a strong base-weak acid and will not be repeated.

A solution made from a salt of a weak acid–weak base is analogous to the previously discussed hydrolysis systems. For example, for an ammonium acetate solution the hydrolysis steps are

$$C_2H_3O_2^- + H_2O \rightleftharpoons HC_2H_3O_2 + OH^-$$

$$NH_4^+ + 2H_2O \rightleftharpoons NH_4OH + H_3O^+$$

The pH of the solution will be determined by which is the least dissociated, the weak acid or weak base. If it is the former, the solution will be basic while if it is the latter, the solution will be acidic. Hence, any calculation must consider both hydrolysis steps by combination of the two previous cases.

For the weak acid–weak base case, it can be shown that

$$[H_3O^+] = \sqrt{\frac{K_w K_a}{K_b}} \tag{8-31}$$

where $K_a$ and $K_b$ are the constants for the weak acid and base participating in the hydrolysis. Hence, it would be concluded that the pH of the solution will be independent of the analytical concentration of the salt used.

Several assumptions limit the use of Eq. (8-31). The principal one is that the concentrations of the cation and anion (ammonium and acetate ions—in the example) must be nearly equal to the analytical concentration of the salt. Or, stated differently, the amount of undissociated weak acid and base formed during hydrolysis must be negligible in concentration.

## BUFFERS

In many solutions little change in pH occurs even though strong acid or strong base is added to the solution. These kinds of solutions tend to resist changes in pH and are said to be buffered.

Buffered solutions are of great significance. For example, many physiological processes require a fixed pH to function properly. Often the permissable pH range is very narrow. To maintain this pH, nature has included buffers into the system, and, frequently, the components of these buffers are the same ones that the chemist uses in the laboratory. Thus, if these same processes are to be studied in the laboratory, the conditions for the buffer must be clearly understood.

The control of pH in experimental chemistry is essential in all sorts of chemical and instrumental applications. Often the success in these applications will be determined by how carefully the pH is controlled and maintained.

The most common type of buffer solution is made by dissolving a weak acid and a salt of the same weak acid in water or by dissolving a weak base and a salt of the same weak base in water. For example, consider the first case where ionization will occur according to the reaction

$$HA + H_2O \rightleftharpoons H_3O^+ + A^- \tag{8-32}$$

If a salt of the weak acid, NaA, is added, the $H_3O^+$ concentration will decrease (pH increases) as predicted by the Le Chatelier principle; that is, the equilibrium is shifted to the left.

If a strong acid or some other source of hydronium ion is introduced into the solution, association with the anion $A^-$, reverse of reaction (8-32), takes place. Since there is a large reservoir of $A^-$ little change in pH is observed.

If a strong base or hydroxide ion is introduced into the solution, neutralization of the hydronium ion in reaction (8-32) occurs. To replace the consumed hydronium ion more HA is ionized and since there is a reservoir of HA the pH changes only slightly.

A buffer made from a weak base (BOH) and its salt (BX) can be described in an analogous way. In this case ionization of the weak base must be considered.

$$BOH \overset{H_2O}{\rightleftharpoons} B^+ + OH^- \tag{8-33}$$

As hydronium ions are added, they are neutralized by the hydroxide ion which are replaced by more ionization of BOH. If hydroxide ions are added, they are consumed by reaction with $B^+$ to form undissociated weak base.

*Example 8–9.*　　Calculate $[H_3O^+]$ and pH for a buffer made from 0.100 mole of $HC_2H_3O_2$ and 0.100 mole of $NaC_2H_3O_2$ diluted to 1 liter of solution.

$$HC_2H_3O_2 + H_2O \rightleftharpoons H_3O^+ + C_2H_3O_2^-$$

$$K_a = 1.76 \times 10^{-5} = \frac{[H_3O^+][C_2H_3O_2^-]}{[HC_2H_3O_2]}$$

If it is assumed that very little of the $HC_2H_3O_2$ and $C_2H_3O_2^-$ ionizes and hydrolyzes, respectively, in reference to their original analytical concentrations, the following approximations can be written:

$$C_{HC_2H_3O_2} = 0.100\ M \cong [HC_2H_3O_2]$$

$$C_{C_2H_3O_2^-} = 0.100\ M \cong [C_2H_3O_2^-]$$

Substitution into the ionization expression gives

$$1.76 \times 10^{-5} = \frac{[H_3O^+][0.100]}{[0.100]} = \frac{[H_3O^+]C_{C_2H_3O_2^-}}{C_{HC_2H_3O_2}} \tag{8-34}$$

$$[H_3O^+] = 1.76 \times 10^{-5}\ M; \quad pH = 4.75$$

*Example 8–10.*　　Calculate the grams of $NaC_2H_3O_2$ that must be added to 100 ml of 0.100 $F$ $HC_2H_3O_2$ to give a buffer solution of pH 4.20. Assume the volume remains constant.

$$pH = 4.20 = -\log[H_3O^+]$$

$$[H_3O^+] = 6.31 \times 10^{-5}\ M$$

$$K_a = 1.76 \times 10^{-5} = \frac{[H_3O^+][C_2H_3O_2^-]}{[HC_2H_3O_2]}$$

$$C_{HC_2H_3O_2} = 0.100\ M \cong [C_2H_3O_2]$$

$$1.76 \times 10^{-5} = \frac{(6.31 \times 10^{-5})C_{C_2H_3O_2^-}}{(0.100)}$$

$$C_{C_2H_3O_2^-} = 0.0279\ M \cong [C_2H_3O_2^-]$$

$$g\ NaC_2H_3O_2 = 100\ ml \times M\ NaC_2H_3O_2 \times \text{formula wt}\ NaC_2H_3O_2$$

$$g\ NaC_2H_3O_2 = 100\ ml \times 0.0279\ \frac{mole}{1000\ ml} \times 82.03\ g/mole = 0.229\ g$$

The approximations leading to (8-34) are reasonable provided the acid is not too strong and the analytical concentrations are not too small. If the acid is too strong, $H_3O^+$ from the ionization of the acid would have to be accounted for, while for very dilute solutions $OH^-$ from water would have to be accounted for.

Equation (8-34) can also be arranged into the form

$$[H_3O^+] = K_a \frac{C_{HC_2H_3O_2}}{C_{C_2H_3O_2^-}} = K_a \frac{\text{Fraction unneutralized}}{\text{Fraction neutralized}} \qquad (8\text{-}35)$$

If the entire equation is multiplied by $-\log$, it becomes

$$\overset{4.2 \quad 4.75}{\text{pH} = \text{p}K_a + \log \frac{\text{Fraction neutralized}}{\text{Fraction unneutralized}}} \qquad (8\text{-}36)$$

In many scientific areas Eq. (8-36) is used in this form and is known as the Henderson–Hasselbalch equation. The main advantage of the Henderson–Hasselbalch equation is that it gives the pH value directly.

*Example 8-11.* Calculate $[H_3O^+]$ and pH for a buffer made from 0.100 mole of $NH_3$ and 0.100 mole of $NH_4Cl$ diluted to one liter of solution.

$$NH_3 + H_2O \rightleftharpoons NH_4^+ + OH^-$$

$$K_b = 1.79 \times 10^{-5} = \frac{[NH_4^+][OH^-]}{[NH_3]}$$

The problem is analogous to Example 8–9. Since the base is not very strong, and the concentrations are not dilute, approximations can be made since ionization ($NH_3$) and hydrolysis ($NH_4^+$) will be small.

$$C_{NH_3} = 0.100\ M \cong [NH_3]$$

$$C_{NH_4^+} = 0.100\ M \cong [NH_4^+]$$

Substitution into the ionization expression gives

$$1.79 \times 10^{-5} = \frac{[0.100][OH^-]}{[0.100]} = \frac{C_{NH_4^+}[OH^-]}{C_{NH_3}}$$

or

$$1.79 \times 10^{-5} = \frac{[0.100]K_w}{[0.100][H_3O^+]} = \frac{C_{NH_4^+}K_w}{C_{NH_3}[H_3O^+]} \qquad (8\text{-}37)$$

$$[H_3O^+] = 5.59 \times 10^{-10}\ M; \quad \text{pH} = 9.25$$

Equation (8-37) written in the Henderson–Hasselbalch form is

$$\text{pH} = \text{p}K_w - \text{p}K_b + \log \frac{\text{Fraction unneutralized}}{\text{Fraction neutralized}} \qquad (8\text{-}38)$$

In terms of $K_a$ for the conjugate acid of the weak base, the equation becomes

$$pH = pK_a^{BH+} + \log \frac{\text{Fraction unneutralized}}{\text{Fraction neutralized}} \tag{8-39}$$

*Example 8–12.* Calculate the molar ratio of $NH_3/NH_4^+$ in a buffered solution of pH 9.30. If the solution contains 10.0 g of $NH_3$, and is exactly 500 ml, calculate the grams of $NH_4Cl$ that must be present.

$$[H_3O^+] = 9.30 = -\log[H_3O^+]$$

$$[H_3O^+] = 5.01 \times 10^{-10}\,M; \quad [OH^-] = 2.00 \times 10^{-5}\,M$$

$$K_b = 1.79 \times 10^{-5} = \frac{[NH_4^+][OH^-]}{[NH_3]}$$

$$1.79 \times 10^{-5} = \frac{[NH_4^+][2.00 \times 10^{-5}]}{[NH_3]}$$

$$[NH_3]/[NH_4^+] = 1.12$$

Or, by using the Henderson–Hasselbalch form [Eq. (8-38)]

$$9.30 = 14 - 4.75 + \log \frac{[NH_3]}{[NH_4^+]}$$

$$\log \frac{[NH_3]}{[NH_4^+]} = 0.05$$

$$[NH_3]/[NH_4^+] = 1.12$$

$$M_{NH_3} = \frac{g\,NH_3}{\text{formula wt } NH_3}$$

$$M_{NH_3} = \frac{10\,g/500\,ml}{17.04\,g/mole \times (1/1000\,ml)} = 1.17\,M$$

$$M_{NH_4Cl} = \frac{M_{NH_3}}{1.12} = \frac{1.17\,M}{1.12} = 1.04\,M$$

$$g\,NH_4Cl = 500\,ml \times M_{NH_4Cl} \times \text{formula wt } NH_4Cl$$

$$g\,NH_4Cl = 500\,ml \times 1.04\,\frac{mole}{1000\,ml} \times 53.49\,g/mole = 27.8\,g$$

*Example 8–13.* Calculate the pH for the solution made by mixing 100 ml of 0.0800 $F$ HCl and 50.0 ml of 0.200 $F$ ethanolamine ($K_b = 4.0 \times 10^{-5}$). Assume that the volume change is additive.

$$HOCH_2CH_2NH_2(B) + HCl \rightarrow HOCH_2CH_2NH_3^+Cl^- (HB^+Cl^-)$$

$$
\begin{array}{l}
50 \text{ ml} \times 0.200\ F = 10.0 \text{ mmoles B} \\
\underline{100 \text{ ml} \times 0.800\ F = \phantom{1}8.00 \text{ mmoles HCl}} \\
150 \text{ ml} \phantom{xxxxxxxxx} 2.00 \text{ mmoles B in excess}
\end{array}
$$

$$C_B = \frac{2.00 \text{ mmoles}}{150 \text{ ml}} = 1.33 \times 10^{-2}\ M \cong [B]$$

From the stoichiometry of the reaction the mmoles of $HB^+$ formed will equal the concentration of acid added.

$$C_{HB^+} = \frac{8.00 \text{ mmoles}}{150 \text{ ml}} = 5.33 \times 10^{-2}\ M \cong [HB^+]$$

In the above, two approximations are made. First, the equilibrium concentrations of $HB^+$ and B are equal to the stoichiometric concentrations and, second, their values are not influenced by the ionization of B.

$$pH = pK_w - pK_b + \log \frac{C_B}{C_{HB^+}}$$

$$pH = 14 - (-\log 4.0 \times 10^{-5}) + \log \frac{1.33 \times 10^{-2}}{5.33 \times 10^{-2}} = 9.00$$

**Buffer Capacity.**    The ability of a buffer to consume acid or base does not continue indefinitely. Each prepared buffer is capable of resisting only a certain amount of added acid or base. A measure of this quantity is referred to as buffer capacity and can be defined as the number of moles of strong base or strong acid required to cause a unit increase or decrease in pH in 1 liter of a buffered solution.

There are two experimental techniques for producing a high buffer capacity, which is usually desirable. First, a large concentration of the buffer components can be used. Or, second, the concentration of the weak acid (base) that is present should equal the concentration of its salt. Hence, maximum capacity is obtained when the dissociation constant for the weak acid ($K_w/K_b$ for a weak base) is equal to the hydronium ion concentration that is desired. Usually, a combination of these two techniques is utilized.

*Example 8-14.*    Calculate the change in pH if $1.000 \times 10^{-3}$ mole of HCl were added to the buffer prepared in Example 8-9. Assume that the volume does not change.

An approximate answer is obtained if it is assumed that the HCl stoichiometrically converts $C_2H_3O_2^-$ to $HC_2H_3O_2$. This is reasonable because the amount of acid that is added, even though it is a strong acid, is very small in comparison to the concentration of the buffer components. Therefore,

$$C_{HC_2H_3O_2} = 0.100\ M + 0.001\ M \cong [HC_2H_3O_2]$$

$$C_{C_2H_3O_2^-} = 0.100\ M - 0.001\ M \cong [C_2H_3O_2^-]$$

Substitution into Eq. (8-35) yields

$$1.76 \times 10^{-5} = \frac{[H_3O^+][0.099]}{[0.101]}$$

$$[H_3O^+] = 1.80 \times 10^{-5}\ M; \quad pH = 4.74$$

The addition of the acid to the buffer solution in Example 8–9 causes the pH to drop by 0.01 pH units. Adding the same amount of HCl to 1 liter of pure water in which no volume change takes place would cause the pH to drop from 7.00 to 3.00, or 4.00 full pH units.

**Preparation of Buffer Solutions.**     The National Bureau of Standards through the years has provided the main leadership in the establishment of standard buffer solutions. Table 8–4 lists the concentrations of salts needed to prepare several standard pH solutions. These solutions, if carefully prepared at a controlled temperature, can be treated as primary standards except for the $KH_3(C_2O_4)_2$ and $Ca(OH)_2$ solutions.

Two other important properties of the pH standard mixtures are listed in Table 8–4. The change in pH that occurs upon 1:1 dilution of the solution with water is expressed as $\Delta pH_{1/2}$ and the temperature coefficient or change in pH with change in temperature in the 25°C region is listed at $dpH/dt$.

Certified samples of the salts needed for the preparation of the buffers listed in Table 8–4 can be obtained from the National Bureau of Standards. No special experimental techniques are needed for the preparation of the solutions. In general, the salt is carefully weighed and diluted to a known volume to yield a buffer of reasonably high buffer capacity.

Several precautions must be taken when handling buffer salts and their solutions.

1. Mold growth in the buffer solution must be prevented during shelf storage. Often a crystal of thymol is added for this purpose.
2. Absorption of carbon dioxide from the atmosphere should be prevented.
3. Salts should be dried or stored according to their hygroscopic properties.
4. Temperature control is necessary.

The solutions described in the previous paragraphs are the primary standards in pH measurements and are used where the most precise pH measurements or pH calibration procedures are required. Thus, they may be used as references in establishing the pH of different solutions, evaluating new hydrogen ion-sensitive electrodes, or calibrating potentiometric cells. The latter

**Table 8–4.** **pH Values for a Series of Primary Standard and Other Useful Buffer Solutions**[a]

| | Secondary standard 0.05 $F$ $KH_3(C_2O_4)_2 \cdot$ 2H$_2$O | Primary standards | | | | Secondary standard Ca(OH)$_2$ (satd. at 25°C) |
| | | $KHC_4H_4O_6$ (satd. at 25°C) | 0.05 $F$ $KHC_8H_4O_4$ | 0.025 $F$ $KH_2PO_4$, 0.025 $F$ $Na_2HPO_4$ | 0.01 $F$ $Na_2B_4O_7 \cdot$ 10H$_2$O | |
| $T$ (°C) | | | | | | |
|---|---|---|---|---|---|---|
| 20 | 1.68 | | 4.00 | 6.88 | 9.22 | 12.63 |
| 25 | 1.68 | 3.56 | 4.01 | 6.86 | 9.18 | 12.45 |
| 30 | 1.69 | 3.55 | 4.01 | 6.85 | 9.14 | 12.30 |
| ΔpH$_{1/2}$, pH units for 1:1 dilution | +0.186 | +0.049 | +0.052 | +0.080 | +0.01 | ca. −0.28 |
| $d$pH/$dt$ at 25°C, pH units | +0.001 | +0.001 | +0.001 | +0.003 | −0.008 | −0.033 |

#### Useful Buffer Solutions

| Composition/100 ml[b] | pH |
|---|---|
| 25 ml 0.2 $F$ KCl + 67.0 ml 0.2 $F$ HCl | 1.0 |
| 25 ml 0.2 $F$ KCl + 6.5 ml 0.2 $F$ HCl | 2.0 |
| 50 ml 0.1 $F$ KHP + 22.3 ml 1 $F$ HCl | 3.0 |
| 50 ml 0.1 $F$ KHP + 0.1 ml 0.1 $F$ HCl | 4.0 |
| 50 ml 0.1 $F$ KHP + 22.6 ml 0.1 $F$ NaOH | 5.0 |
| 50 ml 0.1 $F$ KH$_2$PO$_4$ + 5.6 ml 0.1 $F$ NaOH | 6.0 |
| 50 ml 0.1 $F$ KH$_2$PO$_4$ + 29.1 ml 0.1 $F$ NaOH | 7.0 |
| 50 ml 0.1 $F$ KH$_2$PO$_4$ + 46.1 ml 0.1 $F$ NaOH | 8.0 |
| 50 ml 0.1 $F$ H$_3$BO$_3$ and 0.1 $F$ KCl + 20.8 ml 0.1 $F$ NaOH | 9.0 |
| 50 ml 0.1 $F$ H$_3$BO$_3$ and 0.1 $F$ KCl + 43.7 ml 0.1 $F$ NaOH | 10.0 |
| 50 ml 0.05 $F$ NaHCO$_3$ + 22.7 ml 0.1 $F$ NaOH | 11.0 |
| 50 ml 0.05 $F$ Na$_2$HPO$_4$ + 26.9 ml 0.1 $F$ NaOH | 12.0 |
| 25 ml 0.2 $F$ KCl + 66.0 ml 0.2 $F$ NaOH | 13.0 |

[a] R. G. Bates, "Treatise on Analytical Chemistry," Part 1, Volume 1, Wiley-Interscience, New York, 1959, p. 375. See original for literature references.

[b] Buffers at intermittent pH values can be prepared by using different volumes of HCl or NaOH solution. Composition of other buffer mixtures at fixed ionic strength and those that are useful physiological buffers are listed in chemistry handbooks.

two applications will be considered in detail in later chapters. If the buffer is to be used in living systems the toxic properties of the buffer components must be considered.

One typical compound that provides buffer conditions near the physiological pH is tris(hydroxymethyl)aminomethane $(HOCH_2)_3CNH_2$ (usually referred to as THAM or Tris). This compound, which is a primary standard, has a $K_b$ of $1.26 \times 10^{-6}$ and is routinely used to prepare physiological buffers.

It is not always necessary to use the pH standards designated by the National Bureau of Standards. In fact, in most applications, buffers are prepared from readily available weak acids, bases, and their salts. If the application is pH calibration, such factors as buffer stability, availability of reagents, reactivity of the buffer salts in solution with the components of the potentiometric cell, and the required accuracy of the calibration, should be considered. On the other hand if the application of the buffers is of a chemical nature, the important factors to consider are the reactivity of the buffer components towards the chemical reaction, the accuracy needed, availability of reagents, buffer capacity and concentration, and shelf life of the buffer solution.

## NEUTRALIZATION AS A VOLUMETRIC METHOD

Since a stoichiometric neutralization reaction involves passing from an acidic to basic solution, assuming an acid is being titrated with a base, the progress of the reaction can be followed by determining the pH of the solution as a function of added titrant. Usually, for reasons to be outlined, a strong base or acid is used as the titrant.

A plot of pH vs ml of titrant is called a titration curve. From the curve it is possible to select the volume of titrant needed to neutralize the sample. Since the concentration of the titrant is known the amount of the acidic (or basic) sample present can be calculated. Therefore, in order to understand and use neutralization titrations for analyses it is necessary to understand how the pH of the solution changes quantitatively with added titrant and second, how to detect the stoichiometric point of the neutralization reaction.

Titration curves vary in shape and are influenced by the strength of the acids and bases involved, their concentration, buffer conditions, and the hydrolysis of the salts that are formed (strength of the conjugate acids and bases that are formed). Calculation for each of these kinds of effects have been considered in Examples 8–1 to 8–14. Therefore, the titration curve is a combination of these calculations as a function of concentration changes.

If a color indicator is used for end-point detection, the titrant should be a standard solution of a strong acid or strong base. (For some instrumental end point detection methods a weak acid or base titrant is used because it will provide better accuracy.) This being the case, neutralization titrations can be divided into a strong acid or weak acid titration with strong base, and strong base or weak base with strong acid. The experimental procedure is essentially the same for all the titrations. One precaution, however, is to prevent the basic sample or titrant from undergoing a reaction with atmospheric $CO_2$. For this reason water is often boiled before being used as the solvent and sometimes the titration is performed under $N_2$.

**Strong Acid or Weak Acid–Strong Base Titration.** For purposes of illustration assume that the strong acid, HCl, and the weak acid, $HC_2H_3O_2$, are to be individually titrated with the strong base, NaOH. The two reactions can be summarized in the following way:

Strong acid–strong base

$$HCl + NaOH \xrightarrow{\text{neutralization}} NaCl + H_2O$$
$$+$$
$$H_2O$$
$$\updownarrow$$
$$H_3O^+ + OH^-$$

Weak acid–strong base

$$HC_2H_3O_2 + NaOH \underset{\text{hydrolysis}}{\overset{\text{neutralization}}{\rightleftharpoons}} NaC_2H_3O_2 + H_2O$$
$$+ \qquad\qquad\qquad\qquad +$$
$$H_2O \qquad\qquad\qquad\qquad H_2O$$
$$\updownarrow \qquad\qquad\qquad\qquad \updownarrow$$
$$H_3O^+ + C_2H_3O_2^- \qquad\qquad H_3O^+ + OH^-$$

The essential difference between the two titrations is that throughout the weak acid–strong base titration the effect of ionization in some way influences the change in pH since the product of the reaction is a conjugate base of moderate strength.

For strong acid–strong base reactions, the entire titration curve is determined by calculating the amount of acid present at any point before the stoichiometric point and the amount of base present at any point after the stoichiometric point. Since the two are strong electrolytes, they are completely ionized. In contrast, a weak acid–strong base titration is characterized by three different ionization effects. The initial pH of the weak acid solution is given by the $K_a$ expression (Example 8–1). The instant base is added, and until the stoichiometric point is reached, the solution is a buffer (Example 8–9). At the stoichiometric point a salt of a strong base–weak acid is formed and therefore, the pH of the solution is determined by the presence of a conjugate base of moderate strength which undergoes hydrolysis (Example 8–7). After the stoichiometric point, excess strong base is added and the excess determines the pH of the solution. A flow sheet on pages 158 and 159 provides the equations for the complete calculation of the two types of titration curves.

Figure 8–1 shows a calculated titration curve for the HCl–NaOH titration plotted vs ml of NaOH and percent neutralized. The titration break is very sharp and at the stoichiometric point the pH changes approximately by 8 pH units for a small fraction of added titrant. It is this sharp change that is to be detected as the end point.

The concentration of the strong acid and strong base influences the shape of the titration curve. Figure 8–2 also illustrates this concentration effect As the concentration of the system decreases, the magnitude of the pH break decreases. (The stoichiometric point is always at pH = 7.) Thus, the ability to select the end point becomes impaired. If the titration is being used for analy-

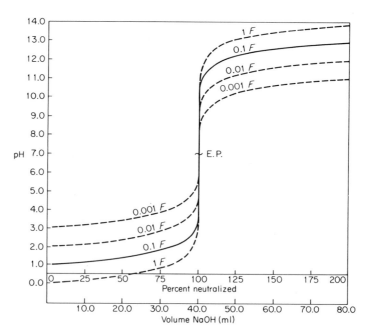

**Fig. 8–1.** Calculated titration curve for the titration of 40.0 ml of 0.100 $F$ HCl with 0.100 $F$ NaOH. The effect of dilution on the titrant and sample is illustrated by the dotted lines.

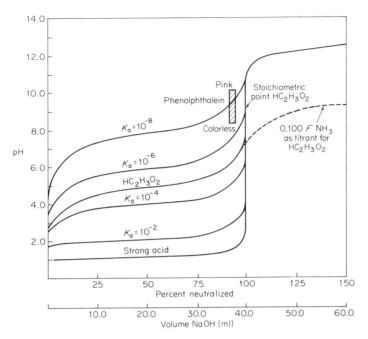

**Fig. 8–2.** Calculated titration curve for the titration of 0.100 $F$ $HC_2H_3O_2$ with 0.100 $F$ NaOH and the effect of $K_a$ on the titration curve.

sis, the lower concentration limits, particularly if a weak acid or a weak base is being titrated, are about 0.001 $F$.

Figure 8–2 illustrates the titration curve for $HC_2H_3O_2$–NaOH. Several different features are obvious when comparing this curve to the strong acid–strong base titration curve. Prior to the stoichiometric point the pH tends to change more rapidly. For the strong acid–strong base titration this portion of the curve is very flat. Therefore, the pH break at the stoichiometric point is less for the weak acid–strong base titration. Another difference is that the end point of the titration occurs on the basic side of the pH scale.

As the strength of the acid decreases, the size of the titration break decreases. Figure 8–2 also illustrates titration curves as a function of $K_a$ in comparison to a strong acid. It should be noted that the stoichiometric point occurs at a more basic pH with a decrease in acidity. This is a direct result of the increase in basicity of the corresponding conjugate bases that are formed in the neutralization. When analysis is the goal, it is not possible to accurately determine acids that have a $K_a$ less than $10^{-8}$ by titration in aqueous solutions.

Dilution effects would lead to the same result as for the strong acid–strong base titrations; that is, a reduction in the pH break occurs. However, it is more difficult to generalize in the weak acid–strong base case because the

# Flow Sheet for the Calculation of Titration Curves for a Strong Acid or Weak Acid with a Strong Base Titrant

Assume that for each titration 40.0 ml of 0.100 F
HCl and HC$_2$H$_3$O$_2$  ($K_a = 1.76 \times 10^{-5}$)  are
titrated with 0.100 F NaOH

| NaOH (ml) | Titration of HCl | Titration of HC$_2$H$_3$O$_2$ |
|---|---|---|
| 0 ml | $C_{HCl} = C_{H_3O^+} = 0.100\ F$ <br> pH $= -\log [H_3O^+]$ <br> pH $= -\log 1.00 \times 10^{-1}$ <br> pH $= 1.00$ | $[H_3O^+] = \sqrt{K_a C_{HC_2H_3O_2}}$ <br> $[H_3O^+] = \sqrt{1.76 \times 10^{-5} \times 0.100}$ <br> $[H_3O^+] = 1.32 \times 10^{-3}\ M$; pH $= 2.88$ <br> (See Example 8-4 for discussion of approximation) |
| Any volume after and before the stoichiometric point volume; as an example consider 10.0 ml (25% neutralized) | mmoles of <br> initial acid: $40.0\text{ ml} \times 0.100\ F = 4.0$  mmole <br> mmoles of <br> base added: $10.0\text{ ml} \times 0.100\ F = 1.00$ mmole <br> amount of <br> acid left in: 50.0 ml $\qquad\qquad$ 3.00 mmole <br><br> $[H_3O^+] = \dfrac{\text{mmoles acid}}{\text{total volume, ml}}$ <br><br> $[H_3O^+] = \dfrac{3.00\text{ mmoles}}{50.0\text{ ml}} = 0.0600\ M$ <br><br> pH $= 1.22$ | Calculate the mmoles of acid left in the strong acid case. Since the solution contains a weak acid and its salt, it is a buffer. Therefore, <br><br> $[H_3O^+] = K_a\,\dfrac{C_{HC_2H_3O_2}}{C_{C_2H_3O_2^-}}$ <br><br> or <br><br> $\text{pH} = pK_a + \log \dfrac{\text{Fraction neutralized}}{\text{Fraction unneutralized}}$ <br><br> (See Example 8-9 for the discussion of the approximations) <br><br> $C_{HC_2H_3O_2} = \dfrac{3.0\text{ mmoles}}{50.0\text{ ml}} = 0.0600\ M$ <br><br> $C_{C_2H_3O_2^-} = \dfrac{1.0\text{ mmoles}}{50.0\text{ ml}} = 0.0200\ F$ <br><br> $[H_3O^+] = 1.76 \times 10^{-5}\ \dfrac{0.0600}{0.0200}$ <br><br> $[H_3O^+] = 5.28 \times 10^{-5}\ M$; pH $= 4.28$ |

**40.0 ml (Stoichiometric point)**

Since no acid or base is present
$$[H_3O^+] = [OH^-] = 1.00 \times 10^{-7}\ M$$
$$pH = 7.00$$

A salt of a strong base–weak acid is present and the $H_3O^+$ is given by the hydrolysis effect

$$[OH^-] = \sqrt{\frac{K_w C_{C_2H_3O_2^-}}{K_a}} \qquad \left([H_3O^+] = \sqrt{\frac{K_a K_w}{C_{C_2H_3O_2^-}}}\right)$$

or

$$[OH^-] = \sqrt{K_b C_{C_2H_3O_2^-} C_{C_2H_3O_2}} \qquad \left([H_3O^+] = \sqrt{\frac{K_w^2}{K_b C_{C_2H_3O_2^-} C_{C_2H_3O_2}}}\right)$$

(See Example 8–7 for a discussion of the approximations)    (Must account for dilution)

$$C_{C_2H_3O_2^-} = \frac{40.0\ \text{ml} \times 0.100\ F}{80.0\ \text{ml}}$$

$$C_{C_2H_3O_2^-} = 0.0500\ F$$

$$[H_3O^+] = \sqrt{\frac{1.76 \times 10^{-5} \times 1 \times 10^{-14}}{0.0500}}$$

$$[H_3O^+] = 1.87 \times 10^{-9}\ M;\ pH = 8.73$$

**50.0 ml (Beyond the stoichiometric point)**

Excess base is being added. Therefore, for any volume past the stoichiometric point the pH is determined by the excess base

| | | |
|---|---|---|
| mmoles of base added: | $50.0\ \text{ml} \times 0.100\ F =$ | $5.00$ mmole |
| mmoles of acid started: | $40.0\ \text{ml} \times 0.100\ F =$ | $4.00$ mmole |
| excess base in: | $90.0$ ml | $1.00$ mmole |

$$C_{OH^-} = \frac{1.00\ \text{mmoles}}{90.0\ \text{ml}} = 1.11 \times 10^{-2}\ F$$

$$[H_3O^+] = 9.01 \times 10^{-13}\ M;\ pH = 12.05$$

For the case of a weak acid titration the hydrolysis due to the presence of the salt of the weak acid–strong base is negligible.

dilution effect will be more noticeable as the acids become weaker. For acetic acid 0.001 $F$ is the limit.

At the midpoint for the titration of $HC_2H_3O_2$ half of the acid has been neutralized and therefore

$$C_{C_2H_3O_2^-} = C_{HC_2H_3O_2}$$

which yields the relationship

$$K_a = [H_3O^+]$$

and

$$pK_a = pH$$

This is very useful since a rapid, moderately accurate determination of the ionization constant of a weak acid is possible by measuring the pH at the midpoint of the titration. As the strength of the acid increases or decreases from the value $K_a = 10^{-5}$, the accuracy for the determination of $K_a$ by this technique falls off significantly. However, by writing a more exact expression it is possible to obtain an accurate $K_a$ even for these stronger or weaker acids.

**Strong Base or Weak Base–Strong Acid Titration.**    A titration of a strong base or a weak base with a strong acid is analogous to the strong acid or weak acid titration. The main differences are that the direction of the pH change is reversed and that the behavior of the weak base is influenced by the lack of complete ionization.

If the primary goal is analysis, the limits for an accurate determination of weak bases are $K_b = 10^{-8}$ and a concentration level of 0.001 $F$. The reasons for these limits are essentially the same as those presented in the section on strong acid or weak acid titrations.

A summary of the two individual reactions taking place in the titrations are as follows:

Strong base–strong acid

$$\text{HCl} + \text{NaOH} \xrightarrow{\text{neutralization}} \text{NaCl} + \text{H}_2\text{O}$$
$$+$$
$$\text{H}_2\text{O}$$
$$\Updownarrow$$
$$\text{H}_3\text{O}^+ + \text{OH}^-$$

Weak base–strong acid

$$\text{NH}_4\text{OH} + \text{HCl} \underset{\text{hydrolysis}}{\overset{\text{neutralization}}{\rightleftharpoons}} \text{NH}_4\text{Cl} + \text{H}_2\text{O}$$

$$\text{H}_2\text{O} \updownarrow \qquad\qquad\qquad + $$
$$\text{H}_2\text{O}$$
$$\Updownarrow$$
$$\text{NH}_4^+ + \text{OH}^- \qquad\qquad \text{H}_3\text{O}^+ + \text{OH}^-$$

A strong base–strong acid titration is calculated by determining whether the acid or base is in excess and at what concentration level. At the stoichiometric point the solution is neutral; therefore, the pH is 7.

For a weak base–strong acid titration the effect of ionization must be considered. Therefore, the initial pH is given by the ionization expression for the weak base, during the titration a buffer system is present, at the stoichiometric point the solution is acidic due to the presence of the neutralization product which is a conjugate acid of moderate strength (hydrolysis), and after the stoichiometric point excess strong acid is added. A flow sheet on pages 162 and 163 provides the equations for calculating the two types of titration curves.

At the midpoint (or 50% neutralized) of the titration of a weak acid the concentrations in this buffer region are

$$C_{NH_4^+} = C_{NH_4OH}$$

Therefore,

$$K_b = [OH^-]; \qquad pK_b = pOH$$

or

$$K_b = \frac{K_w}{[H_3O^+]}; \quad pH = pK_w - pK_b$$

Thus, a good semiquantitative determination of the $pK_b$ is possible by examination of the titration curve. The accuracy limits to a $pK_b$ determination by this method is analogous to the weak acid system.

Figure 8–3 illustrates the calculated titration curves for the titration of a strong base and a weak base. The effect of $K_b$ is also illustrated. It should be noted that the conclusions regarding these systems, strong base–strong acid and weak base–strong acid, are analogous to the titrations of' acids with strong bases. The main difference is the direction of the pH change.

### Weak Acid–Weak Base Titration.

Titration of a weak base with a weak acid or vice versa does not yield a useful pH–titration curve from the standpoint of analysis. Of course, the acid–base reaction still takes place but the pH change during the titration is strongly influenced by hydrolysis and ionization. The various equilibrium steps can be summarized as follows:

$$
\begin{array}{ccc}
& \text{neutralization} & \\
HC_2H_3O_2 + NH_4OH & \rightleftharpoons & NH_4C_2H_3O_2 + H_2O \\
+ \qquad\qquad + & \text{hydrolysis} & + \\
H_2O \qquad\quad H_2O & & H_2O \\
\updownarrow \qquad\quad \updownarrow & & \updownarrow \\
H_3O^+ + C_2H_3O_2^- \quad NH_4^+ + OH^- & & H_3O^+ + OH^-
\end{array}
$$

Qualitatively, the shape of the titration of $HC_2H_3O_2$ with $NH_4OH$ can be

# Flow Sheet for the Calculation of Titration Curves for a Strong Base or Weak Base with a Strong Acid Titrant

Assume that for each titration 40.0 ml of 0.100 $F$ NaOH and NH₃ ($K_b$ 1.79 × 10⁻⁵) are titrated with 0.100 $F$ HCl.

| HCl (ml) | Titration of NaOH | Titration of NH₃ |
|---|---|---|
| 0 ml | $C_{NaOH} = C_{OH} = 0.100\ F$ <br> $pOH = -\log[OH^-]$ <br> $pOH = -\log 1.00 \times 10^{-1}$ <br> $pOH = 1.00;\ pH = 13.00$ | $[OH^-] = \sqrt{K_b C_{NH_3}}$ <br> or <br> $[H_3O^+] = \sqrt{\dfrac{K_w^2}{K_b C_{NH_3}}}$ <br> $[H_3O^+] = \sqrt{\dfrac{(1.00 \times 10^{-14})^2}{1.79 \times 10^{-5} \times 0.100}}$ <br> $[H_3O^+] = 7.48 \times 10^{-12}\ M;\ pH = 11.13$ <br> (See Example 8-6 for a discussion of approximations) |
| Any volume after and before the stoichiometric point volume; as an example consider 10.0 ml (25% neutralized) | mmoles of initial base: $40.0\ \text{ml} \times 0.100\ F = 4.00$ mmole <br> mmoles of acid added: $10.0\ \text{ml} \times 0.100\ F = 1.00$ mmole <br> amount of acid left in: 50.0 ml = 3.00 mmole <br><br> $[OH^-] = \dfrac{\text{mmoles acid}}{\text{total volume, ml}}$ <br> $[OH^-] = \dfrac{3.00}{50.0} = 0.0600\ M$ <br> $pOH = 1.22;\ pH = 12.78$ | Calculate the mmoles of base left as in the strong base case. Since the solution contains a weak base and its salt, it is a buffer. Therefore, <br> $[OH^-] = K_b \dfrac{C_{NH_3}}{C_{NH_4^+}}$ or $[H_3O^+] = \dfrac{K_w}{K_b} \dfrac{C_{NH_4^+}}{C_{NH_3}}$ <br> and <br> $pH = pK_{NH_4^+} + \log \dfrac{\text{Fraction unneutralized}}{\text{Fraction neutralized}}$ <br> or <br> $pH = pK_w - pK_b + \log \dfrac{\text{Fraction unneutralized}}{\text{Fraction neutralized}}$ <br> (See Example 8-11 for the discussion of approximations) <br> $C_{NH_4^+} = \dfrac{1.0\ \text{mmoles}}{50.1\ \text{ml}} = 0.0200\ F$ |

$$C_{NH_3} = \frac{\text{mmoles}}{50.0 \text{ ml}} = 0.0600 \ F$$

$$[H_3O^+] = \frac{1.00 \times 10^{-14} \times 0.0200}{1.79 \times 10^{-5} \times 0.0600}$$

$$[H_3O^+] = 1.86 \times 10^{-10} \ M; \quad pH = 9.73$$

**40.0 ml (stoichiometric point)**

At the equivalence point a salt of a stronger base–weak acid is present. Therefore, the $H_3O^+$ is given by the hydrolysis effect and

$$[H_3O^+] = \sqrt{\frac{K_w}{K_b} C_{NH_4^+}}; \quad [H_3O^+] = \sqrt{K_{NH_4^+} C_{NH_4^+}}$$

(See Example 8-8 for the discussion of approximations)

$$C_{NH_4^+} = \frac{40.0 \text{ ml} \times 0.100 \ F}{80.00 \text{ ml}} = 0.0500 \ F$$

$$[H_3O^+] = \sqrt{\frac{1.00 \times 10^{-14} \times 0.0500}{1.79 \times 10^{-5}}}$$

$$[H_3O^+] = 5.29 \times 10^{-6} \ M; \quad pH = 5.28$$

Since no acid or base is present
$$[H_3O^+] = [OH^-] = 1.00 \times 10^{-7} \ M$$
$$pH = 7.00$$

**50.0 ml (beyond the stoichiometric point)**

Excess acid is being added. Therefore, for any volume past the stoichiometric point the pH is determined by the excess acid

mmoles of acid added: $50.0 \text{ ml} \times 0.10 \ F = 5.00$ mmole
mmoles of base started: $40.0 \text{ ml} \times 0.100 \ F = 4.00$ mmole
excess acid in: $90.0 \text{ ml} = 1.00$ mmole

$$C_{H_3O^+} = \frac{1.00 \text{ mmoles}}{90.0 \text{ ml}} = 1.11 \times 10^{-2} \ F$$

$$[H_3O^+] = 1.11 \times 10^{-2} \ M; \quad pH = 1.96$$

For the case of weak base titration the hydrolysis due to the presence of the salt of the weak base–strong acid is negligible.

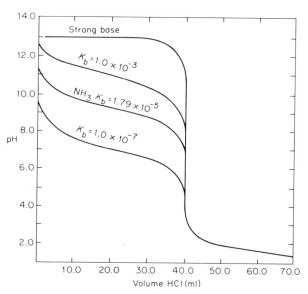

**Fig. 8–3.** Titration curves illustrating the effect of $K_b$. (Calculated curves for the titration of 40.0 ml of 0.100 $F$ weak base with 0.100 $F$ HCl.)

described by considering only the points at 0, 50, 100, 150, and 200% neutralized.

The pH at 0% neutralized is the pH of a solution of a weak acid. If a 0.100 $F$ $HC_2H_3O_2$ solution is used, the pH is 2.88. At 50% neutralization using 0.100 $F$ $NH_3$ titrant

$$pH = pK_a = 4.75$$

Actually, the pH will be slightly higher than this value since the solution contains a weak acid and a salt of a weak acid–weak base which undergoes hydrolysis. The calculations, however, assume hydrolysis is negligible.

The pH at 100% neutralized is calculated from Eq. (8–31)

$$[H_3O^+] = \sqrt{\frac{10^{-14} \times 1.76 \times 10^{-5}}{1.79 \times 10^{-5}}} = 9.92 \times 10^{-8}\ M$$

$$pH = 7.01$$

The solution after the stoichiometric point is a buffer composed of $NH_4C_2H_3O_2$ and $NH_4OH$. If dilution is ignored, an approximate pH at 150% neutralized can be calculated by substitution into the $K_b$ expression for a buffer (see Example 8–11) where

$$\frac{C_{NH_4^+}}{C_{NH_4OH}} \cong \frac{[NH_4^+]}{[NH_4OH]} \cong \frac{2}{1}$$

and, thus,

$$[OH^-] = 8.95 \times 10^{-6}\,M; \quad pH = 8.95$$

At 200% neutralized the ratio

$$\frac{C_{NH_4^+}}{C_{NH_4OH}} \cong \frac{[NH_4^+]}{[NH_4OH]} \cong \frac{1}{1}$$

and

$$K_b = [OH^-] = 1.79 \times 10^{-5}\,M; \quad pH = 9.25$$

An assumption in this calculation is that the hydrolysis of acetate ion does not contribute to the pH. However, this must be considered for a precise calculation.

A plot of the approximate pH values against percent neutralization was shown in Fig. 8–2. The pH break at the stoichiometric point is not well defined and, therefore, an accurate end point can not be detected by ordinary means.

End-point detection by ordinary means in this case implies potentiometry or the use of indicators. Actually, end points for weak acid–weak base titrations are readily and accurately determined by measuring change in conductance, heat change, or change in absorption of radiant energy. Also, by appropriately choosing a solvent other than water many weak acid–weak base titrations can be followed by potentiometry or indicators. Consequently, it is possible to use a weak acid–weak base titration for analysis.

## POLYFUNCTIONAL ACIDS AND BASES

Many acids and bases are capable of furnishing more than one hydronium or hydroxide ion per molecule and are called polyprotic acids or polyfunctional bases. The removal of each proton or hydroxide ion, if the substance is a weak acid or base, respectively, constitutes a separate equilibrium step with its corresponding equilibrium constant. Consequently, any precise calculations dealing with these kinds of systems must include a consideration of all ionization steps. Whether approximations are possible will be determined by the magnitude of each ionization constant, the difference between the stepwise constants, and the purpose for doing the calculation. In general, reasonable answers can often be obtained if the polyprotic systems are considered to be a series of individual steps. Hence, each system is examined and the main equilibrium step is identified and used for the calculation. As the stepwise constants approach each other in value, the error in the calculation becomes larger. Thus, it should be recognized that the answers are not exact and the degree of error can be estimated by inserting the approximate answers into equations which are a more exact description of the multistep equilibrium system.*

A typical weak acid is $H_3PO_4$ and the ionization steps and constants are given by

$$H_3PO_4 + H_2O \rightleftharpoons H_3O^+ + H_2PO_4^-$$

$$H_2PO_4^- + H_2O \rightleftharpoons H_3O^+ + HPO_4^{2-}$$

$$\underline{HPO_4^{2-} + H_2O \rightleftharpoons H_3O^+ + PO_4^{3-}}$$

$$H_3PO_4 + 3H_2O \rightleftharpoons 3H_3O^+ + PO_4^{3-}$$

$$K_{a_1} = \frac{[H_3O^+][H_2PO_4^-]}{[H_3PO_4]} = 7.5 \times 10^{-3}$$

$$K_{a_2} = \frac{[H_3O^+][HPO_4^{2-}]}{[H_2PO_4^-]} = 6.2 \times 10^{-8}$$

$$K_{a_3} = \frac{[H_3O^+][PO_4^{3-}]}{[HPO_4^{2-}]} = 4.8 \times 10^{-13}$$

$$K = K_{a_1}K_{a_2}K_{a_3} = \frac{[H_3O^+]^3[PO_4^{3-}]}{[H_3PO_4]}$$

If the $K_a$'s or $K_b$'s differ by at least a factor of $10^{-3}$ and are not less than $10^{-8}$ to $10^{-9}$, a pH break in the titration curve should be expected for each ionization step. Hence, the calculations for this system are partially simplified since it can be treated as a mixture of three weak acids with the last one being too weak to titrate. It should also be noticed that in the neutralization of $H_3PO_4$ three conjugate bases are produced stepwise with the order of base strength being $PO_4^{3-} > HPO_2^{2-} > H_2PO_3^-$. The two protonated forms are amphoteric and this property will influence the pH of solutions of these salts.

Before proceeding into the calculation of the titration curve, consider the calculation of the pH for solutions of salts of a polyprotic acid.

*Example 8-15.*     Calculate the pH for a 0.100 $F$ solution of $NaH_2PO_4$.

Only the first two equilibrium steps must be considered since further ionization

---

* J. N. Butler, "Ionic Equilibrium A Mathematical Approach," Addison-Wesley, Reading, Massachusetts, 1964.

A. J. Bard, "Chemical Equilibrium," Harper & Row, New York, 1966.

H. Freiser and Q. Fernando, "Ionic Equilibria in Analytical Chemistry," Wiley, New York, 1963.

to $PO_4^{3-}$ is negligible. From the second ionization step

$$[H_3O^+] = [HPO_4^{2-}]$$

However, some of the $H_3O^+$ combines with $H_2PO_4^-$ to form $H_3PO_4$. Therefore, the correct equation is

$$[H_3PO_4] + [H_3O^+] = [HPO_4^{2-}]*$$

Substitution of $K_{a_1}$ and $K_{a_2}$ and rearrangement leads to

$$\frac{[H_3O^+][H_2PO_4^-]}{K_{a_1}} + [H_3O^+] = \frac{K_{a_2}[H_2PO_4^-]}{[H_3O^+]}$$

$$[H_3O^+] = \sqrt{\frac{K_{a_1}K_{a_2}[H_2PO_4^-]}{K_{a_1} + [H_2PO_4^-]}}$$

If $[H_2PO_4^-]$ is large in comparison to $K_{a_1}$, the equation simplifies to

$$[H_3O^+] = \sqrt{K_{a_1}K_{a_2}} \qquad (8\text{-}40)$$

Therefore, the pH for a $NaH_2PO_4$ solution is independent of concentration, provided the above condition is fulfilled and

$$[H_3O^+] = \sqrt{7.5 \times 10^{-3} \times 6.2 \times 10^{-8}} = 2.15 \times 10^{-5}\ M; \quad pH = 4.67$$

---

* Complicated equilibrium problems are often calculated, particularly if exact calculations are desired, through the use of mass balances, charge balances, and proton condition if the equilibrium includes an acid–base system. A mass balance is a mathematical relationship which states that a given atom or group of atoms remains constant throughout the chemical reaction. A charge balance is a electroneutrality relation and is obtained by counting the number of positive charges per unit volume and setting it equal to the number of negative charges per unit volume. The proton condition is obtained by writing a mass balance on $H^+$, a mass balance on $OH^-$, a proton balance, or from the charge balance.

In Example 8–15 the charge balance, and mass balances are

$$[H_3O^+] + [Na^+] = [OH^-] + [H_2PO_4^-] + 2[HPO_4^{2-}] + 3[PO_4^{3-}] \text{ (charge balance)}$$

$$[H_3PO_4] + [H_2PO_4^-] + [HPO_4^{2-}] + [PO_4^{3-}] = 0.10 \text{ (mass balance on P)}$$

$$[Na^+] = 0.10 \text{ (mass balance on Na)}$$

The proton condition is obtained by combining the three equations eliminating $[H_2PO_4^-]$ and hence,

$$[H_3O^+] + [H_3PO_4] = [OH^-] + [HPO_4^{2-}] + 2[PO_4^{3-}],$$

Since the solution must be acidic, the concentration of $[OH^-]$ and $[PO_4^{3-}]$ must be negligible in comparison to the others. This leads to

$$[H_3PO_4] + [H_3O^+] = [HPO_4^{2-}]$$

For a complete discussion of this exact mathematical approach to equilibrium calculations see the references on page 166.

*Example 8–16.*     Calculate the pH for a 0.100F solution of $Na_2HPO_4$. This is analogous to Example 8–14 except that $K_{a_2}$ and $K_{a_3}$ must be considered. Following Example 8–14 the $H_3O^+$ is given by

$$[H_3O^+] = \sqrt{K_{a_2}K_{a_3}} \tag{8-41}$$

provided $[HPO_4{}^{2-}]$ is large in comparison to $K_{a_2}$. Therefore,

$$[H_3O^+] = \sqrt{6.2 \times 10^{-8} \times 4.8 \times 10^{-13}} = 1.40 \times 10^{-10} \, M; \quad pH = 9.85$$

Now consider the titration of 30.0 ml of 0.100 $F$ $H_3PO_4$ with 0.100 $F$ NaOH. On the basis of the $K_a$'s two breaks in the titration curve should be expected. The stoichiometry for the reaction corresponding to the first break is

$$H_3PO_4 + NaOH \rightarrow NaH_2PO_4 + H_2O \tag{8-42}$$

and for the second break is

$$NaH_2PO_4 + NaOH \rightarrow Na_2HPO_4 + H_2O \tag{8-43}$$

The pH at the start of the titration is determined by the first ionization. Thus, the calculation is reduced to the case of a single monoprotic acid.

$$[H_3O^+] \cong [H_2PO_4{}^-]$$

$$C_{H_3PO_4} = 0.100 \cong [H_3PO_4]$$

Substitution into the $K_{a_1}$ expression gives

$$7.5 \times 10^{-3} = \frac{[H_3O^+][H_3O^+]}{(0.100)}$$

$$[H_3O^+] = 2.75 \times 10^{-2} \, M; \quad pH = 1.56$$

Addition of 15.0 ml of titrant will convert half of the $H_3PO_4$ to $NaH_2PO_4$ if it is assumed that the other ionization steps are not involved. Thus,

$$[H_3PO_4] \cong [H_2PO_4{}^-]*$$

Substitution into the $K_{a_1}$ expression gives

$$[H_3O^+] = K_{a_1} = 7.5 \times 10^{-3}; \quad pH = 2.12$$

This is the midpoint for the first pH break. Calculation of other points before reaching the first equivalence point would require calculation of the amounts of $H_3PO_4$ and $H_2PO_4{}^-$ in solution just as if it were a monoprotic weak acid. Also, this part of the titration curve is a buffer region.

At the first stoichiometric point, 30.0 ml of titrant added, the solution con-

---

* A more exact expression must consider the ionization of $H_3PO_4$ and hydrolysis of $H_2PO_4{}^-$.

tains $NaH_2PO_4$. Therefore, its pH is calculated, as in Example 8–15, by

$$[H_3O^+] = \sqrt{K_{a_1}K_{a_2}} = 2.15 \times 10^{-5}; \quad pH = 4.67$$

Continued addition of titrant involves neutralization according to reaction (8–43). The calculation is simplified since only two species, $H_2PO_4^-$ and $HPO_4^{2-}$, are involved. Therefore, the system is a buffer and its $H_3O^+$ concentration is given by the equilibrium constant expression for $K_{a_2}$. If 45.0 ml of 0.100 $F$ NaOH is added, the midpoint of the second break is reached and

$$[H_2PO_4^-] \cong [HPO_4^{2-}] *$$

Substitution into the $K_{a_2}$ expression gives

$$[H_3O^+] = K_{a_2} = 6.2 \times 10^{-8}; \quad pH = 7.21$$

The pH values at other volumes of titrant after the first and before the second break would be calculated just as if the system were a monoprotic acid. Thus, concentrations of $H_2PO_4^-$ and $HPO_4^{2-}$ present would be determined by the amount of base added.

At 60.0 ml the stoichiometric point for the second neutralization step is reached and the solution contains as the major species $HPO_4^{2-}$. If it is assumed that this is the only species, the expression developed in Example 8–16 can be used:

$$[H_3O^+] = \sqrt{K_{a_2}K_{a_3}} = 1.72 \times 10^{-10} \, M; \quad pH = 9.76$$

A third break is not observed because the pH change is gradual as further neutralization takes place. ($K_{a_3}$ is very small.) Consequently, it can be assumed, if dilution is ignored, that the pH will gradually approach a pH of 13. A plot of the calculated pH vs ml of titrant is shown in Fig. 8–4.

Other polyprotic acids can be treated in the same way. Whether approximations can be made or not will be determined by the $K_a$ values and the concentration of the acid. The pH calculations for the titration of polyfunctional bases with strong acid are analogous except that the equations are derived in terms of $K_b$.

**Molecules with Both Acidic and Basic Properties.** There are several different systems which contain both acidic and basic groups. Many of these are organic molecules and are of biological and pharmaceutical interest. Two typical examples are amino acids and sulfonamides. Glycine (**I**), the simplest amino acid, and sulfanilanilide (**II**) are shown below.

**I**                **II**

---

* A more exact expression must consider the dissociation and hydrolysis of each species.

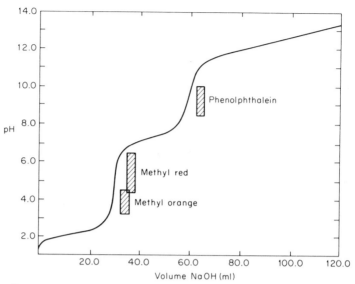

**Fig. 8–4.** Calculated titration curve for the titration of 30.0 ml of 0.100 $F$ $H_3PO_4$ with 0.100 $F$ NaOH.

The procedure for discussing the equilibrium steps and for making calculations of pH for solutions of these compounds is exactly the same as for polyprotic acids. Using glycine as an example, the ionization steps are as follows:

$$^+H_3NCH_2CO_2H + H_2O \rightleftharpoons H_3O^+ + H_2NCH_2CO_2H; \quad K_{a_1} = \frac{[H_3O^+][HG]}{[H_2G^+]}$$
$$\underline{H_2G^+} \qquad\qquad\qquad\qquad\qquad \underline{HG}$$

$$H_2NCH_2CO_2H + H_2O \rightleftharpoons H_3O^+ + H_2NCH_2CO_2^-; \quad K_{a_2} = \frac{[H_3O^+][G^-]}{[HG]}$$
$$\underline{HG} \qquad\qquad\qquad\qquad\qquad \underline{G^-}$$

Hence, the equilibrium constants are tabulated as successive ionization constants of the most highly protonated form and the system is treated as a polyprotic acid.

It should be noted that compounds like glycine, which have acidic and basic properties, will exist at some particular pH value in the neutral form. However, it is observed that their solutions possess ionic character. This results because of internal ionization and systems of this type are called *zwitterions*. For glycine the neutral and internal ionized forms are

*Example 8–17.*     Calculate the pH of a 0.0100 $F$ phenylalanine solution. Phenylalanine is an essential amino acid in human nutrition, and is not synthesized by the human body. Recommended intake for a normal male is about 2.2 g/day.

The equilibria are

$$K_{a_1} = \frac{[H_3O^+][HP]}{[H_2P^+]} = 2.63 \times 10^{-3}$$

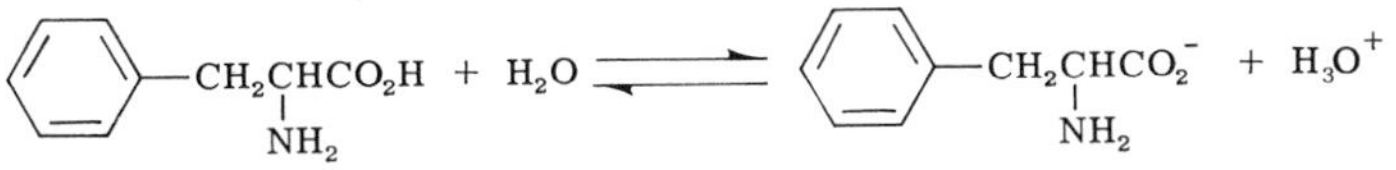

$$K_{a_2} = \frac{[H_3O^+][P^-]}{[HP]} = 5.76 \times 10^{-10}$$

The main phenylalanine species must be HP because of the magnitude of $K_{a_1}$ and $K_{a_2}$. Therefore, as an approximation

$$[HP] \cong C_{HP} = 0.0100 \; M$$

Some of the HP will dissociate to form $P^-$ and some will hydrolyze to form $H_2P^+$; hence, the $H_3O^+$ concentration must be

$$[H_3O^+] = [P^-] - [H_2P^+]$$

This equation includes an approximation and that is that the contribution of $[H_3O^+]$ and $[OH^-]$ from the ionization of water is insignificant. To consider this, $[OH^-]$, which is equal to $K_w/[H_3O^+]$, would be added to right side of the equation. Substitution for $[P^-]$ and $[H_2P]$ by the $K_{a_1}$ and $K_{a_2}$ expressions gives

$$[H_3O^+] = \frac{K_{a_2}[HP]}{[H_3O^+]} - \frac{[H_3O^+][HP]}{K_{a_1}}$$

which upon rearrangement gives

$$K_{a_1}[H_3O^+]^2 = K_{a_1}K_{a_2}[HP] - [H_3O^+]^2[HP]$$

$$[H_3O^+]^2 = \frac{K_{a_1}K_{a_2}[HP]}{K_{a_1} + HP}$$

Inserting the values

$$[H_3O^+]^2 = \frac{(2.63 \times 10^{-3})(5.76 \times 10^{-10})(1.00 \times 10^{-2})}{2.63 \times 10^{-3} + 1.00 \times 10^{-2}} = 4.17 \times 10^{-13}$$

$$[H_3O^+] = 6.46 \times 10^{-7}; \; pH = 6.19$$

## TITRATION OF SALTS

Many salts of weak acids and bases including polyprotic and polyfunctional acids and bases can be quantitatively titrated. The number of breaks in the titration curve will depend on the number of $K_a$'s or $K_b$'s, their magnitude, and their difference. Table 8–5 lists several salts that can be titrated.

A typical example is the titration of $Na_2CO_3$ with standard HCl titrant. The calculation of this titration curve is similar to the previous calculations since carbonate ion is the conjugate base of carbonic acid.* Therefore, the titration curve can be divided into several segments. The initial pH for a carbonate solution is due to the basicity of the conjugate base (hydrolysis). Subsequently, as the HCl titrant is added the titration curve passes through a buffer region composed of the ratio $CO_3^{2-}/HCO_3^-$, a stoichiometric point corresponding to the formation of $HCO_3^-$, a buffer region composed of the ratio $HCO_3^-/H_2CO_3$, and a second stoichiometric point corresponding to the formation of $H_2CO_3$. This is exactly the reverse of the titration of $H_2CO_3$ with NaOH.

Assume that 20.0 ml of 0.100 $F$ $Na_2CO_3$ is titrated with 0.100 $F$ HCl. The ionization expressions are

$$H_2CO_3 + H_2O \rightleftharpoons H_3O^+ + HCO_3^-$$

$$K_{a_1} = \frac{[H_3O^+][HCO_3^-]}{[H_2CO_3]} = 4.47 \times 10^{-7}$$

$$HCO_3^- + H_2O \rightleftharpoons H_3O^+ + CO_3^{2-}$$

$$K_{a_2} = \frac{[H_3O^+][CO_3^{2-}]}{[HCO_3^-]} = 4.68 \times 10^{-11}$$

The initial pH is given by the hydrolysis of $Na_2CO_3$:

$$CO_3^{2-} + H_2O \rightleftharpoons HCO_3^- + OH^-$$

If hydrolysis of $HCO_3^-$ is negligible, the system can be treated as a salt of a weak monoprotic acid (see Example 8–7). Therefore,

$$C_{CO_3^{2-}} = 0.100 \ F$$

$$[H_3O^+] = \sqrt{\frac{K_w K_{a_2}}{C_{CO_3^{2-}}}} = \sqrt{\frac{1 \times 10^{-14} \times 4.68 \times 10^{-11}}{0.100}}$$

$$[H_3O^+] = 2.16 \times 10^{-2} \ M; \quad pH = 11.67$$

---

* Carbonic acid, $H_2CO_3$, does not exist in this form and is actually $CO_2$ dissolved in water. Whenever $H_2CO_3$ is used in this text it actually refers to $CO_2/H_2O$.

Table 8–5.   A Partial List of Acidic and Basic Salts That Can
Be Quantitatively Titrated

---

Basic Salts[a]

| $Na_2CO_3$ | $NaH_2PO_4$ | NaCN |
|---|---|---|
| $NaHCO_3$ | $NaBO_2$ | $Na_3AsO_4$ |
| $Na_2HPO_4$ | | |

Acidic Salts[a]

Salts of the type $RNH^+ X^-$ where $X^-$ is an anion of a strong acid
Pyridinium hydrochloride
Anilinium hydrochloride

---

[a] The general requirement is that the hydrolysis constant must
be $10^{-7}$ or less. For basic salts, $K_H = K_w/K_a < 10^{-7}$; and for acidic
salts, $K_H = K_w/K_b < 10^{-7}$. Therefore, $K_a$ or $K_b$ must be less than
$10^{-7}$.

As the titrant is added, neutralization takes place stepwise.

$$CO_3^{2-} + H_3O^+ \rightarrow HCO_3^- + H_2O \qquad (8\text{-}44)$$

Thus, if hydrolysis and ionization steps are negligible the pH in the buffer
region up to the first stoichiometric point is determined by the ratio of $C_{HCO_3^-}$
to $C_{CO_3^{2-}}$ and $K_{a_2}$. For 10.0 ml of added titrant

$$C_{HCO_3^-} = C_{CO_3^{2-}}$$

and

$$[H_3O^+] = K_{a_2} = 4.68 \times 10^{-11}; \quad pH = 10.33$$

When 20.0 ml of the HCl titrant is added, the first stoichiometric point is
reached. The product of the reaction is the conjugate base, $HCO_3^-$, which has
both acidic and basic properties (amphoteric). Therefore, it is involved in
several equilibrium steps:

$$HCO_3^- + H_2O \rightleftharpoons H_2CO_3 + OH^- \quad \text{(Hydrolysis)}$$

$$HCO_3^- + H_2O \rightleftharpoons H_3O^+ + CO_3^{2-} \quad \text{(Ionization)}$$

Whether the pH of the equilibrium condition is acidic or basic will depend on
the relative magnitude of the equilibrium constants for these processes.

A rigorous expression for the hydronium ion concentration of such a solution must consider these processes. However, an approximate calculation can
be made by using Eq. (8-40). For example, the $H_3O^+$ is calculated to be

$$[H_3O^+] = \sqrt{K_{a_1}K_{a_2}} = \sqrt{4.47 \times 10^{-7} \times 4.68 \times 10^{-11}} = 4.57 \times 10^{-9} \ M$$

and the pH is 8.34.

Continued addition of HCl will lead to the second neutralization step.

$$HCO_3^- + H_3O^+ \rightarrow H_2CO_3 + H_2O \tag{8-45}$$

If 30.0 ml of acid are added, the midpoint of the above reaction has been reached and the hydronium ion concentration during this buffer portion of the curve is determined by the ratio of $C_{H_2CO_3}$ to $C_{HCO_3^-}$. Therefore,

$$C_{H_2CO_3} = C_{HCO_3^-}$$

and

$$[H_3O^+] = K_{a_1} = 4.47 \times 10^{-7} \, M; \quad pH = 6.35$$

The stoichiometric point for the second neutralization step is reached when 40.0 ml of HCl is added to the solution. Since a weak acid is the product of the reaction, the hydronium ion concentration can be calculated by using the expressions describing a solution of a weak acid.

As an approximation assume that the second step is negligible with respect to the first. Thus, $H_2CO_3$ is treated as a monoprotic acid and the hydronium ion concentration can be calculated from the $K_{a_1}$ expression. (See Example 8-4.) The concentration of $H_2CO_3$ is calculated considering the dilution that takes place. Since the $Na_2CO_3$ is converted to $H_2CO_3$, the concentration of $H_2CO_3$ at the stoichiometric point is given by

$$C_{H_2CO_3} = \frac{20.0 \text{ ml} \times 0.100 \, F}{20.0 \text{ ml} + 40.0 \text{ ml}} = 3.33 \times 10^{-2} \, F$$

Thus,

$$[H_3O^+] = \sqrt{K_{a_1}C_{H_2CO_3}} = \sqrt{4.47 \times 10^{-7} \times 3.33 \times 10^{-2}}$$

$$[H_3O^+] = 1.27 \times 10^{-4} \, M; \quad pH = 3.90$$

The pH after the stoichiometric point is determined by the presence of excess strong acid.

These pH values are plotted in Fig. 8-5. In the vicinity of the midpoints for the two breaks, two buffer regions $CO_3^{2-}/HCO_3^-$ and $HCO_3^-/H_2CO_3$, respectively, are present. Therefore, little change in pH occurs in these segments of the titration curve.

The first stoichiometric point can be detected by using phenolphthalein as indicator. For the second, methyl orange (pH 3.2–4.4) is often used. However, methyl red can also be used. From Table 8-8 it can be seen that methyl red changes color in the range of pH 4.8–6.0. It would appear that if this indicator is used the color change would occur prematurely with respect to the equivalence point. Experimentally, the technique is to titrate to the color change and then boil the solution for a minute or two. Boiling will drive out the $H_2CO_3$ (dissolved $CO_2$) and the pH changes toward a neutral value. Thus, the color goes back to the basic side. Continued titration to the acidic color marks the end point. This pH change with heating is shown as a dotted line in Fig. 8–5.

**Physiological Systems.**    In general, the two physiologically important buffer systems are phosphate and carbonate. The $HCO_3^-/H_2CO_3$ buffer system is physiologically important since it is this system which accounts for the transport of $CO_2$ in blood and the control of blood pH. For a healthy adult the arterial blood pH is remarkably constant at a value of 7.38–7.42 (venous blood pH ranges from 7.36 to 7.40). Medically, any change of as little as $\pm0.05$ pH units is an indication of metabolic acidosis or alkalosis or of respiratory acidosis or alkalosis and can be the result of one of several different diseases. Because of this fact the allowable error for an analytical measurement is very small and a useful clinical pH measurement must be made with a tolerance of $\pm0.02$ pH units. (The clinical determination of blood pH is performed with a glass electrode and saturated calomel electrode; these electrodes are described in Chapter 13.)

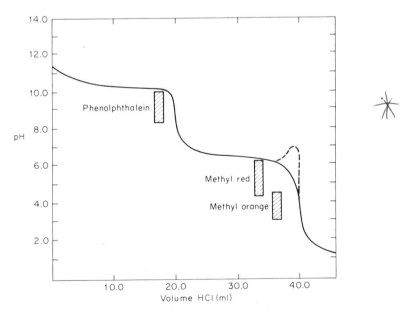

**Fig. 8–5.**   Calculated titration curve for the titration of 20.0 ml of 0.100 $F$ $Na_2CO_3$ with 0.100 $F$ HCl.

The buffer system that contributes to the pH of blood is more complicated than a $HCO_3^-/H_2CO_3$ (dissolved $CO_2$) ratio. In addition to these components, others are hemoglobin/oxyhemoglobin, $H_2PO_4^-/HPO_4^{2-}$, and plasma protein (contains amino acid units which provides both acidic and basic groups).

In general, the clinical determination of blood pH, which is to be used diagnostically, is for the $HCO_3^-/H_2CO_3$ ratio. Since this ratio is the result of the first ionization of $H_2CO_3$, the equilibrium constant expression for $K_{a_1}$ can be rearranged into the form

$$[H_3O^+] = K_{a_1} \frac{[H_2CO_3]}{[HCO_3^-]}$$

and in the Henderson–Hasselbalch form as

$$pH = pK_{a_1} + \log \frac{[HCO_3^-]}{[H_2CO_3]} \tag{8-46}$$

For a blood pH value of 7.40 ($pK_{a_1} = 6.35$), the $[HCO_3^-]/[H_2CO_3]$ ratio is about 11.2. It should be noted, however, that $H_2CO_3$ is dissolved $CO_2$ and the total volume of the gas in the blood is directly proportional to the partial pressure of the gas in the blood and its solubility coefficient. The solubility coefficient of $CO_2$ in blood plasma at 38°C and a $CO_2$ pressure of 760 mm Hg is 0.51 ml $CO_2$ under these conditions. Considering the properties of gases, it can be shown that at 38°C

$$pH = 6.35 + \log \frac{C_{HCO_3^-}}{0.03 \, P_{CO_2}} \tag{8-47}$$

where $C_{HCO_3^-}$ is millimolar concentration and $P_{CO_2}$ is the partial pressure of $CO_2$ in the blood. For normal conditions where the blood pH is 7.40, $P_{CO_2}$ is 40 mm Hg.

In the clinical laboratory, methods are available for the determination of pH, bicarbonate (titration with HCl), and $CO_2$. A manometric (gas measurement) method, known as the Van Slyke method, is used for $CO_2$.

*Example 8–18.*     An enzyme-catalyzed reaction was carried out in a solution containing 0.180 $F$ phosphate buffer at pH $= 7.40$. As a result of the reaction, 0.020 moles/liter of $H_3O^+$ was consumed. (1) What were the concentrations of the conjugate base and acid at the start of the reaction? (2) What were they at the end of the reaction? (3) What was the final pH? (4) What would the final pH be if there had been no buffer? The ionization steps and constants are listed on page 166.

(1) At pH 7.40 and consideration of $K_{a_1}$, $K_{a_2}$, and $K_{a_3}$ the two principle species are $HPO_4^{2-}$ and $H_2PO_4^-$. Writing the $K_{a_2}$ expression in a Henderson–Hasselbalch

form gives

$$pH = pK_{a2} + \log \frac{[HPO_4^{2-}]}{[H_2PO_4^-]}$$

$$7.40 = 7.21 + \log \frac{[HPO_4^{2-}]}{[H_2PO_4^-]}$$

$$[HPO_4^{2-}] = 1.55[H_2PO_4^-] \tag{a}$$

It can also be written that

$$[HPO_4^{2-}] + [H_2PO_4^-] = 0.180 \tag{b}$$

Solving (a) and (b) simultaneously gives

$$[H_2PO_4^-] = 0.071 \text{ moles/liter}$$

$$[HPO_4^{2-}] = 0.109 \text{ moles/liter}$$

(2) As a result of the reaction, the concentrations change according to

$$[H_2PO_4^-] = 0.071 \text{ moles/liter} - 0.020 \text{ moles/liter} = 0.051 \text{ moles/liter}$$

$$[HPO_4^{2-}] = 0.109 \text{ moles/liter} + 0.020 \text{ moles/liter} = 0.129 \text{ moles/liter}$$

(3)
$$pH = pK_{a2} + \log \frac{[HPO_4^{2-}]}{[H_2PO_4^-]}$$

$$pH = 7.21 + \log \frac{0.129}{0.051}$$

$$pH = 7.61$$

(4) If no buffer were present, the 0.020 mole/liter $H_3O^+$ consumed would be the same as if 0.020 mole/liter of $OH^-$ were added to the solution. Hence,

$$pOH = -\log[OH^-] = -\log 2.0 \times 10^{-2} = 1.70$$

$$pH = 14 - pOH = 14 - 1.70 = 12.30$$

It should be noted that in this calculation the approximation was made that at pH 7.40 the only phosphate species in solution are $HPO_4^{2-}$ and $H_2PO_4^-$. This means that ionization to form $PO_4^{3-}$ or hydrolysis to form $H_3PO_4$ are neglected.

If this enzyme reaction were part of a living system the slight change in pH as calculated in part (3), in many cases, would lead to a serious disorder. Certainly, the absence of buffer, as calculated in part (4) would be disastrous. Hence, in the living system, as the enzymatic reaction would take place, other reactions would occur to replenish the buffer components and maintain the pH.

## DIFFERENTIAL TITRATION

Frequently, samples encountered in practical situations contain two or more acidic or basic species. If all of the acids or bases are strong enough

($K_a$ or $K_b$ > $10^{-8}$) the total acidic or basic content of the sample can be determined. However, it would be useful to be able to determine the amount of each of the acidic or basic species in the sample. Whether this is possible or not will depend on the strength of the individual acids or bases.

For the titration of $H_3PO_4$ two distinct breaks corresponding to the neutralization of two protons are observed because the $K_{a_1}$ and $K_{a_2}$ are sufficiently different in magnitude. Similarly the basic salt, $Na_2CO_3$, yields two distinct breaks because of the two different ionization constants involved in this system. These two examples, one acidic and the other basic, are examples of a differential titration. In the first case the two acids that are being differentiated are $H_3PO_4$ and $H_2PO_4^-$ while in the second the two bases being differentiated are $Na_2CO_3$ and $HCO_3^-$. Since $H_2PO_4^-$ and $HCO_3^-$ are the products of the first neutralization, respectively, the volume of titrant needed for the second neutralization will be the same as that needed for the first.

What if the sample contained a mixture of $H_3PO_4$ and $H_2PO_4^-$? Upon titration with NaOH two breaks will be obtained but the volume needed for the second must always be greater than that needed for the first. Similarly, if the mixture contained $Na_2CO_3$ and $NaHCO_3$, the volume required for the second break must be greater than what is needed for the first. This is illustrated in Fig. 8–6, curves (a) and (b).

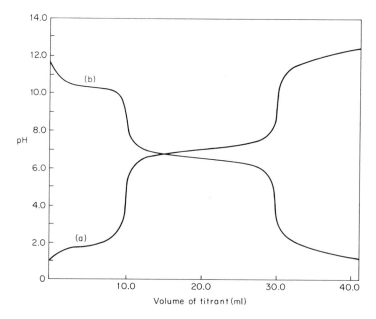

**Fig. 8–6.** Predicted differential titration curves. (a) 10.0 ml of 0.100 $F$ $H_3PO_4$ and 10.0 ml of 0.100 $F$ $NaH_2PO_4$ titrated with 0.100 $F$ NaOH. (b) 10.0 ml of 0.100 $F$ $Na_2CO_3$ and 10.0 ml of 0.100 $F$ $NaHCO_3$ titrated with 0.100 $F$ HCl.

In contrast, samples may contain a mixture of acids or bases that are dissimilar in chemical composition except that they are acidic or basic, respectively. Thus, typical mixtures would be mixtures of strong acids, of strong and weak acids, and of weak acids. Mixtures of bases could be designated in a similar fashion.

In titrating acids the strong basic titrant will seek out the stronger of the acids in the mixture, followed by the neutralization of the second strongest, etc. The completeness of each of these neutralization steps will depend on the difference in strength of the acids. For strong acids no differentiation is possible, at least in water, and thus, the determination of total acidity is only possible. If a strong acid and a weak acid are in the mixture, two breaks are obtained. The first break corresponds to the strong acid and the second to the weak acid. Figure 8–7 illustrates the titration of HCl mixed with a series of successively weaker acids. It should be noted that the weaker the acid the more well-defined the break for the strong acid, HCl.

Whether a mixture of weak acids can be differentiated or not will be determined by their $K_a$ values. In general, if they differ by at least $10^3$, differentiation is possible.

Mixtures of strong bases cannot be differentiated. For strong base–weak base mixtures the strong acidic titrant seeks out the stronger base first and differentiation is possible. As the weak base gets weaker the break for the

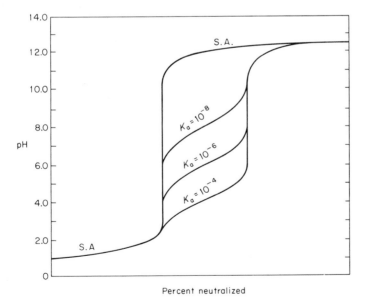

**Fig. 8–7.** Titration curves for HCl and HCl–weak acid mixtures. Effect of $K_a$ on the titration curve for the mixture is illustrated. S.A., strong acid.

stronger base becomes sharper. If in a mixture of weak bases, the bases differ by at least $10^3$ in strength, differentiation is possible.

The ability to differentiate mixtures of acids or bases can be greatly improved by using nonaqueous solvents in place of water. The reasons for this are discussed in the next chapter.

## PRIMARY STANDARDS

Throughout this discussion it has been assumed that standard solutions were available. There are a variety of acidic and basic substances that meet the requirements of a primary standard. Unfortunately only a few of these are strong acids or bases and thus, the acidic and basic primary standards are not often used as titrants. Their main applications are for standardizing acidic and basic titrants, such as HCl and NaOH solutions, and for preparation of buffers.

**Acidic Primary Standards.**    Table 8–6 lists several substances that are acidic primary standards. Of these the first two are routinely used in aqueous solutions for the standardization of basic titrants. The remainder of the acids with the exception of HCl, $H_2SO_4$, and benzoic acid have been recently suggested as standards. The two inorganic acids are prepared as standards according to special distillation procedures while benzoic acid is not very soluble in water.

Of all the acids listed KHP is perhaps the most useful. It meets virtually all of the requirements of a primary standard as outlined previously. Sulfamic acid can be used even though it hydrolyzes in water because the product of the reaction, $HSO_4^-$, is still acidic.

$$H_2NSO_3H + H_2O \rightarrow NH_4HSO_4$$

If hydrolysis takes place with atmospheric water before weighing of the salt, the standardization will be in error.

**Table 8–6.   Common Acidic Primary Standards**[a]

| | |
|---|---|
| 1. Potassium acid phthalate | 7. Potassium hydrogen bis(3,5-dinitrobenzoate) |
| 2. Sulfamic acid | |
| 3. Potassium hydrogen iodate | 8. Benzoic acid |
| 4. Calcium hydrogen malate hexahydrate | 9. Sodium hydrogen diglycolate |
| 5. Di-(m-nitrobenzenesulfonyl)amine | 10. 2,4,6-Trinitrobenzoic acid |
| 6. Cadmium hydrogen ethylenediamine N-hydroxyethyl-N,N',N'-triacetate | 11. Constant boiling HCl |
| | 12. $H_2SO_4$ |

[a] See Table 8–4 for several primary standard salts that are available for buffers.

Table 8–7. Common Basic Primary Standards

1. $Na_2CO_3$ [$CaCO_3$, $Tl_2CO_3$, $NaHCO_3$, potassium bitartrate ($KHC_4H_4O_6$)]
2. Borax ($Na_2B_4O_7 \cdot 10H_2O$)
3. Potassium tetraborate ($K_2B_4O_7 \cdot 4H_2O$)
4. Tris(hydroxymethyl)aminomethane (THAM)
5. 4-Aminopyridine
6. Mercuric oxide

$^a$ See Table 8–4 for several primary standard salts that are available for buffers.

The cadmium salt is an interesting standard for two reasons. First, it has a large molecular weight (388.66). Second and very important, is that it can supply a known amount of cadmium ion and thus serves as a standard for chelometric titrations. The calcium hydrogen malate hexahydrate is also versatile in that it can be used as an acidic, basic, and chelometric standard. It can also be used as a standard for the Karl Fischer titration (titrimetric method for water analysis).

**Basic Primary Standards.**  Table 8–7 contains a list of primary standard basic substances. The first four are the ones that are most often used.

Of all the carbonate salts $Na_2CO_3$ is the most appropriate. The others, although useful, offer no special advantage over $Na_2CO_3$ and in addition, several are of limited solubility. The bitartrate salt after weighing is converted to $Na_2CO_3$ by heating.

Borax and potassium tetraborate are also widely used but care should be exercised because the salts tend to change in hydration. Usually these should be stored in a hygrostat.

THAM is not only useful for standardizing strong acids but also for preparing buffer solutions in the neutral region. This is particularly useful for buffers that are used in biological studies. THAM is often used for standardizing acids in nonaqueous solvents.

## END-POINT DETECTION

The two most common ways of detecting the end point in acid–base titrations are by the use of indicators and by hydronium ion indicating electrodes. The latter technique requires a glass and reference electrode and the potential developed by this cell is measured potentiometrically (see Chapter 13).

The older technique of end point detection is by indicators. Historically, this method also served the purpose of establishing the pH of an unknown solution. In general the most useful types of acid–base indicators are those whose color or fluorescent properties are dependent on the pH of the solution.

**Color Indicators.**    Numerous substances which change color according to the pH of the solution are known. Many of these occur naturally in plants and were recognized historically as substances capable of differentiating between acidic and alkaline solutions.

Typical acid–base indicators are organic molecules, usually of high molecular weights, that possess weakly acidic or basic properties. Thus, they act like any other weak acid or base in water and will ionize in the following way:

$$HIN + H_2O \rightleftharpoons H_3O^+ + IN^- \qquad (8\text{-}48)$$

Acid color                          Base color

$$IN + H_2O \rightleftharpoons OH^- + HIN^+ \qquad (8\text{-}49)$$

Base color                          Acid color

where IN represents an organic structure with an acidic or basic site. To be useful the indicator must have contrasting colors between its acidic and basic form.

Concentration and degree of ionization are the two main factors which determine the color of the indicators. Equilibrium expressions for (8–48) and (8–49) are

$$K_a = \frac{[H_3O^+][IN^-]}{[HIN]} \quad \text{and} \quad K_b = \frac{[HIN^+][OH^-]}{[IN]}$$

Experimental observation has shown that to see the color of one form over the other the concentration of the first should be 10 times the second. Thus, to see the acidic form color $[IN^-]/[HIN] = 1/10$ and to see the basic color $[IN^-]/[HIN] = 10/1$. The contrast of the two colors is also important, but, in general, the tenfold relationship will apply.

If these two concentration ratios are substituted into the equilibrium expression for the indicator the dependency of hydronium ion concentration is demonstrated. For the acid color, the expression simplifies to

$$\frac{[H_3O^+]\,1}{10} = K_a, \quad [H_3O^+] = 10\,K_a, \quad \text{and} \quad pH = pK_a - 1$$

and for the basic color

$$\frac{[H_3O^+]\,10}{1} = K_a, \quad [H_3O^+] = \frac{K_a}{10}, \quad \text{and} \quad pH = pK_a + 1$$

Therefore, the indicator changes color over a 2 pH unit range.

$$pH \text{ range} = pK_a \pm 1 \qquad (8\text{-}50)$$

If the indicator has a $K_a$ of $1 \times 10^{-6}$, the indicator will change color in passing from pH 5 to pH 7. It is also possible for the indicator to have more that one ionization and hence more than one color transition.

These indicators can be used to determine the pH of an unknown solution by comparing the color of the indicator in the unknown solution to the color of the indicator in a series of buffer solutions. For an accurate pH measurement the color should be in the indicator transition stage and the series of the buffer solutions must be of gradual difference in pH values. In the modern laboratory this comparison technique has been replaced by the glass electrode. However, prior to the use of the glass electrode almost all routine hydronium ion concentrations were determined by the indicator method.

There is an indicator for almost any desired pH range. The more common ones are listed in Table 8–8. In choosing an indicator for an acid–base titration, two properties are examined carefully. First, an indicator whose $pK_a$ is the same or very close to the stoichiometric point pH is chosen. Consequently, the indicator transition range coincides with the stoichiometric point pH. Second, the two contrasting colors should be discernible. Often, it is necessary to compromise between these two requirements. Suitable indicators have been suggested in Figs. 8–2, 8–4, and 8–5. Figure 8–8 shows the structure and ionization reactions for several common indicators.

Several other factors, although of lesser importance than $pK_a$ and color contrast, will often affect the color transition. These include temperature, electrolyte concentration, presence of solvents other than water and presence of colloidal particles. Also, it should be apparent that excessive addition of indicator will cause a titration error since the indicator itself is acidic (or basic) and will consume titrant. This error is minimized by keeping the indicator concentrations small and by titrating to matched colors. If the standardization titration and unknown titration are performed to the same indicator color, the indicator error is negligible, provided titrant volumes used are similar. Titrating to the same color also compensates for errors resulting from an indicator transition that occurs just before or just after the stoichiometric point. An alternative way to account for the indicator error is to titrate a solution containing everything but the unknown sample (a blank titration). Thus, the volume of titrant required for the indicator transition is determined and subtracted from future titrations.

Improvement in color transition can be obtained by adding a second colored compound to the indicator solution. Usually, this substance does not undergo a color transition itself but maintains its own color. The advantage of complementary colors is exploited with a mixed indicator solution. For example, the indicator methyl purple is a mixed indicator made from methyl red (red → yellow pH; 4.2 → 6.2) and a blue dye. When mixed, the color change is from purple to green above pH 5.4. There is also an intermediate gray color in the very narrow pH range of 4.8–5.4. Not only is the color change more easily

**Table 8-8. Useful Acid–Base Color Indicators[a]**

| Chemical name | Common name | pH range | $\lambda_{max}$ (nm) | Color change[b] | Preparation |
|---|---|---|---|---|---|
| 2,4,6-Trinitrophenol | Picric acid | 0.6–1.3 | | c–y | 0.04% aq. |
| Thymolsulfonphthalein | Thymol blue | 1.2–2.8 | 544, 430 | r–y | 0.1% alc. |
| 2,4-Dinitrophenol | α-Dinitrophenol | 2.4–4.0 | | c–y | 0.4% aq. |
| Tetrabromophenolsulfonphthalein | Bromophenol blue | 3.0–4.6 | 436, 592 | y–b | 0.1% aq. |
| Dimethylaminoazobenzene-p-sulfonate | Methyl orange | 3.1–4.4 | 522, 464 | r–o | 0.1% aq. |
| Tetrabromo-m-cresolsulfonphthalein | Bromocresol green | 3.8–5.4 | 444, 617 | y–b | 0.1% aq. |
| Dimethylaminoazobenzene-o-carboxylic acid | Methyl red | 4.2–6.3 | 530, 427 | r–y | 0.1% in 60% alc. |
| Dibromo-o-cresolsulfonphthalein | Bromocresol purple | 5.2–6.8 | 433, 591 | y–p | 0.04% aq. |
| Dibromothymolsulfonphthalein | Bromothymol blue | 6.2–7.6 | 433, 617 | y–b | 0.5% aq. |
| Phenolsulfonphthalein | Phenol red | 6.8–8.4 | 433, 558 | y–r | 0.05% aq. |
| o-Cresolsulfonphthalein | Cresol red | 7.2–8.8 | 434, 572 | y–r | 0.05% aq. |
| Thymolsulfonphthalein | Thymol blue | 8.0–9.6 | 430, 596 | y–b | 0.04% aq. |
| Di-p-dioxydiphenylphthalide | Phenolphthalein | 8.3–10.0 | 553 | c–p | 0.05% in 50% alc. |
| Dithymolphthalide | Thymolphthalein | 9.3–10.5 | 598 | c–b | 0.04% in 50% alc. |
| m-Nitrobenzeneazosalicylic acid | Alizarin yellow GG | 10.0–12.0 | | c–y | 0.1% alc. |
| 2,4,6-Trinitrophenolmethylnitramine | Nitramine | 10.8–13.0 | | c–o | 0.01% aq. |

[a] From "Handbook of Analytical Chemistry" (L. Meites, ed.), 1st ed. McGraw-Hill, New York, 1963.
[b] c, colorless; p, purple; b, blue; o, orange; r, red; g, green; y, yellow.

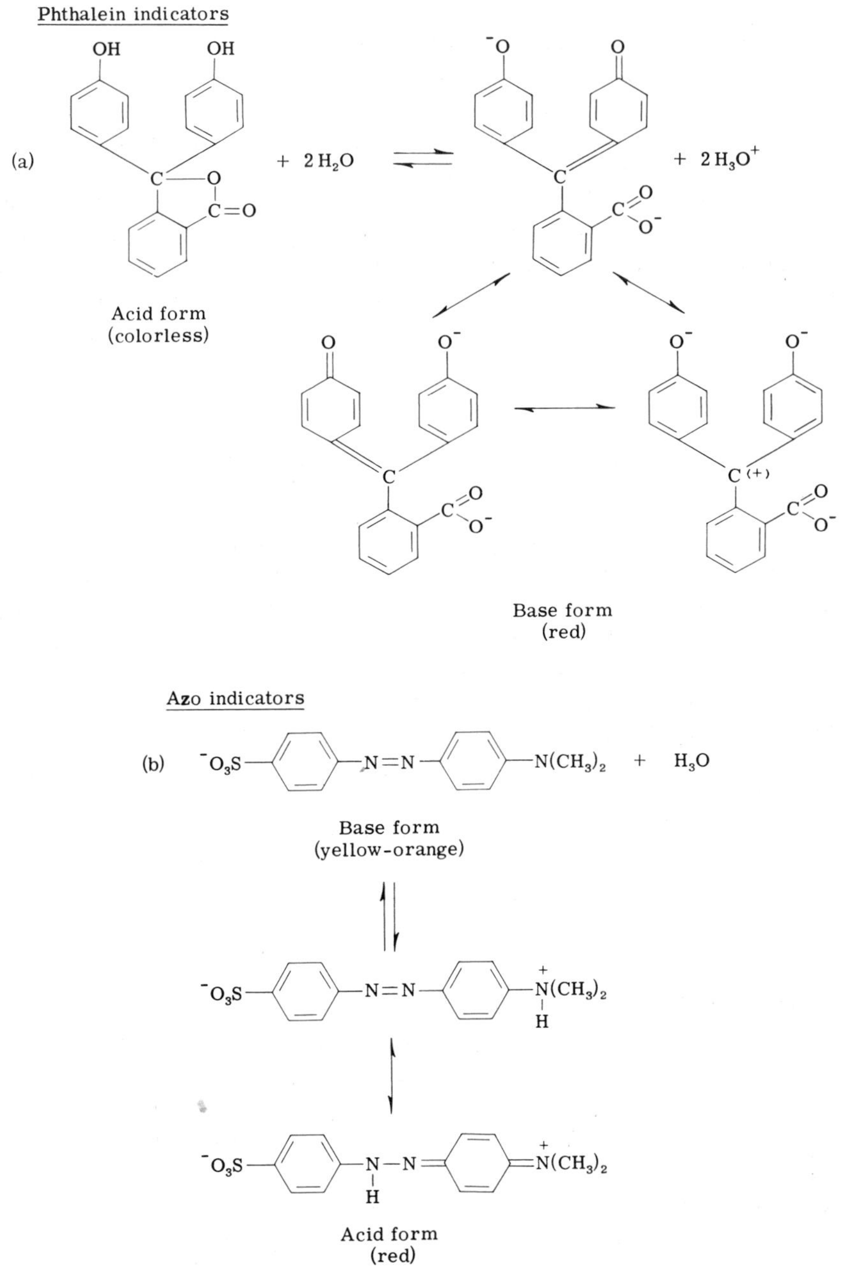
Phthalein indicators
(a)
+ 2 $H_2O$
+ 2 $H_3O^+$
Acid form
(colorless)
Base form
(red)
Azo indicators
(b)
+ $H_2O$
Base form
(yellow-orange)
Acid form
(red)

**Fig. 8–8.** Types of acid–base indicators and an example of each. (a) Phenolphthalein. (b) Methyl orange.

seen but the range of change is reduced. This mixed indicator is commonly used in the titration of soda ash ($Na_2CO_3$) samples with HCl titrant.

The property of fluorescence (emission of radiant energy after first being activated or excited by a specific region of electromagnetic radiation) is also pH-sensitive. Thus, as the pH of the solution changes fluorescence may appear or be quenched at some particular pH value. These types of indicators are used only in special or unusual applications. Since the end point involves the emission or quenching of emitted light it is not necessary that the solution be transparent. Thus, one application of fluorescent acid–base indicators is to use them in highly colored solutions. Although fluorescence change can be used for end-point detection, it has greater utility in other applications (see Chapter 21).

## SUMMARY OF ACID–BASE TITRATION CURVES

In the development of the titration curves in this section several points were illustrated. First approximations were made in order to simplify the calculations. In general the calculated values are within 5% of the true value. Whether these approximations are reasonable or not will depend on the magnitude of the equilibrium constants that are involved and the concentrations of the substances.

These approximations allow a rapid calculation of the titration curve which in turn suggests the feasibility of the titration towards analysis. From the curve the stoichiometric point pH, which must be detected by some means, is established. Furthermore, the properties of hydrolysis and buffer conditions and how they affect pH is illustrated by examining a titration curve. It is also possible to estimate equilibrium constants from titration curves.

## APPLICATIONS

Many industrial, biological, pharmaceutical, and naturally occurring samples contain acids or bases. If the acids or bases in the sample are quantitatively leached out into water or if the sample is soluble itself the acid or base content can be determined by titration with standard titrant. The other requirements for a successful analysis are that the strength of the acids or bases be greater than $K_a = 10^{-8}$ or $K_b = 10^{-8}$, respectively, and that their concentration be in the range of 0.1–0.001 $F$. End-point detection can be by color indicator or one of several different instrumental techniques. Often, by combination of special laboratory techniques and end-point detection techniques, quantitative analysis at the $10^{-3}$–$10^{-4}$ $M$ range is also possible. Industrial cleaning, paint removal, rust removal, and dip solutions, $Na_2CO_3$

content of washing soda, carbonate content of minerals and ores, $CO_2$ in the atmosphere, $HCO_3^-$ in antacid tablets such as Alka Seltzer, acetic acid content of vinegar, caustic soda ($Na_2CO_3$–$NaOH$ mixtures), $NaHCO_3$–$Na_2CO_3$ mixtures, $H_3PO_4$ in commercial orthophosphoric acid, boric acid and borax, and nitrogen analysis are typically determined by acid–base procedures. On the basis of the titration curves already presented several of these analyses are obvious. For this reason only three specific types of applications are discussed in detail.

**Determination of Carbonates, Phosphates, and Borates.** Titrations of soda ash ($Na_2CO_3$), sodium bicarbonate ($NaHCO_3$)$_2$, and mixtures of $Na_2CO_3$–$NaHCO_3$ and $NaOH$–$Na_2CO_3$ with standard acid are important industrial analyses. A mixture of $NaOH$–$NaHCO_3$ will not exist, since the $HCO_3^-$ is acidic enough to undergo a reaction with $NaOH$.

The procedures usually employ either a color indicator or a recording of a pH–titration curve, such as the ones in Figs. 8–5 and 8–6b, for end-point detection. For single components, only one end point is detected, while for the mixtures two end points must be detected. By considering the stoichiometry of the titration and the volume of the titrant required to reach the end point, the % composition can be calculated. This calculation is illustrated in Example 8–23 in the Calculation section at the end of this chapter.

If titration curves are recorded, end points are established from the curve. Typical indicators used are phenolphthalein to detect the first end point and methyl orange or methyl red to detect the second end point (see discussion on page 174). (Mixed indicators can also be used.) This basic procedure will work for the titration of both $Na_2CO_3$–$NaHCO_3$ and $NaOH$–$Na_2CO_3$ mixtures. An improvement in the accuracy for the analysis of the second mixture is achieved by titrating the mixture to the methyl orange end point ($NaOH$–$Na_2CO_3$ total). Another sample is taken, $BaCl_2$ added which quantitatively precipitates the $CO_3^{2-}$, and the $NaOH$ is titrated without filtering to the phenolphthalein end point. The amount of $CO_3^{2-}$ is found by difference.

A procedure employing $BaCl_2$ can be used for a $Na_2CO_3$–$NaHCO_3$ mixtures. However, this one is only used when the sample contains large amounts of $CO_3^{2-}$ and small amounts of $HCO_3^-$.

Phosphoric acid and its salts are not only important industrial compounds but are also routinely used in the preparation of buffers in the laboratory. Neutralization methods for their determination are straightforward and procedures for the titration of $H_3PO_4$ or $Na_2HPO_4$ are similar to the titration of other weak acids, while procedures for the titration of the salts $Na_3PO_4$ and $Na_2HPO_4$ are similar to the titration of carbonate salts. ($BaCl_2$ is not used.) The stoichiometry is determined by the selection of the end point (see Fig. 8–4). Furthermore, mixtures of $H_3PO_4$ and its acidic salts, and $NaOH$ and the basic salts of $H_3PO_4$ can be differentially titrated (see Fig. 8–6a).

Boric acid, $K_a = 5.83 \times 10^{-10}$, is too weak of an acid to be quantitatively titrated with standard base using conventional end-point detection systems. However, it is transformed into a complex species which is acidic enough to titrate by adding organic compounds with hydroxyl groups, such as glycerol, mannitol, or dextrose. Phenolphthalein is usually used as the indicator.

Borax, also known as sodium tetraborate $(Na_2B_4O_7 \cdot 10H_2O)$, in solution can be considered as boric acid which is 50% neutralized or

$$Na_2B_4O_7 + 5H_2O \rightleftharpoons 2NaH_2BO_3 + 2H_3BO_3$$

The $NaH_2BO_3$ is basic enough to be titrated with standard acid; the product of the neutralization being boric acid. The stoichiometry is given by the reaction

$$Na_2B_4O_7 + 2HCl + 5H_2O \rightarrow 4H_3BO_3 + 2NaCl$$

Usually, methyl red or methyl orange (or a mixed indicator) is used to detect the end point.

**Determination of Carbon Dioxide in the Atmosphere.**     For the determination of $CO_2$ in the atmosphere a known volume of air is passed through a standard solution of excess $Ba(OH)_2$.

$$Ba(OH)_2 + CO_2 \rightarrow BaCO_{3(s)} + H_2O$$

(Standardize by titration with standard HCl using phenolphthalein indicator). Since insoluble $BaCO_3$ forms, the excess $Ba(OH)_2$ is titrated with standard HCl (phenolphthalein). Therefore, the amount of $CO_2$ in the sample is found by difference. Since the sample is a gas, caution must be exercised to ensure complete contact of the gas with the $Ba(OH)_2$ solution. The sampling technique also requires a measurement of the temperature and pressure of the sample.

Several other elements can be determined through neutralization methods. Some of these procedures are useful in atmospheric analysis, while others are used in micro determinations. Table 8–9 lists several of these procedures. It should be noted that if the procedure is to be used for elemental analysis, the element must be converted into the acidic or basic species. For example, S in the sample must be converted to $SO_2$ or $SO_3$ prior to employing the neutralization procedure.

**Determination of Salt Content.**     The concentration of a salt in solution can be titrated as an acid after passage through a column of cation exchange resin. As the salt solution passes over the resin, the metal ion is ex-

**Table 8–9.   Acidic and Basic Gases That Can Be Determined by Neutralization Methods**

| Gas | Trapping procedure | Titration procedure |
|---|---|---|
| $CO_2$ | $CO_{2(g)} + Ba(OH)_2 \rightarrow BaCO_{3(s)} + H_2O$ | Excess $Ba(OH)_2$ titrated |
| $NH_3$ | $NH_{3(g)} + HCl \rightarrow NH_4Cl$ | Excess HCl titrated |
| $SO_2$ | $SO_{2(g)} + H_2O_2 \rightarrow H_2SO_{4(aq)}$ | Direct titration of $H_2SO_4$ |
| HCl | $HCl_{(g)} + H_2O \rightarrow HCl_{(aq)}$ | Direct titration of HCl |
| $SiF_4{}^a$ | $SiF_{4(g)} + H_2O \rightarrow H_2SiF_6$ | Direct titration of $H_2SiF_6$ |

$^a$ $SiF_4$ is formed by the reaction of $SiO_2$ and HF and can be used for the determination of F or Si.

changed for hydronium ion stoichiometrically.

$$Resin-SO_3^-H^+ + M^+X^- \rightleftharpoons Resin-SO_3^-M^+ + H^+X^-$$

Alternatively, the salt solution can be passed over a column of anion exchange resin and converted to a base which is titrated with a strong acid titrant.

$$Resin-NR_3^+OH^- + M^+X^- \rightleftharpoons Resin-NR_3^+X^- + M^+OH^-$$

This procedure is often used, particularly in the pharmaceutical industry, for the determination of total salt, tetraalkylammonium salts, and many other organic salts such as amine hydrochlorides, sulfates, and perchlorates.

**Determination of Nitrogen.**     Ammonium salts, nitrate salts, and inorganic and organic nitrogen can be determined by an acid–base titration. Each of these involves the conversion of the sample to ammonia which is titrated.

Ammonium salts are the easiest to handle. In the direct procedure, a solution of the ammonium salt is treated with a NaOH solution and the ammonia is distilled into a solution of standard HCl. A typical distillation apparatus for the distillation and collection of $NH_3$ is shown in Fig. 8–9. The remaining acid is titrated with standard NaOH (methyl red or methyl orange).

$$NH_4^+ + OH^- \rightarrow NH_{3(g)} + H_2O$$

$$NH_{3(g)} + HCl \rightarrow NH_4Cl$$

An indirect procedure requires the boiling of the ammonium salt (except carbonate or bicarbonate) with a standard solution of NaOH. Subsequently, ammonia ceases to be expelled and the remaining NaOH is titrated with standard HCl (methyl red or methyl orange).

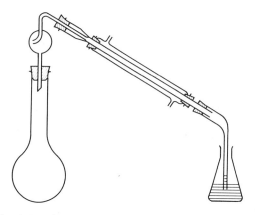

**Fig. 8–9.**   Apparatus for the distillation of ammonia.

Nitrate salts are reduced to ammonia by Al, Zn, and by Devarda's alloy (50Cu, 45Al, 5Zn) in strong alkaline solution.

$$3NO_3^- + 8Al + 5OH^- + 2H_2O \rightarrow 8AlO_2^- + 3NH_2$$

The ammonia produced in the reaction is distilled into acid as previously described. Nitrite salts can also be determined by this procedure.

The determination of nitrogen in inorganic and organic samples by the Kjeldahl method is widely employed in the industrial, clinical, and research laboratory. Typical analyses would be nitrogen in blood and other biological substances, in cereal, and in fertilizer. In this method the nitrogen in the sample is converted into ammonia by digestion with boiling concentrated sulfuric acid.

$$\text{Organic N} \xrightarrow[\text{H}_2\text{SO}_4^- \text{ catalysts}]{\text{oxidation}} NH_4HSO_4$$

Subsequently, the acid is carefully neutralized with strong alkali solution and the ammonia is distilled as before. The digestion process is very slow and various modifications have been suggested to speed it up. Potassium sulfate, which raises the boiling point, is often added. Catalysts, such as Hg, HgO, $CuSO_4$, Se, or $Se$–$FeSO_4$ mixtures, can be used. Nitrogen in proteins, amines, and amides generally do not require special catalysts. However, nitro, azo, hydrazo, and cyano compounds usually require modification of the procedure.

A useful modification of the distillation procedure is to distill the ammonia into a near saturated solution of boric acid rather than into HCl.

$$NH_3 + H_3BO_3 \rightarrow NH_4^+ + H_2BO_3^-$$

The $H_2BO_3^-$ that is formed is titrated with standard HCl back to $H_3BO_3$. At the stoichiometric point of this titration the solution will contain $H_3BO_3$ and $NH_4Cl$. Therefore, an indicator which changes color in the pH range 5–6 is necessary. Bromocresol green or the mixed indicator bromocresol green–methyl red are satisfactory indicators. The main advantage is that only one standard solution (HCl) is required. The exact concentration of the boric acid solution is not required, however, a blank titration of the indicator is essential.

## CALCULATIONS

The basic stoichiometric relationship underlying calculations in analytical chemistry were described in Chapter 3. In the next five examples these principles are applied to neutralization volumetric analysis.

*Example 8-19.* Vinegar is an aqueous solution of acetic acid that is produced by fermentation. A 25.00-ml aliquot of a vinegar sample was taken, diluted, and titrated with 31.75 ml of 0.2550 $F$ NaOH to the phenolphthalein end point. Calculate the g of acetic acid per 100 ml of vinegar.

$$HC_2H_3O_2 + NaOH \rightarrow NaC_2H_3O_2 + H_2O$$

$$ml_{HC_2H_3O_2} \times F_{HC_2H_3O_2} = ml_{NaOH} \times F_{NaOH}$$

$$25.00 \text{ ml} \times F_{HC_2H_3O_2} = 31.75 \text{ ml} \times 0.2550 \text{ } F$$

$$F_{HC_2H_3O_2} = 0.3239 \text{ mmoles/ml}$$

$$0.3239 \text{ mmoles/ml} \times 0.06005 \text{ g/mmole} = 0.01949 \text{ g/ml}$$

$$0.01949 \text{ g/ml} \times 100 = 1.945 \text{ g } HC_2H_3O_2/100 \text{ ml}$$

*Example 8-20.* A soda ash ($Na_2CO_3$) sample weighing 0.1196 g is titrated to the methyl orange end point with 42.45 ml of a HCl titrant. If the titrant is standardized by titrating pure $Na_2CO_3$ (0.1425 g) with the HCl (24.30 ml) to the same end point, calculate the percent $CO_2$ in the sample.

Since the methyl orange end point is used, the reaction is

$$Na_2CO_3 + 2HCl \rightarrow 2NaCl + H_2O + CO_2$$

and the reaction ratio is 1/2.

$$\text{wt of } Na_2CO_3, \text{ mg} = ml_{HCl} \times F_{HCl} \times \text{reaction ratio} \times Na_2CO_3$$

$$142.5 \text{ mg} = 24.30 \text{ ml} \times F_{HCl} \text{ (mmole/ml)} \times 1/2 \times 106.0 \text{ mg/mmole}$$

$$F_{HCl} = 0.1107 \text{ } F$$

$$\%CO_2 = \frac{ml_{HCl} \times F_{HCl} \times \text{reaction ratio} \times CO_2 \times 100}{\text{wt of sample}}$$

$$\%CO_2 = \frac{42.45 \text{ ml} \times 0.1107 \text{ mmole/ml} \times 1/2 \times 44.01 \text{ mg/mmole} \times 100}{119.6 \text{ mg}} = 86.46\%$$

*Example 8-21.*    A sample of milk weighing 0.4750 g is digested with $H_2SO_4$ and catalyst converting the protein to ammonium salts. Excess NaOH is carefully added and the $NH_3$ distilled into 25.00 ml of HCl. The remaining acid required 13.12 ml of 0.07891 *F* NaOH for titration. If 25.00 ml of the same HCl is completely titrated by 15.83 ml of the NaOH, calculate the %N in the sample.

The reactions are

$$NH_3 + HCl \rightarrow NH_4Cl$$

$$HCl + NaOH \rightarrow NaCl + H_2O$$

and the reaction ratio for both reactions is 1/1. Therefore,

$$ml_{NaOH} \times F_{NaOH} = ml_{HCl} \times F_{HCl} \times \text{reaction ratio}$$

$$15.83 \times 0.07891 = 25.00 \times F_{HCl} \times 1/1$$

$$F_{HCl} = 0.04998$$

$$\%N = \frac{(ml_{HCl} \times F_{HCl} - ml_{NaOH} \times F_{NaOH})\text{reaction ratio} \times N \times 100}{\text{wt sample}}$$

$$\%N = \frac{(25.00 \times 0.04998 - 13.12 \times 0.07891)1/1 \times 14.01 \text{ mg/mmole} \times 100}{475.0 \text{ mg}} = 0.6318\%$$

*Example 8-22.*    A milk of magnesia sample [suspension of $Mg(OH)_2$] weighing 0.9322 g required 36.85 ml of 0.1511 *F* HCl for complete neutralization. Calculate percent MgO in the sample.

The reaction is

$$Mg(OH)_2 + 2HCl \rightarrow MgCl_2 + 2H_2O$$

and the reaction ratio is 1/2.

$$\%MgO = \frac{ml_{HCl} \times F_{HCl} \times \text{reaction ratio} \times MgO \times 100}{\text{wt sample}}$$

$$\%MgO = \frac{36.85 \text{ ml} \times 0.1511 \text{ mmole/ml} \times 1/2 \times 40.31 \text{ mg/mmole} \times 100}{932.2 \text{ mg}} = 12.03\%$$

*Example 8-23.*    A mixture containing NaOH, $Na_2CO_3$, and inert material weighing 0.2744 g was dissolved and titrated with 0.1042 *F* HCl. If 34.11 ml of the acid was required to reach the phenolphthalein end point and an *additional* 7.12 ml to reach the methyl orange end point, calculate the percent NaOH and $Na_2CO_3$ in the sample.

The reactions are

1. $NaOH + HCl \rightarrow NaCl + H_2O$ (Phenolphthalein)

2. $Na_2CO_3 + HCl \rightarrow NaHCO_3 + H_2O + NaCl$ (Phenolphthalein)

3. $NaHCO_3 + HCl \rightarrow NaCl + CO_2 + H_2O$ (Methyl orange)

The volume from the phenolphthalein to the methyl orange end point corresponds to the amount of $NaHCO_3$, which on a molar basis is equal to the amount of $Na_2CO_3$ originally present. Remember that the $Na_2CO_3$ has been converted stoichiometrically to $NaHCO_3$; see reaction 2. Therefore, the reaction is 1/1 and

$$\%Na_2CO_3 = \frac{ml_{HCl} \times F_{HCl} \times \text{reaction ratio} \times Na_2CO_3 \times 100}{\text{wt sample}}$$

$$\%Na_2CO_3 = \frac{7.12 \, ml \times 0.1042 \, mmole/ml \times 1/1 \times 106.0 \, mg/mmole \times 100}{274.4 \, mg} = 28.66\%$$

Since 7.12 ml of the acid was required for reaction 3, 7.12 ml must also be required for reaction 2. Therefore, the amount of acid needed to neutralize the NaOH is given by

$$34.11 - 7.12 = 26.99 \, ml$$

and

$$\%NaOH = \frac{ml_{HCl} \times F_{HCl} \times \text{reaction ratio} \times NaOH \times 100}{\text{wt sample}}$$

$$\%NaOH = \frac{26.99 \, ml \times 0.1042 \, mmole/ml \times 1/1 \times 40.00 \, mg/mmole \times 100}{274.4 \, mg} = 41.00\%$$

## Questions

1. Differentiate between Arrhenius and Brønsted–Lowry acids and bases.
2. Give the conjugate acids for the bases: $HCO_3^-$, $F^-$, $C_2H_3O_2^-$, $H_2O$, $C_2O_4^{2-}$, glycine ($H_2NCH_2CO_2H$), and $C_6H_5NH_2$.
3. Give the conjugate bases for the acids: $H_2O$, $HNO_2$, $H_3PO_4$, $H_2PO_4^-$, $H_2NCH_2CO_2H$, $H_2CO_3$, and $HCO_3^-$.
4. Rank the bases in Question 2 in order of increasing basic strength.
5. Rank the acids in Question 3 in order of increasing acidic strength.
6. List several strong acids and strong bases.
7. List several amphiprotic substances.
8. What is the leveling property of a solvent?
9. Show that $pH + pOH = 14$.
10. What does a negative pH value mean?
11. Why is it more practical to plot pH rather than $H_3O$ concentration vs volume of titrant?

12. What is autoprotolysis?
13. Write the equations illustrating the ionization of the following substances in water: $H_2SO_3$, $HNO_3$, $CO_2$, $H_2C_2O_4$, $H_2NCH_2CH_2NH_2$, $HOCH_2CH_2NH_2$, $H_2NCH_2CO_2H$, $PO_4^{3-}$, $NH_4^+$, and $CH_3NH_3^+$.
14. What is the difference between a concentrated acid solution and a strong acid solution?
15. What is hydrolysis?
16. Suggest whether solutions of the following substances are acidic, basic, or neutral. $NaCl$, $Na_2SO_4$, $NaCN$, $NH_4NO_3$, $Na_2CO_3$, $NH_4ClO_4$, $C_6H_5CO_2Na$, $NaHC_2O_4$.
17. What is the difference between a buffer and buffer capacity?
18. List the components needed to prepare a buffer that has a pH of about 6.
19. What special utility does the Henderson–Hasselbalch equation have?
20. Show that Eq. (8-27) is equal to pH $= \frac{1}{2}pK_w + \frac{1}{2}pK_a + \frac{1}{2}\log C$.
21. Show that Eq. (8-30) is equal to pH $= \frac{1}{2}pK_w - \frac{1}{2}pK_b - \frac{1}{2}\log C$.
22. Why is a strong base or strong acid titrant usually used?
23. Why is $HNO_3$ not usually used as a strong acid titrant?
24. Explain why $Na_3PO_4$ is a useful salt for preparing buffers.
25. Predict whether the hydrogen ion concentration and the pH increases, decreases, or remains the same for the following.
    a. Addition of 50 ml of water to 50 ml of 0.1 $F$ $NH_3$.
    b. Addition of 50 ml of water to 50 ml of 0.1 $F$ $HC_2H_3O_2$.
    c. Addition of 50 ml of 0.1 $F$ $NaCl$ to 50 ml of 0.1 $F$ $HC_2H_3O_2$.
    d. Addition of 1 g of $NaCl$ to 50 ml of 0.1 $F$ $HC_2H_3O_2$.
    e. Addition of 1 g of $NaHCO_3$ to 50 ml of 0.1 $F$ $Na_2CO_3$.
    f. Addition of 10 ml of 0.1 $F$ $HCl$ to 50 ml of 0.1 $F$ $HClO_4$.
    g. Bubbling of $CO_2$ through 100 ml of 0.1 $F$ $NaOH$.
    h. Addition of 1 g of $NaF$ to 25 ml of 0.1 $F$ $HF$.
    i. Addition of 1 g of $NaOH$ to 25 ml of 0.1 $F$ $NaHCO_3$.
    j. Addition of 1 g of $NH_4Cl$ to 25 ml of 0.2 $F$ $NH_3$.
    k. Bubbling of $HCl$ gas through 50 ml of 0.1 $F$ $Na_2HPO_4$ solution.
    l. Bubbling of $HCl$ gas through 50 ml of 0.1 $F$ $NaH_2PO_4$ solution.
    m. Bubbling of $HCl$ gas through 50 ml of 0.1 $F$ $H_3PO_4$ solution.
26. Predict whether one, two, etc. pH breaks are observed for the following titrations; also list the order of titration for the mixtures.
    a. $HCl–HC_2H_3O_2$        f. $H_2C_2O_4$
    b. $HCl–H_2SO_4–HC_2H_3O_2$    g. $CH_3NH_2–NH_3$
    c. $HF–HC_2H_3O_2$        h. $H_2NCH_2CH_2NH_2$
    d. $NaOH–Na_3PO_4$        i. $HOCH_2CH_2NH_2–CH_3CH_2CH_2CH_2NH_2$
    e. $Na_2CO_3–Na_3PO_4$      j. Pyridine hydrochloride–aniline hydrochloride
27. Explain how an acid–base indicator works.
28. Why must the indicator concentration be kept at a low value?
29. What effects in titrations would be noticed if distilled water containing dissolved $CO_2$ was used in the preparation of a $HCl$ titrant, of a $NaOH$ titrant?
30. Explain why phenolphthalein is a good indicator for the titration of $HCl$ with $NaOH$ but not for $NaOH$ with $HCl$.
31. Why is it impractical to titrate very weak acids or bases using color indicators to determine the stoichiometric point?

# Problems

1. Calculate the pH for the following.
   a. $1.4 \times 10^{-3}$ F HCl
   b. $6.1 \times 10^{-2}$ F NaOH
   *c. $1.0 \times 10^{-3}$ F $H_2SO_4$
   d. $2.0 \times 10^{-4}$ F NaCl
   e. 2.0 F HCl
   f. $1.0 \times 10^{-8}$ F HCl p13⁹

2. Calculate the pH for the following.
   *a. 0.150 F $HC_2H_3O_2$
   *b. 0.200 F HF
   c. 0.065 F pyridine
   d. 0.10 F lactic acid
   e. 0.10 F $(CH_3CH_2)_2NH$
   f 0.10 F $ClCH_2CO_2H$

3. Calculate the pH for the following.
   a. 40.0 ml of $4.6 \times 10^{-2}$ F HCl.
   *b. A mixture of 10.0 ml of 0.050 F $HNO_3$, 10.0 ml of water, and 45.0 ml of 0.050 F HCl.
   c. A mixture of 25.0 ml of water, 60.0 ml of 0.10 F $HC_2H_3O_2$, and 25.0 ml of 0.10 F NaCl.
   *d. A mixture of 40.0 ml of 0.050 F pyridine and 20.0 ml of water.

4. Calculate the pH for the following.
   *a. 1.15 g of $NaC_2H_3O_2$ per 100 ml.
   *b. 0.05 F $NH_4Cl$.
   c. 0.010 F ethanolamine hydrochloride.
   d. 0.020 F sodium formate.
   e. 0.075 F $NH_4C_2H_3O_2$.
   f. 0.961 g $NH_4Cl$ per 250 ml.

5. A 0.0100 F solution of phenol is 0.05% ionized at 25°C. What is the $pK_a$ of the acid?

6. What concentration of nicotinic acid would have a pH of 4.1?

7.* A 0.8150-g sample of an unknown pure weak monoprotic acid was dissolved in water and titrated with 0.1100 F NaOH. The stoichiometric point was reached at 24.60 ml and the pH was 4.80 when 11.00 ml of the base was added. Calculate the $pK_a$ for the weak acid.

8.* Calculate the pH for the solution made by dissolving 1.00 g of $NaC_2H_3O_2$ and 1.00 g of $HC_2H_3O_2$ per 100 ml of solution.

9. Calculate the pH for the solution made by dissolving 0.150 mole of $NH_3$ and 0.100 mole of $NH_4Cl$ per liter of solution.

10.* In what molar ratio must formic acid and sodium formate be mixed to give a buffer of pH = 4.35.

11.* What is the pH for a solution prepared by mixing 45.0 ml of 0.150 F $NH_3$ and 60.0 ml of 0.100 F $NH_4Cl$?

12.* How many grams of $NaC_2H_3O_2$ must be added to 200 ml of 0.100 F $HC_2H_3O_2$ to give a buffer of pH = 4.40? (Assume that volume changes are negligible.)

13. Calculate the molar proportion of tris (hydroxymethyl)aminomethane and its hydrochloride salt which gives a solution of pH 6.45.

14. Calculate the pH for a solution made by mixing 40.0 ml of 0.040 F lactic acid and 25.0 ml of 0.015 F sodium lactate.

---

* Answers are listed at the end of the book for problems marked with an asterisk.

15. Calculate the pH for the following solutions. (Assume that volume changes are additive.)

*a. 25.5 ml of 0.125 $F$ NaOH is mixed with 41.5 ml of 0.110 $F$ HCl.

b. 14.2 ml of 0.100 $F$ $H_2SO_4$, 10.0 ml of water, 15.0 ml of 0.110 $F$ HCl, and 25.0 ml of 0.120 $F$ NaOH are mixed.

*c. 1.25 g NaOH added to 150 ml of 0.140 $F$ $HC_2H_3O_2$.

d. 10.0 ml of 0.135 $F$ HCl and 20.0 ml of 0.110 $F$ $NH_3$ are mixed.

e. 20.0 ml of 0.150 $F$ NaOH and 30.0 ml of 0.100 $F$ lactic acid are mixed.

16.* Calculate the pH at 0, 10, 25, 50, 75, 90, 100, 125, and 200% neutralized for the titration of 60.0 ml of 0.050 $F$ HCl with 0.0750 $F$ NaOH.

17.* Calculate the pH at 0, 10, 25, 50, 75, 90, 100, 125, and 200% neutralized for the titration of 40 ml of 0.150 $F$ barbituric acid ($K_a = 9.9 \times 10^{-5}$) with 0.200 $F$ NaOH.

18. Nicotinic acid has a $K_a$ of $1.4 \times 10^{-5}$. Calculate the titration curve for this acid if NaOH is used as the titrant.

19. Calculate the pH at 0, 10, 25, 50, 75, 90, 100, 125, and 200% neutralized for the titration of 50.0 ml of 0.100 $F$ tris(hydroxymethyl)aminomethane with 0.200 $F$ HCl.

20.* Calculate the pH for a 0.05 $F$ $H_3PO_4$ solution.

21.* Calculate the grams of NaOH that must be added to a 1 liter solution of 0.100 $F$ $H_3PO_4$ solution to give a pH of 4.35.

22. Calculate the grams of $H_3PO_4$ and $Na_2H_2PO_4$ in 1 liter of solution that has a pH of 2.42.

23. Calculate the molar ratio of $NaH_2PO_4$ to $Na_2HPO_4$ in a solution of pH = 6.5.

24.* What is the pH of a solution containing 0.1 g/liter $Na_2CO_3$ and 0.1 g/liter $NaHCO_3$?

25. What is the pH for a solution made by mixing 40.0 ml of 0.150 $F$ $Na_2CO_3$ and 30 ml of 0.150 $F$ HCl? Or with 40.0 ml of 0.150 $F$ HCl?

26. If arterial blood has a pH of 7.41 and changes to 7.37, calculate the change that takes place in the $HCO_3^-/H_2CO_3$ ratio.

27.* If arterial blood has a pH of 7.40 and is 0.0240 $M$ in $HCO_3^-$, calculate the $M$ concentration of $H_2CO_3$.

28. A plasma sample had a $P_{CO_2}$ of 28 mmHg and $[HCO_3^-] = 0.015$ $M$. Calculate the pH.

29.* A patient swallowed 10 g $NH_4Cl$ and after 1 hour his blood had a pH of 7.35. Calculate the mole ratio of $HCO_3^-/H_2CO_3$ in his blood.

30. Isotonic sodium lactate at pH 7.40 is commonly administered intravenously to combat metabolic acidosis. How many ml of concentrated lactic acid (85% by weight, density 1.20) and how many grams of NaOH would be required to prepare 3 liters of the solution?

31.* What is the pH of a saturated $Mg(OH)_2$ solution in contact with solid $Mg(OH)_2$?

32. What is the pH of a saturated $Mg(OH)_2$ solution that also contains 0.10 $F$ $MgCl_2$?

33.* A 50.00-ml aliquot taken from a saturated $Ca(OH)_2$ solution in contact with $Ca(OH)_2$ solid required 12.93 ml of 0.07550 $F$ HCl. Calculate the $K_{sp}$ for $Ca(OH)_2$.

34.* A 0.4252-g sample of potassium acid phthalate required 24.11 ml of NaOH for neutralization. Calculate the formality of the titrant.

35. A 0.1314-g sample of $Na_2CO_3$ which is 99.5% pure required 23.21 ml of HCl for neutralization to the methyl orange point. Calculate the formality of the titrant.

36.* A 0.1414-g sample of the primary standard sulfamic acid ($NH_2SO_3H$) was dissolved and titrated with 31.25 ml of NaOH. What is the formality of the base?

37. An unknown acid monoprotic sample weighed 0.4165 g. After dissolution, it was titrated with 28.25 ml of 0.09911 $F$ NaOH. Calculate the %H in the sample.

38. A 0.100 $F$ HCl solution has a pH of 1.00, while a 0.100 $F$ $HC_2H_3O_2$ solution has a pH of 2.88. What volume of 0.1225 $F$ NaOH would be required to titrate 50.0-ml aliquots of each acid to their respective end points? Explain your answer.

39.* A sample of pure oxalic acid ($H_2C_2O_4 \cdot 2H_2O$) weighing 0.1950 g required 25.42 ml of KOH for complete neutralization. Calculate the formality of the titrant. If an oxalate unknown weighing 0.2915 g requires 21.50 ml, calculate the percent $H_2C_2O_4$ in the sample.

40.* A sample containing $Na_2CO_3$, $NaHCO_3$, and inert material weighing 0.2905 g is titrated with 0.1141 $F$ HCl. If 22.15 ml is needed to reach the phenolphthalein end point and 46.34 ml needed to reach the methyl orange end point, calculate the percent $Na_2CO_3$ and percent $NaHCO_3$ in the sample.

41. Two 50.00-ml aliquots of a solution containing $H_2SO_4$ and $H_3PO_4$ are titrated with 0.1000 $F$ NaOH. The first requires 26.15 ml to reach the methyl red end point, while the second requires 36.03 ml to reach the phenolphthalein end point. Calculate the grams of each acid per 50 ml of solution.

42. A 0.6900-g sample of NaOH, $Na_2CO_3$, and inert material was dissolved and required 33.51 ml of HCl to reach the phenolphthalein end point. Methyl orange was added, and 7.25 ml more HCl is required to reach the methyl orange end point. If the HCl is 0.2395 $F$ calculate the %NaOH and $Na_2CO_3$ in the sample.

43. An ammonium salt weighing 0.6151 g was heated with NaOH solution and the evolved $NH_3$ was trapped in 75.00 ml of 0.3511 $F$ HCl. The excess acid required 4.15 ml of 0.2100 $F$ NaOH. Calculate the %$NH_4^+$ in the sample?

44. A protein (0.2318 g) was digested and the nitrogen determined by the Kjeldahl method. The ammonia was distilled into 50.00 ml of HCl and the remaining acid required 21.15 ml of 0.1200 $F$ NaOH. If a 40.00-ml aliquot of the HCl required 34.12 ml of the NaOH, calculate the percent N in the sample.

45.* A 4.752-g sample of milk was digested and titrated by the Kjeldahl method. The ammonia was distilled into 40.00 ml of $H_2SO_4$ and the remaining acid required 18.68 ml of 0.1000 $F$ NaOH. If a 40.00-ml aliquot of the $H_2SO_4$ required 22.17 ml of the NaOH, calculate the percent N in the sample.

46. A sample of Chile saltpeter (natural occurring $NaNO_3$) was heated with Devarda's alloy and the $NH_3$ gas was trapped in 50.00 ml of HCl. The excess acid required 14.45 ml of 0.1283 $F$ NaOH. If 25.00 ml of the HCl was neutralized by 27.11 ml of the NaOH and the sample weighed 0.4489 g calculate the %$NaNO_3$ in the sample.

47.* Oxalic acid is widely used in the analytical laboratory, industry, and in veterinarian applications. If a 0.2145-g sample of an oxalic acid sample is taken and 23.87 ml of 0.1410 $F$ NaOH is required for complete neutralization, calculate the percent purity assuming the oxalic acid is a dihydrate.

48. A sample of urban air was passed at controlled flow rate, pressure, and temperature through a solution containing 100.0 ml of 0.02040 $F$ $Ba(OH)_2$. After pre-

cipitation of $BaCO_3$ the excess $Ba(OH)_2$ was titrated with 25.45 ml of 0.03311 $F$ HCl. If the total air sample was 4.125 liters calculate the ppm (ml $CO_2/10^6$ ml air) of $CO_2$. The density of $CO_2$ (at collected conditions) is 1.799 g/liter.

49.* Acetylsalicylic acid (aspirin) can be determined by hydrolyzing the aspirin with a known amount of excess base (boiling for 10 minutes) and then titrating the remaining base with acid.

$$+ \; 2\,NaOH \longrightarrow CH_3COO^- \; Na^+ \; + \; C_6H_4(OH)COO^- \; Na^+$$

If the sample weighed 0.2745 g, 50.00 ml of 0.1000 $F$ NaOH is used, and 11.03 ml of 0.2100 $F$ HCl is required for the excess base (phenol red is usually used), calculate the percent purity of the sample. Discuss the interferences if this method were to be used for aspirin containing pharmaceutical products.

50. Aspirin (acetylsalicylic acid, $K_a = 3.16 \times 10^{-4}$) is absorbed from the stomach in the conjugate acid form. A single 5-grain aspirin tablet will provide 0.3250 g of aspirin. If two tablets are dissolved in 100 ml of water, calculate the weight of aspirin that is in the conjugate acid form. If the solution is pH = 1 calculate the weight of aspirin that is in the conjugate acid form. Discuss the significance of this calculation with respect to the availability of aspirin in the stomach.

# Chapter
# Nine
# Nonaqueous
# Acid–Base Titrations

## SCOPE

Neutralization in an organic reaction is very rapid and quantitative. It is, therefore, not surprising to find that many methods for determining organic compounds involve neutralization. This may be a direct titration, in which the organic compound being acidic or basic, is titrated with a standard base or acid. Or, an indirect titration where the product produced or reactant that is consumed exhibits acidic or basic properties.

Table 9–1 lists many of the common functional groups which exhibit acidic or basic properties. Absent from this list are the many special cases in which the acidic or basic properties are only brought out by the solvent, substituents, or other structural features within the molecule.

Why is it desirable to switch from water, a cheap solvent, in which the concepts of ionization and equilibria are so clearly understood, to nonaqueous solvents, most of which are expensive, toxic, and have an unpleasant odor? Three apparent reasons are: (1) many organic compounds are insoluble in water, (2) acids and bases with $K_a$'s and $K_b$'s of less than $10^{-7}$ can not be quantitatively titrated in water, and (3) the strongest possible acid and base in water is the hydronium ion and hydroxide ion, respectively. Unfortunately, it is difficult to quantitatively derive expressions describing the various equilibria that are involved in nonaqueous solutions and this is one of the major limitations of the nonaqueous acid–base titration.

## SOLVENTS

Analysis in a nonaqueous solvent is controlled by the strength of an acid or base in the solvent, the acid–base properties of a solvent, and the dielectric

Table 9–1.  Acidic and Basic Compounds That Can Be Titrated[a]

| Acids | Bases |
|---|---|
| Aliphatic carboxylic acids | Aliphatic amines |
| Aromatic carboxylic acids | Aromatic amines |
| Phenols | Alkoxides |
| Enols | Quaternary ammonium hydroxides |
| Imides | Polyamines |
| Thioureas | Alkaloids |
| Sulfonamides | Amino acids |
| Sulfonic acids | Antihistamines |
| Phosphonic acids | Xanthates |
| Arsonic acids | Phenothiazines |
| Thiophenols | Schiff bases |
| Barbituates | Hydrazides |
| Hydantoins | Amides |
| Certain salts | Thioamides |
| Inorganic acids | N-, P-, or S- oxides |
|  | Certain salts |
|  | Inorganic bases |

[a] Successful titration, particularly for the weak acids and bases, will depend on the choice of solvent.

constant of a solvent. One method of classifying the solvents is to refer to those solvents with acidic and basic properties as amphiprotic, while those without these properties are called aprotic (inert) solvents. The amphiprotic type can be divided into a group characterized by strong acid properties (protogenic), and those possessing strong basic properties (protophilic). These two groups will still show slight basic and acidic tendencies, respectively.

A summary of the more common solvents is listed in Table 9–2. Values for the dielectric constant and autoprotolysis constant, where known, are also included.

It should be noted that solvents with basic properties such as pyridine, ethers, ketones, and esters are put in the aprotic class. The reason for this is that these solvents do not show acidic properties, or at least the acidic nature is not detectable under normal conditions. Nitromethane, nitroethane, and dimethylsulfoxide are the opposite in that they exhibit slightly acidic and no basic properties. The basic and acidic properties of these solvents, respectively, are considerably less than the protogenic and protophilic solvents.

In part this classification is one of convenience since the distinction between classes is not always clear. For example, it is doubtful whether there are solvents with no acidic and basic properties. They may have such low self-

Table 9-2.  Solvents for Acid–Base Titrations

| Solvent[a] | Dielectric constant | Solvent[a] | Dielectric constant |
|---|---|---|---|
| Amphiprotic, amphoteric | | Aprotic (inert) | |
| Ethylene glycol | 24.3 | Acetonitrile | 36.0 |
| Methanol (16.7) | 32.6 | Acetone | 20.7 |
| Ethanol (19.1) | 24.3 | Methyl isobutyl ketone | 13.1 |
| Isopropanol | 18.3 | Pyridine | 12.5 |
| t-Butanol | | Dimethylformamide | 27.0 |
| Water (14.0) | 78.5 | Nitromethane | 35.9 |
| Amphiprotic, protogenic | | Acetic anhydride | 20.7 |
| Acetic acid (14.45) | 6.13 | Dioxane | 2.21 |
| Formic acid (6.2) | 58.5 | Nitrobenzene | 39.0 |
| Amphiprotic, protophilic | | Benzene | 2.3 |
| Ammonia | 22.0(−33°) | Chloroform | 4.8 |
| Butylamine | 5.3 | | |
| Ethylenediamine | 12.9 | | |

[a] Autoprotolysis constant in parenthesis.

ionization that it is not detected by present means. An alternative way of classification is to divide the aprotic group into acidic, basic, and neutral subgroups as is done in the amphiprotic group.

## EQUILIBRIA IN A NONAQUEOUS SOLVENT

The fact that a solvent has or does not have acidic and basic properties is of vital concern. In an aprotic solvent, such as benzene and carbon tetrachloride, little or no dissociation of a solute takes place. Thus, the acidic or basic properties of the solute are exhibited only upon the addition of a base or acid, respectively.

If the solvent has basic properties, the addition of an acidic solute will result in the following equilibrium where HA is the acid, SH the basic solvent, and $SH_2^+$ the solvated proton:

$$HA + SH \rightleftharpoons SH_2^+ + A^- \tag{9-1}$$

As the basicity of the solvent increases, the degree of dissociation of the acid increases.

In a similar fashion, if the solvent has acidic properties and a basic solute is added, the equilibrium in reaction (9-2) takes place:

$$B + SH \rightleftharpoons BH^+ + S^- \tag{9-2}$$

where B is a base, SH is now a solvent with acidic properties, $S^-$ is the anion of the solvent (lyate ion), and $BH^+$ is the conjugate acid of the base B. As the acidity of the solvent increases, the dissociation of the base increases.

If the solvent has acidic and basic properties, the strongest acid and base possible are $SH^+$ and $S^-$, respectively [see (9-3)], and a measure of this equilibrium is the autoprotolysis constant.

$$2SH \rightleftharpoons SH_2^+ + S^- \tag{9-3}$$

## CHOOSING A SOLVENT

What does all this mean to a nonaqueous titration? From a practical viewpoint it provides a basis for choosing the appropriate solvent for the titration and from a theoretical viewpoint it provides a firm foundation for the understanding of the chemistry taking place in the solvent.

As an illustration, assume that it is desirable to titrate a weak base. The acidic behavior of the titrant and the basic behavior of the solute will be determined by the solvent properties. The strong acid titrant might be made in one of the following solvents. If HA is one of the typical strong inorganic acids, such as $HClO_4$ or $HCl$, the equilibria will lie to the right as the basicity of the solvent increases. Therefore, water being the most basic means that the acid exists as the $H_3O^+$ ion and thus the two acids appear the same in strength. Acetic acid, being the least basic, causes the equilibrium to be shifted to the left. As a result, the acidity of the solution now depends on HA. In this situation $HClO_4$ is a stronger acid than $HCl$; acetic acid is not acting as a leveling solvent.

| Acid | Solvent (B) | | Acid | Base | |
|------|-------------|---|------|------|---|
| $HA +$ | $H_2O$ | $\rightleftharpoons$ | $H_3O^+$ | $+ A^-$ | (9-4) |
| $HA +$ | $CH_3CH_2OH$ | $\rightleftharpoons$ | $CH_3CH_2OH_2^+$ | $+ A^-$ | (9-5) |
| $HA +$ | $HC_2H_3O_2$ | $\rightleftharpoons$ | $H_2^+C_2H_3O_2$ | $+ A^-$ | (9-6) |

If a weak base, B, is dissolved in the same three solvents, reactions (9-7) to (9-9) are possible.

| Base | Solvent (A) | Acid | Base | |
|------|-------------|------|------|---|
| $B +$ | $H_2O$ | $\rightleftharpoons BH^+ +$ | $OH^-$ | (9-7) |
| $B +$ | $CH_3CH_2OH$ | $\rightleftharpoons BH^+ +$ | $CH_3CH_2O^-$ | (9-8) |
| $B +$ | $HC_2H_3O_2$ | $\rightleftharpoons BH^+ +$ | $C_2H_3O_2^-$ | (9-9) |

Reactions (9-7) and (9-8) are not likely since the conjugate acid, $BH^+$, and conjugate base, $OH^-$ or $CH_3CH_2O^-$, are much more acidic and basic, respectively, than the acid, $H_2O$ or $CH_3CH_2OH$, and base, B. This causes the equilibrium to lie to the left. In reaction (9-9), however, the acidity of $BH^+$ is compared to $HC_2H_3O_2$ and the basicity of $C_2H_3O_2^-$ to B with the result being that the equilibrium does not lie to the left. The conclusion is that, of the three solvents, glacial acetic acid would be the most appropriate one while a solvent with basic properties would be a disastrous choice.

A similar situation would exist for the titration of a weak acid; the difference being that a solvent possessing basic properties would be the appropriate one.

**Leveling.**    There is a major difficulty in using acidic or basic solvents. Since these are leveling solvents, they do not normally permit differentiation of mixtures of acids or bases. For example, only total basicity can be determined for a mixture of butylamine and pyridine if glacial acetic acid is used since it is the acetate ion that is being titrated.

$$CH_3CH_2CH_2CH_2NH_2 \ + \ HC_2H_3O_2 \rightleftharpoons CH_3CH_2CH_2CH_2NH_3^+ \ + \ C_2H_3O_2^-$$

$$\langle N \rangle + HC_2H_3O_2 \rightleftharpoons \langle NH^+ \rangle + C_2H_3O_2^-$$

In a similar fashion, basic solvents such as butylamine or ethylenediamine are leveling solvents for acids. As a general rule, the more acidic (or basic) the solvent is the greater its leveling power for bases (or acids). These types of solvents would be used to bring out very weakly acidic (basic solvent) or basic (acidic solvent) properties of a compound.

**Differentation.**    If the solvent is lacking in acidic or basic properties such as an aprotic solvent, the problem of leveling is eliminated and the solvent is useful for differentiation. In Fig. 9–1 titration data for a series of strong acids illustrate the differentiating power of the aprotic solvent methyl isobutyl ketone. In water, the same curves would be obtained for all the inorganic acids and thus, differentiation is not possible. However, in the ketone solvent differentiation is possible and provides a convenient method for the analysis of mixtures of the strong acids. Other interesting features are the separate titration break for each hydrogen of $H_2SO_4$ and the extent of the potential break. For $HClO_4$ the potential break is about 1400 mV (from $-700$ to $+700$). Since there are approximately 60 mV per pH unit, the titration curve corresponds to almost a 23.5 pH unit titration break. In water the titration is about a 12 pH unit break.

The differentiating power of methyl isobutyl ketone is not limited to strong acids as illustrated in Fig. 9–1. Figures 9–2 and 9–3 illustrate differentiating power of other solvents.

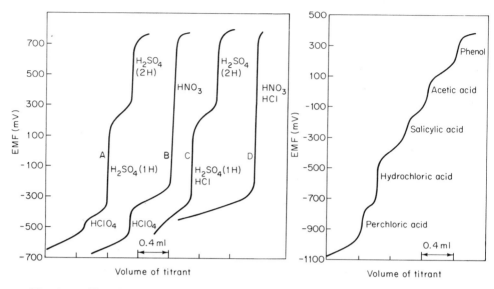

**Fig. 9–1.** Titration curves of strong acids in methyl isobutyl ketone–tetrabutylammonium hydroxide titrant. [L. G. Bruss and G. E. A. Wyld, *Anal. Chem.* **29,** 232(1957).]

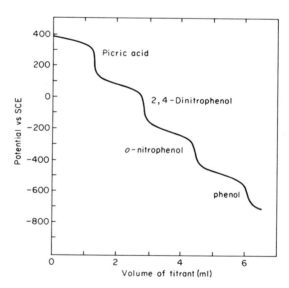

**Fig. 9–2.** Differentiation of a series of acids in *t*-butyl alcohol–tetrabutylammonium hydroxide titrant. [J. S. Fritz and L. W. Marple, *Anal. Chem.* **34,** 921(1962).]

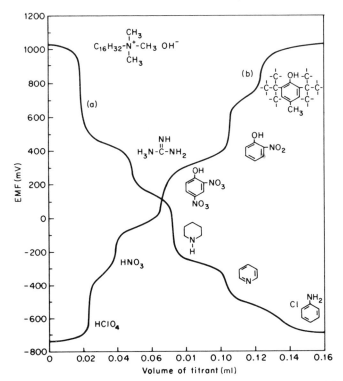

**Fig. 9–3.** Differentiation of acids and bases in 3-methylsulfolane with (a) HClO₄ and (b) tetrabutylammonium hydroxide titrant. [D. H. Morman and C. A. Harlow, *Anal. Chem.* **39,** 1869(1967).]

In summary, from a practical point of view, an inert or nonleveling type of solvent should be used when attempting to titrate a mixture of acids or bases in which little is known about their strengths. If the acids or bases are very weak, a leveling solvent, an acidic one for bases and a basic one for acids, is used to help bring out the acidic and basic properties of the solute. Generally this is done at the expense of differentiation.

The solvent should also satisfy other practical considerations such as the following:

1. Titrant and solute should be readily miscible with the solvent.

2. The solvent should be readily available, of low toxicity, easily purified, and inexpensive.

3. Products of the neutralization should be soluble in the solvent. If it is not, then a crystalline precipitate is desirable.

4. No side reactions should take place.

5. The solvent should have a reasonable dielectric constant, particularly if potentiometry is to be used for end-point detection.

## TITRANTS

Many different titrants employing a variety of solvents have been used in nonaqueous titrations. In general, the stronger acids or bases are used to obtain a well-defined titration curve. If the solvent that is used for the titrant is different than the one used for the sample, a mixed solvent results during the titration. In general, the mixture will impart a leveling property characteristic of the more leveling solvent in the mixture and, therefore, the titration range of the solvent mixture will be reduced. Hence, it is often advantageous to use the same solvent for both the titrant and the sample.

**Acidic Titrants.**     The acidic titrants are listed in Table 9–3. Perchloric acid, by far the most popular of all the common strong acids, is used generally in acetic acid or dioxane as solvent. Although dioxane gives the sharper titration breaks, it is more difficult to use because dioxane is often contaminated by water. Since the source of the $HClO_4$ is a 72% solution, water is automatically introduced into the titrant. In acetic acid, the introduced water is removed by adding a stoichiometric amount of acetic anhydride.

The formation of insoluble perchlorate neutralization salts is sometimes encountered with the $HClO_4$ titrant. If the precipitate is not crystalline, it can interfere in the function of the electrode when potentiometry is used for end-point detection.

Another important factor which accounts for the popular use of $HClO_4$ is that it has outstanding acid strength. Only a few other acids appear to be of similar strength. These are fluorosulfonic acid, trifluoromethylsulfonic acid, 2,4-dinitrobenzenesulfonic acid, and 2,4,6-trinitrobenzenesulfonic acid. The latter two are readily available and easily used.

The primary standards available for standardization are potassium hydrogen phthalate, tris(hydroxymethyl)aminomethane, diphenylguanidine, and sodium carbonate. Their use probably decreases in the order listed.

**Table 9–3.   Titrants for Nonaqueous Titrations**

| Acidic titrants | Basic titrants |
| --- | --- |
| $HClO_4$ in dioxane | Tetraalkylammonium hydroxides or alkoxides |
| $HClO_4$ in glacial acetic acid | Tetrabutylammonium hydroxide or alkoxides |
| Alkylsulfonic acids | $KOCH_3$ or $NaOCH_3$ |
| p-Toluenesulfonic acid | $NaOCH_2CH_2NH_2$ |
| 2,4-Dinitrobenzenesulfonic acid | Tetramethylguanidine |
| 2,4,6-Trinitrobenzenesulfonic acid | Diphenylguanidine |
| Fluorosulfonic acid | $NaC_2H_3O_2$ in $HC_2H_3O_2$ |
| Trifluoromethanesulfonic acid | |

**Basic Titrants.**    Table 9–3 lists the bases that have been used as titrants. Of these, the quaternary ammonium type bases are the most widely used. Solvents usually used for the bases are methanol, ethanol, or isopropanol or a mixed solvent of benzene and one of the alcohols.

The R group may be one or a combination of a wide variety of hydrocarbons; generally R = butyl. Two methods of preparation of the quaternary ammonium bases are available. In the first, tetrabutyl ammonium iodide is shaken with $Ag_2O$ in alcohol.

$$2\ R_4N^+I^- + Ag_2O + CH_3OH \rightarrow R_4N^+OH^- + R_4N^+OCH_3^- + 2\ AgI$$

The AgI is filtered and the solution made to volume with benzene (about 90%). Since methanol is acidic, isopropanol is a better choice, but, unfortunately, the solubility is also lower. The second method involves passing tetrabutylammonium iodide in isopropanol through a strongly basic anion exchange resin charged in the $OH^-$ form.

The problems in using the quaternary base stem from $CO_2$ and decomposition through the Hoffman reaction which produces tertiary amines, olefins, and water. The structure of R in the base can also affect the shape of the titration curve.

The quaternary bases, even though they are strong, are not ideal titrants. In fact there is no real ideal basic titrant. The alkali metal hydroxides cause glass electrode errors and very frequently form sparingly soluble neutralization salts. The quaternary bases, by virtue of the methods of preparation, will always have amphiprotic solvents in their solutions and the other organic bases listed in Table 9–3 are not strongly basic. Often potential response after the end point will be erratic when using the quaternary bases in the very basic solvents. The organic bases are sometimes useful in these situations.

Benzoic acid and phenylcinchoninic acid are suitable primary standards for standardization of the basic titrants. The former is used the most.


# END-POINT DETECTION

**Color Indicators.**    A very useful method of end-point detection is by color change of an acid–base indicator. As in aqueous media, an indicator which changes color at or very close to the stoichiometric point is chosen. A systematic listing of the indicators is done by listing the potential at which the indicator changes color. Thus, by knowing the potential at the stoichiometric point the appropriate indicator can be chosen.

The main limitation to using indicators is that indicator data for a particular solvent do not necessarily apply to all solvents in the same way. Thus, it is often necessary to have the indicator potential data for the solvent being used. As a general guideline, methyl violet or crystal violet, 1-naphtholbenzein, and methyl red are used for titrations of weak bases while thymol blue, azo violet (*p*-nitrobenzeneazoresorcinol), and *o*-nitroaniline are used for weak acids.

**Other Methods for End-Point Detection.**     Potentiometry is probably used as much as visual indicators for end point detection in nonaqueous titrations. This method has the advantage of furnishing a permanent record of the complete titration curve. Often anomalous behavior can be detected by examination of the curve. Response is generally rapid and the method is easily automated. The technique of measuring potential and the type of electrodes required for this measurement are discussed in Chapters 10 to 13 End points in nonaqueous titrations can also be conveniently detected by conductometry spectrophotometry, thermometry, and amperometry.

## APPLICATIONS

Nonaqueous acid–base titrations are a widely used, practical, technique in organic analysis. Many different functional groups can be titrated as acids or bases by choosing the correct solvent and titrant.

The technique of carrying out the titration requires more care and experimental skill than for aqueous acid–base titrations. Dry glassware must be used, and often it is necessary to prepare moisture-free solvents. Solutions and titrants must be protected from atmospheric water and carbon dioxide. In general practice, a water impurity is the main limitation since it reduces the size of the titration break. However, the effect of water on the titration varies according to the strength of the acids or bases being titrated; the stronger they are the smaller the effect. For very weak acids or bases the

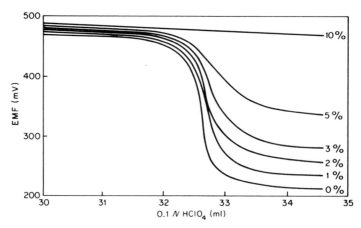

**Fig. 9–4.** The effect of water on the titration of bases in glacial acetic acid with perchloric acid titrant. [C. W. Pifer and E. G. Wollish, *Anal. Chem.* **24,** 300(1952).]

presence of as little as 0.1 to 0.2% water can eliminate the titration break. Figure 9–4 illustrates the effect of water on the nonaqueous titration of bases. Also, side reactions between the sample or titrant and the solvent are possible.

The extra care that must be utilized to obtain a successful titration has not hindered the application of nonaqueous titrations in solving practical problems. In general, the accuracy is better than 1% and often as good as 0.1%; the value will be determined by the solvent and titrant used and by the strength of the acid or base being titrated. In most instances, the titration is applied to analysis in the macro to semimicro range.

Nonaqueous titrations are frequently employed in all types of industrial areas. Quality control of organic products, pharmaceutical preparations, explosives, and biologically important compounds are some of the areas of application.

## INDIRECT ACID–BASE METHODS

Many useful quantitative procedures, which involve an indirect acid–base measurement, are available for the analysis of organic functional groups. In general, the functional group undergoes a reaction with an added reagent. For some, the product of the reaction is acidic or basic, which is titrated. In others the reagent, which is acidic or basic, is used in excess and the remaining amount is titrated. It is not always necessary that the final acid–base titration be carried out in a nonaqueous media. The following sections, which illustrate indirect acid–base methods, describe some of the more useful procedures.

**Hydroxy and Amine Group.**    The hydroxy and primary and secondary amine groups can be determined by an acetylation reaction.

Since the reaction is acid catalyzed (although other acids can be used, $HClO_4$ is the best), the acetylating reagent is a mixture of acetic anhydride and $HClO_4$ in ethyl acetate. A measured aliquot is added to the alcohol sample. After an appropriate reaction time, a water–pyridine mixture is added which hydrolyzes the excess anhydride.

The acid produced and the acid catalyst are titrated with standard base. An aliquot of the acetylating mixture minus the sample is treated in the same way. Therefore, the difference in the two titrations corresponds to the amount of hydroxyl group present in the sample. The acetylation reaction is also base-catalyzed; pyridine is usually used. In most cases acid catalysis provides a much faster reaction time.

**Carbonyl Group.**    A versatile method for the analysis of aldehydes and ketones is by the oxidation reaction.

$$\begin{aligned}\text{>C=O} + \text{H}_2\text{NOH} \rightleftarrows \text{>C=N-OH} + \text{H}_2\text{O}\end{aligned}$$

The problem with this reaction is that it is an equilibrium reaction, free hydroxylamine is readily air oxidized, and the reaction is acid catalyzed. The optimum reaction conditions are provided in the following way. An excess of $NH_2OH \cdot HCl$ (the hydroxylamine hydrochloride does not have to be accurately measured) is added to the carbonyl sample. Then, an accurately measured amount of base (a tertiary amine such as 2-$N,N$-dimethylamino-ethanol is used) is added and the reaction proceeds according to

$$NH_2OH \cdot HCl + B \rightleftarrows BH + Cl^- + NH_2OH$$

$$\text{excess}$$

$$NH_2OH + \text{>C=O} \longrightarrow \text{>C=NOH} + H_2O$$

$$NH_2OH + HClO_4 \longrightarrow NH_3OH + ClO_4^-$$
$$\text{excess} \qquad \text{titrant}$$

The $NH_2OH$ is stoichiometrically produced according to the amount of base added and after the reaction, the remaining $NH_2OH$ is titrated with standard acid. The reaction is carried out in methanol–isopropanol mixture and the acid is in Methyl Cellosolve ($CH_3OCH_2CH_2OH$) which provides a completely nonaqueous system.

**Epoxides.**    Epoxides are very important industrial chemicals that are used in polymer synthesis and as starting materials for related applications. Quantitative conversion to the chlorohydrin takes place when a measured excess of HCl in dioxane is added to the epoxide sample.

$$-C\!-\!\!-\!\!C- + HCl \longrightarrow -C-C-$$
$$\phantom{xxx}O\phantom{xxxxxxxxxxxxx}HO\ \ Cl$$

The remaining acid is titrated with standard base (NaOH in methanol). A more reactive acid, HBr, can be used for a direct titration of epoxides with glacial acetic as solvent.

**Esters.**     Esters, amides, and nitriles can be saponified under alkaline conditions.

$$RCO_2R' + NaOH \rightarrow RCO_2^-Na^+ + R'OH$$

$$RCONH_2 + NaOH \rightarrow RCO_2^-Na^+ + NH_3$$

$$RC{\equiv}N + NaOH + H_2O \rightarrow RCO_2^-Na^+ + NH_3$$

There are several different procedures for esters. However, for a fast, stoichiometric reaction, a large excess of NaOH is required. Therefore, accuracy is lost if a measured aliquot of concentrated NaOH solution is added and the remaining base titrated with standard acid. A better procedure is to treat the ester sample with a large excess of NaOH in a glycol solvent at elevated temperatures. Subsequently, the entire mixture is passed through a strong acid, H-form cation resin. Thus, the NaOH is converted to water and the $RCO_2^-Na^+$ is converted to $RCO_2H$ which is titrated with standard base. In this procedure, it is not required to know the exact amount of NaOH used.

## CALCULATIONS

*Example 9–1.*     An unknown containing a weakly acidic phenol was weighed (0.2213 g) and dissolved in 25 ml of dimethylformamide. The titration to the azo violet indicator end point (or by potentiometry) required 23.29 ml of 0.1010 *F* tetrabutylammonium hydroxide. Calculate the percent OH in the sample.

$$R'{-}OH + R_4NOH \rightarrow R_4N^+O{-}R' \; + HOH$$

(phenol
derivative)

It is assumed that the phenol is monoprotic; therefore, the reaction ratio is 1.

$$\%OH = \frac{ml_{R_4NOH} \times F_{R_4NOH} \times \text{reaction ratio} \times OH \times 100}{\text{weight of sample}}$$

$$\%OH = \frac{23.29 \text{ ml} \times 0.1010 \text{ mmole/ml} \times 1 \times 17.01 \text{ mg/mmole} \times 100}{221.3 \text{ mg}} = 18.07\%$$

*Example 9–2.*     It is common to report acid–base titration results in terms of neutralization equivalents. Calculate the neutralization equivalent for the unknown acid in Example 9–1. The formal concentration for the base must be changed to normality. Thus, since the reaction ratio is 1,

$$0.1010 \; F = 0.1010 \; N$$

and

$$\text{mg sample} = ml_{R_4NOH} \times N_{R_4NOH} \times Eq\ wt$$

$$221.3\ mg = 23.29\ ml \times 0.1010\ mEq/ml \times Eq\ wt$$

$$Eq\ wt = 94.08\ mg/mEq$$

If the exact stoichiometry for the reaction is known (number of replaceable acidic hydrogens per molecule) and the sample is pure, the molecular weight for the compound has been determined.

*Example 9–3.*    Isocyanates, which are used commercially for the preparation of polyurethanes, can be determined by reaction with excess butylamine and titration of the excess in dioxane with $HClO_4$. If 0.2151 g of the isocyanate is combined with 25.00 ml of 0.1157 *F* butylamine and 14.91 ml of 0.1153 *F* $HClO_4$ is required to titrate the excess butylamine, calculate the % SCN in the sample.

$$RSCN + CH_3CH_2CH_2CH_2NH_{2excess} \rightarrow RNHCSNHCH_2CH_2CH_2CH_3$$

$$CH_3CH_2CH_2CH_2NH_2 + HClO_4 \rightarrow CH_3CH_2CH_2CH_2NH_3^+ClO_4^-$$

$$\% \text{ SCN} = \frac{(ml_{BuA} \times F_{BuA} - ml_{HClO_4} \times F_{HClO_4}) \times \text{reaction ratio} \times \text{SCN} \times 100}{\text{wt of sample}}$$

% SCN

$$= \frac{(25.00\ ml \times 0.1157\ F - 14.91\ ml \times 0.1013\ F) \times 1 \times 58.08\ mg/mole \times 100}{215.1\ mg}$$

$$= 37.34\%$$

*Example 9–4.*    A 2.311-g sample of Carbowax was dissolved in a minimum amount of ethyl acetate. Exactly 5.00 ml of an acid catalyzed acetylating mixture $[HClO_4-(CH_3CO)_2O$ in ethyl acetate] is added to the sample and the container is closed. After a 10-minute reaction time a pyridine–water mixture is added which hydrolyzes the remaining acetic anhydride. Subsequently, the acidic species in solution are titrated with 33.45 ml of 0.5105 *F* methanolic NaOH. If a 5.00-ml aliquot of the acetylating mixture after hydrolysis required 42.50 ml of the NaOH titrant, calculate the %OH in the sample.

The reactions are

and consequently the reaction ratio is 1.

$$\%OH = \frac{(ml_{NaOH} \times F_{NaOH} - ml_{NaOH} \times F_{NaOH}) \times \text{reaction ratio} \times OH \times 100}{wt \ sample}$$

$$\%OH = \frac{(42.50 \ ml \times 0.5105 \ mmole/ml - 33.45 \ ml \times 0.5105 \ mmole/ml)}{231.1 \ mg} \times 1 \times 17.01 \ mg/mmole \times 100 = 3.40\%$$

Carbowax is an industrially important polymer and is used as a water-soluble lubricant in molds, textile fibers, metal-forming operations, hair preparations and other cosmetics, paints, polishes, and as an ointment base in pharmaceuticals. Its structure is

$$H-[-O-CH_2CH_2-O-]_n-H$$
Polyethylene glycol

and its molecule weight is in the range from 1000 to 6000. In this procedure the number of terminal hydroxy groups are being determined which can be correlated to molecular weight. Since different molecular weight ranges have different properties and applications, this analysis is very important in industry.

## Questions

1. What advantages does a nonaqueous acid–base titration have over an aqueous titration?
2. List the following functional groups in order of increasing acidity: enol, sulfonic acid, thiophenol, aliphatic carboxylic acid, phenol.
3. Differentiate between amphiprotic, aprotic, protogenic, and protophilic solvents.
4. Under what conditions would a basic solvent be used, an acidic solvent be used, a neutral solvent be used?
5. What is a differentiating solvent, a leveling solvent?
6. Why is HCl not frequently used as a titrant in nonaqueous titrations?
7. Explain why $HClO_4$ is a stronger acid in glacial acetic acid than in water.

8. Select a solvent and titrant for the titration of each of the following:

   a. Caffeine
   b. Sulfa drugs
   c. Amino acids
   d. Alkaloids
   e. Antihistamine hydrochlorides

   f. Fatty acids
   g. Butylamine–pyridine mixture
   h. Benzoic acid–phenol mixture
   i. $H_2SO_4$–HCl mixture
   j. Anhydrides

9. Why does water have to be excluded from nonaqueous titrations?

10. Salts of monosubstituted alkyl esters of sulfonic acids can be hydrolyzed with HCl to the corresponding alcohol and bisulfate ion.

$$\underset{\underset{O}{\overset{O}{\|}}}{ROSO^-} \xrightarrow[H_2O/HCl]{\Delta} ROH + \underset{\underset{O}{\overset{O}{\|}}}{HOSO^-}$$

    Since sodium alkyl sulfates are important anionic surfactants, discuss how this reaction might be useful as a potential method for their analysis.

11. Why is acetone not a suitable solvent for the strong acid titrant $HClO_4$? For the strong base titrant $(CH_3CH_2CH_2CH_2)_4N^+OH^-$?

12. Explain why pyridine is a suitable solvent for the titration of acids but not for bases.

13. What effect does dissolved $CO_2$ have on the titration of acids in methyl isobutyl-ketone? Of bases in acetonitrile?

14. Esters can be determined by saponification according to the reaction

$$RCO_2R' + NaOH \xrightarrow[\text{Solvent}]{\Delta} RCO_2^- + Na^+ + ROH$$

    Outline a procedure employing an acid–base titration which can be used for the determination of esters.

## Problems

1.* A $HClO_4/HC_2H_3O_2$ titrant was standardized against KHP (204.2). If 0.2600 g of KHP was neutralized by 15.11 ml of titrant, calculate the formal concentration of the titrant.

2. An unknown acidic sample weighing 0.3415 g was dissolved in acetone and titrated with 27.54 ml of 0.1100 $F$ $R_4NOH$. Calculate the percent $CO_2H$ in the sample.

3. An unknown amine sample weighing 0.2511 g was dissolved in $HC_2H_3O_2$ and titrated with 10.56 ml of 0.1511 $F$ $HClO_4/HC_2H_3O_2$. Calculate the percent $NH_2$ in the sample.

---

* Answers are listed at the end of the book for problems marked with an asterisk.

4.* A sample of a mixture of sulfathiazole (255.3) and sulfapyridine (249.3) weighing 0.7111 g was dissolved in acetone, and titrated with 0.0959 $F$ $R_4NOH$. The first and second break occurred at 13.12 ml and 21.65 ml, respectively. Calculate the percent of each sulfa in the mixture.

5.* A 0.2273-g epoxide sample was treated with 50.00 ml of a 0.1000 $F$ HCl/dioxane solution. The excess required 14.88 ml of 0.1240 $F$ NaOH. Calculate the percent epoxide ($C_2O$) in the sample.

6.* One of the tests for quality control of esters is the measurement of the "saponification value," which is defined as the number of mg of KOH required to neutralize the fatty acids resulting from the complete hydrolysis of one gram of the sample. A 1.000-g sample of castor oil is heated with 25.00 ml of approximately 0.2 $F$ KOH in ethanol. After cooling the excess KOH required 8.20 ml of 0.2050 $F$ HCl. A blank treated the same way required 24.02 ml of the acid. Calculate the saponification value for the castor oil.

7. A carbonyl sample weighing 2.0110 g was treated with exactly 20 ml of 0.2500 $F$ 2-dimethylaminoethanol and 25 ml of 0.4000 $F$ hydroxylammonium chloride. After an appropriate reaction time the solution was titrated with 14.35 ml of 0.2010 $F$ $HClO_4$. A blank prepared in the same way required 24.05 ml of the acid titrant. Calculate the percent CO in the sample.

8.* Primary amines can be determined by acetylation. A 0.6750-g sample of an amine is dissolved in ethyl acetate and treated with 10.00 ml of an acid catalyzed acetylating mixture. After suitable reaction time and hydrolysis the sample is titrated with 25.76 ml of 0.4107 $F$ NaOH. A 5.00-ml aliquot of the acetylating agent required 19.70 ml of the base after hydrolysis. Calculate the $\%NH_2$ in the sample.

9.* Codeine, an alkaloid, can be determined by titration with $HClO_4$ in acetic acid. If a codeine sample weighed 0.3169 g and 9.75 ml of a 0.1061 $F$ $HClO_4$ titrant is required for neutralization calculate the $\%$ purity of the sample.

10.* p-Aminosalicyclic acid (PAS) which is an antituberculosis drug and prepared in tablets is determined by pulverizing the tablets, extraction with anhydrous acetone, filtration, and titration of the filtrate with $R_4NOH$ titrant. A sample of the powder weighed 0.3123 g and 7.91 ml of 0.1081 $F$ $R_4NOH$ is required for neutralization. Calculate the mg of PAS per tablet if the average weight of a tablet is 1.2141 g.

# Chapter
# Ten
# Oxidation–Reduction
# Equilibria

## DEFINITIONS

An oxidation–reduction reaction (redox) is one in which the reactants undergo changes in oxidation state. In reaction (10-1),

$$Ce^{4+} + Fe^{2+} \rightleftharpoons Ce^{3+} + Fe^{3+} \tag{10-1}$$

which is also a suitable reaction for the determination of iron, cerium changes its oxidation state from $4+$ to $3+$, a gain of one electron, while iron changes from $2+$ to $3+$, a loss of one electron.

A gain of electrons is a reduction process, while oxidation is a loss of electrons. In a redox reaction both processes must occur. Reduction cannot take place in the absence of oxidation or vice versa. In addition, the total number of electrons lost must equal the total number of electrons gained.

The substance that decreases in oxidation state is the oxidizing agent; the substance that increases in oxidation state is the reducing agent. An oxidizing agent causes another substance to be oxidized, while it itself is reduced. In contrast the reducing agent causes another substance to be reduced, while it itself is oxidized.

In reaction (10-1) $Ce^{4+}$ is the oxidizing agent and $Fe^{2+}$ is the reducing agent. The oxidation and reduction steps can be represented by the half-reactions

$$Ce^{4+} + 1e = Ce^{3+} \quad \text{(Reduction)}$$

$$Fe^{2+} = Fe^{3+} + 1e \quad \text{(Oxidation)}$$

or in general by

$$A_{ox} + ne = A_{red}$$

$$B_{red} = B_{ox} + ne$$

In each half-reaction the oxidized and reduced forms can also be referred to as a redox couple.

It is possible that in another reaction $Fe^{3+}$ will be the oxidizing agent and $Fe^{2+}$ will be formed in the reaction. For example, $Fe^{3+}$ is reduced by $Sn^{2+}$

$$2Fe^{3+} + Sn^{2+} = 2Fe^{2+} + Sn^{4+}$$

where the half-reactions are

$$Fe^{3+} + 1e = Fe^{2+}$$

$$Sn^{2+} = Sn^{4+} + 2e$$

Even though the iron half-reaction will take place in both directions, it should not be confused with being an equilibrium reaction. As stated previously, one half-reaction will not take place by itself.

If a suitable oxidizing agent and reducing agent are brought together as in the case of $Ce^{4+}$–$Fe^{2+}$, a reaction takes place providing there is no kinetic problem. The complete reaction is the sum of the two half-reactions

$$Ce^{4+} + 1e = Ce^{3+}$$

$$\frac{Fe^{+2} = Fe^{3+} + 1e}{Ce^{4+} + Fe^{2+} = Ce^{3+} + Fe^{3+}}$$

In this example one electron is gained and one electron is lost. On the other hand, for the $Fe^{3+}$–$Sn^{2+}$ reaction one electron is gained, while two are lost. Since the number of electrons gained and lost must be equal, the $Fe^{3+}$–$Fe^{2+}$ half-cell must be multiplied by 2. Thus, two electrons are gained and two are lost or

$$2Fe^{3+} + 2e = 2Fe^{2+}$$

$$\frac{Sn^{2+} = Sn^{4+} + 2e}{2Fe^{3+} + Sn^{2+} = 2Fe^{2+} + Sn^{4+}}$$

The sum of the two yields the balanced reaction.

**Table of Oxidizing and Reducing Agents.** Oxidizing agents and their reduced forms can be arranged in order according to their ability to gain and lose electrons, respectively. Appendix IV presents such an arrangement for a selected group of half-reactions and is called the Table of Standard Reduction Potentials. This table and the ensuing conventions follow the recommendations set forth by the International Union of Pure and Applied Chemistry (IUPAC) in 1953.

Several important conclusions can be made from the manner in which Appendix IV is written. All the reactions are written as reductions. Thus,

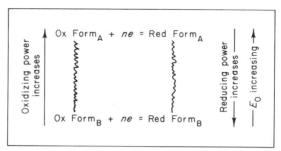

**Fig. 10–1.** Trends in a table of reduction potentials.

the oxidizing agents are on the left and the reducing agents are on the right. Those substances at the top of the oxidized form side (most positive $E$) are the stronger oxidizing agents, while the substances at the bottom of the reduced form side (most negative $E$) are the stronger reducing agents. The stronger the oxidizing and reducing agents are, the more complete the reaction is between the two, providing there are no kinetic factors. On the other hand, as the oxidizing agent and reducing agent become closer to each other in the chart the completeness of the reaction decreases. Figure 10–1 illustrates these conclusions.

From the Table of Standard Reduction Potentials many chemical properties can be calculated and predicted. For example, it is possible to predict if a reaction will occur, calculate the equilibrium constant, calculate the concentrations of the various species at equilibrium, calculate cell potentials that are produced, and balance redox equations. However, it is important to emphasize that it is not possible to predict from Appendix IV whether a reaction is fast or slow.

## ELECTROCHEMICAL CELLS

There are two kinds of electrochemical cells: galvanic and electrolytic cells. A galvanic cell is a cell in which a chemical change takes place spontaneously with the production of electrical energy. With this type of cell it is possible to convert the electrical energy into useful work. A typical example is the ordinary lead storage cell. In contrast an electrolytic cell is a cell in which nonspontaneous electrochemical reactions are forced to take place by impressing an external voltage to the cell. Thus, electrical energy, or work is consumed in order to provide a particular electrode reaction as in the case of chrome plating.

**Galvanic Cell.** Figure 10–2 illustrates a typical simple galvanic cell. The essential parts are a beaker containing a Zn rod in 1.0 $M$ $Zn^{2+}$, a Cu

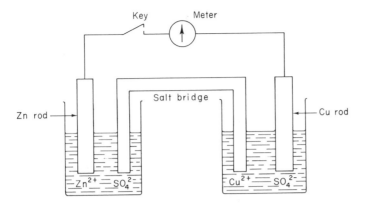

**Fig. 10–2.** A typical galvanic cell. Initial preparation of the cell is such that all concentrations are 1.00 $M$ (unit activity).

rod in 1.0 $M$ $Cu^{2+}$ in a second beaker, 1.0 $M$ $H^+$ in both beakers, and a salt bridge which connects the two solutions. The two rods are connected through a potential reading device. When the key is pressed a pathway for the flow of current through the wire, rods, solutions, and salt bridge is possible. The salt bridge contains KCl suspended in gelatin and provides an electrical pathway between the two solutions without mixing of the $Cu^{2+}$ and $Zn^{2+}$ solutions.

Immediately, as the key is closed, a potential of 1.100 V is observed on the potential readout and evidence of chemical reactions are readily apparent in the two beakers. The accuracy and number of significant figures involved will depend on the quality of the experimental setup. Two possible reactions can occur in each half-cell*:

$$\text{Left cell} \begin{cases} Zn^{2+} + 2e \rightarrow Zn° & \text{(Reduction)} \\ \\ Zn \rightarrow Zn^{2+} + 2e & \text{(Oxidation)} \end{cases}$$

$$\text{Right cell} \begin{cases} Cu^{2+} + 2e \rightarrow Cu° & \text{(Reduction)} \\ \\ Cu° \rightarrow Cu^{2+} + 2e & \text{(Oxidation)} \end{cases}$$

Since both oxidation and reduction must occur, the four reactions can be

---

\* It is also reasonable to speculate about the possibility of $H_2O$ or the anions participating in an electrochemical reaction. The fact that these reactions do not take place in this example will become readily apparent from later sections.

divided into two pairs:

$$Zn^{2+} + 2e \rightarrow Zn$$
$$\underline{Cu \rightarrow Cu^{2+} + 2e}$$
$$Zn^{+2} + Cu \rightarrow Cu^{2+} + Zn \qquad (10\text{-}2)$$

$$Cu^{2+} + 2e \rightarrow Cu$$
$$\underline{Zn \rightarrow Zn^{2+} + 2e}$$
$$Cu^{2+} + Zn \rightarrow Zn^{2+} + Cu \qquad (10\text{-}3)$$

It should be possible chemically to tell which reaction is taking place. For reaction (10-2) the weight of the zinc rod should increase and the copper rod decrease, while the concentration of $Zn^{2+}$ and $Cu^{2+}$ decreases and increases, respectively. The opposite would take place if reaction (10-3) were the case. Experimentally, the observation is that the zinc atoms (rod) are oxidized to zinc(II) ions by the electrons produced at the zinc electrode. These electrons flow through the external wire to the copper electrode and combine with the copper(II) ions. Thus, the copper(II) ions are reduced to Cu metal which plates out on the copper electrode.

It is not possible to measure experimentally the potential for an individual half-cell. Consequently, some particular half-cell must be established as the standard and the others are compared relative to the standard. Any half-cell with any assigned value could be chosen.

**Hydrogen Electrode.**    By agreement, all potentials are referred to the normal hydrogen electrode (NHE) as the reference electrode. This electrode,

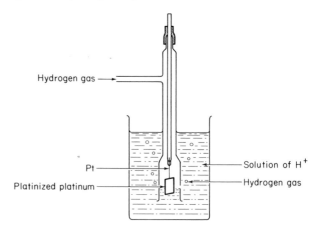

**Fig. 10–3.**  A typical hydrogen electrode at standard conditions. $H^+$ = unit activity; $T = 25°C$; $H_2 = 1$ atm pressure.

provided it meets a series of specified conditions, is arbitrarily assigned a value of zero. Thus, to determine the $Zn^{2+}$–Zn and $Cu^{2+}$–Cu half-cell potentials, each is combined with the NHE and the cell potential is measured. Since the NHE has a fixed, reproducible potential of 0.000 V, by definition, the potential for the other half-cell is readily calculated. Before illustrating how the half-cell potentials are calculated it is necessary to discuss the properties of the NHE.

Figure 10–3 illustrates a typical design of a hydrogen electrode. A spongy Pt or platinized platinum surface which has a black appearance is used rather than a shiny surface because the former is able to respond to the half-cell reaction (10-4) in a reversible manner:

$$2H^+ + 2e \rightleftharpoons H_2 \qquad (10\text{-}4)$$

This special type of platinum electrode is made by electroplating platinum on a platinum electrode from a platinum(IV)–HCl solution.

In addition to reversibility, other requirements are a hydrogen gas pressure of 1 atm, a hydrogen ion activity of unity, and a temperature of 25°C.

If a cell composed of the NHE and the $Zn^{2+}$–Zn half-cells is prepared, a voltage of 0.763 V is measured and two sets of reactions are possible:

$$\begin{array}{l} Zn^{2+} + 2e \rightarrow Zn \text{ (Reduction)} \\ \underline{\phantom{xx} H_2 \rightarrow 2H^+ + 2e \text{ (Oxidation)}} \\ Zn^{2+} + H_2 \rightarrow Zn + 2H^+ \end{array} \qquad (10\text{-}5)$$

or

$$\begin{array}{l} 2H^+ + 2e \rightarrow H_2 \text{ (Reduction)} \\ \underline{\phantom{xx} Zn \rightarrow Zn^{2+} + 2e \text{ (Oxidation)}} \\ 2H^+ + Zn \rightarrow Zn^{2+} + H_2 \end{array} \qquad (10\text{-}6)$$

The latter reaction [Eq. (10-6)], where $H^+$ is being reduced and Zn is being oxidized, is the one that takes place. Since the NHE is 0.000 V, the zinc half-cell as written must be 0.763 V.

$$E^\circ_{H^+, H_2} + E^\circ_{Zn, Zn^{2+}} = 0.763 \text{ V}^*$$

Thus, the following can be written:

$$Zn \rightarrow Zn^{2+} + 2e \qquad E^\circ_{Zn, Zn^{2+}} = 0.763 \text{ V} \qquad (10\text{-}7)$$

---

* Note the nomenclature: $E^\circ_{H^+, H_2}$ means the potential of the half-cell reaction is written as a reduction process, while $E^\circ_{Zn, Zn^{2+}}$ means the potential of the half-cell reaction is written as an oxidation.

and

$$Zn^{2+} + 2e \rightarrow Zn \qquad E^{\circ}_{Zn^{2+},Zn} = -0.763 \text{ V} \qquad (10\text{-}8)$$

For the $Cu^{2+}$, Cu, and NHE cell a potential of 0.337 V would be found and two sets of reactions are possible:

$$
\begin{array}{l}
Cu^{2+} + 2e \rightarrow Cu \quad \text{(Reduction)} \\
\underline{\quad H_2 \rightarrow 2H^+ + 2e \quad \text{(Oxidation)}} \\
Cu^{2+} + H_2 \rightarrow 2H^+ + Cu
\end{array} \qquad (10\text{-}9)
$$

or

$$
\begin{array}{l}
2H^+ + 2e \rightarrow H_2 \quad \text{(Reduction)} \\
\underline{\quad Cu \rightarrow Cu^{+2} + 2e \quad \text{(Oxidation)}} \\
2H^+ + Cu \rightarrow Cu^{2+} + H_2
\end{array} \qquad (10\text{-}10)
$$

The reaction that takes place is (10-9) or the reduction of $Cu^{2+}$ and oxidation of $H_2$. Since the NHE is 0.000 V the copper half-cell must be 0.337 V or

$$E^{\circ}_{Cu^{2+},Cu} + E^{\circ}_{H_2,H^+} = 0.337 \text{ V}$$

$$E^{\circ}_{Cu^{2+},Cu} = 0.337 \text{ V}$$

and, therefore, it can be written that

$$Cu^{2+} + 2e \rightarrow Cu \qquad E^{\circ}_{Cu^{2+},Cu} = 0.337 \text{ V} \qquad (10\text{-}11)$$

or

$$Cu \rightarrow Cu^{2+} + 2e \qquad E^{\circ}_{Cu,Cu^{2+}} = -0.337 \text{ V} \qquad (10\text{-}12)$$

Following the recommended conventions, the half-cell reactions would be expressed as reductions and their potentials are listed in an order relative to the NHE is the Table of Standard Reduction Potentials, Appendix IV. The values for the $Zn^{2+}$–Zn couple (10-8), and for the $Cu^{2+}$–Cu couple (10-11) are included in Appendix IV because standard conditions were used in these half-cells. In summary, the following criteria are used to define standard conditions:

1. Hydrogen ion is unit activity (1.00 $M$).
2. Concentrations of all soluble species are unit activity (1.00 $M$).
3. Partial pressure of all gases are 1.00 atm.
4. Temperature is 25°C.
5. Complexing agents should be absent.

Not all half-cell potentials are determined by direct comparison to the hydrogen electrode. In many cases, which will become apparent in later chapters, the values can and are obtained by calculation. It is also important to remember that it is immaterial whether a potential is referred to the NHE

or any other reference electrode since the difference between electrode potentials is independent of the electrode used as a reference.


## CELL CALCULATIONS

Cell calculations are important to the analytical chemist because this
information can be used to predict the extent of a redox reaction and potential
change as a function of concentration. The following summarizes the basic
principles required for these calculations.

   1. By convention a shorthand representation is used to describe complete
cells. Several general rules facilitate the writing of cells. These are reviewed in
the following:

    a. Molecules, elements, gases, and electrode materials are represented
by the usual chemical symbols. Concentrations of ions and molecules and
partial pressures of gases are cited in parentheses.

    b. A single vertical line ($|$) is used to illustrate a boundary between an
electrode phase and solution phase or between two different solution
phases. The potential across the interface is included in the total potential
of the cell.

    c. A double vertical line ($\|$) represents an electrolytic contact between
the two half-cells, such as a salt bridge, which has zero potential difference (a small potential actually occurs at the contact and is known as the
liquid junction potential) across the phase boundary. At this point in
this introductory discussion the liquid junction potential, which is
usually small and often minimized experimentally, is considered to be
zero.

    d. Arbitrarily, the right electrode in the galvanic cell is assumed to be
the cathode and the left electrode the anode and are therefore, positively
and negatively charged, respectively.

   2. A general equation describing the cell potential is given by

$$E^{\circ}_{\text{right (red)}} + E^{\circ}_{\text{left (ox)}} = E^{\circ}_{\text{cell}} \qquad (10\text{-}13)$$

Alternatively, the cell potential is calculated in terms of reduction potentials
by the equation

$$E^{\circ}_{\text{right (red)}} - E^{\circ}_{\text{left (red)}} = E^{\circ}_{\text{cell}} \qquad (10\text{-}14)$$

The advantage of Eq. (10-13) is that the half-cells and their respective
potentials are written as they occur. Thus, the overall reaction is readily
obtained by addition of the two balanced half-cells.

   3. If the reactants are at their standard states the free energy change for

the cell is given by

$$\Delta G^\circ = -nFE^\circ \tag{10-15}$$

where superscript $^\circ$ refers to standard conditions, $\Delta G^\circ$ is the standard free energy change in joules,* $n$ is the number of electrons taking part in the reaction, $F$ is 1 Faraday (96,487 coulombs), and $E^\circ$ is the cell voltage at standard conditions.

4. For a positive $E_{cell}$ (free energy change is negative) the reaction can take place as written and is said to be spontaneous (galvanic cell). If the $E_{cell}$ is negative (free energy change is positive) the reaction is nonspontaneous; energy must be put into the reaction in order for it to take place as written (electrolytic cell). The sign *does not* indicate whether the reaction actually takes place. This is a kinetic problem and the reaction rate can only be determined by the investigation of the kinetics of the reaction.

5. The potential can be calculated for a half-cell at conditions other than standard conditions provided all concentrations and the standard reduction potential are known. This calculation is possible with the Nernst equation.

*Example 10–1.*    Write the shorthand representation for the cell in Fig. 10–2 as written and for its reverse direction. Calculate the potential for each cell.

The shorthand representation and the reaction for the cell as written according to the conventions are

$$Cu^{2+} + Zn \rightarrow Zn^{2+} + Cu$$

$$-Zn \mid Zn^{2+}(1M) \parallel Cu^{2+}(1M) \mid Cu(+)$$
$$\text{anode oxidation} \qquad \text{cathode reduction}$$

and the cell potential is

$$E^\circ_{\text{right (red)}} + E^\circ_{\text{left (ox)}} = E^\circ_{cell}$$

$$0.337 \text{ V} + 0.763 \text{ V} = +1.100 \text{ V}$$

or by using reduction potentials, it is

$$E^\circ_{\text{right (red)}} - E^\circ_{\text{left (red)}} = E^\circ_{cell}$$

$$0.337 \text{ V} - (-0.763 \text{ V}) = +1.100 \text{ V}$$

Since a positive potential is calculated, the reaction can take place as written with the production of a potential corresponding to 1.100 V.

If the diagram in Fig. 10–2 is reversed or the zinc couple is on the right and the copper couple on the left, the reaction and its cell are

$$Zn^{2+} + Cu \rightarrow Cu^{2+} + Zn$$

$$Cu^\circ \mid Cu^{2+}(1 \text{ } M) \parallel Zn^{2+}(1 \text{ } M) \mid Zn^\circ$$

_____

* 1 cal/mole = 4.184 joules. It is customary to express free energies in cal or kcal/mole rather than in joules.

and the cell potential is

$$E^\circ_{\text{right (red)}} + E^\circ_{\text{left (ox)}} = E^\circ_{\text{cell}}$$

$$-0.763 \text{ V} + (-0.337 \text{ V}) = -1.100 \text{ V}$$

or by using reduction potentials it is

$$E^\circ_{\text{right (red)}} - E^\circ_{\text{left (red)}} = E^\circ_{\text{cell}}$$

$$-0.763 \text{ V} - (+0.337 \text{ V}) = -1.100 \text{ V}$$

The reaction as written is nonspontaneous and energy would have to be supplied from the surroundings to the system in order to make it proceed as written.

*Example 10–2.* Calculate the cell potential for the cell.

$$(-)\text{Pt, H}_2(1 \text{ atm})| \text{ HCl}(1 \ M), \text{AgCl}_{(s)}| \text{ Ag}(+)$$

Anode–Oxidation      Cathode–Reduction

$$\text{H}_2 + 2\text{AgCl} \rightarrow 2\text{HCl} + 2\text{Ag}$$

$$E^\circ_{\text{cell}} = E^\circ_{\text{AgCl,Ag}} - E^\circ_{\text{H}^+,\text{H}_2} = 0.222 \text{ V} - 0 \text{ V} = +0.222 \text{ V}$$

*Example 10–3.* Calculate the cell potential for the cell.

$$(-)\text{Pt} | \text{Ce}^{3+}(1 \ M), \text{Ce}^{4+}(1 \ M), 1 \ F \text{ HClO}_4 \,||\, \text{Fe}^{3+}(1 \ M), \text{Fe}^{2+}(1 \ M)| \text{ Pt}(+)$$

$$\text{Ce}^{3+} + \text{Fe}^{3+} \rightarrow \text{Ce}^{4+} + \text{Fe}^{2+}$$

$$E^\circ_{\text{cell}} = E^\circ_{\text{Fe}^{3+},\text{Fe}^{2+}} - E^\circ_{\text{Ce}^{4+},\text{Ce}^{3+}} = 0.771 \text{ V} - (+1.70 \text{ V}) = -0.93 \text{ V}$$

*Example 10–4.* Calculate the cell potential for the cell.

$$(-)\text{Pb} | \text{PbSO}_{4(s)}, \text{SO}_4^{-2}(1 \ M) \,||\, \text{Cl}^-(1 \ M), \text{AgCl}_{(s)} | \text{ Ag}(+)$$

$$2\text{AgCl}_{(s)} + \text{Pb} + \text{SO}_4^{-2} \rightarrow 2\text{Ag}^\circ + \text{PbSO}_{4(s)} + 2\text{Cl}^-$$

$$E^\circ_{\text{cell}} = E^\circ_{\text{AgCl,Ag}} - E^\circ_{\text{PbSO}_4,\text{Pb}} = 0.222 \text{ V} - (-0.356 \text{ V}) = +0.578 \text{ V}$$

These three examples also illustrate several other useful properties of cells and their design. The cell in Example 10–2 illustrates a cell without a liquid junction potential. Figure 10–4 shows a typical design. Cells of this type have been very useful in establishing precisely hydrogen ion activities in solution. Both half-cells in Example 10–3 are composed of only ionic species, while in Example 10–4 the two half-cells are dependent on the concentration of $\text{SO}_4^{2-}$ and $\text{Cl}^-$ in solution, respectively, even though neither are directly involved in changes in oxidation state. This concluded by examination of the half-cells. For example, for the left half-cell the reaction is

$$\text{PbSO}_{4(s)} + 2e \rightleftharpoons \text{Pb}_{(s)} + \text{SO}_4^{2-}$$

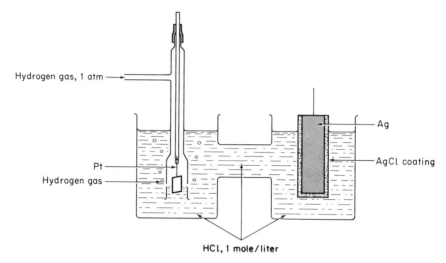

Hydrogen gas, 1 atm

Ag

AgCl coating

Pt

Hydrogen gas

HCl, 1 mole/liter

**Fig. 10–4.** A cell design for Example 10–2 which also illustrates a cell without liquid junction potential.

By applying the Le Chatelier principle any changes in $SO_4^{2-}$ concentration will alter the equilibrium position.

The shorthand cell description duplicates the picture of the drawn cell by describing the chemical reaction and indicating the phase separations and electrolytic contacts (liquid junction potential). It should be apparent that if spontaneous cells are to be devised from the half-cell reduction potentials, the half-cell with the most positive reduction potential automatically becomes the right electrode according to the conventions.

## NERNST EQUATION

In all the previous examples the concentrations (activities) have been expressed as being at standard conditions or 1.00 $M$ (unit activities). Most conditions that would be encountered in the laboratory would be something other than standard conditions. Therefore, an important question is, what is the effect of concentration (activity) on electrode potential?

The quantitative relationship between reduction potentials and concentrations of the reduced and oxidized forms is given by the Nernst equation. For the general reversible half-cell reaction

$$aA + bB + \cdots + ne = cC + dD + \cdots \tag{10-16}$$

The Nernst equation is expressed as

$$E_{\text{ox,red}} = E_{\text{ox,red}}^{\circ} - \frac{RT}{nF} \ln \frac{a_C^c a_D^d}{a_A^a a_B^b} \tag{10-17}$$

where

$E_{\text{ox,red}}$ = reduction potential in volts

$E^{\circ}_{\text{ox,red}}$ = standard reduction potential in volts

$R$ = gas constant and has the value 8.314 joules $^{\circ}K^{-1}$

$T$ = absolute temperature and has the value 298°K (25°C)

$n$ = number of electrons participating in the half-cell reaction

$F$ = Faraday and has the value 96,487 coulombs/equivalent

$\ln$ = natural logarithm and has the value 2.303 $\log_{10}$

$a$ = the activity of each of all species involved in the half cell raised to the power corresponding to the number of moles of each participating in the half cell

Substituting all the values and making the approximation that activities and concentrations are equal, yields the Nernst equation in the form

$$E_{\text{ox,red}} = E^{\circ}_{\text{ox,red}} - \frac{0.0592}{n} \log \frac{[C]^c[D]^d}{[A]^a[B]^b} \qquad (10\text{-}18)$$

For more precise calculations the value 0.05916 should be used. It should be noted that the log portion of the Nernst equation is written as products/reactants following the convention adopted for equilibrium constants. Alternatively it can be written as reactants/products which requires a change of sign in front of the log term or

$$E_{\text{ox,red}} = E^{\circ}_{\text{ox,red}} + \frac{0.0592}{n} \log \frac{[A]^a[B]^b}{[C]^c[D]^d} \qquad (10\text{-}19)$$

If the reactant is a gas, it is expressed as a partial pressure in units of atmospheres. For pure solids or liquids, which are present as a second phase, their concentrations remain constant (activities = 1) and are included in the value $E^{\circ}$. Since water is the solvent, its concentration is also invariant and its contribution is included in the $E^{\circ}$ value. If water is not the solvent and $H_2O$ appears in the half-cell it would be considered in the Nernst expression for the half-cell.

*Example 10-5.* Write the Nernst equation for the half-cell

$$MnO_4^- + 8H^+ + 5e = Mn^{2+} + 4H_2O$$

$$E_{MnO_4^-,\, Mn^{2+}} = E^{\circ}_{MnO_4^-,\, Mn^{2+}} - \frac{0.0592}{5} \log \frac{[Mn^{2+}]}{[MnO_4^-][H^+]^8}$$

**Formal Potentials.**　　Frequently, calculated potentials do not agree with experimentally observed potentials. Two principal reasons account for this difference. First, concentrations are used in the Nernst equation even though the measurements in the laboratory are activities. Since electrochemical cells tend to be modestly concentrated rather than dilute, interactions between ions, solvent molecules, and with each other can be significant. Hence, the approximation that activities being equal to concentration can often be in error. In general, this error is small and can be overlooked except when dealing with concentrated solutions, with fundamental studies, or in studies that require the utmost accuracy.

The second contributing factor to the observed differences in the calculated and experimental potentials is that the species in the half-cells do not exist in solution as simple ions or molecules as written in the half-cell. Often they participate in a series of competing equilibria. For example, in the cell in Fig. 10–2 both the $Zn^{2+}$ and $Cu^{2+}$ concentrations would be affected by the presence of $Cl^-$, since this ion is capable of complexing with the two cations. It is also possible for the two metal ions to exist in a hydrolyzed form. In the cell in Fig. 10–4 the solution in contact with the Ag/AgCl electrode could contain the species $Ag^+$, $AgCl_{(s)}$, $AgCl_{(aq)}$, $AgCl_2^-$, $AgCl_3^{2-}$, and $AgCl_4^{3-}$. The net effect of these equilibria is to reduce the equilibrium concentration of the free metal ion, thus affecting the electrode potential. If the existence of the competing equilibria are known and if the corresponding equilibrium constants are available then the effect of the competing equilibria can be accounted for. Since these data are often lacking these corrections can not be routinely computed.

To overcome these problems a secondary type of reduction potential can be defined. This potential is referred to as the *formal reduction potential* and differs from the *standard reduction potential* in that the oxidant and reductant are at unit formal concentrations in a specified electrolytic solution. The numerical values and sign are still relative to the NHE. Appendix IV lists formal potentials for several different systems. It should be emphasized that a formal potential for a specified set of experimental conditions is applicable only to that specific system and should not be applied to a different set of experimental conditions.

If a formal reduction potential is used rather than a standard reduction potential, the symbol $E^\circ_{ox,red}$ becomes $E^f_{ox,red}$ and concentrations are expressed in formal units.

*Example 10–6.*　　Calculate the potential at a platinum wire in a half-cell containing $1.00 \times 10^{-4}$ *F* Fe(III) and $1.00 \times 10^{-2}$ *F* Fe(II) in 1 *F* HCl.

$$Fe^{3+} + 1e \rightleftharpoons Fe^{2+}$$

$$E = E^f_{Fe^{3+},Fe^{2+}} - \frac{0.0592}{1} \log \frac{[Fe^{2+}]}{[Fe^{3+}]}$$

$$E = +0.700 \text{ V} - \frac{0.0592}{1} \log \frac{[0.01]}{[0.0001]}$$

$$E = +0.700 \text{ V} - 0.118 \text{ V} = +0.582 \text{ V}$$

**Example 10-7.** Calculate the potential at a platinum wire in a half-cell containing $1.00 \times 10^{-2} F$ Ce(IV) and $1.00 \times 10^{-3} F$ Ce(III) in $1 F$ HCl.

$$\text{Ce}^{4+} + 1e \rightleftharpoons \text{Ce}^{3+}$$

$$E = E_{\text{Ce}^{4+},\text{Ce}^{3+}} - \frac{0.0592}{1} \log \frac{[\text{Ce}^{3+}]}{[\text{Ce}^{4+}]}$$

$$E = 1.28 \text{ V} - \frac{0.0592}{1} \log \frac{[0.001]}{[0.01]}$$

$$E = +1.28 \text{ V} - (-0.0592 \text{ V}) = 1.339 \text{ V}$$

**Example 10-8.** Calculate the cell potential for the cell made by combining the half-cells in Examples 10-6 and 10-7 in the following way:

Pt $|$ Fe$^{2+}$ (0.0100 $F$), Fe$^{3+}$ (0.000100 $F$), 1 $F$ HCl $||$ Ce$^{4+}$ (0.0100 $F$), Ce$^{3+}$ (0.00100 $F$),

$$1 \ F \text{ HCl} \ | \ \text{Pt}$$

For this cell the chemical reaction is

$$\text{Fe}^{2+} + \text{Ce}^{4+} \rightarrow \text{Fe}^{3+} + \text{Ce}^{3+}$$

$$E_{\text{cell}} = E_{\text{Ce}^{4+},\text{Ce}^{3+}} - E_{\text{Fe}^{3+},\text{Fe}^{2+}}$$

$$E_{\text{cell}} = E'_{\text{Ce}^{4+},\text{Ce}^{3+}} - \frac{0.0592}{1} \log \frac{[\text{Ce}^{3+}]}{[\text{Ce}^{4+}]} - \left[ E'_{\text{Fe}^{3+},\text{Fe}^{2+}} - \frac{0.0592}{1} \log \frac{[\text{Fe}^{2+}]}{[\text{Fe}^{3+}]} \right]$$

$$E_{\text{cell}} = 1.28 \text{ V} - \frac{0.0592}{1} \log \frac{[0.00100]}{[0.0100]} - 0.700 \text{ V} + \frac{0.0592}{1} \log \frac{[0.0100]}{[0.000100]}$$

$$E_{\text{cell}} = 1.28 \text{ V} - (-0.0592 \text{ V}) - 0.700 \text{ V} + 0.118 \text{ V} = 0.757 \text{ V}$$

**Example 10-9.** Write the reaction and calculate the potential for the cell

Pt $|$ Fe$^{2+}$ (0.0100 $F$), Fe$^{3+}$ (0.00100) $F$), HCl (1 $F$) $||$ Cr$_2$O$_7^{2-}$ (0.0200 $F$), Cr$^{3+}$ (0.00500 $F$),

$$\text{HCl} \, (0.100 \ F) \ | \ \text{Pt}$$

Reaction:

$$1 \, (\text{Cr}_2\text{O}_7^{2-} + 14\text{H}^+ + 6e \rightarrow 2\text{Cr}^{3+} + 7\text{H}_2\text{O})$$

$$\frac{6 \, (\text{Fe}^{2+} \rightarrow \text{Fe}^{3+} + 1e)}{\text{Cr}_2\text{O}_7^{2-} + 14\text{H}^+ + 6\text{Fe}^{2+} \rightarrow 2\text{Cr}^{3+} + 6\text{Fe}^{3+} + 7\text{H}_2\text{O}}$$

Cell potential:

$$E_{cell} = E_{Cr_2O_7{}^{2-},Cr^{3+}} - E_{Fe^{3+},Fe^{2+}}$$

$$E_{cell} = E^f_{Cr_2O_7{}^{2-},Cr^{3+}} + \frac{0.0592}{6} \log \frac{[Cr^{3+}]^2}{[Cr_2O_7{}^{2-}][H^+]^{14}} - \left[ E^f_{Fe^{3+},Fe^{2+}} - \frac{0.0592}{1} \log \frac{[Fe^{2+}]}{[Fe^{3+}]} \right]$$

$$E_{cell} = 0.93 \text{ V} + \frac{0.0592}{6} \log \frac{[0.00500]^2}{[0.0200][0.100]^{14}} - 0.700 \text{ V} + \frac{0.0592}{1} \log \frac{[0.0100]}{[0.00100]}$$

$$E_{cell} = 0.93 \text{ V} - 0.109 \text{ V} - 0.700 \text{ V} + 0.0592 \text{ V} = 0.180 \text{ V}$$

*Example 10–10.*    Write the reaction and calculate the potential for the cell

Ag | AgCl$_{(s)}$, KCl (0.0400 $M$) || H$^+$ (0.00700 $M$), H$_2$ (0.400 atm) | Pt

Reaction:

$$2H^+ + 2e = H_2$$

$$\underline{2(Ag + Cl^- = AgCl_{(s)} + 1e)}$$

$$2Ag + 2H^+ + 2Cl^- = H_2 + 2AgCl$$

Cell potential:

$$E_{cell} = E_{H^+,H_2} - E_{AgCl,Ag}$$

$$E_{cell} = E^o_{H^+,H_2} - \frac{0.0592}{2} \log \frac{P_{H_2}}{[H^+]^2} - \left[ E^o_{AgCl,Ag} - \frac{0.0592}{1} \log \frac{[Cl^-]}{1} \right]$$

$$E_{cell} = 0 \text{ V} - \frac{0.0592}{2} \log \frac{[0.400]}{[7.00 \times 10^{-3}]^2} - 0.222 \text{ V} + \frac{0.0592}{1} \log 4.00 \times 10^{-2}$$

$$E_{cell} = 0 \text{ V} - 0.116 \text{ V} - 0.222 \text{ V} - 0.0828 \text{ V} = -0.421 \text{ V}$$

Since the cell potential is negative, the reaction is nonspontaneous. This means that over 0.421 V would have to be applied to the cell to initiate the reaction.

## EQUILIBRIUM CONSTANT

The equilibrium constant for a redox reaction will reveal how complete the reaction is at equilibrium. Consequently, this information is extremely valuable in predicting whether a particular redox reaction is useful for quantitative analysis.

In Chapter 6 it was shown that the equilibrium constant is related to $\Delta G°$, the standard free energy, by the equation

$$\Delta G° = -RT \ln K \qquad [\text{see Eq. (6-31)}] \qquad (10\text{-}20)$$

Combining Eqs. (10-15) and (10-19) yields

$$-nFE° = -RT \ln K$$

and

$$\ln K = \frac{nFE^\circ_{\text{cell}}}{RT} \qquad (10\text{-}21)$$

Substitution for $F$, $R$, $T$, and conversion to log form leads to

$$\log K = \frac{nE^\circ_{\text{cell}}}{0.0592} \qquad (10\text{-}22)$$

*Example 10–11.* Calculate the equilibrium constant for the reaction between Fe(II) and Ce(IV).

The two half-cells, standard reduction potentials, final reaction, and equilibrium constant expression are as follows:

$$Ce^{4+} + 1e \rightarrow Ce^{3+}; \qquad E^\circ_{Ce^{4+},Ce^{3+}} = 1.70 \text{ V}$$

$$Fe^{3+} + 1e \rightarrow Fe^{2+}; \qquad E^\circ_{Fe^{3+},Fe^{2+}} = 0.771 \text{ V}$$

$$Ce^{4+} + Fe^{2+} \rightarrow Ce^{+3} + Fe^{3+}$$

$$K = \frac{[Ce^{3+}][Fe^{3+}]}{[Ce^{4+}][Fe^{2+}]}$$

$$\log K = \frac{(E^\circ_{\text{right (red)}} - E^\circ_{\text{left (red)}})^*}{0.0592}$$

$$\log K = \frac{(1)(1.70 - 0.771)}{0.0592} = 15.69$$

$$K = 10^{15.69} = 4.90 \times 10^{15}$$

*Example 10–12.* Calculate the solubility product of AgCl from the data

$$Ag^+ + 1e = Ag_{(s)} \qquad\qquad E^\circ_{Ag^+,Ag} = +0.799 \text{ V}$$

$$AgCl_{(s)} + 1e = Ag_{(s)} + Cl^- \qquad E^\circ_{AgCl,Ag} = +0.222 \text{ V}$$

$$\begin{array}{c} AgCl_{(s)} + 1e = Ag_{(s)} + Cl^- \\ Ag_{(s)} = Ag^+ + 1e \\ \hline AgCl_{(s)} = Ag^+ + Cl^- \end{array}$$

$$\log K_{\text{sp}} = \frac{1[0.222 - (+0.799)]}{0.0592}$$

$$K_{\text{sp}} = 1.82 \times 10^{-10}$$

---

* This can also be written as $\log K = (E^\circ_{\text{right (red)}} + E^\circ_{\text{left (ox)}})/0.0592$.

*Example 10–13.*      A $2.00 \times 10^{-2}$ $F$ solution of NaI saturated with AgI is made part of the following cell after reaching equilibrium.

$$\text{Ag} \mid \text{AgI}_{(s)}, \text{NaI} (0.0200 \ M) \mid\mid \text{H}^+ (1 \ M), \text{H}_2 (1 \text{ atm.}) \mid \text{Pt}$$

If the cell potential is 0.0480 V, calculate the $K_{sp}$ for AgI. The half-cell reactions are

$$2\text{H}^+ + 2e = \text{H}_{2(g)}$$

$$\text{Ag}^+ + 1e = \text{Ag}$$

$$E_{cell} = E_{\text{H}^+,\text{H}_2} - E_{\text{Ag}^+,\text{Ag}}$$

$$E_{cell} = E^{\circ}_{\text{H}^+,\text{H}_2} - \frac{0.0592}{2} \log \frac{P_{\text{H}_2}}{[\text{H}^+]^2} - \left[ E^{\circ}_{\text{Ag}^+,\text{Ag}} - \frac{0.0592}{1} \log \frac{1}{[\text{Ag}^+]} \right]$$

$$0.0480 \text{ V} = 0.00 \text{ V} - \frac{0.0592}{2} \log - \frac{1}{[1]^2} - 0.799 \text{ V} + \frac{0.0592}{1} \log \frac{1}{[\text{Ag}^+]}$$

$$[\text{Ag}^+] = 4.16 \times 10^{-15}$$

$$\text{AgI}_{(s)} \rightleftharpoons \text{Ag}^+ + \text{I}^-$$

$$[\text{I}^-] = 4.16 \times 10^{-15} \ M + 2.00 \times 10^{-2} \ M \cong 2.00 \times 10^{-2} \ M$$

$$K_{sp} = [\text{Ag}^+][\text{I}^-]$$

$$K_{sp} = [4.16 \times 10^{-15}][2.00 \times 10^{-2}] = 8.32 \times 10^{-17}$$

## MEASUREMENT OF POTENTIAL

If a direct-current voltmeter is used, as shown in Fig. 10–2, a small current must be passed through the meter in order for it to be operable. Since current is drawn from the cell, a variation in the concentrations of the reacting species will take place which leads to a change in the cell voltage. An additional factor, because of internal resistance in the cell, is the development of an ohmic potential drop which opposes the potential due to the two electrodes. Thus, to correctly measure the potential of the cell it is necessary to be able to make the measurement with an insignificant passage of current. The instrument which is used for the accurate measurement of potentials is a potentiometer.

A diagram of a simple potentiometer is shown in Fig. 10–5. The instrument can be conveniently divided into two parts: a voltage divider and the galvanic cell portion which is enclosed in dotted lines in Fig. 10–5.

A voltage divider is a device that supplies a continuously variable voltage from zero (point C is at A) to the total output (point C is at B) of the battery. The essential parts are a battery connected to the resistance AB upon which a sliding contact C passes. A voltmeter from A to the contact C permits the voltage measurement. Since a current $I$ flows and using Ohm's law, it can be readily shown that

$$E_{AC} = E_{AB} \frac{R_{AC}}{R_{AB}} \qquad (10\text{--}23)$$

which states that a variable voltage is obtained.

The galvanic cell is connected to AC in such a way that its output opposes the output of the working battery which must have a potential equal to or greater than what is being measured. Also, connected in parallel with the cell is a galvanometer for current measurement and a tapping key which permits intermittent completing and disrupting the circuit.

Three possibilities exist if the potential divider is adjusted to a position of $E_{AC}$ and the key is pressed. Two of these are when $E_{AC}$ is greater than $E_{cell}$ or $E_{AC}$ is less than $E_{cell}$. For the former, electrons flow from right to left, while for the latter, the opposite occurs. The third case is where $E_{AC}$ is equal to $E_{cell}$, and for this situation no current flows through the galvanometer and galvanic cell.

In practice the key is tapped and C is moved until the galvanometer indicates no current flow. When this balance is found, the voltage of $E_{cell}$ is read from the voltmeter providing that the slide wire has been calibrated. It should be emphasized that, at balance, the current needed by the volt-meter is supplied by the battery and not by the galvanic cell (only infinitesimally small currents are drawn from the galvanic cell). For successful operation of the potentiometer, as shown in Fig. 10–5, the positive and negative

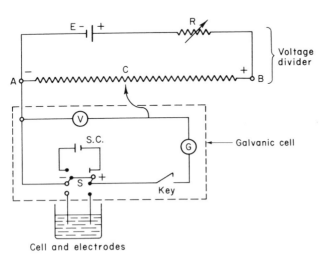

**Fig. 10–5.** A simple potentiometer. E, source of voltage; R, variable resistance; C, variable slide wire; V, voltmeter; G, galvanometer; S, switch; S. C., standard cell.

electrodes of the cell must be attached opposite the positive and negative arms of the potentiometer.

Standardization of the potentiometer is accomplished with a Weston cell which produces a reproducible, accurately known electromotive force. This cell, shown in Fig. 10–6, is a typical galvanic cell and is represented by

$$Cd(Hg) \mid CdSO_4 \cdot \tfrac{8}{3}H_2O_{(s)}, \; Hg_2SO_{4(s)} \mid Hg$$

where the reactions are

$$Cd_{(s)} + SO_4^{2-} \rightarrow CdSO_{4(s)} + 2e$$

$$Hg_2SO_{4(s)} + 2e \rightarrow 2Hg_{(l)} + SO_4^{2-}$$

At 25°C the potential of the cell is 1.0183 V.

In Fig. 10–5 the Weston cell is included in the potentiometer as S.C. (standard cell). Thus, to calibrate the slide wire so that it will read directly in volts, the potentiometer is first connected to the standard cell. The resistance R is adjusted until the voltage response is 1.0183 V. If the test cell is then put in the circuit and the measurement procedure carried out as described before, the voltage of the cell is determined.

The potential for the Weston cell is dependent on temperature and will change its potential about 0.04 mV for each degree increase in temperature. This effect is largely due to temperature effects on the solubility of $CdSO_4$ and $Hg_2SO_4$. The Weston cell is also very sensitive to the passage of current.

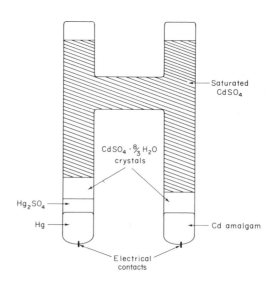

**Fig. 10–6.** A saturated standard Weston cell.

Thus, the cell should be in the completed circuit for only very brief periods of time. Excessive passage of a current will result in concentration changes in the cell which in turn changes its potential.

## REFERENCE-INDICATOR ELECTRODES

A potentiometer can be used to measure the potential of the Zn–Cu cell (Fig. 10–2). If either or both $Zn^{2+}$ and $Cu^{2+}$ concentrations are changed, a new cell potential is measured because

$$E_{cell} = 0.337 \text{ V} - \frac{0.0592}{2} \log \frac{1}{[Cu^{2+}]} - \left[ -0.763 \text{ V} - \frac{0.0592}{2} \log \frac{1}{[Zn^{2+}]} \right]$$

Since an infinite number of concentration combinations will produce the same potential, it is not possible to deduce the concentration of the two ions from the cell potential. If one of them is known, then the concentration of the other can be calculated.

In general, the analytical chemist is interested in measuring the potential or potential changes of one half-cell. Since it is only possible to measure the voltage of a complete cell, the potential at one half-cell can only be measured if the other retains a fixed, reproducible voltage and is insensitive to changes of the solution composition. This type of electrode is a *reference electrode*. The NHE is a reference electrode and was used in this capacity early in this chapter to evaluate the standard reduction potentials for $Cu^{2+}/Cu$ and $Zn^{2+}/Zn$. However, it is too cumbersome to use in routine applications.

The second electrode in combination with the reference electrode is called an *indicator electrode* and its response should be dependent upon changes in concentration of the species of interest. This change should be reversible and follow the Nernst equation. In general, the indicator electrode should collect electrons in the potential-determining step in the absence of interfering side reactions.

**Indicator Electrodes.**    There are several different kinds of indicator electrodes. Several metals, such as silver, copper, lead, cadmium, and mercury, will participate in a reversible electron exchange and can serve as indicator electrodes for their ions. In this list mercury is perhaps the most valuable (see Chapter 29). In general, most other common metals, except noble metals, are not satisfactory as indicator electrodes in ordinary applications, usually, because of oxide coatings on the surface and other surface properties that hinder electron exchange. Noble metals are chemically inert and can conveniently act as collector electrodes for half-cell reactions that involve charged species or gases. Of all the noble metals platinum and gold are used the most.

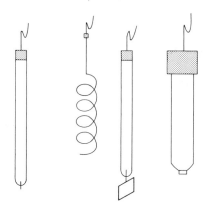

**Fig. 10–7.** Typical indicator electrode designs.

Usually, all metal electrodes are used as a wire, strip, or button set in plastic or glass (see Fig. 10–7).

Several metals can serve as indicator electrodes for anions that form slightly soluble precipitates with the cation of the metal. A typical example is the use of a silver electrode for indicating chloride ion. Once the solution is saturated with the sparingly soluble AgCl, the silver wire becomes coated with AgCl and responds to $Cl^-$ concentration through the following half-reaction

$$AgCl_{(s)} + 1e = Ag_{(s)} + Cl^-$$

There are several other indicator electrodes. Many of these have become so valuable to the analytical chemist in modern potentiometry that they are treated separately under the heading "Ion-Selective Electrodes" in Chapter 13.

**Reference Electrodes—Saturated Calomel Electrode.**      Several electrodes are useful as reference electrodes. Of these the one used the most is the calomel electrode. Figure 10–8 shows a typical laboratory and commercial design. The half-cell reaction for this electrode is

$$Hg_2Cl_{2(s)} + 2e \rightleftharpoons 2Hg_{(l)} + 2Cl^-$$

and its potential is given by

$$E = E^\circ_{Hg_2Cl_2,Hg} - \frac{0.059}{2} \log [Cl^-]^2$$

where the chloride ion concentration is expressed in molar units rather than

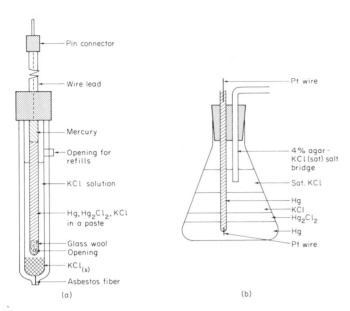

**Fig. 10–8.** A typical (a) commercial and (b) laboratory-made saturated calomel reference electrode.

activity. If the solution is saturated with KCl, the electrode is called the saturated calomel electrode (SCE). Since the potential of the electrode is determined by the chloride ion concentration, it can also be used at other KCl concentrations (see Table 10–1). Regardless of the level of KCl used the solution is always saturated with $Hg_2Cl_2$. The SCE is the easiest to prepare and is used the most. However, because of the saturated condition, it will undergo a greater potential change with change in temperature in comparison to the other KCl–calomel reference electrodes.

**Table 10–1.   Potentials for Reference Electrodes**

| Electrode | Conditions | Potential (volts) vs NHE at 25°C |
|---|---|---|
| Mercury–mercurous chloride | KCl (sat.) | +0.2412 |
| | 1.0 $M$ KCl | +0.2801 |
| | 0.10 $M$ KCl | +0.3337 |
| Silver–silver chloride | KCl (sat.) | +0.199 |
| | 1.0 $M$ KCl | +0.237 |
| | 0.10 $M$ KCl | +0.290 |
| Mercury–mercurous sulfate | $K_2SO_4$(sat.) | +0.64 |
| | 0.05 $M$ $H_2SO_4$ | +0.68 |

**Other Reference Electrodes.**    Two other electrodes which are useful as reference electrodes are the *silver–silver chloride* and *mercury–mercurous sulfate* electrode. The half-cells for these electrodes are, respectively,

$$AgCl_{(s)} + 1e \rightarrow Ag_{(s)} + Cl^-$$

$$Hg_2SO_{4(s)} + 2e \rightarrow 2Hg_{(l)} + SO_4{}^{2-}$$

If Nernst expressions are written, it is readily apparent that these electrodes are similar to the SCE in that their potentials are determined by the $Cl^-$ and $SO_4{}^{2-}$ concentration, respectively. Both of these are used at saturated conditions. A typical design of the Ag/AgCl electrode which is second in application to the SCE is shown in Fig. 10–9. The potentials are listed in Table 10–1.

**Criteria for Reference Electrodes.**    To be useful as a reference electrode several criteria should be met. The most important ones are:

1. The electrode should be easily prepared from readily available materials.
2. An accurate, reproducible potential should be rapidly attained.
3. Potential of the electrode should remain constant over a long period of storage time.
4. The electrode should not suffer from a thermal hysteresis.
5. The voltage change for the electrode per degree of temperature change should be known and reproducible. In some experimental operations the electrode may be heated or cooled. If the electrode is eventually taken back to its original temperature, the original potential should also be obtained.
6. The electrode should be able to withstand the passage of small amounts of current for short periods of time without change in its potential. If this criterion is met, the electrode is said to have a low polarizability.

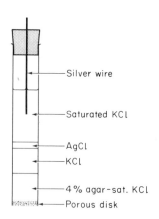

**Fig. 10–9.**  A typical silver–silver chloride reference electrode.

In many electrochemical applications commercial reference electrodes cannot be used and a reference electrode must be made in the laboratory. Therefore, it is important to consider these requirements carefully when preparing a reference electrode.

## BIOLOGICAL SYSTEMS

The principles of oxidation–reduction are vital to the understanding of many biological processes. For example, biological oxidation can be viewed in terms of removal of hydrogen or loss of electrons. Perhaps the main difference between the biological systems and the types described throughout this chapter is that enzymes provide catalytic activity in the former systems. Therefore, biological redox studies must also include studies on the properties of enzymes.

Two examples are cited here to illustrate biological oxidation–reduction. An enzyme containing organically bound iron is responsible for the catalytic oxidation of organic foodstuffs, in a manner similar to catalytic action of metal ions in the oxidation of organic compounds. A mechanism illustrating this action is

$$\text{Enzyme} \cdot \text{Fe(III)} + \text{metabolite} \rightarrow \text{enzyme} \cdot \text{Fe(II)} + 2\text{H}^+ + \text{oxidized metabolite}$$

$$\text{Enzyme} \cdot \text{Fe(II)} + \text{O}_2 \rightarrow \text{enzyme} \cdot \text{Fe(II)} \cdot \text{O}_2 \text{ (activated complex)}$$

$$\text{Enzyme} \cdot \text{Fe(II)} \cdot \text{O}_2 + 2\text{H}^+ \rightarrow \text{enzyme} \cdot \text{Fe(III)} + \text{H}_2\text{O (or H}_2\text{O}_2)$$

The enzyme $\cdot$ Fe(II) is reoxidized to enzyme $\cdot$ Fe(III) by reaction with molecular oxygen and this latter enzyme becomes available again for oxidation of more of the metabolite. In this mechanism electron transfer occurs with the hydrogen atom being converted to a hydrogen ion and the metal ion reduced to a lower oxidation state. The enzyme and molecular oxygen are the electron acceptors, while the metabolite and reduced enzyme are the electron donors.

The consumption of ethanol in the liver involves oxidation–reduction in which the alcohol is oxidized to acetaldehyde. Reaction (10–24) illustrates this conversion:

$$\text{NAD}^+ \quad + \quad \text{CH}_3\text{CH}_2\text{OH} \quad \underset{\substack{\text{alcohol} \\ \text{dehydrogenase}}}{\rightleftharpoons} \quad \text{CH}_3\overset{\text{O}}{\overset{\|}{\text{CH}}} \quad + \quad \text{H}^+ \quad + \quad \text{NADH} \tag{10–24}$$

where $\text{NAD}^+$ and NADH are nicotinamide–adenine dinucleotide and its reduced form, respectively. Not shown in the reaction is how the enzyme participtates in the electron transfer. The biological reactions do not stop at this point. Acetaldehyde is oxidized in presence of another enzyme catalyst to acetic acid. Also, $NAD^+$ is replenished through a series of known and complex steps involving enzyme catalysts with the final result being oxidation of NADH to $NAD^+$ and oxygen being reduced to water.

Half-reactions of biological interest can be tabulated according to their reduction potentials. An abbreviated list of reduction processes is shown in Table 10–2. From data of this type it is possible to predict whether reactions are spontaneous, calculate equilibrium constants, calculate potentials as a function of concentration and pH (Nernst equation) and determine cell potentials. The only difference between these calculations and those described throughout this chapter is that biological reactions are now of interest.

It is not possible to review this very important part of biochemistry in this short space. However, on the basis of these brief comments it should be

**Table 10–2.   Standard Reduction Potentials for Several Biological Systems**[a]

| System | $E°$ (pH 0 at 30°C) (volts) | $E$ (pH 7.0 at 30°C) (volts) |
|---|---|---|
| $\frac{1}{2}O_2 + 2H^+ + 2e \rightarrow H_2O$ | +1.229 | +0.816 |
| $Fe^{3+} + e \rightarrow Fe^{2+}$ | +0.771 | +0.771 |
| $Br_2 + 2e \rightarrow 2Br^-$ | +0.652 | +0.652 |
| $I_2 + 2e \rightarrow 2I^-$ | +0.536 | +0.536 |
| Cytochrome-$a$ $Fe^{3+} + e \rightarrow$ cytochrome-$a$ $Fe^{2+}$ | +0.290 | +0.290 |
| Cytochrome-$c$ $Fe^{3+} + e \rightarrow$ cytochrome-$c$ $Fe^{2+}$ | | +0.250 |
| 2,6-Dichlorophenolindophenol + $2H^+ + 2e^- \rightarrow$ reduced 2,6-dichlorophenolindophenol | | +0.22 |
| Dehydroascorbate + $2H^+ + 2e \rightarrow$ ascorbate | +0.390 | +0.060 |
| Fumarate + $2H^+ + 2e \rightarrow$ succinate | +0.433 | +0.031 |
| Methylene blue$^+$ + $2e + 2H^+ \rightarrow$ leuco methylene blue $H^+$ | +0.532 | +0.011 |
| FAD + $2H^+ + 2e \rightarrow$ FAD2H | | −0.06 |
| Oxalacetate + $2H^+ + 2e \rightarrow$ malate | +0.330 | −0.102 |
| Pyruvate + $2H^+ + 2e \rightarrow$ lactate | +0.224 | −0.190 |
| $(Cyst-S)_2 + 2H^+ + 2e \rightarrow$ 2 cysteine-SH | | −0.22 |
| $(Glutathione-S)_2 + 2e + 2H^+ \rightarrow$ 2 glutathione-SH | | −0.23 |
| Safranine-T + $2e \rightarrow$ leucosafranine-T | −0.235 | −0.289 |
| Acetoacetate + $2H^+ + 2e \rightarrow$ L-$\beta$-hydroxybutyrate | | −0.293 |
| $(C_6H_5S)_2 + 2H^+ + 2e \rightarrow 2C_6H_5SH$ | | −0.30 |
| $DPN^+ + 2H^+ + 2e \rightarrow DPNH(H^+)$ | −0.107 | −0.320 |
| Xanthine + $2H^+ + 2e \rightarrow$ hypoxanthine + $H_2O$ | | −0.371 |
| $H^+ + e \rightarrow -H_2$ | +0.000 | −0.420 |
| Gluconate + $2H^+ + 2e \rightarrow$ glucose + $H_2O$ | | −0.45 |

[a] In the health sciences standard conditions are usually defined as pH = 7.

obvious that the fundamental concepts of oxidation and reduction are necessary in the understanding of certain aspects of biological reactions.


# *Questions*


1. Define the terms oxidation, reduction, oxidizing agent, reducing agent, electrolysis, galvanic cell, and half-cell.
2. Compare standard reduction potential to formal reduction potential.
3. What parameters define the standard reduction potential?
4. List the following in order of increasing oxidizing power: $Zn^{2+}$, $Ce^{4+}(HClO_4)$, $Cr_2O_7^{2-}$, $Ag^+$, $I_2$, $H^+$, and $Pb^{2+}$.
5. List the following in order of increasing reducing power: Cd, Ag, $Cr^{2+}$, $H_2$, K, $Br^-$, $Mn^{2+}$, $Sn^{2+}$, and $I^-$.
6. Why is it not possible to measure absolute potential?
7. What is the significance of the sign for a calculated cell voltage?
8. What is the significance of a voltage of zero for a cell reaction?
9. What is the significance of adopting a set of conventions for writing cells, their reactions, and calculating their potentials?
10. What is the significance of the Nernst equation?
11. Write the Nernst expression for the following half-cells.
    a. $Sn^{4+} + 2e = Sn^{2+}$
    b. $AgBr_{(s)} + 1e = Ag_{(s)} + Br^-$
    c. $Cl_{2(g)} + 2e = 2Cl^-$
    d. $AsO_4^{3-} + 2H^+ + 2e = AsO_3^{3-} + H_2O$
    e. $DPN^+ + 2H^+ + 2e = DPNH(H^+)$
    f. Gluconate $+ 2H^+ + 2e =$ glucose $+ H_2O$
12. Explain why the potential of the $Cr_2O_7^{2-}/Cr^{3+}$ half-cell is affected by acidity.
13. How does acidity affect the oxidizing power of $I_2$?
14. How does acidity affect the reducing power of $As_2O_3$?
15. What affect does basicity have on the reducing power of $Na_3AsO_3$?
16. How does an increase in temperature affect the standard reduction potential?
17. What is a salt bridge and what purpose does it serve?
18. What happens to the reduction potential for the half-cell, $Ag^+ (1\ M)\ |\ Ag$, if NaI is added to the half-cell.
19. Using the half-cells in Example 10–9 prove that

$$\log K = \frac{n(E^f_{Cr_2O_7^{2-},Cr^{3+}} - E^f_{Fe^{3+},Fe^{2+}})}{0.059} = \frac{6(0.93 - 0.700)}{0.059}$$

20. Criticize the statement, "The reaction between $S_2O_8^{2-}$ and $H_2C_2O_4$ is a useful analytical reaction for the determination of $H_2C_2O_4$ by titration since the equilibrium constant is very large."
21. Describe how a simple potentiometer works.

22. What is a Weston cell?
23. Differentiate between a reference and indicator electrode.
24. Explain why a Fe wire is not a suitable indicator electrode for sensing the $Fe^{3+}/Fe^{2+}$ half-cell.
25. Select a book on biochemistry in the library and find two biological systems that undergo oxidation–reduction.

## Problems

1. Write the balanced reaction and calculate the potential for the following cells which are at *standard conditions*.
    *a. $Zn \mid Zn^{2+} \parallel Ni^{2+} \mid Ni$
    b. $Pt, H_2 \mid H^+ \parallel Br_2, Br^- \mid Ag$
    c. $Pt \mid Sn^{2+}, Sn^{4+} \parallel Cr_2O_7^{2-}, Cr^{3+}, H^+ \mid Pt$
    d. $Pt \mid I^-, I_2 \parallel MnO_4^-, Mn^{2+}, H^+ \mid Pt$
    *e. $Pt \mid Br^-, Br_2 \parallel Fe^{3+}, Fe^{2+} \mid Pt$
    f. $Pt \mid S_2O_3^{2-}, S_4O_6^{2-} \parallel I_2, I^- \mid Pt$

2. Calculate the reduction potential developed at a Pt wire for each of the following half-cells. Assume that volumes are additive.
    *a. 25.0 ml of $0.0100 \ F \ Fe^{2+}$ and 40.0 ml of $0.100 \ F \ Fe^{3+}$ are mixed.
    b. 30.0 ml of $0.0400 \ F \ Cr_2O_7^-$ and 10.0 ml of $0.0100 \ F \ Cr^{3+}$ are mixed at constant pH = 2.0.
    *c. $H_2$ at 1.8 atm is passed through a pH 5.45 solution.
    d. 30.0 ml of $0.100 \ F \ Ti^{3+}$ in $4 \ F \ H_2SO_4$ and 60.0 ml of $0.0750 \ F \ Ti^{4+}$ in $4 \ F \ H_2SO_4$ are mixed.
    *e. 30.0 ml of $0.0100 \ F \ Ce^{4+}$ in $1 \ F \ H_2SO_4$ and 65.0 ml of $0.00100 \ F \ Ce^{3+}$ in $1 \ F \ H_2SO_4$ are mixed.

3. Calculate the half-cell reduction potential developed at a Ag wire immersed in each of the following:
    a. $0.00600 \ F \ AgNO_3$.
    *b. $0.0300 \ F \ NaI$ that is saturated with AgI.
    c. $0.0015 \ F \ Na_2SO_4$ that is saturated with $Ag_2SO_4$.

4. Write the balanced reaction, calculate the cell potential, calculate the equilibrium constant, and predict whether the reaction is spontaneous or not for each of the following.
    *a. $Pb^\circ \mid Pb^{2+}(0.01 \ M) \parallel I_2(1 \ M), I^-(0.001 \ M) \mid Pt$
    b. $Pt \mid H_2(0.5 \ atm) \mid H^+(0.01 \ M) \parallel Hg_2SO_{4(s)} \mid Hg_{(1)}, SO_4^{2-}(10^{-4} \ M)Pt$
    *c. $Pt \mid Tl^+(0.1 \ M), Tl^{3+}(0.001 \ M) \parallel MnO_4^-(0.001 \ M), Mn^{2+}(0.1 \ M), H^+(pH = 2) \mid Pt$

---

* Answers are listed at the end of the book for problems marked with an asterisk.

d. $Ag° \mid Ag^+(0.1\ M) \| Sn^{4+}(0.1\ M),\ Sn^{2+}(0.001\ M),\ Pt$

e. $Pt \mid Fe^{2+}(0.1\ M),\ Fe^{3+}(0.01\ M),\ 5.0\ F\ HCl \| Ce^{4+}(0.01\ M),\ Ce^{3+}(0.1\ M),$
$1\ F\ HCl \mid Pt$

f. $Pt \mid Ti^{3+}(0.01\ M),\ Ti^{4+}(0.1\ M),\ 5\ F\ H_3PO_4 \| Ti^{4+}(0.01\ M),\ Ti^{3+}(0.0001\ F),$
$4\ F\ H_2SO_4 \mid Pt$

5. How many grams of $FeCl_3$ must be added to 400 ml of a 0.0400 $F$ $Fe^{2+}$ which is also 10 $F$ in HCl so that the potential at a Pt wire immersed in the solution will have a reduction potential of $+0.450$ V?

6. How many grams of $CrCl_3$ per liter must be added to a half-cell containing 0.1 $F$ $K_2Cr_2O_7$ and 1 $F$ HCl so that its potential is 1.049 V?

7.* Calculate the equilibrium constants for the cells in Question 1.

8. Calculate the potential for the cell

$$Pt \mid H_2(1\ atm),\ H^+(0.100\ M) \| Ag^+(0.100\ M) \mid Ag$$

after each of the following changes. Assume that the volume of each half-cell is 1 liter.

a. Addition of 0.100 mole NaOH to the left cell.

*b. Addition of 0.100 mole NaCl to the right cell.

c. 1 g of Ag powder is added to the right cell.

d. The $H_2$ gas is reduced to 0.1 atm pressure.

e. Addition of 0.100 mole $AgNO_3$ to the right cell.

f. Addition of 500 ml $H_2O$ to the left cell.

9.* Calculate the solubility product of ZnS from the data

$$Zn^{2+} + 2e = Zn_{(s)} \qquad\qquad E°_{Zn^{2+},Zn} = -0.763\ V$$

$$ZnS_{(s)} + 2e = Zn_{(s)} + S^{2-} \qquad\qquad E°_{ZnS,Zn} = -1.44\ V$$

10. Calculate the solubility product of AgBr from the data

$$Ag^+ + 1e = Ag_{(s)} \qquad\qquad E°_{Ag^+,Ag} = +0.799\ V$$

$$AgBr_{(s)} + 1e = Ag_{(s)} + Br^- \qquad\qquad E°_{AgBr,Ag} = +0.070\ V$$

11. Calculate the formation constant for $Ag(CN)_2^-$ from the data

$$Ag^+ + 1e = Ag_{(s)} \qquad\qquad E°_{Ag^+,Ag} = +0.799\ V$$

$$Ag(CN)_2^- + 1e = Ag_{(s)} + 2CN^- \qquad E°_{Ag(CN)_2^-,Ag} = -0.310\ V$$

12.* The voltage for the following cell is 0.250 V.

$$Pt \mid H_2(1\ atm),\ HA[2.00 \times 10^{-2}] \| H_2(1\ atm),\ H^+(1.00\ M) \mid Pt$$

Calculate the $K_a$ for the weak acid HA.

13. Calculate the final concentration of $Fe^{2+}$ in a solution prepared by mixing 40.0 ml each of 0.0500 $F$ $Fe^{2+}$ and 0.0500 $F$ $Ce^{4+}$. Assume acidity is constant at 1 $F$ HCl.

14.* The potential for the following cell is 0.463 V. Calculate the concentration of $Ag^+$ in its half-cell

$$Cu_{(s)} \mid Cu^{2+}(0.0100 \ M)\|Ag^+ \mid Ag_{(s)}$$

15. Predict the qualitative effect of each of the following on the cell.

$$Pt \mid Fe^{2+}(0.01 \ M), \ Fe^{3+}(0.001 \ M), \ 1 \ F \ HCl \| Cr_2O_7{}^{2-}(0.01 \ M),$$

$$Cr^{3+}(0.01 \ M), \ 1 \ F \ HCl \mid Pt$$

    a. Addition of 2 g of $FeCl_2$ to the left cell.
  *b. Addition of 2 g of $FeCl_3$ to the left cell.
    c. Addition of 2 g of $K_2Cr_2O_7$ to the right cell.
  *d. Addition of 2 g of $CrCl_3$ to the right cell.
    e. Neutralizing 99% of the acid in the right cell.
    f. The addition of 100 ml of water to the left cell.
    g. The addition of 100 ml of water to the right cell.

16.* Calculate the standard free energy change for the oxidation of diphosphopyridine nucleotide (DPN) by flavin adenine dinucleotide (FAD) at pH = 7.

$$DPNH(H^+) + FAD \rightarrow DPN^+ + FADH_2$$

17. Calculate the equilibrium constant and predict whether the following occurs spontaneously at pH = 7.

$$\text{succinate} + 2 \text{ cytochrome-}c \ Fe^{3+} \rightarrow \text{fumarate} + 2 \text{ cytochrome-}c \ Fe^{2+}$$

# Chapter Eleven
# Oxidation–Reduction Titrations

## INTRODUCTION

If a redox reaction is to be used for a titration, it must meet the same general requirements that apply to other successful titration procedures. Consequently, the reaction should be rapid, go to completion, be stoichiometric, and a means for end-point detection should be available. Many different inorganic species exist in more than one stable oxidation state and frequently can be determined through a redox titration. Similarly, certain organic functional groups are quantitatively oxidized or reduced and can, therefore, be analyzed by a titration procedure. In this chapter the feasibility of redox titrations and end-point detection by color indicators and potentiometry are considered.

In general, redox reactions tend to be slow and often the redox reaction only becomes useful after a suitable catalyst becomes available. For example, the best way to standardize $Ce^{4+}$ solutions is by titrating primary standard $As_2O_3$ [tris-1,10-phenthroline iron(II) is used as indicator]. However, poor results are obtained because the reaction rate is too slow. If $OsO_4$ is added as a catalyst, the reaction proceeds conveniently and an accurate standardization is obtained. It is difficult to generalize in regard to useful catalysts for redox reactions. Some are catalyzed by acid, others by base, and still others by metal ions. Also, there are reactions, such as $Fe^{2+}-Ce^{4+}$, $I_2-S_2O_3^{2-}$, $BrO^--Br^-$, and others, that are rapid and do not require catalysts.

## TITRATION CURVE

A redox titration curve describes the change in concentration of the species of interest as a function of the titrant. Since potential is related to concentration through the Nernst equation, a redox titration curve is a plot of potential vs milliliter of titrant (reducing or oxidizing agent).

It is possible to predict the shape of a titration curve prior to actually carrying out the titration in the laboratory. This is important since it provides useful information about the quantitative aspects of the reaction.

In practice the cell potential is calculated with the Nernst equation as concentration changes during the course of the titration. For convenience the standard $H^+/H_2$ half-cell is used as the reference half-cell. In the laboratory, a saturated calomel reference electrode is usually used for the experimentally determined titration curve.

A convenient way to illustrate these calculations is to use the Fe(II)–Ce(IV) reaction as an example. If the titration of Fe(II) with Ce(IV) is carried out in 1 $F$ $H_2SO_4$, the equilibrium constant for the reaction is

$$\log K = \frac{n(E^f_{Ce^{4+}, \, Ce^{3+}} - E^f_{Fe^{3+}, \, Fe^{2+}})}{0.0592}$$

$$\log K = \frac{1 \times (1.44 \text{ V} - 0.68 \text{ V})}{0.0592}$$

$$K = 10^{12.84} = 6.92 \times 10^{12}$$

and since this is a very favorable equilibrium constant, the reaction should go to completion.

For purposes of illustration assume that 40.0 ml of 0.100 $F$ $Fe^{2+}$ is titrated with 0.100 $F$ $Ce^{4+}$ in 1 $F$ $H_2SO_4$. Also, as previously mentioned, assume that the reference electrode is the NHE. Thus, the cell potential, since the NHE is zero, will be equal to the potential developed at the indicator electrode.

**Initial Potential.**     The solution initially is composed only of Fe(II). Perhaps there is a finite quantity of $Fe^{3+}$ present as a result of air oxidation. However, since the $Fe^{3+}$ concentration is so small, a calculation of the initial potential has no real significance.

**Potential during the Titration.**     With the addition of the $Ce^{4+}$ titrant the reaction takes place to form a stoichiometric amount of Fe(III) and Ce(III). When the system reaches equilibrium, the potential at the indicator electrode, $E_{ie}$, is given by

$$E_{ie} = E^f_{Fe^{3+}, \, Fe^{2+}} - \frac{0.0592}{1} \log \frac{[Fe^{2+}]}{[Fe^{3+}]} = E^f_{Ce^{4+}, \, Ce^{3+}} - \frac{0.0592}{1} \log \frac{[Ce^{3+}]}{[Ce^{4+}]}$$

This relationship can be used to calculate the potential at any point during the titration or for any point after passage of the stoichiometric point. Since the equilibrium constant is favorable ($10^{12.84}$), the concentration of Ce(IV) is very small in comparison to the concentration of Ce(III) and for this reason the potential at the indicator electrode is calculated from the ratio of $[Fe^{2+}]/[Fe^{3+}]$.

A typical calculation would be as follows, assuming 10.00 ml of the titrant are added:

$$\begin{array}{ll} 40.0 \text{ ml} \times 0.100\ F = 4.00 \text{ mmoles of } Fe^{2+} \text{ started} \\ 10.0 \text{ ml} \times 0.100\ F = 1.00 \text{ mmoles of } Ce^{4+} \text{ added} \\ \hline 50.0 \text{ ml} \qquad\qquad\quad 3.00 \text{ mmoles of } Fe^{2+} \text{ left} \end{array}$$

Since the reductant and oxidant undergo a reaction with a reaction ratio of 1, 3.00 mmoles of $Fe^{2+}$ must remain in a total volume of 50.0 ml. Also, the 50.0 ml contains 1.00 mmoles of Fe(III). Therefore, the concentrations of $Fe^{2+}$ and $Fe^{3+}$ are

$$[Fe^{2+}] = \frac{3.00 \text{ mmoles}}{50 \text{ ml}} = 0.0600\ M; \quad [Fe^{3+}] = \frac{1.00 \text{ mmoles}}{50 \text{ ml}} = 0.0200\ M$$

$$E_{ie} = E'_{Fe^{3+},\ Fe^{2+}} - \frac{0.0592}{1} \log \frac{[Fe^{2+}]}{[Fe^{3+}]}$$

$$E_{ie} = 0.68 - \frac{0.0592}{1} \log \frac{[0.0600]}{[0.0200]} = 0.652 \text{ V}$$

**Midpoint of Titration.** If 20.0 ml of $Ce^{4+}$ are added, the calculation would be

$$\begin{array}{ll} 40.0 \text{ ml} \times 0.100\ F = 4.00 \text{ mmoles of } Fe^{2+} \text{ started} \\ 20.0 \text{ ml} \times 0.100\ F = 2.00 \text{ mmoles of } Ce^{4+} \text{ added} \\ \hline 60.0 \text{ ml} \qquad\qquad\quad 2.00 \text{ mmoles of } Fe^{3+} \text{ formed} \end{array}$$

Although the concentration of the $Fe^{3+}$ and $Fe^{2+}$ can be calculated, it should be apparent that the $Fe^{2+}$ and $Fe^{3+}$ are equal in concentration. Thus,

$$[Fe^{2+}] = [Fe^{3+}]$$

and

$$E_{ie} = 0.681 - \frac{0.0592}{1} \log 1 = 0.68 \text{ V}$$

Potentials for other titrant additions prior to the end point can be calculated in the same way.

**Stoichiometric Point Potential.** Since the sample at the start was 40.0 ml of 0.100 F $Fe^{2+}$ and the reaction ratio is 1, the stoichiometric point is reached when 40.0 ml of 0.100 $M$ $Ce^{4+}$ are added. Thus, $Fe^{3+}$ and $Ce^{3+}$ are stoichiometrically formed. However, even though the equilibrium constant is favorable it does not mean that the reaction is entirely complete or that $[Fe^{2+}] = 0$ and $[Ce^{4+}] = 0$ at the stoichiometric point. These concentrations can be predicted on the basis of $K$ to be very small but they are not zero. What can be stated on the basis of the reaction is that at the stoichiometric point

$$[Fe^{2+}] = [Ce^{4+}] \rightarrow 0 \quad \text{and} \quad [Fe^{3+}] = [Ce^{3+}]$$

At the stoichiometric point of the reaction the potential developed at the indicator electrode for each half-cell is given by

$$E_{SP} = E^f_{Ce^{4+},\, Ce^{3+}} - \frac{0.0592}{1} \log \frac{[Ce^{3+}]}{[Ce^{4+}]}$$

$$E_{SP} = E^f_{Fe^{3+},\, Fe^{2+}} - \frac{0.0592}{1} \log \frac{[Fe^{2+}]}{[Fe^{3+}]}$$

where $E_{SP}$ is the stiochiometric point potential. If these two expressions are added and $[Ce^{4+}]$ and $[Ce^{3+}]$ substituted for $[Fe^{2+}]$ and $[Fe^{3+}]$, respectively, the equation becomes

$$2E_{SP} = E^f_{Ce^{4+},\, Ce^{3+}} + E^f_{Fe^{3+},\, Fe^{2+}} - \frac{0.0592}{1} \log \frac{[Ce^{3+}]}{[Ce^{4+}]} - \frac{0.0592}{1} \log \frac{[Ce^{4+}]}{[Ce^{3+}]}$$

and simplifies to

$$2E_{SP} = E^f_{Ce^{4+},\, Ce^{3+}} + E^f_{Fe^{3+},\, Fe^{2+}}$$

$$E_{SP} = \frac{E^f_{Ce^{4+},\, Ce^{3+}} + E^f_{Fe^{3+},\, Fe^{2+}}}{2} = \frac{1.44 \text{ V} + 0.68 \text{ V}}{2} = 1.06 \text{ V}$$

In this example the stoichiometric point potential is the average of the potentials for the two half-cells. Thus, for the general half-cells

$$A_{ox} + n_a e = A_{red}$$

$$B_{ox} + n_b e = B_{red}$$

where the reaction is

$$A_{ox} + B_{red} = A_{red} + B_{ox}$$

the equation for the stoichiometric point potential can be shown to be

$$E_{SP} = \frac{n_a E^\circ_{A_{ox},\, A_{red}} + n_b E^\circ_{B_{ox},\, B_{red}}}{n_a + n_b} \tag{11-1}$$

Equation (11–1) is limited to those cases in which the electron change involves a small number of electrons and the half-cells are not complex. For more complex reactions the equation must be derived.

*Example 11–1.*    Derive an expression for the stoichiometric point potential for the reaction

$$5Fe^{2+} + MnO_4^- + 8H^+ = 5Fe^{3+} + Mn^{2+} + 4H_2O$$

The half-cells are

$$MnO_4^- + 8H^+ + 5e = Mn^{2+} + 4H_2O$$

$$Fe^{3+} + 1e = Fe^{2+}$$

At the stoichiometric point

$$[Fe^{2+}] = 5[MnO_4^-] \rightarrow 0 \quad \text{and} \quad [Fe^{3+}] = 5[Mn^{2+}]$$

and the stoichiometric point potential is given by either

$$E_{sp} = E^{\circ}_{MnO_4^-, Mn^{2+}} - \frac{0.0592}{5} \log \frac{[Mn^{2+}]}{[MnO_4^-][H^+]^8}$$

or

$$E_{sp} = E^{\circ}_{Fe^{3+}, Fe^{2+}} - \frac{0.0592}{1} \log \frac{[Fe^{2+}]}{[Fe^{2+}]}$$

The $MnO_4^-/Mn^{2+}$ half-cell is multiplied by 5 and the two are added to give

$$6E_{sp} = 5E^{\circ}_{MnO_4^-, Mn^{2+}} + E^{\circ}_{Fe^{3+}, Fe^{2+}} - 0.0592 \log \frac{[Fe^{2+}][Mn^{2+}]}{[Fe^{3+}][MnO_4^-][H^+]^8}$$

Substitution gives

$$E_{sp} = \frac{5E^{\circ}_{MnO_4^-, Mn^{2+}} + E^{\circ}_{Fe^{3+}, Fe^{2+}}}{6} - \frac{0.0592}{6} \log \frac{5[MnO_4^-][Mn^{2+}]}{5[Mn^{2+}][MnO_4^-][H^+]^8}$$

$$E_{sp} = \frac{5E^{\circ}_{MnO_4^-, Mn^{2+}} + E^{\circ}_{Fe^{3+}, Fe^{2+}}}{6} - \frac{0.0592}{6} \log \frac{1}{[H^+]^8} \qquad (11-2)$$

In Example 11–1 the stoichiometric point potential is dependent on the acidity of the solution and when the $H^+$ ion concentration is 1 $M$, Eq. (11-2) simplifies to Eq. (11-1). It should be noted that for any reaction in which the stoichiometry is not 1 the equation defining the stoichiometric point potential will show a dependence on the concentration of one of the reactants.

As pointed out, for more complex reactions the stoichiometric point potential equation must be derived. However, depending on the application, a reasonable qualitative answer is readily obtained by applying Eq. (11-1) to all systems.

**Calculations beyond the Stoichiometric Point.**     When   excess   $Ce^{4+}$
titrant is added, the molar concentration of $Fe^{3+} \ggg Fe^{2+}$, since the equilibrium constant is favorable for the reaction. Although the potential at the
indicator electrode could be calculated from the $[Fe^{2+}]/[Fe^{3+}]$ ratio, it is
more convenient to make the calculation from the ratio of $[Ce^{3+}]/[Ce^{4+}]$. It
should be noted that the reasoning here is identical to the one used for suggesting the use of the $[Fe^{2+}]/[Fe^{3+}]$ ratio and not the $[Ce^{3+}]/[Ce^{4+}]$ ratio
in calculations before the stoichiometric point.

An example of the calculation is illustrated in the following for the point
where 50.00 ml of $Ce^{4+}$ titrant have been added (10.00 ml in excess).

For example, if 50.0 ml of $Ce^{4+}$ are added, the following can be written:

$$
\begin{array}{ll}
50\ \text{ml} \times 0.100\ F = 5.00\ \text{mmoles of } Ce^{4+} \text{ added} \\
40\ \text{ml} \times 0.100\ F = 4.00\ \text{mmoles of } Fe^{2+} \text{ started} \ (= Ce^{3+} \text{ formed}) \\
\hline
90\ \text{ml} \qquad\qquad\qquad 1.00\ \text{mmoles of } Ce^{4+} \text{ in excess}
\end{array}
$$

Thus, in molar units

$$
[Ce^{4+}] = \frac{1.00\ \text{mmoles}}{90.0\ \text{ml}} = 0.0111\ M; \qquad [Ce^{3+}] = \frac{4.00\ \text{mmoles}}{90.0\ \text{ml}} = 0.0444\ M
$$

and

$$
E_{ie} = E'_{Ce^{4+},\ Ce^{3+}} - \frac{0.0592}{1} \log \frac{[Ce^{3+}]}{[Ce^{4+}]}
$$

$$
E_{ie} = 1.44 - \frac{0.0592}{1} \log \frac{[0.0444]}{[0.0111]} = 1.40\ V
$$

When 80 ml of $Ce^{4+}$ are added,

$$
[Ce^{3+}] = [Ce^{4+}]
$$

and

$$
E_{ie} = E'_{Ce^{4+},\ Ce^{3+}} - \frac{0.0592}{1} \log \frac{[Ce^{3+}]}{[Ce^{4+}]}
$$

$$
E_{ie} = 1.44\ V
$$

The complete titration curve is plotted in Fig. 11–1. Note that in the vicinity
of the stoichiometric point the potential changes very abruptly. The potentials
calculated for this curve were calculated vs the NHE. If they were calculated
vs some other reference electrode the entire curve would be shifted upward or
downward depending on the relationship of the electrode toward the hydrogen
electrode. For example, if the saturated calomel electrode is used, the curve
is shifted downward (dotted curve in Fig. 11–1) by +0.241 V.

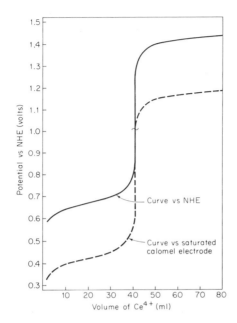

**Fig. 11–1.** Titration curve for the titration of 40.0 ml of 0.100 $F$ $Fe^{2+}$ with 0.100 $F$ $Ce^{4+}$ titrant in 1.0 $F$ $H_2SO_4$.

Usually redox titrations are carried out in 0.1 $F$ solutions. If more dilute solutions are used it is readily shown that the potential is the same as if 0.100 $F$ solutions were used. Thus, although concentrations are changed, the ratio in the log term remains constant and the potential is the same or independent of dilution. In practice, however, redox titration methods used for analysis are generally not done in lower than $10^{-3}$ to $10^{-4}$ $F$ concentration levels.

**Completion of Reaction.**     The degree of completion for a redox reaction can be calculated from the equilibrium constant. This is illustrated for the $Fe^{2+}$–$Ce^{4+}$ titration in the following example.

*Example 11–2.*     Calculate the mg of $Fe^{2+}$ remaining at the stoichiometric point for the titration of 40.0 ml of 0.100 $F$ $Fe^{2+}$ with 0.100 $F$ $Ce^{4+}$ at constant 1 $F$ $H_2SO_4$.

$$\log K = \frac{n\left(E^{f}_{Ce^{4+},Ce^{3+}} - E^{f}_{Fe^{3+},Fe^{2+}}\right)}{0.0592}$$

$$\log K = \frac{1(1.44 \text{ V} - 0.68 \text{ V})}{0.592}$$

$$K = 6.89 \times 10^{12}$$

$$Fe^{2+} + Ce^{4+} = Ce^{3+} + Fe^{3+}$$

$$K = \frac{[Ce^{3+}][Fe^{3+}]}{[Ce^{4+}][Fe^{2+}]}$$

At the stoichiometric point

$$[Fe^{2+}] = [Ce^{4+}]$$

$$[Fe^{3+}] = [Ce^{3+}] \simeq 40.0 \text{ ml} \times 0.100 \ F/(40.0 \text{ ml} + 40.0 \text{ ml}) = 0.0500 \ M$$

Substituting into the ionization constant expression

$$6.98 \times 10^{12} = \frac{[0.0500][0.0500]}{[Fe^{2+}][Fe^{2+}]}$$

$$[Fe^{2+}] = 3.63 \times 10^{-16} \ M$$

$$80.0 \text{ ml} \times 3.63 \times 10^{-16} \ M = 2.90 \times 10^{-14} \text{ mmoles}$$

$$2.90 \times 10^{-14} \text{ mmoles} \times 55.8 \text{ mg/mmole} = 1.62 \times 10^{-12} \text{ mg}$$

**Examples of Titration Curves.**    A very good qualitative prediction of the shape for a titration curve can be made by calculations at three points: potential at 50% reacted, stoichiometric point potential or at 100% reacted, and potential at 200% reacted. The potential at the 50% point for any redox reaction is given by the standard (formal) reduction potential of the system being titrated. The stoichiometric point is calculated as a weighted average from the two standard (formal) reduction potentials that are involved, while the 200% point is given by the standard (formal) reduction potential of the titrant. Thus, all that is required is the two standard (formal) reduction potentials, since the magnitude and sharpness of the titration break will increase as the difference between the two potentials increases. If needed, the equilibrium constant can also be calculated from this minimum amount of data.

Prediction of redox titration curves from a minimum number of calculations is illustrated for the following titrations:

(a) $Ce^{4+} + Fe^{2+} \xrightarrow{\text{1 F HClO}_4} Ce^{3+} + Fe^{3+}$

(b) $Ce^{4+} + Fe^{2+} \xrightarrow{\text{1 F HNO}_3} Ce^{3+} + Fe^{3+}$

(c) $MnO_4^- + 5Fe^{2+} + 8H^+ \xrightarrow{\text{H}^+ = 1M} Mn^{2+} + 5Fe^{3+} + H_2O$

(d) $Ce^{4+} + Fe^{2+} \xrightarrow{\text{1 F H}_2\text{SO}_4} Ce^{3+} + Fe^{3+}$

(e) $2Ce^{4+} + AsO_2^- + H_2O \xrightarrow{\text{1 F HClO}_4} 2Ce^{3+} + AsO_3^- + 2H^+$

(f) $2Ce^{4+} + C_2O_4^{-2} \xrightarrow{1\,F\,HClO_4} 2Ce^{3+} + 2CO_2$

(g) $2Ce^{4+} + 2I^- \xrightarrow{1\,F\,HClO_4} I_2 + 2Ce^{3+}$

(h) $Br_2 + 2I^- \rightarrow I_2 + 2Br^-$

(i) $Fe^{3+} + 2I^- \rightarrow I_2 + Fe^{2+}$

Table 11–1 lists the data and Fig. 11–2 illustrates the S-shaped curves that can be drawn through the three points.

Several important points are illustrated in Fig. 11–2. For example, in Fig. 11–2A the effect of increasing oxidizing power of the titrant, which leads to a large and sharper potential break, is shown. In addition, it is important to notice how the experimental conditions can affect the titration. For example, the oxidizing power of cerium(IV) varies with the type of acid used and changes in the order $HClO_4 > HNO_3 > H_2SO_4 > HCl$.

If the Nernst equation is written for the $Ce^{4+}/Ce^{3+}$ couple, there is no indication that acidity should affect the potential of the half-cell. However, if some other reaction takes place which reduces the $Ce^{4+}$ concentration, the half-cell potential will decrease. This is exactly what happens in that $Cl^- > SO_4^{-2} > NO_3^- > ClO_4^-$ in complexing power with $Ce^{4+}$, which acts as a means of decreasing the $Ce^{4+}$ concentration.

In Fig. 11–2B the effect of reducing power is shown. As predicted, the stronger the reducing agent, the larger the potential break.

In Fig. 11–2C the reverse is illustrated or the effect of increasing oxidizing

**Table 11–1. Equilibrium Constant and Potentials at Three Points for Several Titrations by Calculation**

| Titra-tion | $E^{\circ}_{titrant}$ | $E^{\circ}_{reactant}$ | $K$ | Potential (V) | | |
|---|---|---|---|---|---|---|
| | | | | 50% | 100% [a] | 200% |
| a | +1.70; $Ce^{4+}$ | +0.771; $Fe^{2+}$ | $10^{15.8}$ | 0.771 | 1.24 | 1.70 |
| b | +1.60; $Ce^{4+a}$ | +0.771; $Fe^{2+}$ | $10^{14.1}$ | 0.771 | 1.19 | 1.60 |
| c | +1.51; $MnO_4^-$ | +0.771; $Fe^{2+}$ | $10^{62.8}$ | 0.771 | 1.39 | 1.51 |
| d | +1.70; $Ce^{4+}$ | +0.68; $Fe^{2+a\,b}$ | $10^{17.3}$ | 0.680 | 1.19 | 1.70 |
| e | +1.70; $Ce^{4+}$ | +0.577; $HAsO_2^b$ | $10^{38.1}$ | 0.577 | 0.951 | 1.70 |
| f | +1.70; $Ce^{4+}$ | −0.49; $H_2C_2O_4$ | $10^{74.4}$ | −0.490 | 0.240 | 1.70 |
| g | +0.536; $I^-$ | +1.70; $Ce^{4+}$ | $10^{39.5}$ | 1.70 | 1.12 | 0.536 |
| h | +0.536; $I^-$ | +1.09; $Br_2$ | $10^{18.8}$ | 1.09 | 0.813 | 0.536 |
| i | +0.536; $I^-$ | +0.771; $Fe^{3+}$ | $10^{0.796}$ | 0.771 | 0.654 | 0.536 |
| j | +0.771; $Fe^{2+}$ | +1.70; $Ce^{4+}$ | $10^{15.8}$ | 1.70 | 1.24 | 0.771 |

[a] These are calculated as approximations using Eq. (11–1).
[b] Potentials are formal reduction potentials, while others are standard reduction.

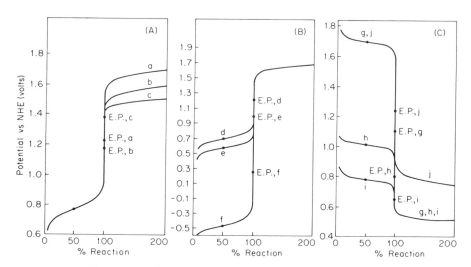

**Fig. 11–2.** Calculated titration curves. (See Table 11–1.)

| (A) | | (B) | | (C) | |
|---|---|---|---|---|---|
| Titrant | Reactant | Titrant | Reactant | Titrant | Reactant |
| (a) $Ce^{4+}$, $HClO_4$ | $Fe^{2+}$ | (d) $Ce^{4+}$ | $Fe^{2+}$, $H_2SO_4$ | (g) $I^-$ | $Ce^{4+}$, $HClO_4$ |
| (b) $Ce^{4+}$, $HNO_3$ | $Fe^{2+}$ | (e) $Ce^{4+}$ $HClO_4$ | $HAsO_2$ | (h) $I^-$ | $Br_2$ |
| (c) $MnO_4^-$ | $Fe^{2+}$ | (f) $Ce^{4+}$, $HClO_4$ | $H_2C_2O_4$ | (i) $I^-$ | $Fe^{3+}$ |
| | | | | (j) $Fe^{2+}$ | $Ce^{4+}$ |

power. For the same titrant, $I^-$, a larger break is observed as the oxidizing power of the sample increases. Also shown is the effect of reducing power of the titrant. For the titration of $Ce^{4+}$ a larger break is found when using a stronger reducing agent (compare $I^-$ and $Fe^{2+}$ titrants).

Reactions (a) to (f) can be carried out quantitatively by a titration technique. In some cases catalysts are used to compensate for slow reaction rates. Although reactions (g) to (i) are favorable and stoichiometric, they are normally not carried out by direct titration with iodide ion. In practice, an excess of $I^-$ is placed in the sample solution. A stoichiometric amount of $I_2$ is produced which is conveniently titrated with thiosulfate solution. The chemistry is described by the reactions given below:

$$\text{Oxidant} + 2I^-_{(excess)} \rightarrow \text{reductant} + I_2$$
$$I_2 + S_2O_3^{2-} \rightarrow S_4O_6^{2-} + 2I^- \tag{11-3}$$

The $I_2/I^-$ and $S_2O_3^{2}/S_4O_6^-$ reactions are more familarly known under the title *iodimetry* and the methods of analysis involving these two reactions are referred to as *iodometric methods*.

# DIFFERENTIAL TITRATIONS

All the previous examples have been cases where the sample contains only one reducing agent (or one oxidizing agent). If two or more are present the titration is still possible. However, a titration break for each will be obtained providing the reduction potentials of the two species differ by at least 0.2 V. This is illustrated in Fig. 11–3 where a differential titration curve of titanium(III) and iron(II) with permanganate titrant in acid solution is shown.

The half-reactions that are involved are the following:

$$MnO_4^- + 8H^+ + 5e \rightleftharpoons Mn^{2+} + 4H_2O \quad E^\circ_{MnO_4^-, Mn^{2+}} = +1.51 \text{ V}$$

$$TiO^{2+} + 2H^+ + 1e \rightleftharpoons Ti^{3+} + 3H_2O \quad E^\circ_{TiO^{2+}, Ti^{3+}} = +0.1 \text{ V}$$

$$Fe^{3+} + 1e \rightleftharpoons Fe^{2+} \quad E^\circ_{Fe^{3+}, Fe^{2+}} = +0.771 \text{ V}$$

Since titanium (III) is the stronger reducing agent, it will react first with the permanganate titrant. After this the iron(II) is titrated.

It is possible to calculate the titration curve for this system. Prior to

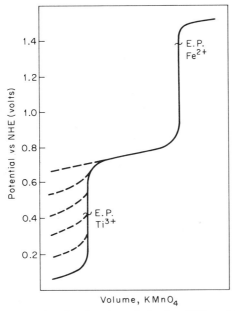

**Fig. 11–3.** Titration curve for the titration of titanium(III) and iron(II) with permanganate ion. (Dotted curves represent a successive increase in the reduction potential of the species titrated first.)

reaching the stoichiometric point for $Ti^{3+}$ the potential at any point on the curve is calculated from the Nernst equation.

$$E = E^{\circ}_{TiO^{2+},\ Ti^{3+}} - \frac{0.0592}{1} \log \frac{[Ti^{3+}]}{[TiO^{2+}][H^+]^2} \tag{11-4}$$

This portion of the curve, as shown in Fig. 11–3, is the same as if titanium (III) was being titrated by itself.

The stoichiometric point potential for the titration of titanium (III) cannot be calculated by expression (11-1), because as the $TiO^{2+}$ concentration increases ($Ti^{3+}$ concentration decreases), the potential also becomes influenced by the presence of the iron (II) concentration. Thus, the potential of the two half-cells at the stoichiometric point will be equal after equilibrium is reached. This is represented by the reaction

$$TiO^{2+} + 2H^+ + Fe^{2+} \rightleftharpoons Ti^{3+} + Fe^{3+} + H_2O$$

and upon adding the two half-cells the following expression is obtained:

$$2E = E^{\circ}_{TiO^{2+},\ Ti^{3+}} + E^{\circ}_{Fe^{3+},\ Fe^{2+}} - \frac{0.0592}{1} \log \frac{[Ti^{3+}][Fe^{2+}]}{[TiO^{2+}][H^+]^2[Fe^{3+}]} \tag{11-5}$$

To complete the calculation it is necessary to make some approximations. Assume that the acidity of the solution is 1 $F$. It can also be assumed that

$$[Fe^{3+}] \cong [Ti^{3+}]$$

Substitution into Eq. (11-5) leads to

$$E = \frac{0.871}{2} - \frac{0.0592}{2} \log \frac{[Fe^{2+}]}{[TiO^{2+}]} \tag{11-6}$$

Thus, the stoichiometric point potential will be determined by the concentration of the titanium (IV) species and iron (II) in the solution. If they are equal, the stoichiometric point potential will be at 0.436 V.

As the titration is continued, iron (II) is oxidized by the permanganate titrant. Prior to reaching the next stoichiometric point, the potential at any point on the curve is given by the expression

$$E = E^{\circ}_{Fe^{3+},\ Fe^{2+}} - \frac{0.0592}{1} \log \frac{[Fe^{2+}]}{[Fe^{3+}]}$$

Thus, this portion of the curve is as if iron (II) were being titrated by itself.

The stoichiometric point potential can be calculated as shown in Example 11–1, since no additional reducing agent is present.

$$E_{SP} = \frac{5E^{\circ}_{MnO_4^-, \, Mn^{2+}} + E^{\circ}_{Fe^{3+}, \, Fe^{2+}}}{6} - \frac{0.0592}{6} \log \frac{1}{[H^+]^8}$$

$$E_{SP} = \frac{5(1.51 \text{ V}) + (0.771 \text{ V})}{6} - \frac{0.0592}{6} \log \frac{1}{[1]^8} = 1.39 \text{ V}$$

Whether the two breaks are as sharply defined as those shown in Fig. 11–3 will depend on the difference in their respective reduction potentials. For the example in Fig. 11–3 the difference is substantial. As pointed out previously, a difference of at least 0.2 V is needed. If the reduction potential of the titanium(IV)–titanium(III) half-cell were more positive a smaller break would be obtained. To illustrate this, a series of dotted lines are shown in Fig. 11–3 which represents a successive positive increase in the reduction potential of the first titrated species. When this potential equals the reduction potential for the second component only one break is obtained. The end point would, therefore, correspond to the titration of both species.

## END-POINT DETECTION

**Potentiometry.**    A classic oxidation–reduction titration procedure, which is very useful for the determination of iron, is the titration of $Fe^{2+}$ with $Ce^{4+}$. Briefly, the procedure involves the adjustment of the iron in solution to the 2+ oxidation state, addition of acid (usually 0.5 $F$ or higher), insertion of a Pt indicator and SCE reference electrode, addition of aliquots of the $Ce^{4+}$ titrant, and recording the potential change as a function of added titrant. Near the stoichiometric point, small increments of titrant are added since a sharp potential change occurs. A plot of the titration curve is like Fig. 11–1 and the end point is selected by using one of the procedures outlined in the chapter on laboratory techniques.

If potentiometry is to be used only for end-point detection for a redox titration, the main goal is to determine the number of milliters of titrant required to reach the end point (stoichiometric point), which is indicated by a sharp change in potential. It is not necessary to know the potential precisely at the beginning, stoichiometric point, or after the stoichiometric point since it is the potential change which allows the selection of the end point. Therefore, the potential may be adjusted arbitrarily at the beginning of the titration.

Potentiometric titrations can be used for a wide variety of inorganic and organic oxidation–reduction reactions in routine quantitative analysis. In some cases the instrumentation is designed to follow the titrations automatically in which the titration curve, (potential vs ml of titrant), or its first or second derivative are plotted or observed on an oscilloscope.

A useful, typical determination in the pharmaceutical industry is the potentiometric titration of sulfanilamide with $NaNO_2$.

$$H_2N\!-\!\!\langle\bigcirc\rangle\!-\!SO_2NH_2 \ + \ NO_2^- \ + \ 2\,H^+ \ \longrightarrow \ {}^+N_2\!-\!\!\langle\bigcirc\rangle\!-\!SO_2NH_2 \ + \ 2\,H_2O \quad (11\text{-}7)$$

The titration is based on the above stoichiometry and the progress of the titration can be followed potentiometrically with a calomel–Pt electrode pair. The reaction is not specific for sulfanilamide since $NO_2^-$ undergoes a reaction with many other types of aromatic primary amines. Hence, this titration can be used for the determination of several other pharmaceutically useful compounds including other sulfa drugs, p-aminophenol, and certain alkaloids.

A change in oxidation state does not have to take place in order to follow the titration potentiometrically. In practice, if the indicator electrode responds to the concentration of the species of interest in the solution, any reaction which removes this species or changes its equilibrium concentrations will lead to a change in potential at the indicator electrode. Hence, neutralization (Chapters 8 and 9), precipitation (Chapter 7), and complexometric (Chapter 15) titrations can be followed potentiometrically.

For example, in the titration of silver ion with chloride ion, the silver ion concentration in solution changes as titrant is added. Although no change in oxidation state occurs in the reaction, the $Ag^+$ concentration can be indicated by a silver wire inserted in the solution as an indicator electrode, because the redox couple $Ag^+/Ag$ is present in the solution. This couple will then be defined by the Nernst equation or

$$E_{Ag^+,\,Ag} = E^{\circ}_{Ag^+,\,Ag} - \frac{0.0592}{1}\log\frac{1}{[Ag^+]}$$

Consequently, as the $Ag^+$ concentration changes through precipitation as $AgCl$, the potential at the silver wire must also change. Insertion of a reference electrode completes the cell and the change in cell potential is easily measured by a potentiometer.

A successful potentiometric titration, assuming the reaction is fast and has a favorable equilibrium constant, requires the selection of the appropriate indicator electrode. The reference electrode is usually a saturated calomel electrode. In modern applications of potentiometric titrations, indicator electrodes are either noble metals (Pt and Au), insoluble metal salt/metal electrode ($AgCl/Ag$), or ion-selective electrodes. The first two were described in Chapter 10 and the last type will be described in Chapter 13.

The noble metal electrodes are primarily used to detect potential changes for systems where both the oxidized and reduced species are soluble in the solvent being used. Several typical titrations that could be followed potentiometrically (second species is the titrant) are $Fe^{2+}-Cr_2O_7^{2-}$, $Fe^{2+}-Ce^{4+}$, $Fe^{3+}-Ti^{3+}$, $Sn^{2+}-Ce^{4+}$, $Mn^{2+}-MnO_4^-$, $I_2-S_2O_3^{2-}$, $Co^{2+}-Fe(CN)_6^{3-}$, $Fe(CN)_6^{3-}-Hg^+$, $I^--MnO_4^-$, and $U^{4+}-Ce^{4+}$.

One major advantage of the potentiometric titration is that it is a convienient and accurate technique for following differential titrations. Hence, individual components of a mixture can be determined, provided they are sufficiently different in oxidizing or reducing power. Furthermore, potentiometric end-point detection is readily automated.

**Redox Color Indicators.** Another convenient way to detect an end point in a redox titration is by the use of a redox color indicator. A visual redox indicator can be defined as some substance which is added to the solution and undergoes a physical change that is seen by the eye, such as a color or fluorescence change, as the stoichiometric point is passed.

The vast majority of redox indicators are oxidizing or reducing agents themselves. Perhaps the principal exception is starch which forms a blue complex with $I_3^-$ ion and is frequently used as the indicator in iodometric methods.

A useful visual redox indicator should possess the following properties.

1. It should change color at the stoichiometric point potential for the titration.
2. The color change should be sharp and involve intense, contrasting colors.
3. The indicator should be soluble, stable, and undergo reversible changes.

Perhaps the one single factor which is least often attained is the one of reversibility. Reversibility, as applied here, means that if the end point is exceeded (for example, titration with an oxidizing agent) and a color change is observed, a reverse in the titration (titration with a reducing agent) should cause the indicator to change back to its original color at the exact point as before. What often happens is that the indicator products of the oxidation (or reduction) undergo additional reactions to other products and thus, the indicator will act irreversibly.

A half-reaction for an indicator process leading to a color change can be written as

$$IN_{ox} + nH_3O^+ + ne \rightleftharpoons H_nIN_{red} + nH_2O \qquad (11\text{-}8)$$
$$\text{Color I} \qquad\qquad \text{Color II}$$

where reversibility is assumed. The corresponding Nernst equation for the indicator half-reaction is

$$E = E^\circ_{IN_{ox},\ HIN_{red}} - \frac{0.0592}{n} \log \frac{[H_nIN]}{[IN_{ox}][H_3O^+]^n} \qquad (11\text{-}9)$$

From experience, it can be concluded that one color will be seen over the other color when the concentration of the first is about ten times the concentration of the second. Thus, to see color I

$$\frac{[IN_{ox}]}{[H_nIN_{red}]} \geqq \frac{10}{1}$$

and to see color II

$$\frac{[IN_{ox}]}{[H_nIN_{red}]} \leqq \frac{1}{10}$$

If these are substituted into the Nernst equation, the following expression is obtained:

$$E = E^{\circ}_{IN_{ox}, HIN_{red}} \pm \frac{0.0592}{n} \log \frac{1}{[H_3O^+]^n} \qquad (11\text{-}10)$$

For the indicator in which $H_3O^+$ is not involved or if $H_3O^+$ is controlled at $1\ M$ the expression simplifies to

$$E = E^{\circ}_{IN_{ox}, HIN_{red}} \pm \frac{0.059}{n} \qquad (11\text{-}11)$$

Consequently, it is possible, if $n$ is known, to qualitatively predict the potential range over which the indicator changes color. In choosing an indicator for a particular titration the indicator that has an $E^{\circ}_{IN_{ox},HIN_{red}}$ at or close to the equivalence point potential of the titration is chosen. Of course the other indicator properties mentioned before must also be considered in the selection. Table 11–2 lists some typical redox indicators.

As an illustration consider the titration of $Fe^{2+}$ with $Ce^{4+}$ in $1\ F\ H_2SO_4^-$. In the calculation of the titration curve the stoichiometric point potential was found to be $+1.06$ V (vs NHE). From Table 11–2, the most appropriate indicator would be the 1,10-phenanthroline–ferrous complex which has an $E^{\circ}_{ox,red}$ at $+1.14$ V.

Experimentally, a few drops of about $0.03\ F$ indicator solution are added to the sample solution followed by the Ce(IV) titrant. In the course of adding the titrant the solution retains a reddish color (color or reduced indicator form) and as the stoichiometric point is reached the indicator passes through its color transition to a pale blue (color of oxidized indicator form). The sharper the color transition, the easier it is to detect the end point and thus the analysis is more accurate. Excessive amounts of indicator must be avoided to prevent large indicator blanks.

In differential titrations, if the difference in potential between the two breaks is 0.4 V or larger, it is possible to use indicators to detect both end points. The difficulty is in choosing indicators whose colors do not conflict with each other. If the potential difference is less than 0.4 V, potentiometry should be used.

**Table 11-2. Several Useful Redox Indicators**[a]

| $E$ (volts vs NHE) | Indicator | Colors of | | Preparation |
|---|---|---|---|---|
| | | Reduced form | Oxidized form | |
| +0.36 | Methylene blue | Colorless | Blue | 2 g/liter $H_2O$ |
| +0.54 | 1-Naphthol-2-sulfonic acid indophenol | Colorless | Red | — |
| +0.76 | 4'-Ethoxy-2,4-diaminoazo-benzene | Red | Pale yellow | — |
| +0.8 | Diphenylamine or diphenyl-benzidine | Colorless | Violet or green | 1% sol in 18 $M$ $H_2SO_4$ |
| +0.81 | $N$-Methyldiphenylamine-$p$-sulfonic acid | Colorless | Red, 510 nm | — |
| +0.85 | Diphenylaminesulfonic acid | Colorless | Violet or green | 3 g Ba salt/liter $H_2O$ |
| +1.06 | $p$-Nitrodiphenylamine | Colorless | Violet | — |
| +1.08 | $N$-Phenylanthranilic acid | Colorless | Pink | 1 g/liter 0.01 $M$ $Na_2CO_3$ |
| +1.14 | $o$-Phenanthroline ferrous complex (ferroin) | Red, 510 nm | Pale blue | 1.5 g dye/100 ml 0.025 $M$ FeSO$_4$ |
| +1.31 | Nitro-$o$-phenanthroline ferrous complex (nitroferroin) | Violet red, 510 nm | Pale blue | 1.7 g dye/100 ml 0.025 $M$ FeSO$_4$ |

[a] L. Meites, ed., "Handbook of Analytical Chemistry," 1st ed., McGraw-Hill, New York, 1963.

**1,10-Phenanthroline Derivatives.** The tris(1,10-phenanthroline) iron (II) complex (ferroin), **I**, and related compounds can be used analytically

**I**

in a variety of ways. One of its applications is as a redox indicator. The redox reaction shown as an oxidation is written as

$$\mathrm{Fe(C_{12}H_8N_2)_3^{2+}} \rightleftharpoons \mathrm{Fe(C_{12}H_8N_2)_3^{3+}} + 1e \qquad (11\text{-}12)$$
$$\text{red} \qquad\qquad \text{pale blue}$$

The red color is very intense, the color change is very sharp, and the indicator action is reversible. Although acidity does not enter into the half-cell, the acidity will affect the indicator. For example, the indicator in its reduced form slowly decomposes in strongly acidic solution. The complex also decomposes in the presence of $Co^{2+}$, $Cu^{2+}$, $Ni^{2+}$, $Zn^{2+}$, and $Cd^{2+}$ since these ions will compete with $Fe^{2+}$ for the coordination sites.

Reduction potentials for the complex can be changed by adding substituents on the phenanthroline ring system or using different metal ions in the complex. Thus, a convenient redox indicator table can be devised which is entirely based on 1,10-phenanthroline derivatives and their complexes. In some cases, a change in fluorescence as well as a color change is observed. For example, tris(5,6-dimethyl-1,10-phenanthroline)ruthenium(II) undergoes a sensitive change from a deep red fluorescence to no fluorescence in the oxidized forms. Other indicators possessing fluorescent changes are also available.
change from a red fluorescence to nonfluorescent compounds in the oxidized forms. Other indicators possessing fluorescent changes are also available.

**Diphenylamine.**      Diphenylamine and diphenylamine-4-sulfonic acid function essentially the same as redox indicators. The main difference is that the latter compound is much more soluble in water. These indicators are particularly suited to the titration of $Fe^{2+}$ with $Cr_2O_7^{2-}$.

The mechanism for the indicator oxidation has been suggested to be

Diphenylamine                                    Diphenylbenzidine

(11-13)

Diphenylbenzidine violet

First, the diphenylamine is oxidized irreversibly to diphenylbenzidine; the latter compound could also be used as the indicator. Further reversible oxidation, as a result of the added titrant and reaching the appropriate potential, to the diphenylbenzidine violet species results in a color change of colorless to violet. The actual color shades, of course, will depend on the colors of the other species in solution. The indicator can also be used for the titration of oxidizing agents such as $Ce^{4+}$, $MnO_4^-$, $Cr_2O_7^{2-}$, and $VO_3^-$ with $Fe^{2+}$ (color change violet to colorless). However, diphenylbenzidine is oxidized to other products and is thus not stable for long periods of time in an oxidizing solution.

## Questions

1.   List the requirements which a redox reaction must satisfy if it is to be used in a volumetric titration.

2. Suggest possible mechanisms which might explain why catalysts are usually required in redox reactions.
3. Predict which of the pairs would give the largest titration break. The second component in each pair is the titrant.
   a. $Fe^{2+}$–$Ce^{4+}$ (HCl) and $Fe^{2+}$–$Ce^{4+}$ ($HNO_3$)
   b. $Fe^{2+}$–$MnO_4^-$ and $Fe^{2+}$–$Cr_2O_7^{2-}$
   c. $Ti^{3+}$–$Fe^{3+}$ and $Sn^{2+}$–$Fe^{3+}$
   d. $Fe^{3+}$–$Ti^{3+}$ and $AsO_3^{3-}$–$I_2$
   e. $AsO_3^{3-}$–$MnO_4^-$ and $AsO_3^{3-}$–$I_2$
4. Describe how the $E$'s change for $A_{ox}$ and $B_{red}$ the instant a solution of $A_{ox}$ and $B_{red}$ are mixed.
5. What is a differential potentiometric titration?
6. Explain why the measurement of the actual potential is not required when determining titration end points by potentiometry.
7. Explain, using reactions, half-cells, and the Nernst equation why a Ag-wire can be used for the potentiometric titration of $S^{2-}$ with $AgNO_3$.
8. Compare a redox indicator to an acid–base indicator with respect to how they function.

## *Problems*

1. Calculate the potential for the following systems vs NHE.
   *a. 60.0 ml of 0.100 $F$ $Ce^{4+}$ is mixed with 20.0 ml of 0.200 $F$ $Fe^{2+}$ at constant 1 $F$ $H_2SO_4$.
   b. 35.0 ml of 0.100 $F$ $Fe^{2+}$ is mixed with 45.0 ml of 0.100 $F$ $Cr_2O_7^{2-}$ at constant 1 $F$ HCl.
   c. 25.0 ml of 0.100 $F$ $Fe^{2+}$ is mixed with 25.0 ml of 0.010 $F$ $MnO_4^-$ at a constant pH of 1.
   d. 45.0 ml of 0.0500 $F$ $Ti^{3+}$ is mixed with 37.5 ml of 0.0600 $F$ $Fe^{3+}$ at constant 1 $F$ $H_2SO_4$.
   e. 55.0 ml of 0.0200 $F$ $MnO_4^-$ is mixed with 55.00 ml of 0.0200 $F$ $C_2O_4^{2-}$ at constant 1 $F$ $H^+$.

2.* Calculate the potential in Problem 1a–e vs the SCE.

3. Calculate the potential vs NHE at 25%, 50%, 100%, 125%, and 200% reacted for the following titrations.
   *a. 50.0 ml of 0.0500 $F$ $Fe^{2+}$ titrated with 0.100 $F$ $Ce^{4+}$ constant 1 $F$ $H_2SO_4$.
   b. 40.0 ml of 0.100 $F$ $Tl^{3+}$ titrated with 0.100 $F$ $Fe^{2+}$ at constant 1 $F$ $H_2SO_4$.
   c. 40.0 ml of 0.0750 $F$ $Fe^{2+}$ titrated with 0.0100 $F$ $Cr_2O_7^{2-}$ at constant 1 $F$ HCl.

4. Calculate the volume of titrant required to reach the stoichiometric point for each of the following titrations.
   a.* 41.0 ml of 0.0200 $F$ $Ti^{3+}$ titrated with 0.0300 $F$ $Fe^{3+}$ at constant 1 $F$ $H_2SO_4$.

---

* Answers are listed at the end of the book for problems marked with an asterisk.

    b. 65.0 ml of 0.0500 $F$ $Fe^{2+}$ titrated with 0.0100 $F$ $Cr_2O_7^{2-}$ at constant 1 $F$ HCl.

    c. 56.5 ml of 0.0420 $F$ ascorbic acid (vitamin C—see Table 10–2) is titrated with 0.0375 $F$ $Ce^{4+}$ at constant 1 $F$ $H_2SO_4$.

5.* Derive an expression which can be used to calculate the stoichiometric point potential for each titration in Question 4.

6.* Calculate the mg of ascorbic acid remaining in solution at the stoichiometric point for the titration in Question 4c.

7. A 35.0-ml aliquot of 0.150 $F$ HCl is titrated with 0.175 $F$ NaOH. If a hydrogen electrode is put together such that the $H_2$ gas is maintained at 1 atm during the titration, calculate the potential at the hydrogen electrode vs SCE at 0, 20, 40, 80, 90, 99, 100, 101, 110, and 120% reaction. Plot a graph of $E$ vs milliliters of titrant, $[H^+]$ vs milliliters of titrant, and pH vs milliliters of titrant.

8.* Calculate the molar ratio of the two forms of the ferroin indicator if the potential of the solution registers +1.04 V vs NHE; +1.24 V vs NHE.

9. The standard reduction potential was to be determined for the half-cell

$$\text{Dehydroascorbate} + 2H^+ + 2e \rightleftharpoons \text{ascorbic acid (vitamin C)}$$

If the following potentials were recorded for the titration of 0.2640 g of pure ascorbic acid in solution with 0.1000 $F$ $Ce(SO_4)_4$ in 1 $F$ $H_2SO_4$, calculate the $E°$ for the above half-cell.

$$\text{Ascorbic acid} + 2Ce^{4+} \rightarrow 2Ce^{3+} + \text{Dehydroascorbate} + 2H^+$$

| Titrant (ml) | Potential vs SCE (V) |
|---|---|
| 5.00 | +0.145 |
| 10.00 | +0.158 |
| 15.00 | +0.169 |
| 20.00 | +0.171 |
| 25.00 | +0.186 |

# Chapter
# Twelve
# Oxidizing and Reducing
# Agents as Titrants in
# Analytical Chemistry

## INTRODUCTION

Many elements have more than one stable oxidation state. Thus, a titration with an oxidizing agent or reducing agent is a very useful general method of analysis. However, for success the species being titrated must be present quantitatively at its lower oxidation state if it is to be titrated with an oxidizing agent. Similarly, the species should be present quantitatively at its higher oxidation state if it is to be titrated with a reducing agent. Consequently, most procedures for inorganic samples, which are based on a redox titration, require as a preliminary step the adjustment of the oxidation state of the sample. This oxidation change must be performed quantitatively and the reagents or conditions used for this purpose must not interfere in the titration.

Many organic functional groups can also be titrated with oxidizing agents or reducing agents. Normally, pretreatment leading to an adjusted oxidation state is not necessary for the determination of organic functional groups.

If optimum conditions are used, redox titrations will yield, in general, a precision and accuracy of $\pm 1\%$ or better. Since many species do not change oxidation state easily, a certain degree of selectivity is obtained by redox methods. In general, the redox titration is used in macro analysis, however, when coupled with the general field of electrochemistry the concentration range and scope is broadened tremendously.

The frequency of application of reducing titrants is less than with oxidizing titrants. The reason for the difference is stability. Virtually all useful reducing agents are prone to air oxidation; the better the reducing agent the better the reaction with oxygen. This problem can be rectified by standardizing the titrant frequently and, for the more severe cases, the titrant can be protected from oxygen (air) by storage under a nitrogen atmosphere.

*265*

## OXIDIZING TITRANTS

The common oxidizing agents are listed in Table 12–1. Several are available as primary standards and can be used for the preparation of standard oxidizing titrants and for the standardization of solutions of reducing titrants.

**Potassium Permanganate.**   Potassium permanganate is a powerful, versatile oxidizing titrant. It has specific applications depending on whether it is used in acidic, neutral, or alkaline solutions, where it exhibits different oxidizing power.

$$MnO_4^- + 8H^+ + 5e = Mn^{2+} + 4H_2O \qquad E^f = 1.51 \text{ V}; [H_3O^+] = 1\ M$$

$$MnO_4^- + 4H^+ + 3e = MnO_{2(s)} + 2H_2O \qquad E^f = 1.69 \text{ V, neutral}$$

$$MnO_4^- + 2H_2O + 3e = MnO_{2(s)} + 40H^- \qquad E^f = 0.51 \text{ V, basic}$$

Another reason for its wide use is that it acts as a self-indicator. The first excess of $MnO_4^-$ passed the stoichiometric point produces a distinct pink coloration.

The disadvantages of $KMnO_4$ titrant are that it is not a primary standard and preparation of the solution will always cause formation of solid $MnO_2$. Therefore, the titrant must be filtered before it is used and also after long

**Table 12–1.   A List of Oxidizing Agents Used as Standards and Titrants**

| Reagent | Conditions | Half-reaction | $E^f$ [a] (volts) |
|---|---|---|---|
| $KMnO_4$ | Acidic | $MnO_4^- + 8H^+ + 5e \rightleftharpoons Mn^{2+} + 4H_2O$ | +1.5 |
| $KMnO_4$ | Neutral | $MnO_4^- + 4H^+ + 3e \rightleftharpoons MnO_2 + 2H_2O$ | +1.7 |
| $K_2Cr_2O_7$[b] | Acidic | $Cr_2O_7^{2-} + 14H^+ + 6e \rightleftharpoons 2Cr^{3+} + 7H_2O$ | +1.3 |
| $(NH_4)_2Ce(NO_3)_6$[b] | Acidic | $Ce^{4+} + 1e \rightleftharpoons Ce^{3+}$ | +1.4–1.7 |
| $Ce(SO_4)_2$ | Acidic | $Ce^{4+} + 1e \rightleftharpoons Ce^{3+}$ | +1.4–1.7 |
| $H_5IO_6$ | Weakly acidic | $IO_6^{5-} + 6H^+ + 2e \rightleftharpoons IO_3^- + 3H_2O$ | +1.5 |
| $KIO_3$[b] | Acidic | $IO_3^- + 6H^+ + 2Cl^- + 4e \rightleftharpoons ICl_2^- + 3H_2O$ | +1.2 |
| $I_2$[b] | Acidic–neutral–basic | $I_2 + 2e \rightleftharpoons 2I^-$ | +0.6 |
| $KBrO_3$[b] | Acidic + KBr | $BrO_3^- + Br^- + 6H^+ \rightarrow 3Br_2 + 3H_2O$ | — |
| | | $Br_2 + 2e \rightleftharpoons 2Br^-$ | +1.1 |
| $NaOCl$ | Neutral–basic | $OCl^- + H_2O + 2e \rightleftharpoons Cl^- + OH^-$ | +0.9 |
| $FeCl_3$ | Acidic | $Fe^{3+} + 1e \rightleftharpoons Fe^{2+}$ | +0.8 |
| $H_2O_2$ | Acidic–neutral | $H_2O_2 + 2H^+ + 2e \rightleftharpoons 2H_2O$ | +1.8 |

[a] Potential will depend on pH and other experimental conditions.

[b] Available as primary standard.

**Table 12-2.   Applications of Potassium Permanganate**

| Ion determined | Product |
|---|---|
| $KMnO_4$ in acidic solution: $MnO_4^- + 8H + 5e \rightleftharpoons Mn^{2+} + 4H_2O$ | |
| $H_2C_2O_4$ | $CO_2$ |
| $MC_2O_4$ where M = $Mg^{2+}$, $Ca^{2+}$, $Zn^{2+}$, $Co^{2+}$, $La^{3+}$, $Th^{4+}$, $Ba^{2+}$, $Sr^{2+}$, $Ce^{4+}$, $Ag^+$, $Pb^{2+}$ | $CO_2$ |
| $HNO_2$ | $HNO_3$ |
| $I^-$ | ICN (in $CN^-$) |
| $As^{3+}$ | $AsO_4^{3-}$ |
| $Br^-$ | $Br_2$ |
| $Sn^{2+}$ | $Sn^{4+}$ |
| $H_2O_2$ | $O_2$ |
| $Fe^{2+}$ | $Fe^{3+}$ |
| $Sb^{3+}$ | $SbO_4^{3-}$ |
| $Fe(CN)_6^{4-}$ | $Fe(CN)_6^{3-}$ |
| $VO^{2+}$ | $VO_3^-$ |
| $Mo^{3+}$ | $MoO_4^{2-}$ |
| $U^{4+}$ | $UO_2^{2+}$ |
| $Ti^{3+}$ | $Ti^{4+}$ |
| $Nb^{3+}$ | $Nb^{5+}$ |
| $KMnO_4$ in neutral solution: $MnO_4^- + 4H^+ + 3e \rightleftharpoons MnO_2 + 2H_2O$ | |
| $Mn^{2+}$ | $MnO_2$ |
| $KMnO_4$ in basic solution: $MnO_4^- + e \rightleftharpoons MnO_4^{2-}$ | |
| $IO_3^-$ | $IO_4^-$ |
| $I^-$ | $IO_4^-$ |
| $CN^-$ | $CNO^-$ |
| $SO_3^{2-}$ | $SO_4^{2-}$ |
| $HS^-$ | $SO_4^{2-}$ |

periods of storage. Storage in the dark is preferred since light catalyzes the formation of $MnO_2$. Formation of $MnO_2$ is also hastened because of the presence of organic matter. For these reasons the $KMnO_4$ solution should be standardized frequently. Oxalic acid dihydrate ($H_2C_2O_4 \cdot 2H_2O$), sodium oxalate ($Na_2C_2O_4$), potassium tetraoxalate ($KHC_2O4 \cdot H_2C_2O_4 \cdot 2H_2O$), or $As_2O_3$ are usually used for standardization.

Table 12-2 lists the more important applications of $KMnO_4$ titrant. Probably, the most useful applications are for the determination of Fe, Mn, peroxides, and metal ions through oxalate formation.

In the determination of iron, the oxidation state of iron is first adjusted to Fe(II). Since chloride ion will be partially oxidized by $MnO_4^-$, chloride ion must be absent. However, the better dissolution procedures for iron ore and adjustment of oxidation state will introduce chloride ion into the sample. Therefore, an alternate procedure is to add $Mn(II)$–$H_2SO_4$ solution to the $Fe(II)$ sample. The effect of the presence of $Mn(II)$ is to reduce the rate at which $MnO_4^-$ oxidizes $Cl^-$ and thereby minimizes the chloride error. Adjustment of the oxidation state of iron is usually by $SnCl_2$ or the zinc reductor. Other possible choices are $H_2S$, $SO_3^{2-}$, or $Ti(III)$.

Although $Mn(II)$ can be titrated directly with $KMnO_4$ in neutral solution (see Table 12–2), this procedure is not used too often because of empirical stoichiometry. There are better procedures, which can be applied to the determination of Mn in the mineral pyrolusite ($MnO_2$) and in steel. The $MnO_2$ is reduced with excess $H_3AsO_3$ in $H_2SO_4$ solution, using the catalyst KI, to $Mn(II)$. Subsequently, the remaining $As(III)$ is titrated with standard $KMnO_4$. Solutions of $Fe(II)$ or $H_2C_2O_4$ can also be used instead of the $As(III)$. After dissolving the steel in $HNO_3$, $NaBiO_3$ is added which oxidizes the $Mn(II)$ to $MnO_4^-$. The excess $NaBiO_3$ is filtered and the filtrate ($MnO_4^-$) is treated with an excess of a $Fe(II)$ solution. The remaining $Fe(II)$ is titrated with standard $KMnO_4$.

The procedure for the determination of $H_2O_2$ solution involves a direct titration. A sample is carefully weighed, acidified with $H_2SO_4$, and titrated with standard $KMnO_4$. Peroxide salts are treated in the same way. Acidifying the peroxide solution should be done carefully with a cold sulfuric acid solution to prevent loss of active oxygen. Frequently, boric acid is added so that the more stable perboric acid is formed. These modifications are particularly important in the determination of peroxide salts. Hydrogen peroxide solutions are important in industrial and pharmaceutical applications. Generally, they are prepared as 10, 20, 40, and 100 volume concentrations (refers to amount of oxygen produced); each can be determined by $KMnO_4$ titration.

Several different metal ions form sparingly soluble oxalates (see Table (12–2). Since the oxalate anion is easily titrated with $KMnO_4$, this titration procedure can also be adapted for the determination of the metal ion. The reactions, using Ca as an example, are the following:

$$Ca^{2+} + C_2O_4^{2-} \xrightarrow{NH_3} CaC_2O_{4(s)}$$

$$CaC_2O_{4(s)} + H_2SO_4 \longrightarrow CaSO_4 + H_2C_2O_4$$

$$5H_2C_2O_4 + 2KMnO_4 + 6H^+ \longrightarrow 2Mn^{2+} + 2K^+ + 10CO_2 + 8H_2O$$

A quantitative precipitation as the oxalate is required. Usually, this is done by the addition of $(NH_4)_2C_2O_4$ and $NH_3$. The precipitate, which suffers from coprecipitation, is filtered, washed, redissolved with $H_2SO_4$, and the $H_2C_2O_4$ titrated with $KMnO_4$ titrant. This method is particularly useful for the determination of calcium in salts and minerals such as $CaCO_3$. Although this method provides good accuracy, calcium is titrated more often with EDTA (see Chapter 15) in recent years because this latter method is faster and does not suffer from coprecipitation errors.

Potassium permanganate can also be used as an oxidizing agent for the determination of poly- and hydroxycarboxylic acids, uric acid, formic acid, formaldehyde, ascorbic acid (vitamin C), polyphenols, and olefins. Most of those involve a back-titration rather than a direct titration procedure.

**Cerium(IV) Titrant.**    Cerium(IV) is most often used as the sulfate salt in $H_2SO_4$. Best results are obtained when the acid concentration is 0.5 $F$ or higher. Cerium(IV) can not be used in basic media because it precipitates as the hydroxide salt. Although cerium(IV) solutions are intensely yellow, it is not usually used as a self indicator. A redox color indicator, potentiometer, or other end-point detection system is used.

The cerium sulfate solution has the advantages of being made from a primary standard, its solutions are stable almost indefinitely, its stability is not affected by $H_2SO_4$ concentration and the reaction involves the production of only one product, Ce(III), which is colorless. Cerium(IV) can be employed in many of the same titrations as $KMnO_4$. Table 12–3 reviews many of the applications. Solutions of the titrant can be standardized with $As_2O_3$ (osmic acid or iodide ion as catalyst), sodium oxalate, pure iron (dissolve and adjust to $Fe^{2+}$), and iron(II) ammonium sulfate. Color redox indicators usually used are ferroin, 5,6-dimethylferroin, or $N$-phenylanthranilic acid (see Chapter 11).

Cerium(IV) sulfate is also useful for the analysis of 1,2-diols. The products of the reactions are formic acid and ketones if there is no hydrogen attached

**Table 12–3.    Applications of Cerium(IV) Sulfate**

Ce(IV) in Acid Solution: $Ce^{4+} + e \rightleftharpoons Ce^{3+}$

| Ion determined | Product | Ion determined | Product |
|---|---|---|---|
| $H_2C_2O_4$ | $CO_2$ | $S_2O_8^{2-}$ | $SO_4^{2-}$ |
| $Fe^{2+}$ | $Fe^{3+}$ | $U^{4+}$ | $UO_2^{2+}$ |
| $NO_2^-$ | $NO_3^-$ | $As^{3+}$ | $AsO_4^{3-}$ |
| $Cu^+(HCl)$ | $Cu^{2+}$ | $Ti^{3+}$ | $Ti^{4+}$ |
| $Mo^{5+}$ | $MoO_4^{2-}$ | $Fe(CN)_6^{4-}$ | $Fe(CN)_6^{3-}$ |
| $Te^{4+}$ | $TeO^{2-}$ | $Cr^{2+}$ | $Cr^{3+}$ |
| $Ce^{3+}$ | $Ce^{4+}$ | $VO^{2+}$ | $VO_3^-$ |
| $H_2O_2$ | $O_2$ | | |

**Table 12–4. Applications of Potassium Dichromate**

$K_2Cr_2O_7$ in acid solution:
$$Cr_2O_7{}^{2-} + 14H^+ + 6e \rightleftharpoons 2Cr^{3+} + 7H_2O$$

| Ion determined | Product |
|---|---|
| $Fe^{2+}$ | $Fe^{3+}$ |
| $Cr^{3+a}$ | $Cr_2O_7{}^{2-}$ |
| $ClO_3{}^{-a}$ | $Cl^-$ |

[a] Indirect titration.

to one or both of the hydroxylated carbon atoms. For example, glycerol is cleaved according to

$$\underset{\underset{HO}{|}}{H_2C}-\underset{\underset{OH}{|}}{CH}-\underset{\underset{OH}{|}}{CH_2} + 8\,Ce^{4+} + 3\,H_2O \longrightarrow 3\,HCO_2H + 8\,Ce^{3+} + 8\,H^+$$

Certain $\alpha$- and $\beta$-diketones, malonic esters, and malic acid are also quantitatively cleaved. Usually, excess Ce(IV) is added to a solution of the sample, heated, and the remaining Ce(IV) titrated with Fe(II) using ferroin as indicator.

**Potassium Dichromate.** Potassium dichromate is a weaker oxidizing agent than $KMnO_4$ or Ce(IV). However, it is a primary standard and its solutions have long lasting stability in acid and are stable to light, to most organic matter, and to chloride ion. It is always used in acid solutions. The principal disadvantage is that both the reactant $Cr_2O_7{}^{2-}$ and product $Cr^{3+}$ are highly colored, orange and green, respectively.

Potassium dichromate is used mostly for the analysis of iron (see Table 12–4). Either $SnCl_2$ or the zinc reductor is used for the adjustment of the iron to the 2+ oxidation state. Sodium diphenylamine sulfonate ($H_3PO_4$ must be present) or 5,6-dimethylferroin and $N$-phenylanthranilic acid can be used as indicators. If needed, a $K_2Cr_2O_7$ solution can be standardized against pure iron.

**Halogens.** The half-reaction for the reversible $I_2/I^-$ couple is very useful in analysis.

$$I_2 + 2e = 2I^-$$

The reduction potential for the system is about $+0.53$ V and occupies a middle of the road position in the Table of Reduction Potentials. For example, $I_2$ is a strong enough oxidizing agent so that it will undergo a reaction with strong to moderately strong reducing agents. On the other hand, $I^-$ will reduce strong oxidizing agents.

Iodine can be used in a direct titration of reducing agents. Or, it can be used in excess to react with a reducing sample and back-titrated with a sodium thiosulfate solution. Iodide ion is used in reactions with oxidizing agents and the iodine produced is titrated with standard sodium thiosulfate. The reactions describing these three titration techniques which are more familarily known as iodimetry (first two) and iodometry (third case) are the following:

*Direct titration:* $I_2$ added as titrant in stoichiometric amount.

$$Sn^{2+} + I_2 \rightarrow Sn^{4+} + 2I^-$$

*Back-titration:* $I_2$ added to reducing sample in excess and back-titrated.

$$Excess \ I_2 + 2S_2O_3{}^{2-} \rightarrow 2I^- + S_4O_6{}^{2-}$$

*Indirect titration:* $2Cu^{2+} + 4I^- \rightarrow 2CuI_{(s)} + I_2$

$$I_2 + \ S_2O_3{}^{2-} \rightarrow 2I^- + S_4O_6{}^{2-}$$

The iodine titrant is used as a solution containing iodide ion. In this solution iodine is present as the $I_3{}^-$ ion and the half-cell is correctly written as

$$I_3{}^- + 2e = 3I^-$$

For convenience, $I_2$ will be used in the discussion in this chapter.

Although the $I_2/I^-$ couple does not involve hydrogen ion, its utility is pH sensitive. Above pH of 8, iodine slowly disproportionates to $I^-$ and $IO_3{}^-$, while in acidic solution, iodide ion is slowly oxidized by oxygen to iodine. The oxidation is catalyzed by light and several different metal ions. Iodine is volatile enough so that losses through evaporation can be appreciable. Its solubility is low but in the presence of iodide ion the solubility is greatly increased due to the formation of $I_3{}^-$ anion. Even with these limitations, iodimetry and iodometry (see page 275) techniques are widely used.

A dilute iodine solution is pale yellow and with practice the appearance of this color, when the iodine is just in excess, can be used as an indicator for the titration. However, starch will undergo a reaction with iodine to produce an intensely blue colored complex. This color is easily detected even at very low concentrations and for this reason the starch indicator is preferred.

Table 12–5 lists the ions that can be determined through a direct titration procedure. The acidity of the solution must be carefully controlled even though the $I_2/I^-$ couple does not involve hydrogen ion. The reason for this is that the reducing power of the ions in Table 12–5 will depend on the pH. The stronger reducing agents, such as $Sn^{2+}$, $SO_2$ (or $H_2SO_3$), $H_2S$, and $Na_2S_2O_3$, can be titrated in acid solution, while the others are titrated in neutral to slightly alkaline solution.

Although iodine can be prepared as a primary standard, its solutions are usually standardized. Standardization is usually against $As_2O_3$ or $Na_2S_2O_3$

**Table 12–5.   Applications of Iodine**

Iodine (direct titration): $I_2 + 2e \rightleftharpoons 2I^-$

| Ion determined | Product |
|:---:|:---:|
| $As^{3+}$ | $AsO_4^{3-}$ |
| $Sb^{3+}$ | $SbO_4^{3-}$ |
| $Sn^{2+}$ | $Sn^{4+}$ |
| $H_2S$ | $S$ |
| $SO_2$ | $SO_4^{2-}$ |
| $S_2O_3^{2-}$ | $S_4O_6^{2-}$ |
| $N_2H_4$ | $N_2$ |
| Metal sulfide | $S$ |

solution which has been standardized against iodine (derived from $KIO_3/I^-$ or $KBrO_3/I^-$, see page 277).

Bromine, in addition to iodine, is very useful in analysis; however, its main applications as an oxidizing agent are for the determination of organic functional groups. For example, olefins will add bromine stoichiometrically according to the reaction

$$\begin{array}{c} \diagdown \\ \diagup \end{array} C = C \begin{array}{c} \diagup \\ \diagdown \end{array} + Br_2 \longrightarrow \begin{array}{c} \diagdown \\ \diagup \end{array} \underset{Br}{\overset{|}{C}} - \underset{Br}{\overset{|}{C}} \begin{array}{c} \diagup \\ \diagdown \end{array}$$

Chlorine or iodine is not useful in this analysis because of interferring side reactions. The reactivity of the olefin varies according to its structure; Lewis acids such as $AlBr_3$, $Hg(II)$, and $Ag(I)$ are often used as catalysts.

Phenols and aromatic amines can be determined by reaction with $Br_2$.

The ortho and para positions are brominated unless these positions are occupied by other groups.

Because of reactivity of the organic compound and the corrosive, volatile nature of bromine, solutions of bromine are not used. In general, the best procedure is to generate bromine *in situ*. This is done by adding an aliquot of standard $KBrO_3$ solution and excess solid KBr to a solution of the sample which generates bromine by the reaction

$$BrO_3^- + 5Br^- + 6H^+ \rightarrow 3Br_2 + 3H_2O$$

After the bromination is complete, the remaining bromine is determined by adding KI

$$Br_2 + I^- \rightarrow 2Br^- + I_2$$

and the iodine produced is titrated with standard $Na_2S_2O_3$ solution. The procedure is repeated minus the sample and the difference between the two allows the calculation of the amount of functional group in the sample. Bromine can also be generated electrochemically (see Chapter 28).

Bromine and iodine can also be used for the analysis of certain organosulfur compounds. For example, for a mercaptan the reaction with $I_2$ is

$$2RSH + I_2 \rightarrow RSSR + 2HI$$

A direct titration is not used. Excess $I_2$ is added and the remaining iodine titrated with sodium thiosulfate.

Dialykyl sulfides and disulfides are determined by bromination using the $BrO_3/Br^-/I^-/S_2O_3^{2-}$ procedure.

$$R_2S + 2Br_2 + 2H_2O \rightarrow R_2SO_2 + 4HBr$$

$$RSSR + 5Br_2 + 4H_2O \rightarrow 2RSO_2Br + 8HBr$$

Bromine can also be used for the determination of sulfonamides and ascorbic acid while iodine is also useful for ascorbic acid, trivalent organo-arsenic compounds, uric acid, RMgI type reagents, certain methyl ketones and acetyaldehyde.

**Karl Fischer Titration.** The Karl Fischer reagent is an extremely useful titrant for the analysis of small amounts of water. The reagent is a mixture of iodine and sulfur dioxide dissolved in a pyridine–methanol mixture. Reaction with water takes place in the following way:

The oxidation of $SO_2$ by $I_2$ takes place in the first step only in the presence of

Table 12–6.  Applications of Potassium Iodate

KIO$_3$ in Acid Solution: $IO_3^- + 6H^+ + 2Cl^- + 4e \rightleftharpoons ICl_2^- + 3H_2O$

| Ion determined | Product | Ion determined | Product |
|---|---|---|---|
| As$^{3+}$ | AsO$_4$$^{3-}$ | Fe$^{2+}$ | Fe$^{3+}$ |
| Sb$^{3+}$ | SbO$_4$$^{3-}$ | N$_2$H$_4$ | N$_2$ |
| I$^-$(Cl$^-$) | ICl$_2$$^-$ | CNS$^-$ | SO$_4$$^{2-}$ + CN$^-$ |
| I$_2$(Cl$^-$) | ICl$_2$$^-$ | SO$_3$$^{2-}$ | SO$_4$$^{2-}$ |
| Sn$^{2+}$ | Sn$^{4+}$ | S$_2$O$_3$$^{2-}$ | SO$_4$$^{2-}$ |
| Tl$^+$ | Tl$^{3+}$ | S$_4$O$_6$$^{2-}$ | SO$_4$$^{2-}$ |
| Hg$_2$Cl$_2$ | HgCl$_2$ | | |

water to produce the pyridine–sulfur trioxide product which undergoes further reaction with methanol to form pyridinium methyl sulfate (step 2). Therefore, the reaction ratio is one I$_2$ per H$_2$O.

The Karl Fischer reagent reacts rapidly with water and can be used in a direct titration procedure. Since the first excess of iodine imparts a yellowish color to the solution, the titrant can be used as a self indicator. However, this end point detection is difficult for many individuals to use and usually requires practice. An alternate instrumental end point detection method is by a modified amperometric technique (see Chapter 29).

The methanolic Karl Fischer reagent is unstable and must be frequently standardized. A more stable reagent is prepared by substituting ethylene glycol monomethyl ether (CH$_3$OCH$_2$CH$_2$OH) for the methanol. Pure water or sodium tartrate dihydrate in methanol solvent are useful primary standards. In general, a water blank must be measured for the methanol. Waters of hydration, water in organic solvents, absorbed water, and water in many other samples are readily determined by the Karl Fischer titration.

**Potassium Iodate and Potassium Bromate.**    These two salts are primary standards, stronger than iodine as oxidizing agents, stable to organic matter such as filter paper, organic acids, and alcohols, and their aqueous solutions are stable indefinitely. Table 12–6 lists the ions that can be determined with potassium iodate. Some of the methods require careful control of the acid concentration, while others have a slow rate of reaction near the equivalence point.

## REDUCING TITRANTS

The common reducing titrants are listed in Table 12–7. Those that are primary standards can be used to prepare standard solutions of the reducing agent or for standardization of oxidizing titrants.

**Table 12–7. A List of Reducing Agents Used as Standards and Titrants**

| Reagent | Conditions | Half-reaction | $E^a$ (volts) |
|---|---|---|---|
| $FeSO_4 \cdot (NH_4)_2SO_4 \cdot$ $6H_2O^b$ | Acidic | $Fe^{3+} + 1e \rightleftharpoons Fe^{2+}$ | +0.8 |
| $Fe^b$ | Acidic | $Fe^{3+} + 1e \rightleftharpoons Fe^{2+}$ | +0.8 |
| $FeSO_4$ | Acidic | $Fe^{3+} + 1e \rightleftharpoons Fe^{2+}$ | +0.8 |
| $As_2O_3{}^b$ | Acidic | $H_3AsO_4 + 2H^+ + 2e \rightleftharpoons H_3AsO_3 + H_2O$ | +0.6 |
| $Na_2S_2O_3 \cdot 5H_2O$ | Neutral | $S_4O_6{}^{2-} + 2e \rightleftharpoons 2S_2O_3{}^{2-}$ | +0.1 |
| $Cr^{2+}$ (prepared) | Acidic | $Cr^{3+} + 1e \rightleftharpoons Cr^{2+}$ | −0.4 |
| $Ti^{3+}$ (prepared) | Acidic | $Ti^{4+} + 1e \rightleftharpoons Ti^{3+}$ | +0.1 |
| $KI^b$ | Acidic–neutral– basic | $I_2 + 2e \rightleftharpoons 2I^-$ | +0.6 |
| $K_4Fe(CN)_6{}^b$ | Acidic | $Fe(CN)_6{}^{3-} + 1e \rightleftharpoons Fe(CN)_6{}^{4-}$ | +0.4 |
| $Na_2C_2O_4{}^b$ | Weakly acidic– neutral | $2CO_2 + 2e \rightleftharpoons C_2O_4{}^{2+}$ | −0.5 |
| $SnCl_2$ | Acidic | $Sn^{4+} + 2e \rightleftharpoons Sn^{2+}$ | +0.1 |
| $H_2S$ | Acidic | $S + 2H^+ + 2e \rightleftharpoons H_2S$ | +0.1 |

$^a$ Potential will depend on pH and other experimental conditions.

$^b$ Available as primary standard.

**Iodometry.** In iodometry, an oxidizing agent is treated with a large excess of iodide ion in acid or neutral solution. The oxidant is quantitatively reduced liberating an equivalent amount of iodine which is titrated with a standard solution of sodium thiosulfate. A summary of the types of oxidizing agents that can be determined this way are listed in Table 12–8.

Starch is usually used as the indicator. However, before it is added the iodine is titrated with thiosulfate until the brown coloration is reduced to a yellow coloration. At this point a small amount of free iodine is left when the starch indicator is added and upon addition of more thiosulfate a color change of blue to colorless is observed. There are also several useful instrumental ways of detecting the end point.

The more important experimental problems in working with iodide and iodine were outlined on page 271. In iodometry the stoichiometry of the $I_2$–$S_2O_3{}^{2-}$ reaction and the handling of the $S_2O_3{}^{2-}$ solution is important.

The stoichiometry is based on the conversion of thiosulfate to tetrathionate according to the reaction

$$S_2O_3{}^{2-} \rightleftharpoons S_4O_6{}^{2-} + 2e$$

provided the solution during the titration of iodine is neutral to weakly acidic. If the solution is too basic, some sulfate will be formed:

$$S_2O_3{}^{2-} + 10OH^- \rightleftharpoons 2SO_4{}^{2-} + 5H_2O + 8e$$

**Table 12–8.  Applications of Reducing Titrants**

Iodine (indirect titration)

Reaction: $2I^- + (oxidant) \rightleftharpoons I_2 + (product)$

Titration: $I_2 + 2S_2O_3^{2-} \rightleftharpoons 2I^- + S_4O_6^{2-}$

| Ion determined | Product | Ion determined | Product |
|---|---|---|---|
| $IO_4^-$ | $I_2$ | $Fe(CN)_6^{3-}$ | $Fe(CN)_6^{4-}$ |
| $IO_3^-$ | $I_2$ | $MnO_4^-$ | $Mn^{2+}$ |
| $BrO_3^-$ | $Br^-$ | $Ce^{4+}$ | $Ce^{3+}$ |
| $ClO_3^-$ | $Cl^-$ | $Cr_2O_7^{2-}$ | $Cr^{3+}$ |
| $HClO$ | $Cl^-$ | $Fe^{3+}$ | $Fe^{2+}$ |
| $Cl_2$ | $Cl^-$ | $Cu^{2+}$ | $Cu^+$ |
| $Br_2$ | $Br^-$ | $O_3$ | $O_2$ |
| $I^-$ | $I_2$ | $H_2O_2$ | $H_2O$ |
| $NO_2^-$ | $NO$ | $MnO_2$ | $Mn^{2+}$ |
| $AsO_4^{3-}$ | $AsO_3$ | Metal | $Cr^{3+}$ |
| $SbO_4^{4-}$ | $SbO_3$ | chromates | |

Cr(II) in acidic solution: $Cr^{2+} \rightleftharpoons Cr^{3+} + e$

| Ion determined | Product |
|---|---|
| $Cu^{2+}$ | $Cu$ |
| $Fe^{3+}$ | $Fe^{2+}$ |

Ti(III) in acidic solution: $Ti^{3+} \rightleftharpoons Ti^{4+} + e$

| Ion determined | Product |
|---|---|
| $Fe^{3+}$ | $Fe^{2+}$ |

In acid solution, decomposition takes place via the reaction

$$S_2O_3^{2-} + 2H^+ \rightleftharpoons H_2SO_3 + S$$

and, although sulfurous acid reacts with iodine, the stoichiometry is different in comparison to the thiosulfate reaction:

$$H_2SO_3 + H_2O \rightleftharpoons SO_4^{2-} + 4H^+ + 2e$$

Sodium thiosulfate, $Na_2S_2O_3 \cdot 5H_2O$, is readily obtained in a state of high purity with respect to thiosulfate content. However, the exact water content is uncertain. For this reason it is unsuitable as a primary standard and solutions of the salt must be standardized. Standardization is usually done with potassium iodate or potassium bromate. Samples of either of the two salts are carefully weighed, dissolved in water, acidified with $H_2SO_4$, and excess KI added.

$$IO_3^- + 5I^- + 6H^+ \rightarrow 3I_2 + 3H_2O$$

$$BrO_3^- + 6I^- + 6H^+ \rightarrow Br^- + 3I_2 + 3H_2O$$

The stoichiometric amount of $I_2$ liberated is titrated with the $Na_2S_2O_3$ solution using starch indicator. Potassium dichromate, pure copper metal, pure iodine, standard $KMnO_4$ solution, and ceric sulfate can also be used but they offer no advantages over $KIO_3$ or $KBrO_3$.

Iodide ion is useful for the analysis of organic peroxides which includes peracids, diacyl and dialkyl peroxides, and alkyl hydroperoxides. For benzoyl peroxide the reaction is

The liberated iodine is titrated with standard sodium thiosulfate.

**Other Reducing Titrants.** Ti(III), Cr(II), and Sn(II) (Table 12–8) are powerful reducing agents that can be used as titrants. However, they are difficult to use experimentally because they readily react with oxygen. Solutions must be stored and used under nitrogen and standardized frequently. Often they are prepared just before they are used. For these reasons these titrants are not used except under special circumstances.

Titanium(III) and chromium(II) can be used for the analyses of nitro, nitrate ester, nitroso, and azo groups. The main difficulties experienced in these procedures are in stoichiometry, rate of reaction, and reactivity of the functional group. For the nitroso and nitro group the reactions are

$$RNO + 6Ti^{3+} + 6H^+ \rightarrow RNH_2 + 6Ti^{4+} + 2H_2O$$

$$RNO_2 + 6Cr^{2+} + 6H^+ \rightarrow RNH_2 + 6Cr^{3+} + 2H_2O$$

One reducing agent which is used frequently, but not as a titrant is a solution of Fe(II). Usually, an aliquot of the Fe(II) solution is added to the oxidant and the remaining Fe(II) titrated with a standard oxidizing titrant. Table 12–9 lists several oxidants that can be determined by this technique.

Other reducing conditions are useful not as titrants but produce a gas as a product or use a gas as a reactant. The analysis is completed by measuring the volume (pressure and temperature must also be measured or held constant)

**Table 12–9.   Applications of Iron(II)**

Reaction: Oxidant $+$ Fe(II)$_{excess}$ $\rightarrow$
                product $+$ Fe(III)

Titration: Fe(II)$_{excess}$ $+$ oxidizing
                titrant $\rightarrow$ Fe(III) $+$ titrant
                product

| Ion determined | Product |
|---|---|
| $ClO_3^-$ | $Cl^-$ |
| $NO_3^-$ | $NO$ |
| $H_2O_2$ | $H_2O$ |
| $VO_3^-$ | $VO^{2+}$ |
| $Ce^{4+}$ | $Ce^{3+}$ |
| $MnO_4^-$ | $Mn^{2+}$ |
| $Cr_2O_7^{2-}$ | $Cr^{3+}$ |

of the reactant or product. For example, hydrogenation can be used for un-saturation:

$$\begin{matrix} \diagup \\ C \\ \diagdown \end{matrix} = \begin{matrix} \diagup \\ C \\ \diagdown \end{matrix} \quad + \quad H_{2\,(g)} \quad \longrightarrow \quad \begin{matrix} \diagup \\ C \\ | \\ H \end{matrix} - \begin{matrix} \diagup \\ C \\ | \\ H \end{matrix}$$

Other multiple bond groups, such as acetylenes, aromatic compounds, azo groups, unsaturated acids, and conjugated dienes, can also be determined. Usually, finally divided Pt, Pd, or Ni is used as catalyst.

## ADJUSTMENT OF OXIDATION STATE

**Adjustment to Lower Oxidation State.**    As previously pointed out, the sample must first be reduced to a lower oxidation state before it can be titrated with the oxidizing agent. Consequently, the general procedure after dissolving the sample is to add excess reducing agent to the solution which lowers the oxidation state of the sample. However, excess reducing agent must be removed since it could react with the oxidizing titrant.

Table 12–10 lists some of the more common systems (compounds and metal reductors) which are used to adjust the sample to a lower oxidation state. The method for the removal of the excess reagent is also listed. Since the reactions with the sample is an ordinary redox reaction, it should be possible to predict if the reaction can take place by comparison of the approximate reduction potential for the systems. In this list zinc provides the most powerful reducing conditions.

Table 12–10.   Reducing Conditions

| Reducing agent | Approximate reduction potential[a] (volts) | Conditions for removal of excess |
|---|---|---|
| Ag | +0.8 | Filter |
| Bi | +0.3 | Filter |
| $Na_2SO_3$ or $SO_2$ | +0.2 | Boil in acid |
| $NH_2OH \cdot HCl$ | — | — |
| $SnCl_2$ | +0.15 | Oxidize with $HgCl_2$ |
| $H_2S$ | +0.14 | Boil |
| $Na_2S_2O_4$ | — | Boil |
| Pb | −0.13 | Filter |
| Cd | −0.4 | Filter |
| Zn | −0.77 | Filter |

[a] Depends on experimental conditions.

Hydroxylamine hydrochloride and $SnCl_2$ are two compounds that are particularly useful for the reduction of Fe(III) to Fe(II). Hydroxylamine hydrochloride can also be used for the reduction of Cu(II) to Cu(I). Since $NH_2OH \cdot HCl$ is a moderately strong reducing agent, it must be removed in titration procedures. This is not easily done and, therefore, is not frequently used in titration procedures. However, it is conveniently used in other procedures requiring the adjustment of the oxidation state of Fe and Cu provided excess $NH_2OH \cdot HCl$ can be tolerated.

In contrast, $SnCl_2$ is removed by the addition of $HgCl_2$. The reactions, where iron is being reduced, are

$$2Fe^{3+} + Sn^{2+} \rightarrow 2Fe^{2+} + Sn^{4+}$$
$$\text{Excess}$$

$$Sn^{2+} + 2HgCl_2 \rightarrow Hg_2Cl_{2(s)} + Sn^{4+} + 2Cl^-$$
$$\text{Excess}$$

The mercurous chloride is insoluble enough so that it does not interfere in the titration of Fe(II) with an oxidizing titrant. If there is a large excess of Sn $Cl_2$, another reaction is possible,

$$Hg_2Cl_{2(s)} + Sn^{2+} \rightarrow 2Hg + Sn^{4+} + Cl^-$$

and its presence is indicated by the formation of a gray or black precipitate. The mercury will interfere if permanganate or dichromate is used as the titrant. Although not indicated, the $SnCl_2$ reaction with Fe(III) is carried out in an HCl solution and the presence of chloride must be considered when choosing the oxidizing titrant.

Table 12–11.  **Metal Reductors**

| Metal ion | Product of reduction | |
|---|---|---|
| | $Zn(H_2SO_4)$ | $Ag(HCl)$ |
| $Fe^{3+}$ | $Fe^{2+}$ | $Fe^{2+}$ |
| $Ti^{4+}$ | $Ti^{3+}$ | No reaction |
| $Cr_2O_7^{2-}$ | $Cr^{2+}$ | $Cr^{3+}$ |
| $MnO_4^-$ | $Mn^{2+}$ | $Mn^{2+}$ |
| $MoO_4^{2-}$ | $Mo^{3+}$ | $Mo^{5+}$ |
| $VO_3^-$ | $V^{2+}$ | $VO^{2+}$ |
| $UO_2^{2+}$ | $U^{3+}$ and $U^{4+}$ | $U^{4+}$ |
| $Cu^{2+}$ | $Cu^0$ | $Cu^+$ (as chloro complex) |
| $Ag^+$ | $Ag^0$ | No reaction |
| $Al^{3+}$ | No reaction | No reaction |

**Metal Reductors.**    The metal reductors are very versatile, easily used and removed from the system, and can be used for the preparation of titrants as well as for adjusting oxidation states of samples. Table 12–11 lists the reductors. They can be used as free metals or as amalgams (Zn–Hg and Ag–Hg). The applications are the same; however, the amalgams provide complete reduction with less metal (therefore it is faster), the amalgams can be used repeatedly, and a blank titration is not necessary. (The Jones reductor can introduce Fe into the sample depending on the quality of the Zn.) Also included in Table 12–11 is a comparison of the different amalgamated metals.

The procedure involves filling a glass tube with the amalgamated metal shot and passing the sample, which is usually acidified with HCl or $H_2SO_4$, through the column at a rate not exceeding 25 ml/minute. An acid solution is used to wash the sample through the column and the entire effluent is collected.

**Zinc Amalgam.**    The zinc amalgam is a powerful reductant. It can also be used for the preparation of solutions of Ti(III) and Cr(II) both of which are useful as reducing titrants. The zinc reductor will remove Cu and Ag from the solution since they are reduced to the metal.

Nitric acid must be absent since it will be reduced to $NH_2OH$ which will react with oxidizing titrants. Certain organic matter and acetate must be absent. Both of these are removed by heating the sample to fumes of $H_2SO_4$ prior to passing the sample through the reductor.

Many of the products of the Zn reduction listed in Table 12–11 are very reactive, for example Ti(III) and Cr(II). These are either collected under $N_2$ or passed into an intermediate solution. Usually the intermediate solution is a Fe(III) solution. For example, if Ti is being determined, a solution of the Ti is passed through the reductor directly into the Fe(III) solution. The Ti(III) reduces Fe(III) stoichiometrically to Fe(II) which is titrated with an appropriate oxidizing titrant. Hence, the very reactive Ti(III) is never isolated.

**Silver Reductor.** The silver reductor is compared to the zinc reductor in Table 12–11 and it can be concluded that the silver reductor is the weaker of the two. Since the silver is usually coated with AgCl, HCl is preferred over $H_2SO_4$ as the acid medium.

**Adjustment to Higher Oxidation State.** Before titration with a reducing agent the oxidation state must be quantitatively adjusted to a higher state. Generally, this is done either as part of the dissolution of the sample or after the sample is put into solution. In many cases oxidizing conditions are needed to dissolve the sample. Therefore, the oxidation state is raised during dissolution. In the latter, excess oxidizing agent is added, the reaction allowed to proceed, and some means is used to remove the remaining oxidizing agent. This last step is necessary since it is likely to undergo a reaction with the reducing titrant.

Table 12–12 lists some of the more common oxidizing conditions for adjustment of oxidation state. In general, a system is adjusted to its higher stable oxidation state. Approximate reduction potentials and techniques for the removal of the excess oxidizing agent are also given.

**Table 12–12. Oxidizing Conditions**

| Oxidizing agent | Approximate reduction potential[a] (volts) | Conditions for removal of excess |
|---|---|---|
| $O_3$ | 2.1 | Boil; decomposes |
| $K_2S_2O_8$ | 2.0 | Boil; decomposes to $SO_4^{2-}$–$SO_3$ |
| $H_2O_2$ | 1.8 | Boil; decomposes |
| $PbO_{2(s)}$ | 1.5 | Filter |
| $NaBiO_{3(s)}$ | | Filter |
| $KClO_3$ | 1.5 | Boil; decomposes in presence of acid |
| $KMnO_4$ | 1.5 | Add $NaN_3$ or boil with added HCl or $NaNO_2$ |
| $HClO_4$ (hot; concentrated) | 1.4 | Cool and dilute with water |
| $KIO_3$ | 1.2 | Precipitate as $Hg_5(IO_6)_2$ |

[a] Reduction potential depends on experimental conditions.

**Summary.** In surveying the methods in Tables 12–1 to 12–9 no attempt was made to list the exact experimental conditions. In many cases catalysts, elevated temperatures, careful control of acid or base concentration, or other specialized experimental conditions are required. Often a back-titration procedure is involved. Fortunately, these many procedures have been carefully documented.

## CALCULATIONS

Calculations in redox reactions require the balanced reaction and the reaction ratio both of which are provided by balancing the electrons that are gained and lost. Once these facts are established the procedure for completing the calculation is the same as for neutralization calculations (see Chapters 3 and 7). Unlike neutralization reactions, redox reactions often involve more complex ratios.

In redox reactions the reaction ratio, $a/b$, is determined by the balancing of the electrons. As defined in Chapter 3, $a$ refers to the sample, while $b$ refers to the titrant. In general form, the coefficients $a$ and $b$ are illustrated in the following:

$$n_2(A + n_1 e \rightarrow \text{products})$$

$$\frac{n_1(B \rightarrow \text{products} + n_2 e)}{n_2 A + n_1 B \rightarrow \text{products}}$$

If A is the titrant and B is the sample, the reaction ratio $(a/b)$ is $n_1/n_2$. One additional complication is that the values $n_2$ and $n_1$ must be on a per ion or molecule basis.

Although normality units can be used for the calculations, formality is stressed because of the similarity to neutralization and because molar (formal) units are required in calculations with equilibrium constants. Before proceeding with the examples cited here the section on calculations in Chapter 3 should be reviewed.

*Example 12–1.* A solution of $KMnO_4$ which is to be used for the determination of the iron content of an iron ore was prepared and standardized by titrating a weighed sample (0.2112 g) of dried $As_2O_3$ dissolved in acidic solution. Calculate the formality of the $KMnO_4$ solution if 36.42 ml of the titrant is required.

$$2 (MnO_4^- + 8H^+ + 5e = Mn^{2+} + 4H_2O)$$

$$\frac{5 (AsO_3^{3-} + H_2O = AsO_4^{3-} + 2H^+ + 2e)}{2MnO_4^- + 6H^+ + 5AsO_3^{3-} = 2Mn^{2+} + 5AsO_4^{3-} + 3H_2O}$$

From the balanced reaction, the reaction ratio is established as 5/2. However, the arsenic is weighed as $As_2O_3$ and since one $As_2O_3$ leads to two $AsO_3^{3-}$

$$As_2O_3 + 3H_2O \rightarrow 2AsO_3^{3-} + 6H^+$$

the reaction ratio based on $As_2O_3$ must be 5/4. That is, two $AsO_3^{3-}$ represent one $As_2O_3$, therefore, the ratio is $5/2 \times 1/2$ or 5/4.

The formality is calculated by

$$\text{wt } As_2O_3 = ml_{KMnO_4} \times F_{KMnO_4} \times \text{reaction ratio} \times As_2O_3$$

$$211.2 \text{ mg} = 36.42 \text{ ml} \times F_{KMnO_4} \times 5/4 \times 197.8 \text{ mg/mmole}$$

$$F_{KMnO_4} = 0.02346 \ F$$

*Example 12–2.* A sample of iron ore (0.5598 g) was fused with $K_2S_2O_7$ in a porcelain crucible over a meeker burner (1 hour) and after cooling was dissolved in water The Fe(III) was reduced to Fe(II) by passing the acidified solution ($H_2SO_4$) through a zinc reductor. After adjusting the acidity in the effluent the iron(II) is titrated with the $KMnO_4$ from Example 12–1, 36.42 ml being required to reach the stoichiometric point. Calculate the iron in the sample as $\%Fe_2O_3$.

$$1 \ (MnO_4^- + 8H^+ + 5e = Mn^{2+} + 4H_2O)$$

$$\frac{5 \ (Fe^{2+} = Fe^{3+} + 1e)}{MnO_4^- + 5Fe^{2+} + 8H^+ = Mn^{2+} + 5Fe^{3+} + 4H_2O}$$

In balancing the reaction, the reaction ratio is concluded to be 5/1. Therefore,

$$\%Fe = \frac{ml_{KMnO_4} \times F_{KMnO_4} \times \text{reaction ratio} \times Fe \times 100}{\text{wt of sample}}$$

$$\%Fe = \frac{36.42 \text{ ml} \times 0.02346 \text{ mmole/ml} \times 5/1 \times 55.85 \text{ mg/mmole} \times 100}{559.8 \text{ mg}} = 42.62\%$$

The reaction ratio must be changed if the $\%Fe_2O_3$ is calculated. Each $Fe_2O_3$ provides 2Fe, therefore the ratio is $5/2 \times 1/2$ or 5/2. Hence,

$$\%Fe_2O_3 = \frac{36.42 \text{ ml} \times 0.02346 \text{ mmole/ml} \times 5/2 \times \overset{159.7}{\cancel{143.7}} \text{ mg/mmole} \times 100}{559.8 \text{ mg}}$$

$$\%Fe_2O_3 = \cancel{54.83}\% \ 60. \ \%$$

*Example 12–3.* A weighed sample of pure Cu wire (0.1105 g) is dissolved (hot $HNO_3$) and an excess of KI is added. The liberated iodine is titrated with 39.42 ml of thiosulfate solution to the starch end point. A sample of Cu ore (0.2129 g) was treated the same way, 28.42 ml of thiosulfate being required for the titration. Calculate the %Cu in the ore.

The reactions are

$$a = 2\ (Cu^{2+} + 1e = Cu^+)$$
$$\underline{1\ (2I^- = I_2 + 2e)}$$
$$2Cu^{2+} + 2I^- = 2Cu^+ + I_2$$

$$\left|\ +2I^-\right.$$

$$\longrightarrow 2CuI\ _{(s)}$$

Copper(I) forms an insoluble iodide salt in the presence of excess iodide ion. Therefore, in balancing the reaction two iodide ions are added to both sides of the reaction.

$$2Cu^{2+} + 4I^- \to 2CuI_{(s)} + I_2$$

The reaction for the titration is

$$2\ (I_2 + 2e = 2I^-)$$
$$\underline{b = 2\ (2S_2O_3^{2-} = S_4O_6^{2-} + 2e)}$$
$$I_2 + 2S_2O_3^{2-} = 2I^- + S_4O_6^{2-}$$

and the reaction ratio is 2/2. Consequently,

$$\text{wt Cu} = ml_{S_2O_3^{2-}} \times F_{S_2O_3^{2-}} \times \text{reaction ratio} \times Cu$$

$$110.5\ mg = 39.42\ ml \times F_{S_2O_3^{2-}} \times 2/2 \times 63.54\ mg/mmole$$

$$F_{S_2O_3^{2-}} = 0.04411\ F$$

$$\%Cu = \frac{ml_{S_2O_3^{2-}} \times F_{S_2O_3^{2-}} \times \text{reaction ratio} \times Cu \times 100}{\text{wt sample}}$$

$$\%Cu = \frac{28.42\ ml \times 0.04411\ mmole/ml \times 2/2 \times 63.54\ mg/mmole \times 100}{212.9\ mg}$$

$$\%Cu = 37.41\%$$

*Example 12–4.*    Calcium in blood or urine can be determined in the clinical laboratory through an oxidation–reduction titration. The basis for the method is to precipitate Ca(II) as $CaC_2O_4$, filter, redissolve, and titrate the $C_2O_4^{2-}$ with $MnO_4^-$ in strong acid solution. A 24-hour urine sample was carefully evaporated to a small volume (1.000 g) and treated appropriately to isolate the $CaC_2O_4$. Titration of $CaC_2O_4$ with 0.08554 $F$ $KMnO_4$ required 27.50 ml to reach the end point. Calculate the mg Ca excreted in a 24-hour period.

$$2\ (MnO_4^- + 8H^+ + 5e = Mn^{2+} + 4H_2O)$$
$$\underline{5\ (C_2O_4^{2-} = 2CO_2 + 2e)}$$
$$2MnO_4 + 5C_2O_4^{2-} + 16H^+ \to 2Mn^{2+} + 10CO_2 + 8H_2O$$

$$\text{wt Ca} = ml_{KMnO_4} \times F_{KMnO_4} \times \text{reaction ratio} \times Ca$$

$$\text{mg Ca} = 27.50\ ml \times 0.08554\ mmole/ml \times 5/2 \times 40.08\ mg/mmole = 235.7\ mg$$

Since the original sample was collected over a 24-hour period, the amount of calcium per 24 hours is 235.7 mg/24 hour. A normal adult on a normal diet excretes in the urine an average of 100–300 mg of Ca per 24 hours.

*Example 12–5.*    The determination of oxidizable material in water is an important environmental analysis used in evaluating both treated as well as polluted water. Depending on the water sample, the oxidizable material is either organic, inorganic, or both. One procedure that has been used for a long time is the measurement of the consumption of an oxidizing titrant ($KMnO_4$) solution under a precise set of conditions. Although this is an established and widely used procedure, the oxidation that takes place is by no means quantitative and many, especially industrial, pollutants are not determined. The results are usually reported as permanganate consumption in units of mg $KMnO_4$/liter of water.

A 100-ml sample of water, 5 ml of 25% $H_2SO_4$, and 15.00 ml of 0.002410 $F$ $KMnO_4$ were mixed in an Erlenmeyer flask equipped with a condenser. The mixture was brought to a boil and gently boiled for exactly 10 minutes. Subsequently, 15.00 ml of 0.005084 $F$ oxalic acid solution was added to the hot solution and the excess oxalic acid was back-titrated with 5.14 ml of 0.002110 $F$ $KMnO_4$. The end point was marked by a persistent pink color. Calculate the mg $KMnO_4$ consumed per liter of water.

mg $KMnO_4$ added $= ml_{KMnO_4} \times F_{KMnO_4} \times KMnO_4$

mg $KMnO_4$ added $= 15.00$ ml $\times 0.002410$ mmole/ml $\times 158$ mg/mmole $= 5.718$ mg

mg $KMnO_4$ left $= (ml_{H_2C_2O_4} \times F_{H_2C_2O_4} - ml_{KMnO_4} \times F_{KMnO_4} \times$ reaction ratio$)$

$$\text{reaction ratio} \times KMnO_4$$

The reaction ratios are determined from the reaction

$$2MnO_4^- + 16H^+ + 5C_2O_4^{2-} \rightarrow 2Mn^{2+} + 10CO_2 + 8H_2O$$

mg $KMnO_4$ left $= (15.00$ ml $\times 0.005084$ $F - 5.14$ ml $\times 0.002110$ $F \times 5/2)$

$$\times 2/5 \times 158 \text{ mg/mmole}$$

mg $KMnO_4$ left $= 3.106$ mg

5.718 mg $- 3.106$ mg $= 2.612$ mg $KMnO_4$/100 ml $H_2O = 26.12$ mg $KMnO_4$/liter $H_2O$

*Example 12–6.*    Calculate the percent methacrylic acid in a drum of methacrylic acid [$CH_2{=}C(CH_3)CO_2H$] (M.W. 86.09) which is to be eventually used for the preparation of a methacrylate based polymer.

A statistical sampling of the liquid in the drum was taken. From this a 0.2100-g sample was accurately weighed and dissolved in water. (Other solvents are used depending on the solubility of the sample.) A bromination flask is evacuated and a 25.00-ml aliquot of a standard $KBrO_3$ solution containing excess KBr is introduced. (The $BrO_3^-$ should ideally be about a 10–15% excess over the olefin content.) Some $H_2SO_4$ is introduced followed by the solution of the sample and several washings. Bromine is produced when the system is acidified. The flask is wrapped with a dark cloth or placed in the dark and shaken for at least 7 minutes. (The reaction time depends on the sample.) A NaCl–KI solution is added and the liberated iodine required

16.14 ml of 0.1951 $F$ $Na_2S_2O_3$ (starch indicator). The entire procedure was repeated minus the sample and 40.42 ml of the $Na_2S_2O_3$ titrant was required.

The reactions are

$$\begin{array}{r} 1\ (2BrO_3^- + 12H^+ + 10e = Br_2 + 6H_2O) \\ 5\ (2Br^- = Br_2 + 2e) \\ \hline BrO_3^- + 6H^+ + 5Br^- \rightarrow 3Br_2 + 3H_2O\ (\div 2) \end{array}$$

$$\underset{/}{\overset{\backslash}{C}}{=}\underset{\backslash}{\overset{/}{C}} + Br_2 \rightarrow -\underset{|}{\overset{|}{C}}-\underset{|}{\overset{|}{C}}-$$
$$\qquad\qquad\qquad Br\ \ Br$$

$$Br_2 + 2I^- \rightarrow 2Br^- + I_2$$

$$I_2 + 2S_2O_3^- \rightarrow 2I^- + S_4O_6^{2-}$$

$$\%CH_2{=}C(CH_3)CO_2H = \frac{(ml_{S_2O_3^{2-}} \times F_{S_2O_3^{2-}} - ml_{S_2O_3^{2-}} \times F_{S_2O_3^{2-}})}{\underset{\times\ \text{reaction ratio} \times CH_2{=}C(CH_3)CO_2H \times 100}{}}{wt\ sample}$$

$$\%CH_2{=}C(CH_3)CO_2H = \frac{(40.42\ ml \times 0.1951\ mmole/ml - 16.14\ ml}{\times\ 0.1951\ mmole/ml \times 1/2 \times 86.09\ mg/mmole \times 100}{210.0\ mg}$$

$$\%CH_2{=}C(CH_3)CO_2H = 92.81\%$$

*Example 12-7.* Chromium in an ore can be determined by dissolving the ore and oxidizing the chromium to chromium(VI). After acidifying the solution (forms $Cr_2O_7^{2-}$), an excess of standard iron(II) solution is added and the excess is titrated with standard $K_2Cr_2O_7$. A 0.2801-g sample of ore is taken and treated as described. Exactly 75.00 ml of 0.1010 $F$ $FeSO_4$ is added and 16.85 ml of 0.02507 $F$ $K_2Cr_2O_7$ is required for the back titration. Calculate the %Cr in the sample.

$$Cr_{ore} \rightarrow Cr^{+6} \qquad 2CrO_4^{2-} + 2H^- \rightleftharpoons Cr_2O_7^{2-} + H_2O$$

$$\begin{array}{r} 1\ (Cr_2O_7^{2-} + 14H^+ + 6e = 2Cr^{3+} + 7H_2O) \\ 6\ (Fe^{2+} = Fe^{3+} + 1e) \\ \hline Cr_2O_7^{2-} + 14H^+ + 6Fe^{2+} = 2Cr^{3+} + 6Fe^{3+} + 7H_2O \end{array}$$

Therefore,

$$\%Cr = \frac{(mmoles\ taken - mmoles\ found \times reaction\ ratio) \times reaction\ ratio \times Cr \times 100}{wt\ sample,\ mg}$$

$$\%Cr = \frac{(ml_{Fe^{2+}} \times F_{Fe^{2+}} - ml_{Cr_2O_7^{2-}} \times F_{Cr_2O_7^{2-}} \times 6/1) \times 2/6 \times 52.00 \times 100}{wt\ sample}$$

$$\%Cr = \frac{(75.00\ ml \times 0.1010\ mmole/ml - 16.85\ ml \times 0.02507\ mmole/ml \times 6/1)}{\times\ 2/6 \times 52.00\ mg/mmole \times 100}{280.1\ mg}$$

$\%Cr = 31.19\%$

*Example 12-8.* Menadione (2-methyl-1,4-naphthoquinone) (**I**), which is useful for treatment of hypoprothrombinemia and in veterinarian applications can be determined by direct titration with Ti(III). A 0.2114-g sample is dissolved in a mixture of 3:2 $HC_2H_3O_2$:$CH_3CH_2OH$. Dry sodium carbonate and sodium tartrate is added, and under a blanket of $N_2$ or $CO_2$ the menadione is titrated with $TiCl_3$ solution. If 32.13 ml of 0.7573 $F$ $TiCl_3$ is required for the titration, calculate the percent purity of the sample.

$$\% \text{ Menadione} = \frac{ml_{Ti^{3+}} \times F_{Ti^{3+}} \times \text{reaction ratio} \times C_{11}H_8O_2 \times 100}{\text{wt sample}}$$

$$\% \text{ Menadione} = \frac{32.13 \text{ ml} \times 0.7573 \text{ mmole/ml} \times 1/2 \times 172.2 \text{ mg/mmole} \times 100}{211.4 \text{ mg}}$$

$$\% \text{ Menadione} = 99.05\%$$

# Questions

1. Write a balanced reaction for each of the following (include acid, base, or neutral conditions).
   a. Standardization of $KMnO_4$ with $Na_2C_2O_4$.
   b. Standardization of $KMnO_4$ with $As_2O_3$.
   c. Standardization of $Ce(SO_4)$ with $As_2O_3$.
   d. Standardization of $K_2Cr_2O_7$ with ferrous ethylenediammonium sulfate.
2. What are the advantages of $KMnO_4$ over $Ce^{4+}$ as an oxidizing titrant?

3. Suggest why the following oxidants are not used as titrants.

    a. Hot $HNO_3$               d. $Na_2O_2$

    b. Cold $HNO_3$             e. $H_2S$

    c. $6\ F\ H_2SO_4$             f. $Br_2$

4. Explain why oxidizing titrants are more widely used than reducing titrants.

5. Explain why a back titration technique is frequently used in titrations employing reducing agents.

6. What happens to the formality of a $KMnO_4$ solution if decomposition to $MnO_2$ takes place?

7. Does the oxidizing power of $KMnO_4$ change as the result of $MnO_2$ formation?

8. List several primary standards that can be used to standardize reducing titrants.

9. Compare the oxidizing power of $Ce^{4+}$ in $HClO_4$, $HNO_3$, $HCl$, and $H_2SO_4$.

10. Compare the advantages and disadvantages of determining $Ca^{2+}$ gravimetrically as $CaC_2O_4$ to the titration of $CaC_2O_4$ with $KMnO_4$.

11. What is an iodometric method?

12. List several primary standard reducing agents that are used to standardize oxidizing titrants.

13. Why is iodine useful as an oxidizing agent even though it has modest oxidizing power?

14. What role does KI play in a $I_2$–KI solution?

15. Compare the strength of Ag as a reducing agent in the presence and absence of HCl.

16. Explain why $Fe^{2+}$-phenanthroline(ferroin) can be used as an indicator in $Ce^{4+}$ titrations but not in $K_2Cr_2O_7$ titrations.


# Problems

1.* If 40.00 ml of a solution of $H_2C_2O_4$ can be titrated with 30.00 ml of 0.4000 $F$ NaOH and if 40.00 ml of the same $H_2C_2O_4$ solution requires 65.00 ml of a $KMnO_4$ solution, what is the formality of the $KMnO_4$ solution?

2. A solution of $KMnO_4$ was standardized by weighing 0.2145 g $As_2O_3$ and after proper treatment was titrated with 42.44 ml of the $KMnO_4$ titrant. Calculate the formality of the $KMnO_4$ solution.

3. How many milliliters of $K_2Cr_2O_7$ solution containing 24.00 g of pure salt per liter would react with 3.315 g of $FeSO_4 \cdot 7H_2O$ in dilute acid solution?

4.* How many milligrams of $H_2O_2$ will react with 40.00 ml of a 0.03100 $F$ $KMnO_4$ solution in acidic solution?

---

\* Answers are listed at the end of the book for problems marked with an asterisk.

5.* A 0.1521-g sample of pure $KIO_3$ was dissolved, acidified, and excess KI added. If 24.12 ml of a thiosulfate solution was required for titration, calculate its formal concentration.

6. A 0.2018-g sample of pure $KBrO_3$ was dissolved, acidified, and excess KI added. If 36.15 ml of a thiosulfate solution was required for titration, calculate its formal concentration.

7.* A 0.2005-g sample of sodium tartrate (230.1) dihydrate was used to standardize a Karl Fischer reagent. If 14.12 ml of titrant were required, calculate the titer of the titrant in terms of milligrams of water per milliliter of Karl Fischer titrant.

8. A 0.3155-g iron sample was dissolved and after suitable reduction required 42.15 ml of a $K_2Cr_2O_7$ solution which contained 4.250 g of pure $K_2Cr_2O_7$ per 1000 ml. Calculate the %Fe in the sample.

9.* A 0.2010-g calcium sample is dissolved and precipitated as $CaC_2O_4$. After proper treatment the $C_2O_4^{2-}$ was titrated with 18.55 ml of 0.01000 $F$ $KMnO_4$. Calculate the %CaO in the sample.

10. An impure 0.1580-g L-cysteine sample, $HSCH_2CHNH_2CO_2H$ (121.2), was treated with 25.00 ml of an iodine solution. The excess iodine required 12.10 ml of 0.01050 $F$ thiosulfate solution. A blank treated in the same way required 27.12 ml of the titrant; calculate the percent purity of the L-cysteine.

11.* What is the percent purity of a sample of impure $Fe_2O_3$ if a 1.101-g sample required 33.46 ml of a 0.1010 $F$ $Ce^{4+}$ solution?

12. A 20.0-ml aliquot of a $H_2O_2$ solution was diluted to 250 ml. A 25.00-ml aliquot of this solution after adjustment of the acidity required 33.12 ml of a 0.5110 $F$ $KMnO_4$ solution. Calculate the grams of $H_2O_2$ per 100 ml of original solution.

13. The calcium in a 10.00-ml serum sample was precipitated as $CaC_2O_4$ which was dissolved and titrated with 9.68 ml of 0.001010 $F$ $KMnO_4$. Calculate the milligrams of Ca/ml of serum. $\left(\#\, moles = \dfrac{volume}{22.4\, l/mole} \#15\right)$

14.* An impure peroxide sample weighing 0.4112 g was dissolved and treated with excess KI. Calculate the %$O_2$ in the sample if 13.45 ml of 0.0845 $F$ thiosulfate was required to titrate the liberated $I_2$.

15.* Calculate the volume of $H_2$ gas at STP required to hydrogenate 0.4121 g of 1-butene (56.10).

16. A 0.2050-g sample of a copper ore is dissolved and after proper treatment excess KI is added and the liberated $I_2$ required 15.26 ml of 0.0400 $F$ $Na_2S_2O_3$ to reach the starch end point. Calculate the %Cu in the sample.

17.* A 0.1880-g sample of pure $K_2Cr_2O_7$ was dissolved, acidified and treated with an excess of KI. The liberated $I_2$ required 41.15 ml of a thiosulfate solution. Calculate the formality of the thiosulfate solution.

18.   Arsenic insecticides can be analyzed for As content by treating the sample with boiling HCl in the presence of a reducing agent. This produces $AsCl_3$ which is collected and titrated with iodine solution. If 25.42 ml of 0.1121 $F$ $I_2$ was required for a 0.4115-g insecticide sample, calculate the %As in the sample.

19.*   A 0.5125-g sample of bleaching powder was dissolved, conditions adjusted, excess KI added, and the liberated $I_2$ titrated with 31.44 ml of 0.2110 $F$ thiosulfate. Calculate the %$Cl_2$ available in the bleach.

# Chapter
# Thirteen
# Ion-Selective
# Electrodes

## INTRODUCTION

In redox methods an indicator electrode is used to sense the presence or change in concentration of the oxidized and reduced forms of a redox couple. Usually, the indicator electrode is an inert noble metal, such as Pt, and the potential of the cell is measured vs a reference electrode. In this type of cell, the Pt does not participate in an actual electrochemical half-cell reaction but acts as a collector of electrons that are part of the half-cell reaction. However some substances, not only collect the electrons, but also participate in the half-cell. For example, a zinc rod responds to $Zn(II)$ concentration, a copper rod responds to $Cu(II)$, and mercury to $Hg(II)$. These, and several other metals, can act as "ion-selective" electrodes toward their own ions.

It would be very convenient to be able to dip an electrode pair (ion-selective electrode and reference electrode) into a solution of the substance to be determined and obtain the sample's concentration from the observed potential. The problems with using metals are that in many cases electrode response is slow, the Nernst equation is not followed, electron change is not well defined, and the metal electrodes change potential due to alteration of the electrode surface. Although there are some useful ion selective metal electrodes ($Zn$, $Cu$, $Hg$), the vast majority suffer from some combination of the aforementioned problems.

There are also many ions that are of analytical interest that do not participate in a half-cell that includes a metal. A typical example is hydronium ion. An accurate measurement of hydronium ion is very important in a wide variety of scientific disciplines. Consequently, it is desirable to be able to make this measurement accurately, and routinely under a variety of solution conditions and concentration levels. There are also many other ions, such as $F^-$,

$SO_4^-$, $NH_4^+$, $Na^+$, $K^+$, and others, which are not part of a redox couple involving a metal or a easily handled metal.

Many different ion selective electrodes have been investigated and shown to be very useful. Most of these are not based on redox half-cells, like the $Zn^{2+}/Zn$ half-cell, but involve membrane or exchange potentials. This chapter considers the modern ion selective electrodes and their applications with the main emphasis directed toward the measurement of pH.

## ELECTRODES FOR HYDRONIUM ION

Historically, almost all effort was toward the development of hydronium ion-indicating electrodes since the determination of pH or acidity of a solution has been important throughout all phases of chemistry and biochemistry. For example, the production of nylon as well as other modern fibers depends on rigid pH control. The pH of blood is normally controlled to within a few tenths of a pH unit by our body chemistry. Proper skin pH is essential for a healthy complexion. The pH of one's stomach directly affects the digestive process, while in soil, the pH regulates the availability of nutrients for plant growth as well as the activity of bacteria. Efficient production of food products depends on pH control. Maintaining proper ecological balance in a river or lake requires pH control, since the pH of the water directly affects the physiological functions and nutrient utilization by plant and animal life. Industry must therefore also control pH of waste waters to maintain the ecological balance. In the laboratory, an accurate pH measurement is often necessary in the study of chemical processes, provides the necessary condition for an analysis, or determines the proper reaction conditions. It is probably a fair statement to say that the measurement of pH or acidity is one of the most frequently made measurements.

At present, virtually all pH measurements are made with the glass electrode. This electrode comes in a variety of shapes and designs, and its development can be credited, more than any other single factor, for the widespread use of pH measurement as a control in research and industry. Other pH electrodes are also available but are of limited use. These will only be briefly described.

Theoretically, any half-cell involving hydronium ion should be capable of acting as an indicator electrode for hydronium ion. For example, hydrogen ion concentration will determine the potential for the $Cr_2O_7^{2-}/Cr^{3+}$ half-cell provided $Cr_2O_7^{2-}$ and $Cr^{3+}$ concentrations are held constant. This is readily seen from the Nernst expression for the reaction

$$Cr_2O_7^{2-} + 14H^+ + 6e = 2Cr^{3+} + 7H_2O$$

$$E = E_{Cr_2O_7^{2-},\ Cr^{3+}} - \frac{0.0592}{6} \log \frac{[Cr^{3+}]^2}{[Cr_2O_7^{2-}]} - \frac{0.0592}{6} \log \frac{1}{[H^+]^{14}}$$

and

$$E = \text{constant} - 7/3(0.0592) \text{ pH}$$

where $Cr_2O_7^{2-}$ and $Cr^{3+}$ are held constant. Thus, as the pH of the solution changes the half-cell potential must also change. Examination of the Table of Standard Reduction Potentials reveals that many other reactions should have the same capabilities.

These reactions, however, are not feasible for two reasons. First, it is not easy to prevent changes in concentration of the other components of the half-cell. In the example cited, the equilibrium concentrations of $Cr_2O_7^{2-}$ and $Cr^{3+}$ will be affected differently at various levels of acidity because of association and hydrolysis

$$Cr_2O_7^{2-} + H^+ \rightleftharpoons HCr_2O_7^-$$

$$Cr_2O_7^{2-} + 2OH^- \rightleftharpoons 2CrO_4^{2-} + H_2O$$

$$Cr^{3+} + OH^- \rightleftharpoons Cr(OH)^{2+}$$

Thus, the half-cell potential would be also influenced by the altered equilibrium concentrations of the chromium species.

The second problem is electrode construction. A durable, rapidly responsive, easily constructed, inexpensive, sensitive, accurate, and versatile electrode is desired. In general, electrodes meeting these requirements are not readily put together in which half-cells such as the dichromate–chromium(III) couple are used. However, there are a few half-cells which can be exploited for pH measurements. In addition, no alternative route to this measurement, or at least one that is easily carried out, has been developed. Consequently, the determination of pH is a type of potentiometric measurement, and it can be stated that this measurement is the most important practical application of potentiometry.

**Hydrogen Electrode.**    The hydrogen electrode is best described as an oxidation–reduction electrode at which equilibrium is established between electrons on a noble metal, hydrogen ions in solution, and dissolved molecular hydrogen. Experimentally, the activity of the dissolved hydrogen gas is fixed by maintaining equilibrium with a known partial pressure of hydrogen. A typical electrode design was shown previously in Fig. 10–3.

The hydrogen electrode, in addition to responding to hydrogen ion activity, is universally adopted as the primary standard with which all other electrodes are compared. This electrode is capable of a high degree of reproducibility and according to many workers it is comparatively easy to prepare and use.

The expression for the half-cell and electrode potential is

$$2H_3O^+(\text{aq soln}) + 2e \rightleftharpoons H_2(\text{aq soln}) + 2H_2O \tag{13-1}$$

$$E = E^\circ_{H^+,\, H_2} - \frac{0.0592}{2} \log \frac{p_{H_2}}{a^2_{H_3O^+}} \tag{13-2}$$

where $E^\circ_{H^+,\, H_2} = 0.000$ V and $H_2$ is expressed in terms of pressure.

The exchange equilibrium, as shown in reaction (13-1), is not established in the solution phase. Thus, when a metal is placed in the solution it will not assume a potential defined by the equilibrium unless it is able to act as a catalyst. In order for this to happen, the metal must absorb the hydrogen atoms and a better statement of the half-cell is

$$2H_3O^+(\text{aq soln}) + 2e \rightleftharpoons 2H(\text{adsorbed on metal}) \rightleftharpoons H_2(\text{aq soln}) \tag{13-3}$$

If the hydrogen electrode is used as an indicator electrode, it is combined with a reference electrode such as the saturated calomel electrode. The whole cell and its potential are given by

$$\text{Pt} \mid H_2(1 \text{ atm}), H_3O^+_{(\text{unk})} \parallel KCl_{(s)}, Hg_2Cl_{2(s)}, \mid Hg \tag{13-4}$$

$$E_{\text{cell}} = E_{\text{SCE}} + E_j - \left( E^\circ_{H^+,\, H_2} - \frac{0.0592}{2} \log \frac{P_{H_2}}{a^2_{H_3O^+}} \right) \tag{13-5}$$

Since $E_{H_3O^+, H_2} = 0$, $P_{H_2} = 1$, and $E_{\text{SCE}} + E_j$ (the junction potential) $= k$, the $E_{\text{cell}}$ simplifies to

$$E_{\text{cell}} = k - \frac{0.0592}{2} \log a^2_{H_3O^+}$$

$$E_{\text{cell}} = k + 0.0592 \text{ pH}$$

$$\text{pH} = \frac{E_{\text{cell}} - k}{0.0592} \tag{13-6}$$

To measure the pH, the system must be calibrated by using buffers of known pH. This procedure will be discussed in the section on glass electrodes.

The main advantages of the hydrogen electrode are that it is useful over the entire pH range, free of salt errors, high accuracy, low internal resistance, and negligible electric leakage errors. Its principal uses are to check the accuracy and stability of reference buffer solutions, determine $Na^+$ errors in glass electrodes, check the accuracy of other pH electrodes, and serve as the primary standard for pH measurements. Its principal disadvantage is that it is more awkward to use in most practical situations in comparison to other pH electrodes (glass electrode).

**Other Half-Cells.**     Two other half-cells that can be used for indicating hydronium ion are the quinhydrone electrode and antimony electrode.

Quinhydrone is a molecular species composed of one $p$-quinone(Q) and one $p$-hydroquinone($H_2Q$) held together by hydrogen bonding.

Its half-cell is given by

$$Q + 2H_3O^+ + 2e = H_2Q + 2H_2O \tag{13-7}$$

and when used with a SCE the cell and its potential are given by

$$Pt \mid H_2Q_{(sat)}, Q_{(sat)}, H_3O^+_{(unk)} \parallel KCl_{(s)}, Hg_2Cl_{2(s)} \mid Hg \tag{13-8}$$

$$E_{cell} = E_{SCE} + E_j - \left( E^\circ_{Q, H_2Q} - \frac{0.0592}{2} \log \frac{a_{H_2Q}}{a_Q a^2_{H_3O^+}} \right) \tag{13-9}$$

The main advantages of the quinhydrone electrode are low internal resistance, rapid response, high accuracy, simplicity, free of errors due to the presence of nonreducing gases, and free of salt errors. Major limitations are that it contaminates the solution, cannot be used to monitor a flowing solution, the solution must be free of strong oxidizing and reducing agents, and it is limited to a pH range of 1 to 9. In alkaline solution, the weakly acidic $H_2Q$ is neutralized and is subject to oxidation by air and dissolved oxygen.

The main application of the quinhydrone electrode is in nonaqueous solvents where reproducible results are obtained in a variety of solvents. The tetrachloro derivative (chloranil electrode) is even more useful.

The antimony electrode is constructed from a rod of high purity, electrolytic antimony the surface of which is coated with a very thin film of oxide. Its half-cell is

$$Sb_2O_{3(s)} + 6H_3O^+ + 6e \rightleftharpoons 2Sb_{(s)} + 9H_2O \tag{13-10}$$

If used in combination with a reference electrode, such as the SCE, the cell and its potential are given by

$$Sb_{(s)}, Sb_2O_{3(s)}, H_3O^+_{(unk)} \parallel KCl_{(s)}, Hg_2Cl_{2(s)} \mid Hg \tag{13-11}$$

$$E_{cell} = E_{SCE} + E_j - \left( E^\circ_{Sb_2O_3, Sb} - \frac{0.0592}{6} \log \frac{1}{a^6_{H_3O^+}} \right) \tag{13-12}$$

The advantages of this electrode are that it is not easily broken, has low resistance, is adaptable to continuous measurement, and is useful in turbid

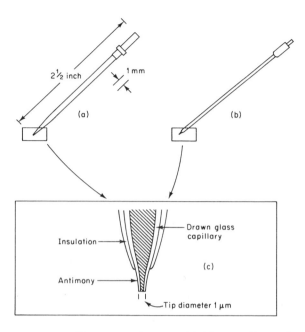

**Fig. 13-1.** Micro antimony pH-sensitive electrode. (a) Glass capillary substrate; (b) metal or glass rod substrate; (c) magnification of electrode tip. (Transidyne General Corporation, Ann Arbor, Michigan.)

and viscous solutions. Limitations of the electrode are that the error is large (about ±0.2 pH units), it suffers from a salt error, calibration is needed for each specific application, oxidizing and reducing agents will interfere, the electrode is poisoned by traces of metals such as copper, silver, and other metals below antimony in the electromotive series, and interference is found by certain complexing agents if present in the solution. The electrode is useful only in the range pH 1 to 10 because of its amphoteric properties.

A recent development is the modification of the antimony electrode into a microelectrode. The tip configuration which involves a vacuum deposition of a very pure antimony film is shown in Fig. 13-1. Such a miniaturized electrode has been suggested to have improved properties over the conventional antimony electrode. Providing the limitations are recognized, the electrode in its micro form should be extremely useful for registering pH information in studies in microbiology and physiology. For example, *in vivo* measurement of pH level in blood is possible.

It can be shown that for both the quinhydrone and antimony electrode the cell potential is given by Eq. (13-6). In practice, therefore, these two electrodes must also be calibrated by standard buffers.

**pH Titrations.**     Since the pH of the solution changes during an acid–base titration, a convenient way to follow the titration is by potentiometry with a glass electrode–reference electrode pair. The cell potentials are measured by a pH meter (see "Measurements with a pH Meter" in this chapter). Plotting of these data, which are read in pH units, vs ml of titrant yields the titration curve for the acid–base system. The end point in the titration curve can be determined by selecting the point of greatest inflection or by plotting the first or second derivative (see Chapter 30 on Experimental Techniques).

Excellent agreement between calculated and experimental acid–base titration curves are found throughout the acidic region and up to about pH 10. At higher pH values a sodium error in the glass electrode is observed; the magnitude of this error depends on the type of glass membrane used.

Since the pH of solutions can be measured with ease, accuracy, precision, speed, and versatility, it is not surprising to find that this measurement is a routine laboratory procedure. From pH measurements, it is possible to calculate $K_a$ and $K_b$ values, buffer ratios, hydrolysis effects, and to obtain additional useful information about the system.

In health related areas pH measurements are very valuable. For example, the pH of arterial blood is remarkably constant ranging from 7.38 to 7.42. From a physiological viewpoint a change of about 0.05 pH units is very significant. Therefore, the precision and accuracy of this determination must be excellent. The major components responsible for the pH are $H_2CO_3$ and $HCO_3^-$. Measurement of the pH of a blood sample allows the calculation of the ratio of $HCO_3^-$ to $H_2CO_3$ ($H_2O$–$CO_2$). Also, it aids the diagnosis of alkalosis or acidosis due either to metabolic or respiratory ailments. (See Chapter 8, page 175.)

# NON-HALF-CELL ION-SELECTIVE ELECTRODES

In general, the potential determining mechanism in non-half-cell types of electrodes appears to involve an ion exchange process and the electrodes can be divided into three general classes: glass membrane electrodes, solid state or precipitate electrodes, and liquid–liquid membrane electrodes. The glass membrane electrode is primarily used for indicating hydronium ion and several other monovalent cations, while the others are used for ions other than hydronium ion. A recent development is fabrication of gas-sensing and enzyme electrodes. These electrodes still belong to the first of the aforementioned classes. The former differs by the fact that a second membrane is involved with the sole purpose of being selectively permeable to a specific gas. The latter differs by having an enzyme immobilized on the ion-sensitive membrane.

Ion-selective electrodes measure single ion activity according to the Nernst equation

$$E = E_c - \frac{0.0592}{n} \log \frac{1}{a_{ion}} \qquad (13\text{-}13)$$

where $E$ is the total potential of the system, $E_c$ is the portion of the total potential due to the reference electrode and internal solutions used, and $n$ is the charge excluding the sign of the ion being detected. For a tenfold change in ionic activity, the electrode potential at 25°C changes by 59.16 mV (monovalent ion), 29.58 mV (divalent ion), etc. In dilute solutions, the activity of an ion approaches the concentration. Therefore, in many cases, the activity is proportional to concentrations and the electrode can be calibrated in terms of concentration.

Many different ion-selective electrodes are available commercially. They are rugged, easy to use, available in different sizes and shapes, provide rapid response, and are of modest cost. Usually, they are used with a saturated calomel electrode and the potentials that are developed are measured by ordinary potentiometric–pH instrumentation. The electrodes can be used to follow titrations or to determine ion concentrations in static or flowing samples.

## GLASS MEMBRANE ELECTRODE

The glass electrode was first described at the turn of the century. It was erroneously thought that the potential developed at the glass membrane was due to the membrane showing preferential permeability to hydronium ion. Consequently, the early development of the glass membrane electrode was mostly empirical in nature. Once this notion was disproved, the understanding of the glass electrode and development of other ion-selective electrodes was rapid.

When a thin membrane of glass is placed between two solutions, a potential difference is observed which depends on the type of cations in the solution. The key to the response is the composition of the glass membrane. If the electrode is to respond to $Na^+$ rather than $H_3O^+$, for example, a particular glass composition suitable for $Na^+$ response must be selected.

The experimental evidence is now overwhelming that the glass membrane functions as a cation exchanger and shows a particular selectivity order for cations. The selectivity and response is altered by the type and concentration of oxides and lattice-modifying additives present in the membrane. This is illustrated in Table 13–1. It is still very difficult to predict electrode selectivity by considering glass composition and this part of the technology of glass membranes still tends to be trial and error. One other property of a glass membrane,

**Table 13–1.  Response Properties for Recommended Cation-Sensitive Glasses[a]**

| Principal cation to be measured | Glass composition | Selectivity characteristics |
|---|---|---|
| Li$^+$ | 15% Li$_2$O–25% Al$_2$O$_3$–60% SiO$_2$ | $K_{\mathrm{Li^+/Na^+}} \approx 3$, $K_{\mathrm{Li^+/K^+/K^+}} > 1000$ |
| Na$^+$ | 11% Na$_2$O–18% Al$_2$O$_3$–71% SiO$_2$ | $K_{\mathrm{Na^+/K^+}} \approx 2800$ at pH 11 |
| | | $K_{\mathrm{Ha^+/K^+}} \approx 300$ at pH 7 |
| | 10.4% Li$_2$O–22.6% Al$_2$O$_3$–67% SiO$_2$ | $K_{\mathrm{Na^+/K^+}} \approx 10^5$ |
| K$^+$ | 27% Na$_2$O–5% Al$_2$O$_3$–68% SiO$_2$ | $K_{\mathrm{K^+/Na^+}} \approx 20$ |
| Ag$^+$ | 28.8% Na$_2$O–19.1% Al$_2$O$_3$–52.1% SiO$_2$ | $K_{\mathrm{Ag^+/H^+}} \approx 10^5$ |
| | 11% Na$_2$O–18% Al$_2$O$_3$–71% SiO$_2$ | $K_{\mathrm{Ag^+/Na^+}} > 1000$ |

[a] G. A. Rechnitz, *Chem. Eng. News* **43** (25), 146 (1967) Reprinted, with permission of the copyright owner, the American Chemical Society.

which appears to be essential, is that it undergo hydration. Nonhygroscopic glasses produce little or no electrode response.

In designing the hydronium ion-indicating electrode a thin-walled glass bulb is filled with a solution of known pH (see Fig. 13–2). The potential across the membrane that will develop is determined by measuring the potential difference between two reference electrodes placed on opposite sides of the membrane. This can be represented as

$$\left.\begin{array}{c}\text{reference}\\\text{electrode}\end{array}\right|\ a'_{\mathrm{H_3O^+}}\ \left|\ \begin{array}{c}\text{glass}\\\text{membrane}\end{array}\ \right|\ a''_{\mathrm{H_3O^+}}\ \left|\ \begin{array}{c}\text{reference}\\\text{electrode}\end{array}\right. \tag{13-14}$$

The potential for this cell will follow, according to experimental observation, the expression

$$E_{\mathrm{cell}} = k - 0.0592 \log \frac{a'_{\mathrm{H_3O^+}}}{a''_{\mathrm{H_3O^+}}} \tag{13-15}$$

where $k$ is a constant. The constant includes the difference in junction poten-

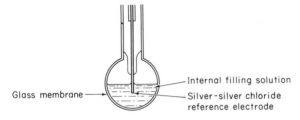

Fig. 13–2.  A typical glass electrode.

tials between the reference electrodes and solution and the asymmetry potential for the glass membrane. The asymmetry potential is due to the difference in the surface of the inner and outer layers of the glass membrane and at most corresponds to about 2–3 mV. If any difference in potential exists between the two reference electrodes, this is also incorporated into $k$.

Assume that $a_{H^+}''$ represents the activity of a fixed, standard solution. For this case $a_{H^+}'$ is an unknown solution and expression (13-15) becomes

$$E_{cell} = k_1 - 0.0592 \log a_{H^+}'$$

which leads to Eq. (13-6) or

$$E_{cell} = k_1 + 0.0592 \, pH \qquad (13\text{-}16)$$

and where $k_1$ now includes the constant factor related to $a_{H^+}''$. It appears, therefore, that the glass electrode suitable for hydronium ion acts just as if it were a hydrogen electrode in terms of potential response [see Eqs. (13-1) to (13-6)].

In practice, a typical commercially available glass and saturated calomel electrode, which is illustrated in Fig. 13–3, are inserted into a test solution The membrane is a small, thin, curved disk having an approximate size of 0.5 cm diameter and 50 μm thickness. A Ag–AgCl reference electrode is the internal reference electrode and makes contact with a ml or so of a 0.1 $F$ HCl or buffer solution (fixed activity) inside the membrane. The outside solution

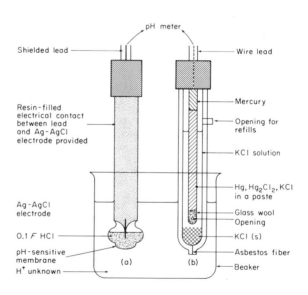

**Fig. 13–3.** Typical cell employing commercial electrodes for the pH measurement. (a) Glass electrode; (b) saturated calomel electrode.

is the test solution. The complete cell is represented by

$$\text{Ag} \mid \text{AgCl}_{(s)}, \text{HCl } (0.1 \ F) \ \left| \ \begin{array}{c} \text{glass} \\ \text{membrane} \end{array} \ \right| \ \begin{array}{c} \text{test} \\ \text{solution} \end{array} \ \right| \ \text{Hg}_2\text{Cl}_{2(s)}, \text{KCl}_{(s)} \mid \text{Hg}$$

$$(13\text{-}17)$$

The membrane potential is determined by measuring the potential difference between the two reference electrodes with a pH meter. An ordinary potentiometer cannot be used because of the high internal resistance of the glass electrode.

Upon examining the complete cell shown in (13-17) it is apparent that the observed cell EMF consists of six sources of electromotive force:

1. Potential of the internal reference Ag/AgCl electrode
2. Potential at the inner surface of the membrane
3. Potential at the outer surface of the membrane
4. Junction potential at the external reference electrode
5. Potential of the external reference Hg/Hg$_2$Cl$_2$ electrode
6. Asymmetry potential

If the test solution is changed, potentials 3, 4, and 6 are altered with the major change being due to 3. Thus, expression (13-16) applies where $k_1$ now represents all of the constant and near constant sources of potentials. It should be noted that expression (13-16) was derived for the hydrogen electrode, quinhydrone electrode, and antimony electrode. Thus, the calibration procedure, as outlined in the next few paragraphs, applies to these electrodes as well as to the glass electrode.

If the value of $k_1$ could be directly and easily evaluated, the use of Eq. (13-16) and the measurement of pH would be straightforward. However, this is not the case, and even if $k_1$ were determined it would not be very precise. It is for these reasons that the measurement must involve a calibration step.

First the test solution is comprised of a standard buffer solution with the pH precisely known. Thus, for the standard

$$(E_{cell})_s = k_1 + 0.0592 \ (\text{pH})_s$$

This is followed by measurement of the unknown solution where

$$(E_{cell})_u = k_1 + 0.0592 \ (\text{pH})_u$$

Elimination of $k_1$ is possible and the pH of the unknown is given by

$$(\text{pH})_u = (\text{pH})_s + \frac{(E_{cell})_u - (E_{cell})_s}{0.0592} \qquad (13\text{-}18)$$

In using Eq. (13-18), the value of $k_1$ should not change and the pH of the standard buffer should be accurately known. As mentioned previously in the section on buffers, the National Bureau of Standards has played a major role in this development. Variations in $k_1$ are minimized by choosing a standard buffer that has a pH similar to that of the unknown solution. With minimum effort, ordinary instruments, and buffers, an error in pH measurement is about $\pm 0.02$ pH units.

The same equations apply for glass membranes sensitive to metal ions. Thus, in Eqs. (13-14) to (13-18) $H_3O^+$ is replaced by $M^+$. In general, glass membranes are suitable for the detection of monovalent cations (see Table 13-1) in addition to hydronium ion. No glass membrane has been designed that shows a potential response toward anions.

A clear understanding of the glass membrane requires consideration of more than just its composition. The details of the membrane can be represented by

| Internal solution | Hydrated gel layer | Dry glass layer | Hydrated gel layer | External solution |
|---|---|---|---|---|

There are probably additional·layers of gradual change on both sides of the dry membrane.

The main bulk of the membrane is the dry glass layer (about 50 $\mu$m thick). When the membrane is placed in water, it tends to swell as external hydration occurs. Simultaneously, the hydrated layer will dissolve slightly while additional dry glass is hydrated. Thus, a steady state is reached and the hydrated thicknesses are about 50–1000 Å. The dissolution rate is closely related to the electrode life expectancy and is very dependent on glass composition.

The hydrated glass surface undergoes cation exchange and involves more than one kind of cation in a concentration profile in the hydrated layer. Often the selectivity properties of the glass membrane are similar to the selectivities found for ion exchange resins (see Chapter 26). Thus, the overall potential of the glass membrane is the sum of the diffusion potential (travel in the hydrated layer) and ion exchange potential. At the solution-hydrated layer interface charge is transferred by the ion exchange process, while in the layer itself diffusion accounts for the current transport. In the glass framework, current is transported by the lowest charged cations. These cations, which are usually sodium ions, do not diffuse through the glass but move perhaps a few atomic diameters and transfer the charge from one sodium ion to another.

The glass electrode, as pointed out, can be designed for the detection of $Li^+$, $Na^+$, $K^+$, and $Ag^+$ in addition to detection of hydronium ion. As a pH electrode, the glass electrode is amazingly versatile. It is not influenced by oxidizing or reducing agents or by the presence of metals. It can be used in nonaqueous, aqueous, dense, and turbid solutions. The response is rapid, accurate, and precise without contaminating the solution or disturbing the solubility of dissolved gases. The main limitations are that the membrane surface tends to absorb ions and undissociated molecules, the electrode has a high internal resistance, it is breakable and sensitive to temperature, and the potential response is affected by sodium ion.

The magnitude of the sodium error, which causes the pH to be lower than its actual value, can be modified by changing the composition of the membrane. For the standard glass electrode the sodium error is encountered in basic solutions (pH > 9); the more alkaline the solution the greater the error. An error is also observed in very strong acid solutions (pH < 1); however, the reasons for this error are not clearly understood.

## SOLID STATE AND PRECIPITATE ELECTRODES

Solid state and precipitate electrodes are similar in that a solid membrane separates the standard and test solution. In the former electrode the membrane is a single crystal doped with another salt. In contrast, the precipitate electrode has a membrane impregnated with or made from some insoluble crystalline salt. An example of the former is the fluoride electrode (manufactured by Orion Research Inc.), which consists of a single crystal of $LaF_3$ doped with an Eu(II) salt, while an example of the latter is a pressed silver halide pellet (also made by Orion). The response mechanism is similar to the ion exchange type mechanism for the glass electrode. Only in the exact details are there principal differences.

Fortunately, several crystalline materials are known which have an ionic conductivity at room temperature. Usually, the lattice ion with the smallest ionic radius and charge is the one which participates in conduction. By their nature, crystalline materials are mechanically stable and often are chemically inert and of low solubility

Conduction in a crystal phase occurs by a lattice defect mechanism. In this mechanism mobile ions move into adjacent vacancy defects. Experimentally, it is possible to control the vacancy with respect to size, shape, and charge, and thus, restrict its availability only to certain mobile ions. Consequently, other ions are not able to contribute to conduction. In this way the crystal membrane demonstrates a particular selectivity toward an ion.

These type of electrodes are perhaps the simplest, since foreign ions are virtually eliminated from entering the crystal phase. Thus, it always performs in a Nernstian manner. If interference occurs, it is usually because of chemical reactions at the crystal surface.

**Solid State Electrodes.**   Solid state electrodes have a doped single crystal membrane. A typical design is shown in Fig. 13–4. The electrodes are frequently characterized by low resistivities and this property is further reduced

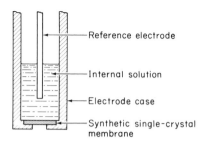

Reference electrode

Internal solution

Electrode case

Synthetic single-crystal membrane

**Fig. 13–4.** A typical solid state electrode.

by doping. Typical electrode lifetimes at room temperature are from 1 to 2 years. Selectivity and lower limits of detection vary with the type of solid electrode, however, the upper limit of detection is the saturated solution. Normally, the electrodes are not used at this level and a practical upper limit is about 1 $M$.

The $LaF_3$ electrode, which was first introduced in 1966, is one of the most useful solid state electrodes. This unique single crystal membrane responds to fluoride ion and virtually no other anion or cation. Nernstian response is from 1 $M$ to below $10^{-5}$ $M$ $F^-$ and only $OH^-$ interferes.

A typical cell employing the saturated calomel electrode and the fluoride electrode containing a $Ag/AgCl$ reference is the following:

$$Hg; Hg_2Cl_{2(s)} \mid KCl_{(sat)} \mid\mid F^-_{(unk)} \mid LaF_{3(s)} \mid NaF \ (0.1 \ M),$$

$$NaCl \ (0.1 \ M) \mid AgCl_{(s)}; Ag \quad (13\text{-}19)$$

The potential of the cell is given by

$$E_{cell} = E_{AgCl} - E_{SCE} + \frac{2.303RT}{F} \log \frac{1}{a_{F^-}} + E_a + E_j \quad (13\text{-}20)$$

where $E_{AgCl}$, $E_{SCE}$, $E_a$, and $E_j$ are constant potentials representing internal reference electrode, external reference electrode, asymmetry properties, and liquid junctions, respectively. Calibration in a solution of known fluoride activity eliminates the need of accurately knowing these constants. It has been demonstrated that the presence of $Na^+$, $K^+$, $Mg^{2+}$, $NO_3^-$, $Cl^-$, or $SO_4^{2-}$ ions cause no interference.

Some typical applications of this electrode have been the determination of fluoride in bone, air and stack gas samples, chromium plating baths, minerals, water, and toothpastes. Without question, this electrode has in a short period of time become a valuable analytical tool.

Fluoride ion can be titrated using the fluoride electrode. Figure 13–5 illustrates titration curves using $Th^{4+}$, $La^{3+}$, and other titrants. If the medium is 60–80 vol% ethanol, an improvement in the titration is observed. The stoi-

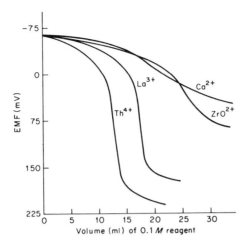

**Fig. 13–5.** Titration of fluoride ion using $LaF_3$ electrode and different titrants. [T. S. Light and R. F. Mannion, *Anal. Chem.* **41**, 107 (1969).]

chiometry of the reaction $F^-/La^{3+}$ is 3/1 and the point of maximum slope does not correspond to the true stoichiometric point. However, by use of fluoride standards the stoichiometric point potential can be determined. It is also possible to use the electrode for measuring $La^{3+}$ content.

The second most useful solid state electrode is based on a $Ag_2S$ membrane. Silver sulfide, like $LaF_3$, is an ionic conductor of low resistance, particularly when doped with other silver salts. The mobile ions are the silver ions. Other important membrane properties are its low solubility, resistance toward oxidizing and reducing agents, and ease in making it into pellets. The electrode reaches equilibrium rapidly and it appears to be superior to a silver metal electrode for detecting silver ion. It has been suggested that free sulfide

**Table 13–2.   Silver–Silver Sulfide Type Electrodes**

| Ion determined | Membrane | Principal interferences |
|---|---|---|
| $Cl^-$ | $AgCl/Ag_2S$ | $Br^-$, $I^-$, $S^{2-}$, $NH_3$, $CN^-$ |
| $Br^-$ | $AgBr/Ag_2S$ | $I^-$, $S^{2-}$, $NH_3$, $CN^-$ |
| $I^-$ | $AgI/Ag_2S$ | $S^{2-}$, $CN^-$ |
| $SCN^-$ | $AgSCN/Ag_2S$ | $Br^-$, $I^-$, $S^{2-}$, $NH_3$, $CN^-$ |
| $S^{2-}$, $Ag^+$ | $Ag_2S$ | $Hg^{2+}$ |
| $CN^-$ | $AgI/Ag_2S$ | $I^-$, $S^{2-}$ |
| $Cu^{2+}$ | $CuS/Ag_2S$ | $Hg^{2+}$, $Ag^+$ |
| $Pb^{2+}$ | $PbS/Ag_2S$ | $Hg^{2+}$, $Ag^+$, $Cu^{2+}$ |
| $Cd^{2+}$ | $CdS/Ag_2S$ | $Hg^{2+}$, $Ag^+$, $Cu^{2+}$ |

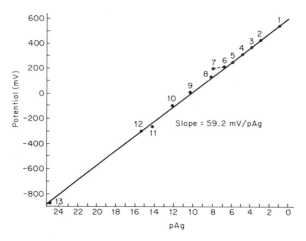

**Fig. 13–6.** Response of silver sulfide electrode to silver activity. [R. A. Durst, "Ion Selective Electrodes," National Bureau of Standards Special Publications 314, R. A. Durst, ed., Washington, D. C., p. 375 (1969).]

| Point | Solution composition | ~pAg (calc) | $E$ (mV) |
|-------|---------------------|-------------|----------|
| 1 | $10^{-1}$ $M$ AgNO$_3$ | 1.1 | +550 |
| 2 | $10^{-3}$ $M$ AgNO$_3$ | 3 | +438 |
| 3 | $10^{-4}$ $M$ AgNO$_3$ | 4 | +385 |
| 4 | $10^{-5}$ $M$ AgNO$_3$ | 5 | +323 |
| 5 | $10^{-6}$ $M$ AgNO$_3$ | 6 | +260 |
| 6 | $10^{-7}$ $M$ AgNO$_3$ | 7 | +225 |
| 7 | $10^{-8}$ $M$ AgNO$_3$ | 8 | +213 |
| 8 | sat'd AgI | 8.2 | +150 |
| 9 | sat'd AgI + $10^{-6}$ $M$ KI | 10.3 | +21 |
| 10 | sat'd AgI + $10^{-4}$ $M$ KI | 12.3 | −91 |
| 11 | sat'd AgCl + 1 $M$ Na$_2$S$_2$O$_3$ | 14.2 | −256 |
| 12 | sat'd AgCl + 0.1 $M$ KI | 15.5 | −298 |
| 13 | 0.1 $M$ Na$_2$S + 1 $M$ NaOH | 24.9 | −872 |

ion levels as low as $10^{-19}$ $M$ are detectable in acidic solutions while levels of Ag$^+$ down to $10^{-20}$ $M$ are detected in titrations. Figure 13–6 illustrates the Ag$^+$ response.

The electrode and its modifications, mixed halide–sulfide and mixed metal ion–silver sulfide systems, can be used for the detection of a wide variety of ions. Nine electrodes and their properties are listed in Table 13–2.

**Precipitate Electrodes.** Precipitate electrodes contain membranes which are made by taking the membrane material, usually an inert binder, and impregnating it with the potential active ingredient. The resulting mixture is then shaped, cut, or pressed into a suitable form.

**Table 13–3.  Several Typical Precipitate Membrane Electrodes**

| Active material | Matrix | Ion sensitivity |
|---|---|---|
| Calcium stearate | Paraffin | $Ca^{2+}$ |
| Potassium tetraphenylborate | Polystyrene–gauze | $K^+$ |
| Barium sulfate | Paraffin | $Ba^{2+}$ |
| Silver halide | Paraffin or silicone rubber | $Ag^+$, halide |
| Silver sulfide | Silicon rubber | $Ag^+$, $S^{2-}$ |
| Calcium fluoride | Silicon rubber | $F^-$ |
| Titanium dioxide | Polyethylene | $H^+$, $OH^-$ |
| Lead tungstate | Paraffin | $Pb^{2+}$, $WO_4^{2-}$ |

Several binders, such as paraffin wax, collodion, polyvinyl chloride; polystyrene, polyethylene, and silicon rubber, have been used. Although its principal purpose is to provide an inert matrix, other properties such as adhesion, strength, swelling, and availability must also be considered. Currently, silicon rubber appears to be the most useful.

Table 13–3 lists a number of precipitate-type membrane electrodes employing sparingly soluble metal salts and chelates. Many of these electrodes are not easily obtained commercially. Consequently, almost all of the data in the literature on this type of electrode are from electrodes made in the research laboratory. Because of the different methods of preparation, different materials, and other factors which cannot be controlled, there appears, in some instances, to be experimental contradictions. Several general quantitative conclusions can be made, however. For example, electrode properties are very sensitive to grain size, crystalline form, precipitation conditions, and solubility product for the active ingredient. It is also frequently necessary to condition the electrode by soaking in appropriate solutions. Fortunately, the literature contains the necessary directions for making the electrodes.

It is reasonable to state, however, that the electrodes exhibit selectivity and can be used in a practical way. The most reproducible type electrodes are the silver halide impregnated silicone rubber electrodes.


## LIQUID–LIQUID ELECTRODE

A liquid ion exchanger in its simplest form can be pictured as a water-immiscible fluid possessing exchange properties placed between two aqueous solutions. Electric properties of such a system were actually studied in 1908. An obvious limiting factor in the design of a liquid–liquid electrode is a mechanical one. It is necessary that the liquid ion exchanger participate in electrolytic contact with a sample solution, but mixing of the two liquid

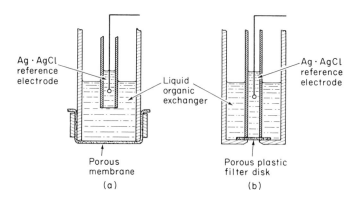

**Fig. 13–7.** Typical liquid membrane electrodes.

phases or dissolving of the ion exchanger in the sample solution must be negligible. Electrodes can be made that are responsive to cations or anions.

Figure 13–7 illustrates two ways of overcoming the mechanical problem. In Fig. 13–7a the liquid membrane is held inside a glass tube, sealed at its end by a dialysis cellulose membrane. The latter membrane is permeable to all ions but not the liquid ion exchanger. A reference electrode, such as a Ag/AgCl electrode, is placed in an agar gel and dipped into the liquid ion exchanger. This type of electrode setup suffers from a high resistance (actual membrane is the distance from dialysis membrane to the reference electrode) and a long response time.

The electrode shown in Fig. 13–7b is more typical of commercial electrodes. In this design the liquid ion exchanger is held in the pores of a thin and porous (very small pore diameter) disk. The area above the disk contains more liquid ion exchanger which is able to pass into the pores of the disk when needed. A reference electrode, Ag/AgCl, is also inserted into the system. In this design internal resistances are much lower since the actual membrane is very thin and response times are rapid.

In a liquid membrane electrode, the dissolved, essentially undissociated salts (acid, etc.), are able to move through the membrane phase. At the membrane interface with the test solution, ion exchange between the ions in the sample solution and mobile ions in the ion exchanger takes place according to the particular selectivity shown by the liquid ion exchanger. Thus, the electrode's selectivity is directly determined by the liquid ion exchanger's selectivity.

In attempting to devise a useful electrode, it is necessary to search for or synthesize a substance possessing the required selectivity as well as the necessary properties. One guideline that has been suggested is the stability constant for various ligands. For example, if the ratio of the stability constant

**Table 13–4.   Typical Liquid Membrane Electrodes**

| Ion measured | Exchange site | Ion measured | Exchange site |
|---|---|---|---|
| $Ca^{2+}$ | $(RO)_2PO_2^-$ | $ClO_4^-$ | |
| $Ca^{2+}$ and $Mg^{2+}$ | $(RO)_2PO_2^-$ | $Cl^-$ | $R_4N^+$ |
| $Cu^{2+}$ | $R-S-CH_2-COO^-$ | $BF_4^-$ $\}$ | |
| $Pb^{2+}$ | $R-S-CH_2-COO^-$ | $NO_3^-$ | |

for an ion of interest to the stability constant of an interfering ion in aqueous solution is large, it can be anticipated that the ligand-salt will also show good selectivity for the ion of interest in the membrane phase. Of course it may be necessary to modify the liquid ion exchanger to increase its solubility in the membrane phase. If mixed complexes involving $H^+$ are formed, a large hydrogen ion interference should be expected. Normally, the solvent for the liquid ion exchanger is not critical with respect to selectivity.

Table 13–4 lists several of the commercially available liquid membrane selective electrodes, their exchange sites, and their selectivities. Figure 13–8 illustrates calibration curves for the $Ca^{2+}$ and $Cu^{2+}$ electrodes. The calcium electrode is particularly useful because of its favorable $Ca^{2+}$ selectivity over $Na^+$, $K^+$, and $Mg^{2+}$. Therefore, it finds great use in water analysis and biological research.

Phosphoric acid esters, which are used in the calcium electrode, are easily synthesized with long hydrocarbon chains. They have appropriate solubility characteristics and form stable complexes with calcium and not other cations. If a diester is used, a hydrogen ion interference (mixed complex) is avoided.

In Figure 13–8, the $Ca^{2+}$ curve was measured with an electrode made from a 0.1 $F$ calcium salt of didecyl phosphoric acid in dioctylphenylphosphonate. A Nernst behavior was observed in the range $10^{-1}$ to $10^{-5}$ $M$; the lower limit is determined by the low solubility of the calcium salt of didecyl phosphoric acid.

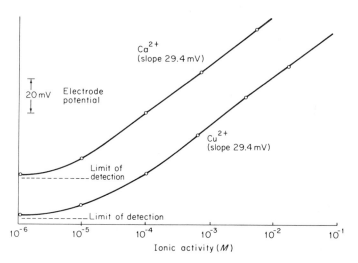

**Fig. 13–8.**   Electrode response for cupric and calcium ion-selective electrodes as a function of ionic activity. [J. W. Ross, Jr., "Ion Selective Electrodes," National Bureau of Standards Special Publication 314, R. A. Durst, ed., Washington, D. C., p. 57 (1969).]

## GAS-  AND  ENZYME-SENSING  ELECTRODES

Figure 13–9 illustrates a typical design of a gas-sensing electrode. Examination of the electrode reveals that it is actually a cell composed of a reference electrode and an indicator electrode with the potential of the latter being determined by the selective passage of a gas through the permeable membrane. Several different gas-sensing electrodes are now commercially available. The nitrogen oxide electrode is briefly described in the following; the others, such as the $SO_2$, $CO_2$, $H_2S$, and $NH_3$ electrode function in a similar manner.

The thin, microporous, gas-permeable membrane in the gas-sensitive electrode is permeable to a gas, while preventing water and electrolytes from entering and passing through the membrane. Consider the gas $NO_x$, where $NO_x$ is a mixture of $NO_2$ and $NO$, which is in equilibrium with an acidified solution of $HNO_2$. As the gas enters the membrane, it will reach an equilibrium with the internal solution. The final equilibrium position is therefore determined by three separate equilibria:

$$(1) \quad NO_{x(aq)}^{external} \rightleftharpoons NO_{x(g)}^{membrane}$$

$$(2) \quad NO_{x(g)}^{membrane} \rightleftharpoons NO_{x(aq)}^{internal}$$

$$(3) \quad NO_{x(aq)}^{internal} + H_2O \rightleftharpoons H^+ + NO_2^-$$

As the $NO_x$ concentration level in the sample changes, the internal $H^+$ concentration is altered and a new potential develops at the indicator electrode.

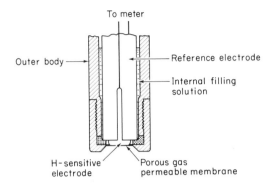

**Fig. 13–9.** A nitrogen oxide gas-sensing electrode.

This electrode potential is essentially Nernstian with respect to $NHO_2$ concentration (expressed as $NO_2$ usually) and it can be shown that

$$pNO_2 = \frac{E_{obs} - k}{0.0592} \qquad (13\text{-}21)$$

even though the indicator electrode is responding to $H^+$. Therefore, like the glass electrode, it is calibrated (see pages 301–302) to eliminate $k$ by using solutions of known $NO_2$ concentration.

The $NO_2$ electrode provides a rapid response, does not suffer from many interferences, and will provide a sensitivity of $NO_2^-$ in water down to 0.02 ppm. It is widely used for the determination of nitrite in foods, food brines, sewage and waste waters, and industrial atmosphere. Figure 13–10 illustrates how nitrates, nitrites, and oxides of nitrogen in the environment can be converted to either $NO_3^-$, for determination with the nitrate ion-selective electrode or $HNO_2$ for determination with the nitrogen oxide gas-sensing electrode.

Several variations of enzyme electrodes have been devised. In general, the enzyme is immobilized and is in contact with the indicator electrode much like the gas-permeable membrane. The substance to be determined is converted through a reaction catalyzed by the immobilized enzyme into a species which influences the potential of the indicator electrode. Since enzymes are highly selective, the electrode will respond to only certain species.

One of the early examples was an electrode in which urease enzyme was immobilized on a cation-sensitive glass electrode. In the presence of urea the enzyme catalyzes the reaction

$$\underset{\overset{\|}{\underset{H_2NCNH_2}{O}}}{} + 2\,H_3O^+ \xrightarrow{\text{urease}} CO_2 + 2\,NH_4^+ + 2\,H_2O \qquad (13\text{-}22)$$

The indicator electrode responds to the level of $NH_4^+$ and shows a Nernstian

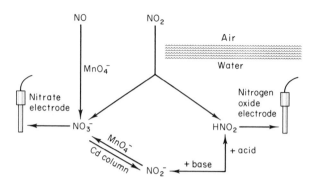

**Fig. 13–10.** Flow diagram for the determination of oxides of nitrogen and nitrates by suitable indicator electrodes.

response at urea concentrations from $10^{-4}$ to approximately $10^{-1}\,M$ urea. In general, enzyme electrodes have been designed to meet quantitative needs in the clinical laboratory and in the physiological and biochemical laboratory. Although presently useful they are not as reliable as other ion-selective electrodes, however, their development is still in the infancy stage.

## CALIBRATION

As previously shown for the hydrogen ion-sensitive electrodes [see Eq. (13-6)], the potential of the indicator electrode and the activity of the ion are related through a Nernstian-like equation where

$$- \log a_{\mathrm{M}} = \mathrm{pM} = \frac{E_{\mathrm{cell}} - k}{0.0592/n} \tag{13-23}$$

Depending on the ion-selective electrode a plot of potential vs activity will be linear over a wide concentration range as shown in Fig. 13–8. If the approximation is made that activity and concentration are equal the error may be significant since over such a large concentration range the activity coefficient will not be constant. Thus, if the change in activity coefficient is not taken into account, the resulting plot of potential vs concentration will be nonlinear and the curvature will be the greatest in the more concentrated solutions. It is this region then where the error due to the approximation will be the largest. A comparison of calibration curves of potential vs activity and potential vs concentration for fluoride ion as determined with the flouride electrode is shown in Fig. 13–11.

Electrode calibration, where potentials are measured for a series of standards, is a desirable procedure for several reasons. Many procedures are available to prepare standard solutions, the measurements are simple and

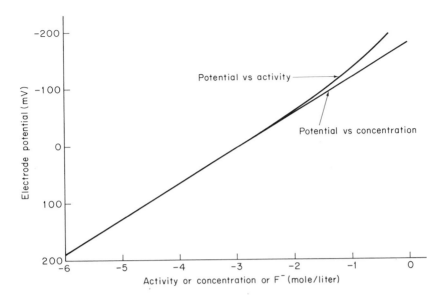

**Fig. 13–11.** Response of a fluoride ion electrode vs activity and concentration of fluoride ion.

straightforward, and response is rapid enough to use in continuous monitoring. To minimize the error due to activity effects, standards and samples are swamped with excess inert electrolyte. The concentration of the electrolyte should be large enough so that the different standards and unknown samples do not differ significantly in ionic strength. This type of empirical calibration is widely used with ion-selective electrodes.

A second procedure that is used is a standard addition method. In this procedure the potential is measured before and after the addition of a known volume of a standard to a known volume of the unkown. Because of the small addition it is assumed that the ionic strength is not affected, and consequently the activity coefficient remains constant.

Assuming the unknown is a monovalent ion $(n = 1)$, activity is equal to concentration, and the potentials $E_1$ and $E_2$ are potentials before and after addition of the standard, respectively, Eq. 13–23 can be written, where $n = 1$, as

$$- \log C_x = \frac{E_1 - k}{0.0592} \tag{13-24}$$

and

$$- \log \frac{C_x V_x + C_s V_s}{V_x + V_s} = \frac{E_2 - k}{0.0592} \tag{13-25}$$

where $C_x$ and $V_x$ are concentration and volume of unknown and $C_s$ and $V_s$ are for the standard. Solving the two equations simultaneously gives an equation which allows the calculation of $C_x$ from the experimental data or

$$C_x = \frac{C_s V_s}{V_x + V_s}\left(10^{-(E_1 - E_2/0.0592)} - \frac{v_x}{v_x + v_s}\right)^{-1} \qquad (13\text{-}26)$$

The standard addition method is a technique that is used with many different instrumental methods and is not specific to measurements with ion-selective electrodes.

## APPLICATIONS

Ion-selective electrodes, with the exception of the pH-sensitive glass electrode, were virtually unknown ten years ago. The potential of these electrodes in solving many practical problems is now being realized. For example, they are widely used in clinical, biological, water, air, oceanographic, and pharmaceutical research and in routine analytical determinations. At present there are reliable, commercially available electrodes for indicating $H^+$, halides ($F^-$, $Cl^-$, $Br^-$, $I^-$), $Cd^{2+}$, $Cu^{2+}$, $CN^-$, $BF_4^-$, $Pb^{2+}$, $NO_3^-$, $ClO_4^-$, $K^+$, $Ag^+$, $S^{2-}$, $Na^+$, and $SCN^-$, for $NH_3$, $H_2S$, $SO_2$, $CO_2$, $NO$ and $NO_2$ gases, and for several different enzymes.

The electrodes can be used for individual measurements, for flowing systems (see Fig. 13–12), and in titrations. Several typical applications are briefly cited below.

Lead poisoning is of considerable public interest and obviously a quick, accurate method of analyzing lead is needed. It is possible to determine lead in blood and urine samples by atomic absorption or ashing the sample and using a colorimetric reagent for the lead in the residue. In both cases the sample is destroyed.

With a $PbS/Ag_2S$ crystal membrane electrode, it is possible to measure the lead directly in a blood or urine sample where the normal Pb levels are approximately 40 $\mu g/100$ ml and 5–10 $\mu g/100$ ml, respectively. No pretreatment or separations are necessary and the analysis is complete in about 10 minutes.

Unlike other halogens, microfluorine quantities are very difficult to determine. Fluoride ion is readily titrated with $Th(NO_3)_4$ (0.005 $F$) in 80% ethanol using a fluoride electrode to indicate the end point (see Fig. 13–5). Samples can be in the range of 1–10 mg and an absolute accuracy of $\pm 0.3\%$ can be expected.

The chloride electrode has been used for the determination of chloride ion in a variety of industrial and physiological samples. Of particular importance is the rapid, accurate clinical determination of chloride ion in sweat. These data are then used in diagnosing cystic fibrosis.

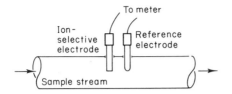

**Fig. 13–12.** Monitoring a flowing stream with an ion-selective electrode.

The ammonia electrode can be used to replace the distillation and titration procedure in the Kjeldahl method (see Chapter 8). After conversion of nitrogen to ammonium ion, the solution is made basic and the ammonia concentration determined with the ammonia electrode. The electrode is suitable down to the $10^{-6}$ $M$ level.

Calcium ion has been determined in beer, boiler water, soil, feedstuffs, flour, minerals, milk, sea water, serum and biological fluids, sugar, pulping liquor, and wine with the calcium electrode. Not all of these are direct measuring procedures. Some involve titration techniques.

Ionized calcium is a physiological active species and a broad range of important physiological processes are known to be critically dependent on calcium ion activity. It has been stated, "there is little doubt that $Ca^{2+}$ is one of the most important electrolytes in human physiology." Successful measurements of $Ca^{2+}$ in biological fluids and related samples have been made with the calcium liquid ion exchange electrode and a specially designed calcium flow through electrode. The latter electrode is ideally suited for serum and other biological fluids because of increased selectivity of $Ca^{2+}$ over $Na^+$ and $K^+$. For example, ionized calcium in serum and simultaneously drawn heparinized whole blood can be determined with this electrode setup. The electrodes have also been used for measuring stability constants of $Ca^{2+}$ complexes and in some cases to follow the kinetics of complex formation.

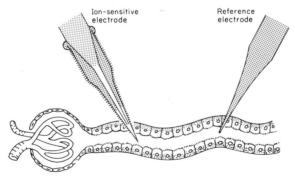

**Fig. 13–13.** An indicator and a reference microelectrode in the lumen of a kidney proximal tubule. [R. N. Khuri, "Ion Selective Electrodes," National Bureau of Standards Special Publication 314, R. A. Durst, ed., Washington, D. C., p. 287 (1969).]

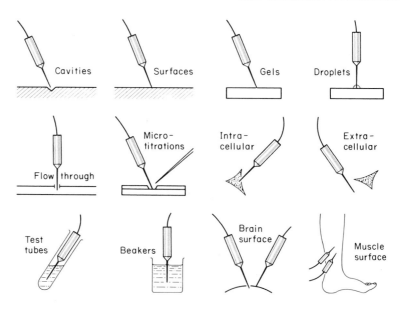

**Fig. 13–14.** Typical applications of micro-, pH, and ion-selective electrodes.

One last application is shown in Fig. 13–13 where the pH or cation activities of luminal fluid in kidney tubules are detected *in situ*. The tubules are of microscopic size and contains the preurine. In this application micro-tipped electrodes such as the one shown in Fig. 13–1 were used. Many of the ion-selective electrodes are available as microelectrodes and are routinely used in a variety of applications. Figure 13–14 summarizes many applications of these microelectrodes.

## MEASUREMENTS WITH A pH METER

One of the limitations of the glass, solid state, and liquid–liquid membrane electrodes is that they have very large internal resistances (megaohm range). For this reason, the simple potentiometric circuit (see Chapter 10) must be modified before it can be used with these electrodes. However, the simple circuit can be used with the hydrogen, quinhydrone, and antimony electrodes.

In order to measure a cell potential, a current, $i$, must flow through the cell. A voltage, $iR$, is produced, which is opposite the potential of the cell, since the cell offers the resistance $R$. The $iR$ value must be below 1 mV; this corresponds to a voltage error of below 1 mV. A typical glass membrane has a resistance of 10 M$\Omega$. Therefore, from Ohm's law the maximum current which can flow and still maintain only 1 mV is $10^{-10}$ A.

$$i = \frac{E}{R} = \frac{0.001 \text{ V}}{10 \times 10^6 \text{ ohm}} = 10^{-10} \text{ A}$$

Generally, galvanometers of the type used in simple potentiometers require more current to produce a detectable deflection. As the resistance of the ion-selective electrode increases, the possibility of using the simple potentiometer is further diminished.

Two types of meters have been designed to overcome this problem. These are the direct reading pH meter and the null-detector potentiometric pH meter. In general, the direct reading pH meter has an accuracy of $\pm 0.1$ pH units, while the null instrument is capable of an accuracy of $\pm 0.01$ pH units.

Both types of instrumental designs can be further modified electronically to take care of temperature change, a switch to allow direct reading of potential or pH values, connections for feeding the data into a recorder, asymmetry potential balancing, and for digital readout. A very precise measurement is possible (0.001 pH unit) in which a full meter scale corresponds to 0.5, 1, or 2 pH units. This type of instrument, which is much more expensive, includes a vibrating reed electrometer or choppers for amplification of the signal.

## *Questions*

1. Differentiate between an indicator electrode, ion selective electrode, and a reference electrode.
2. Discuss whether the following half-cell can be used to indicate hydrogen ion concentration.

$$MnO_4^- + 8H^+ + 5e = Mn^{2+} + 4H_2O$$

3. What is a hydrogen electrode and what special conditions must be satisfied when it is used?
4. Write the cell made between a hydrogen electrode and a Ag–AgCl reference electrode.
5. Derive an expression for the pH of the cell in Question 4.
6. Criticize the statement, "Quinhydrone is an equimolar mixture of *p*-quinone and *p*-hydroquinone."
7. List four basic types of electrodes that are used for the determination of hydrogen ion.
8. Write the basic reaction which leads to a potential for each electrode in Question 7.
9. List the electrodes in question 7 that are *not* applicable for hydrogen ion measurement under the following experimental conditions.
   a. 0.0001 $F$ NaOH.
   b. A sample containing $O_2$ which will be analyzed for $O_2$ by polarography.
   c. A slightly basic solution containing 3 $F$ NaCl.
   d. A solution of equiformal tin(II) and tin(IV).
   e. A very viscous solution.
   f. Checking of the reliability of a new experimental hydrogen ion indicating electrode.

10. Explain how a glass electrode can be used in a volumetric acid–base titration. What are its limitations in this application?
11. It is often necessary to measure continuously the pH of a flowing stream. Suggest how this might be done and the limitations of your procedure.
12. Differentiate between the three types of ion-selective electrodes.
13. What does Nernstian response mean?
14. What are the main properties of a glass membrane that determines a Nernstian response?
15. What does selectivity imply when used in a discussion of ion selective electrodes?
16. What are the sources of potential in a glass electrode–SCE cell? Which are the most significant?
17. Why is it necessary to calibrate ion selective–reference electrode cells?
18. Explain how the glass electrode–SCE cell is calibrated.
19. Suggest an experiment that would demonstrate that the potential across a glass membrane is not due to diffusion of ions through the membrane.
20. Differentiate between a solid state and precipitate ion selective electrode.
21. Explain how the fluoride ion-selective electrode-cell is calibrated.
22. What is the main difference between a pH meter and a potentiometer?

## *Problems*

1. A hydrogen electrode and SCE were used to determine the pH for a series of buffer solutions. Calculate the voltage for each of these solutions if the pH values are the following. Assume the junction potential is negligible and the hydrogen electrode is at standard conditions.
   * a. 4.6                 b. 8.0                 c. 11.4                 d. 2.3

2. Using the same electrode system in Question 1 calculate the pH values for the following potentials.
   * a. 0.315 V         b. 0.814 V         c. 0.463 V         d. 0.711 V

3.* The voltage for the following cell is 0.481 V. Calculate the $K_a$ for the weak acid.

   $$Pt \mid H_2(1 \text{ atm}), HA[1.0 \times 10^{-2}] \parallel Hg_2Cl_{2(s)}, KCl_{(s)} \mid Hg$$

4. The voltage for the following cell is 0.518 V. Calculate the $K_a$ for the weak acid.

   $$Pt \mid H_2(1 \text{ atm}), HA[1.0 \times 10^{-2}], A^-[1.0 \times 10^{-2}] \parallel Hg_2Cl_{2(s)}, KCl_{(s)} \mid Hg$$

5. Show that pH $= (E_{cell} - k)/0.0592$ for the antimony/antimony oxide–SCE electrode pair.

6. Using the calibration curve in Fig. 13–11, calculate the following, assuming temperatures and activity effects for all solutions are the same.

---

* Answers are listed at the end of the book for problems marked with an asterisk.

*a. If a naturally occurring water sample contains 18 ppm fluoride, what potential would be recorded on the meter?

b. If a 1-gallon water sample registers +25.0 mV on the meter calculate the amount of fluoride in the sample as ppm F and mg F?

7. The potential of a 25.0-ml aliquot of an unknown fluoride solution was tested with a fluoride electrode and registered a potential of −14.0 mV vs SCE. A 25.0-ml aliquot of a $4.15 \times 10^{-3}$ F NaF solution was added to the unknown and the potential was now −16.5 mV. Calculate the mg of fluoride in the 25.0-ml sample.

8.* A 8.172-g sample of a soil was treated with a pH 8.2 sodium acetate solution, centrifuged, and the supernate containing calcium was removed and diluted to 100.0 ml. A 50-ml aliquot of this registered a potential of +20.0 MV on a calcium electrode vs the SCE. A 25-ml aliquot of $3.85 \times 10^{-2}$ M was added and the new potential was 0.0 mV. Calculate the % of Ca and the mg Ca/g soil.

9. Chloride can be determined in salted tomato juice with the chloride ion-selective electrode. A 10-ml aliquot of tomato juice gave a potential of −17.2 mV. A 100-ml acidified ($HNO_3$) aliquot of $2.00 \times 10^{-3}$ M NaCl was added and the potential was −34.6 mV. Calculate the mole/liter and mg/liter of $Cl^-$ in the tomato juice.

10. The following potentiometric data were obtained for the titration of a fluoride sample (50.0 ml) with 0.115 F La ($NO_3$)$_3$ titrant at pH of 5 using a fluoride and SCE electrode pair.

| ml titrant | mV | ml titrant | mV |
|---|---|---|---|
| 0.0 | −110 | 25.0 | −50.0 |
| 5.0 | −105 | 27.0 | −40.0 |
| 10.0 | −98 | 29.0 | +5.0 |
| 15.0 | −85 | 29.5 | +30.0 |
| 20.0 | −77.0 | 30.0 | +50.0 |
| | | 35.0 | +110 |

Calculate mg and ppm F in the sample.

# Chapter
# Fourteen
# Precipitation
# Titrations

## INTRODUCTION

A precipitation titration is one in which a substance is titrated with a standard solution of a precipitating agent. At the completion of the precipitation, which is defined by the stoichiometry of the reaction, either the appearance of excess titrant or the disappearance of the reactant is detected. Detection of the stoichiometric point can be accomplished by color indicators as well as by instrumental methods. Of the latter, the principal technique is potentiometric measurement using indicator or ion-selective type electrodes.

The idea of a precipitation titration is a very old one. For example, Gay-Lussac in 1832 determined silver ion with chloride. Mohr and Volhard, whose names identify specific precipitation procedures involving silver ion and halides, made their contributions in 1856 and 1874, respectively. In the early 1900s the formation of turbidity (AgCl) in a titration procedure was successfully used to determine the atomic weight of silver, chlorine, and several metals isolated as pure metal chlorides.

## TITRATION REQUIREMENTS

The main difference between a precipitation titration and other volumetric methods is that a precipitate forms during the course of the titration. Therefore, the titration requirements include the formation of a stoichiometric precipitate as well as those typical to most volumetric procedures.

A stoichiometric relationship between the titrant and sample leading to precipitation must be possible. Even though the precipitate is not isolated the adsorptive nature of the precipitate toward the titrant can lead to large errors. For this reason many reactions involving precipitate formation, such as the titration of metal ions with a hydroxide or sulfide titrant, are not useful.

Since the titrant is added rapidly, equilibrium between the precipitate and its ions in solution must be reached quickly. A slow attainment of equilibrium will lead to an overtitration. In addition to a rapid reaction rate, it is necessary that the equilibrium constant ($K_{sp}$) be favorable. That is, the $K_{sp}$ should be small, which then indicates a low solubility for the product (the precipitate) of the titration.

The last and, perhaps, determining factor is the determination of the stoichiometric point of the titration. Although several different indicator processes are possible, in general, the best way for detecting the end point in precipitation titrations is by an instrumental technique.

Even though many precipitation reactions are known, not all are suitable for a precipitation titration. The reason being that one or more of the aforementioned requirements are not satisfied. Probably the most used precipitation titrations are the titration of halides ($Cl^-$, $Br^-$, $I^-$) and pseudohalides ($S^{2-}$, $HS^-$, $RS^-$, $CN^-$, $SCN^-$), with silver ion (or reverse) and the titration of sulfate ion with barium ion (or reverse).

## TITRATION CURVE

Assume that $AgNO_3$ is being titrated with NaCl to yield the insoluble product AgCl:

$$Ag^+ + Cl^- \rightarrow AgCl_{(s)} \tag{14-1}$$

As the titrant is added the silver ion concentration gradually decreases, since AgCl precipitates. When the stoichiometric point is reached, a stoichiometric amount of NaCl titrant has been added to the $AgNO_3$ solution. However, the silver ion concentration cannot be zero since the AgCl precipitate is in equilibrium with its ions. The amount of silver ion in solution will be determined by the solubility product for AgCl and the stoichiometric point for the titration is reached when this silver ion concentration is reached. If more NaCl titrant is added, the silver ion concentration must decrease further because of the common ion effect.

This entire process is described by a titration curve which is constructed by plotting change in silver ion concentration as a function of the volume of NaCl titrant. If $Cl^-$ is being titrated with $Ag^+$, a plot of chloride ion concentration as a function of the volume of $AgNO_3$ titrant is made.

*Example 14-1.* Calculate the change in silver and chloride ion concentration during the titration of 40.0 ml of 0.100 $F$ $AgNO_3$ with 0.100 $F$ NaCl.

*Initial pAg.* At the outset the silver ion concentration is 0.100 $F$. Therefore, the pAg is

$$pAg = -\log [Ag^+] = -\log 10^{-1} = 1$$

Since no chloride ion has been introduced at this point, the pCl cannot be defined.

*During the Titration.* At 25% titrated 10.0 ml of 0.100 $F$ NaCl has been added. Therefore, the amount of $Ag^+$ left in solution is calculated as follows:

$$\text{mmole } Ag^+ \text{ started} = 40.0 \text{ ml} \times 0.100 \text{ } F = 4.00 \text{ mmole}$$

$$\frac{\text{mmole } Cl^- \text{ added}}{\text{mmole } Ag^+ \text{ left}} = \frac{10.0 \text{ ml} \times 0.100 \text{ } F}{50.0 \text{ ml}} = \frac{1.00 \text{ mole}}{3.00 \text{ mmole}}$$

$$[Ag^+] \cong C_{Ag^+} = \frac{3.00 \text{ mmole}}{50.0 \text{ ml}} = 0.0600 \text{ } M$$

$$pAg = -\log[0.0600] = 1.22$$

In the calculation it should be noted that the concentration used for the silver ion does not contain that minute amount contributed by the solubility of AgCl. This is a reasonable approximation since the silver ion concentration from the AgCl is negligible in comparison to the amount of silver ion (0.0600 $M$) that is in excess of the chloride ion.

For the calculation of pCl, the common ion effect must be considered, since there is an excess of silver ion in the solution.* Therefore,

$$K_{sp} = 1.8 \times 10^{-10} \ (M)^2 = [Ag^+][Cl^-]$$

$$[Cl^-] = S$$

$$[Ag^+] \cong C_{Ag^+} = 0.0600 \text{ } M$$

$$1.80 \times 10^{-10} = [0.0600]S$$

$$S = 3.00 \times 10^{-9} M$$

$$pCl = -\log [3.00 \times 10^{-9}] = 8.52$$

Again, the $Ag^+$ from the solubility of AgCl is not considered as part of the silver ion concentration. If it were, the following would be written:

$$Ag^+ = 0.0600 + S$$

and upon insertion into the solubility product expression the chloride concentration is given by

$$1.80 \times 10^{-10} \ (M)^2 = [0.0600 + S]S$$

---

* It should be noted that the equation

$$pK_{sp} = pAg + pCl$$

can be derived from the $K_{sp}$ expression. Hence, at any point in the titration, if pAg is calculated pCl can be calculated (or vice versa) by this equation.

Solving by the quadratic expression still leads to $3.00 \times 10^{-9}\ M$ for chloride concentration. Hence, the approximation is reasonable.

Other points during the titration would be calculated in the same manner. However, the approximation is the least precise for pCl in the early part of the titration and for pAg just before reaching the equivalence point.

*Stoichiometric Point.* The stoichiometric point is reached when 40.0 ml of NaCl is added to the solution. At this point the solution is a saturated solution of AgCl in contact with solid AgCl. Consequently, the concentration of silver and chloride ion in solution can be calculated from the solubility product.

$$[Ag^+] = S = [Cl^-]$$

$$1.8 \times 10^{-10}\ (M)^2 = S \cdot S$$

$$S = 1.34 \times 10^{-5}\ M$$

$$pAg = pCl = -\log[1.34 \times 10^{-5}] = 4.87$$

*After the Stoichiometric Point.* When 50.0 ml of NaCl is added, excess NaCl is being introduced into the solution. The silver ion concentration in the solution is, therefore, influenced by the common ion effect.

The chloride ion concentration and pCl is calculated by

$$\text{mmole of } Cl^- \text{ added} = 50.0 \times 0.100\ F = 5.00 \text{ mmole}$$

$$\underline{\text{mmole of } Ag^+ \text{ initial} = 40.0 \times 0.100\ F = 4.00 \text{ mmole}}$$

$$\text{mmole of } Cl^- \text{ in excess} \quad 90.0 \text{ ml} \qquad 1.00 \text{ mmole}$$

$$[Cl^-] \cong C_{Cl^-} = \frac{1.00 \text{ mmole}}{90.0 \text{ ml}} = 0.0111\ M$$

In this calculation the chloride ion concentration contributed by the solubility of the AgCl is considered negligible. Such an approximation can be shown to be reasonable and only at points just past the stoichiometric point is this approximation subject to concern.

$$pCl = -\log 1.11 \times 10^{-2} = 1.95$$

For pAg

$$[Ag^+] = S$$

$$[Cl^-] = C_{Cl^-} \cong 0.0111\ M$$

A more precise expression for $[Cl^-]$ is

$$[Cl^-] = C_{Cl^-} = 0.0111 + S$$

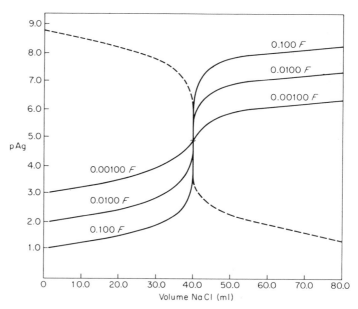

**Fig. 14–1.** Calculated titration curves for the titration of 40.0 ml of 0.100 $F$ AgNO$_3$ with 0.100 $F$ NaCl. Effect of dilution is shown by assuming 0.0100 and 0.00100 $F$ conditions for titrant and sample. The dotted titration curve is a plot of pCl vs volume of AgNO$_3$.

Substituting into the $K_{sp}$ expression and using the approximation for [Cl$^-$] gives

$$1.8 \times 10^{-10} \ (M)^2 = S[0.0111]$$

$$S = 1.62 \times 10^{-8} \ M$$

$$pAg = -\log [1.62 \times 10^{-8}] = 7.79$$

A typical calculated titration curve is plotted in Fig. 14–1. In the vicinity of the stoichiometric point the silver ion concentration changes very rapidly. Since the stoichiometric point concentration is determined only by the solubility product, this point is independent of concentrations employed in the titration. However, other parts of the curve are affected by the concentration level of titrant and sample. This effect is also illustrated in Fig. 14–1 where 0.01 $F$ and 0.001 $F$ titrant and sample are used. As the concentration of the system decreases, the abrupt change in the silver ion concentration becomes less pronounced. With this decrease in abrupt change, it should be expected that the detection of the end point will become more difficult.

The size of the titration break will also be affected by the solubility product. As the solubility of the precipitate formed during the titration decreases

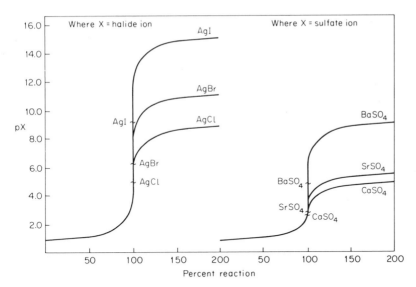

**Fig. 14–2.** Predicted titration curves for several titrations. Although not shown in the figure the slope of the break will become steeper as the solubility of the product of the reaction increases.

| Ion | Titrant | $K_{sp}$ |
|-----|---------|----------|
| $Cl^-$ | $Ag^+$ | $1.8 \times 10^{-10}$ |
| $Br^-$ | $Ag^+$ | $5.2 \times 10^{-13}$ |
| $I^-$ | $Ag^+$ | $8.3 \times 10^{-17}$ |
| $SO_4^{2-}$ | $Ca^{2+}$ | $1.2 \times 10^{-7}$ |
| $SO_4^{2-}$ | $Sr^{2+}$ | $3.2 \times 10^{-7}$ |
| $SO_4^{2-}$ | $Ba^{2+}$ | $1.3 \times 10^{-10}$ |

(smaller $K_{sp}$), the size of the break will increase. This is illustrated in Fig. 14–2 where a series of theoretical titration curves as a function of $K_{sp}$ are shown. From these curves it can be concluded, provided there are no kinetic or side reaction problems, that each of the halides can be quantitatively titrated with silver ion, sulfate can be quantitatively titrated with barium ion, semiquantitatively with strontium ion, and not quantitatively with calcium ion. It can also be predicted that excellent success is possible in differentiating mixtures of $I^-$–$Br^-$ and $I^-$–$Cl^-$ and reasonable success for $Br^-$–$Cl^-$ using a silver titrant.

**End-Point Detection.** Three classical methods, known as the Mohr, Volhard, and Fajans method, utilize color indicators for the end point. In general, these three methods involve a silver ion halide or pseudohalide titration. Thus, the selection of chemical reagents as indicators are for these reactions and cannot be arbitrarily applied to other precipitation titrations. However, the basis of end-point detection in these three methods, specifically, formation of a colored precipitate and formation of a color homogeneously or on a surface, can be applied to other methods.

**Mohr Method.**     In the Mohr method a colored precipitate forms at the end point. For the titration of chloride ion with silver ion, the reactions are

$$\text{Titration:} \qquad Ag^+ + Cl^- \rightarrow AgCl_{(s)} \qquad K_{sp} = 1.8 \times 10^{-10} \qquad (14\text{-}2)$$

$$\text{End point:} \qquad 2Ag^+ + CrO_4^{2-} \rightarrow Ag_2CrO_4 \qquad K_{sp} = 1.1 \times 10^{-12} \qquad (14\text{-}3)$$

Sodium chromate, which acts as the indicator, is added to the chloride sample solution. The chromate ion, if it is to be a useful indicator, must not react with silver ion until the stoichiometric point silver ion concentration is reached. In other words, the silver ion must precipitate the chloride ion first followed by the formation of the red $Ag_2CrO_4$. This is predicted by comparison of the $K_{sp}$ for AgCl and $Ag_2CrO_4$. Unfortunately, the entire titration is affected by contrasting colors. Chromate ion is yellow, AgCl precipitate is white, and the $Ag_2CrO_4$ precipitate is red. Thus, to see a red color over the yellow color, the eye requires that the $CrO_4^{2-}$ concentration be small. Because of these competing factors the Mohr procedure includes a titration error which must be compensated for. Since other end-point detection systems are available, the Mohr method is not used at present as extensively as in the past. In general, the Mohr method can be applied to the determination of chloride and bromide but not for iodide and thiocyanate.

**Volhard Method.**     In the Volhard method, silver ion is titrated directly with a standard thiocyanate solution and the end point is indicated by the formation of a soluble, highly colored complex. Ferric ion, which serves as the indicator, is added to the system. Under the correct conditions and as soon as excess $SCN^-$ is added, a deep red coloration forms due to complexation of Fe(III).

$$\text{Titration:} \quad Ag^+ + SCN^- \rightleftharpoons Ag\,SCN \qquad (14\text{-}4)$$

$$\text{End point:} \quad Fe^{3+} + SCN^- \rightleftharpoons \underbrace{Fe\,(SCN)^{2+}}_{red} \qquad (14\text{-}5)$$

Other iron–thiocyanate complexes can form but require greater $SCN^-$ concentration than is present at the end point. The equilibrium constant for reaction (14-5) is not particularly favorable. Fortunately, the color for $Fe(SCN)^{2+}$ is very intense and, thus, small concentrations of this complex are readily detected. It has been estimated experimentally that the minimum concentration of $Fe(SCN)^{2+}$ in solution that is detectable by the eye is about $6.4 \times 10^{-6} \ M$.

Halides can also be titrated. However, the procedure for their analysis is based on an indirect titration. A measured excess of standard $Ag^+$ solution is pipeted into a solution of the halide sample. Subsequently, the remaining $Ag^+$ is then titrated with standard $SCN^-$ using $Fe^{3+}$ as the indicator.

If the solubility product for AgSCN and the silver halides are compared, it is concluded that AgCl is the most soluble.

$$K_{sp}(AgSCN) = 1.0 \times 10^{-12} \quad K_{sp}(AgBr) = 5.2 \times 10^{-13}$$

$$K_{sp}(AgCl) = 1.8 \times 10^{-10} \quad K_{sp}(AgI) = 8.3 \times 10^{-17}$$

Hence, in the indirect titration of $Cl^-$, some of the AgCl precipitate will dissolve in favor of the formation of AgSCN precipitate upon titrating the excess $Ag^+$ with $SCN^-$. For the indirect titration of $Br^-$ and $I^-$, no error should be expected since these are more insoluble as silver salts in comparison to AgSCN.

The Volhard method can still be used, however, for the titration of chloride ion providing one of several different experimental techniques are employed. In the first, the precipitated AgCl is filtered out of the solution leaving the excess standard silver ion behind. Once the AgCl precipitate is removed, the titration with standard $SCN^-$ solution is straightforward.

A second useful technique is to add nitrobenzene to the solution containing the AgCl–excess $Ag^+$. The nitrobenzene coats the AgCl precipitate and the net effect is as if the precipitate had been filtered away. Upon titration with standard $SCN^-$ solution, the $SCN^-$ is unable to penetrate the thin nitrobenzene layer on the AgCl. Since filtration is not involved, this technique is generally faster. If filtration is to be used, some digestion is usually required in order to increase the size of the AgCl crystals and improve its filtering characteristics.

The last experimental modification that can be employed is to use a large excess of $Fe^{3+}$. It has been found that if the $Fe^{3+}$ is at about $0.7 \ F$, satisfactory titrations and stable end points are obtained. By using very large $Fe^{3+}$ concentrations, the $SCN^-$ finds itself being involved in two competing equilibria: (1) dissolving the silver chloride to form silver thiocyanate precipitate, and (2) being complexed by the $Fe^{3+}$. The absence of a titration error can be shown through calculation.

Other anions that precipitate as silver salts can be determined by the Volhard method. However, successful analysis is possible only if the silver salt of the anion being determined is more insoluble than the AgSCN precipitate or if the procedure is modified to overcome a reverse solubility.

**Fajans Method.** The indicator action in the Fajans method is based on the appearance or disappearance of a color on the surface of the precipitate.

Since the process involves adsorption or desorption of the indicator, the indicators are often called adsorption indicators.

An adsorption indicator will have acid–base properties and participates in the equilibrium

$$HIN + H_2O \rightleftharpoons H_3O^+ + IN^- \tag{14-6}$$

The direction of this equilibrium will be determined by the pH of the solution and the $K_a$ for the adsorption indicator. At the optimum experimental conditions, fluorescein, which is the most frequently used indicator for the $Ag^+$–$Cl^-$ titration, will be in the solution in the $IN^-$ form.

In the Fajans method chloride ion is titrated with a standard $Ag^+$ solution. Prior to reaching the stoichiometric point, the surface of the AgCl precipitate will have a primary layer of chloride ion and therefore have a negatively charged surface.

$$\text{(AgCl)} + Cl^- = \text{(AgCl)} Cl)^-$$

Thus, the solution takes on the greenish-yellow fluorescent color of the anion form of fluorescein since it is not adsorbed on the AgCl precipitate.

As the stoichiometric point is passed, the surface charge changes since the AgCl precipitate is in the presences of silver ions.

$$\text{(AgCl)} + Ag^+ = \text{(AgCl)} Ag)^+$$

Immediately, the fluorescein anion acts as the counterion and is adsorbed on the AgCl surface.

$$\text{(AgCl)} Ag)^+ IN^-$$

This causes the precipitate to take on a pinkish-red color.

The indicator action is reversible. If enough chloride ion solution is added to the above solution to go back to the other side of the stoichiometric point, the pinkish-red adsorbed layer gives way to the formation of the greenish-yellow solution. In addition, the process is clearly a surface phenomenon involving adsorption, since the solubility of the silver salt of the fluorescein anion is not exceeded.

A successful use of an adsorption indicator requires an experimental procedure which promotes surface adsorption. This is in contrast to ordinary gravimetry where every attempt is made to minimize surface adsorption. Consequently, several factors favoring indicator adsorption can be cited.

1. The precipitate should be of small particle size and in a highly dispersed state. Ideally, the suspended particles should stay in this state uniformly throughout the titration. Often compounds are added which help in maintaining a highly dispersed condition. In the $Ag^+$–$Cl^-$ titration addition of dextrin serves this purpose. The presence of electrolytes at high concentration will tend to promote coagulation of the dispersed particles.

2. Principal primary ions involved in the adsorption should be the same kind of ions that make up the precipitate.

3. The adsorption indicator should be strongly adsorbed.

4. The salt formed between the indicator anion (or cation) and primary ion should be sufficiently soluble.

5. Since the adsorption indicator is a weak acid or base the pH of the solution and the $K_a$ of the indicator will determine the concentration of the adsorbing form of the indicator. For this reason the pH must be carefully controlled during the titration.

It is possible to determine sulfate by titration with barium ion in which the end point is detected by an adsorption indicator. At a pH of 3.5 and in water–alcohol solvent mixture the barium sulfate that is formed is a highly adsorptive precipitate that tends to remain in a suspended state. The titration before the stoichiometric point can be represented by

$$Ba^{2+} + SO_4^{2-} \rightarrow \boxed{(BaSO_4}\; SO_4^{2-}) \; Na^+$$

and after the stoichiometric point by

$$Ba^{2+} + BaSO_4 \rightarrow \boxed{(BaSO_4}\; Ba^{2+}) \; 2X^-$$

Two indicators that can be used are Alizarin red S (**I**) and Thorin (**II**). Each one will function as an anionic indicator which replaces $X^-$ in the above reaction. The end point, which is easily seen, changes from a yellow solution to a pink coloration on the surface of the precipitate.

I                                                     II

The main emphasis in this discussion has been the titration of halide ion with silver ion, or the reverse, by the Mohr, Volhard, and Fajans method. A variety of other precipitation reactions are also useful in analysis. Several of these are listed in Tables 14–1 and 14–2. Closely related reactions are those in which the product that is formed is undissociated but remains in solution. These systems can be treated in much the same way that precipitation reactions are handled. However, in calculations the equilibrium constant that is required is not a solubility product but a dissociation constant.

Even though there are indicators available for the various silver titrations and titrations listed in Tables 14–1 and 14–2 the indicator itself is one of the main limitations in precipitation titrations. The indicator must undergo a visible change when a certain pX is reached. If the change occurs too soon

Table 14–1.  Application of Argentometric Precipitation Methods[a]

| Ions | Remarks |
|---|---|
| Volhard | |
| $AsO_4^{3-}$, $Br^-$, $I^-$, $CNO^-$, $SCN^-$ | In presence of Ag salt |
| $CO_3^{2-}$, $CrO_4^{2-}$, $CN^-$, $Cl^-$, $C_2O_4^{2-}$, $PO_4^{3-}$, $S^{2-}$ | Removal of silver salt is required |
| $BH_4^-$ | Involves reduction of $Ag^+$ to Ag |
| $K^+$ | Based on reaction with $NaB(C_6H_5)_4$ |
| Mohr | |
| $Br^-$, $Cl^-$ | Direct titration using $NaCrO_4$ |

[a] Instrumental methods of detecting the end point are not included.

Table 14–2.  Absorption Indicators[a]

| Indicator | Ion | Titrated with |
|---|---|---|
| Alizarin red S | $Fe(CN)_6^{4-}$, $MoO_4^{2-}$ | $Pb(NO_3)_2$ |
| | $F^-$ | $Th(NO_3)_4$ |
| Bromophenol blue | $SCN^-$, $Cl^-$, $Br^-$, $I^-$ | $AgNO_3$ |
| | $Br^-$ | $Hg_2(NO_3)_2$ |
| | $Hg_2^{2+}$ | $CNS^-$, $Cl^-$, or $Br^-$ |
| Congo red | $SCN^-$, $Cl^-$, $Br^-$, $I^-$ | $AgNO_3$ |
| Dibromofluorescein | $HPO_4^{2-}$ | $Pb(OAc)_2$ |
| Dichlorofluorescein | $Cl^-$, $Br^-$, $I^-$ | $AgNO_3$ |
| | $BO_2^-$ | $Pb(OAc)_2$ |
| Diphenylcarbazone | $CN^-$, $Br^-$, $SCN^-$ | $AgNO_3$ |
| Solochrome red B′ | $Mo_7O_{24}^{6-}$ and $Fe(CN)_6^{4-}$ | $Pb^{2+}$ |
| Tartrazine | $Ag^+$ | $SCN^-$, halides |
| Tetrachlorofluorescein | $Cl^-$ and $Br^-$ | $Ag^+$ |
| Bromocresol purple | $SCN^-$ | $Ag^+$ |
| Bromothymol blue | $SCN^-$ | $Ag^+$ |
| Eosine | $Br^-$, $I^-$, $SCN^-$, | $Ag^+$ |
| Fluorescein | Halides, $SCN^-$, and $Fe(CN)_6^{4-}$ | $Ag^+$ |
| | $SO_4^{2-}$ | $Ba(OH)_2$ |
| | $C_2O_4^{2-}$ | $Pb(OAc)_2$ |
| Diiododimethylfluorescein | $I^-$ | $Ag^+$ |
| Dibromofluorescein | $SCN^-$, $Cl^-$, $Br^-$, and $I^-$ | $Ag^+$ |
| Dimethylfluorescein | $Cl^-$ | $Ag^+$ |
| Rose bengal | $I^-$ | $Ag^+$ |
| Diphenylamine blue | $Cl^-$, $Br^-$ | $AgNO_3$ |

[a] Taken from D. D. Perrin, "Organic Complexing Reagents," p. 132. Interscience, New York, 1964. See also J. F. Coetzee, "Treatise on Analytical Chemistry," Part I, Vol. 1, p. 784. Interscience, New York, 1959.

or too late, a titration error occurs. This error can often be accounted for in the standardization procedure. However, this requires careful control and reproducibility in handling of the reagents and experimental procedure. For these reasons more recently developed instrumental methods such as potentiometry are usually preferred for detecting the end point.

**Potentiometry.**     The stoichiometric point for many precipitation reactions can be determined potentiometrically even though there is no change in oxidation state during the titration. Ion-selective electrodes or other type indicator electrodes in combination with a reference electrode provide excellent working cells for following the titration.

Consider the titration of silver ion with chloride ion or

$$Ag^+ + Cl^- \rightarrow AgCl_{(s)}$$

As the titration is performed, the $Ag^+$ concentration will decrease in the solution. If a Ag wire is inserted into the solution the redox couple, $Ag^+/Ag$, is established and the potential at the silver wire will change as the $Ag^+$ concentration decreases according to the Nernst equation

$$E_{Ag^+,Ag} = E^{\circ}_{Ag^+,Ag} - \frac{0.0592}{1} \log \frac{1}{[Ag^+]} \qquad (14\text{-}7)$$

In order to make potentiometric measurements, a reference electrode, such as the saturated calomel electrode, is added to complete the cell. Therefore, as titrant is added the potential of the cell changes (defined by the Nernst equation) and a typical S-shaped titration curve is obtained if the cell potential is plotted vs the volume of titrant.

A typical cell arrangement for the $Ag^+$–$Cl^-$ titration is shown in Fig. 14–3.

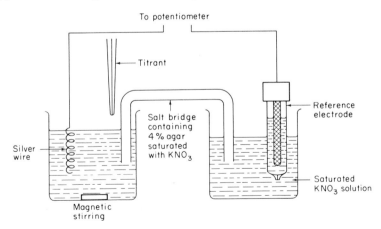

**Fig. 14–3.**   A potentiometric cell for a silver–halide titration.

For the most precise work a salt bridge is used to connect the reference compartment to the indicating component since if the SCE were to be placed directly in the sample solution, slight leakage of KCl into the solution would take place. However, for routine applications in which macro amounts of silver are being determined the SCE can be inserted directly into the sample solution. The titration curve that is obtained experimentally will follow the calculated curve shown in Fig. 14–1, providing exact potentials are measured.

The same cell can be used for the reverse titration, where chloride ion is titrated with silver ion. However, in this case the Ag wire becomes coated with AgCl as the titration is started and the potential is determined by the couple AgCl/Ag or

$$AgCl_{(s)} + 1e = Ag_{(s)} + Cl^-$$

and given by the Nernst equation

$$E_{AgCl,Ag} = E^\circ_{AgCl,Ag} - \frac{0.0592}{1} \log \frac{[Cl^-]}{1} \tag{14-8}$$

Thus, as the chloride concentration changes during the titration the cell potential (a reference electrode must be added to complete the cell) also changes and a S-shaped titration curve is obtained upon plotting cell potential vs volume of titrant.

A silver salt/silver electrode can be used potentiometrically in titrations of anions that form insoluble silver salts. For example, quantitative results can be obtained for individual titrations of $I^-$, $Br^-$, $Cl^-$, $SCN^-$, $S^-$, $SH^-$, $N_3^-$, $(C_6H_5)_4B^-$, $Fe(CN)_6^{3-}$, $Fe(CN)_6^{4-}$, $PO_4^{3-}$, $CN^-$, $C_2O_4^{2-}$ and others employing $AgNO_3$ as titrant.

Theoretically, any metal electrode should respond to a precipitation reaction in which metal ions of the electrode metal are removed from solution by precipitation. In practice, however, only a few provide the rapid, reversible response that is required for a successful potentiometric titration. In general, silver and mercury electrodes (see Chapter 10) have the widest utility.

Ion-selective electrodes, which were described in Chapter 13, are a recent development, and broaden the scope of potentiometric end-point detection in precipitation reactions. For example, by using the fluoride electrode, fluoride can be titrated in alcoholic solution with either $La(NO_3)_3$ or $Th(NO_3)_4$ to produce the precipitate $LaF_3$ or $ThF_4$ (see Fig. 13–5). Other halides can be determined individually or in mixtures using a halide ion-selective electrode and $AgNO_3$ as titrant. The lead ion-selective electrode can be used for the determination of lead where $Na_2C_2O_4$ is the titrant. If a dioxane–water solvent mixture is used, sulfate can be titrated using the sulfate ion-selective electrode with $Pb(NO_3)_2$ as titrant.

## CALCULATIONS

*Example 14-2.* A 0.1755-g dissolved sample of a silver alloy required 20.92 ml of a 0.07101 $F$ potassium thiocyanate solution (Volhard titration). Calculate the percent Ag in the alloy.

Since the reaction is

$$Ag^+ + SCN^- \rightarrow AgSCN_{(s)}$$

the reaction ratio is 1/1 and therefore

$$\%Ag = \frac{ml_{SCN} \times F_{SCN} \times \text{reaction ratio} \times Ag \times 100}{\text{wt sample, mg}}$$

$$\%Ag = \frac{20.92 \text{ ml} \times 0.07101 \text{ mmole/ml} \times 1/1 \times 107.9 \text{ mg/mmole} \times 100}{175.5 \text{ mg}} = 91.33\%$$

*Example 14-3.* A normal adult excretes in urine 75–200 mmole of chloride in a 24-hour period. There are several clinical methods for determining the chloride in urine. One of these is based on the Volhard method. An aliquot of standard silver nitrate is added to the chloride sample and the remaining silver is titrated with potassium thiocyanate.

A urine sample is collected for 24 hours, evaporated and diluted to 1000 ml in a volumetric flask. A 25.00-ml aliquot is taken and combined with 50.00 ml of 0.1241 $F$ $AgNO_3$. The remaining silver is titrated with 21.22 ml of 0.1211 $F$ KSCN. Calculate the amount of chloride excreted in 24 hours.

The reactions are

$$Cl^- + Ag^+_{(excess)} \rightarrow AgCl_{(s)}$$

$$Ag^+_{(excess)} + SCN^- \rightarrow AgSCN_{(s)}$$

where the reaction ratio for both reactions is 1/1. Therefore,

$$mgCl = (ml_{Ag} \times F_{Ag} - ml_{SCN} \times F_{SCN}) \times \text{reaction ratio} \times Cl$$

$$mgCl = (50.00 \times 0.1241 - 21.22 \times 0.1211) \times 1/1 \times 35.45$$

$$mgCl = 128.9 \text{ mg}$$

$$\text{mmole Cl} = mgCl \div Cl \text{ mg/mmole}$$

$$\text{mmole Cl} = 128.9 \div 35.45 = 3.636 \text{ mmole Cl}$$

Since 25 ml were taken, this is the number of mmole in 25 ml. The number of mmole in 1000 ml which is equal to the number of mmole in the 24-hour period is given by

$$3.636 \text{ mmole} \times \frac{1000 \text{ ml}}{25 \text{ ml}} = 145.4 \text{ mmole Cl/24 hours}$$

This method can be applied to the analysis of chloride in many physiological samples. However, if proteins are present, they must be removed first, since proteins will precipitate as silver salts.

*Example 14–4.*     A compound which has known therapeutic value was to be used as a standard and was submitted for C, H, and S analysis. The C and H was determined as described in Chapter 7. A combustion procedure, whereby a micro amount of sample was burned in a fast steam of oxygen

$$\text{organic S} \xrightarrow{\text{O}_2} \text{SO}_2 \text{ and SO}_3$$

and the oxidation products are collected in a dilute solution of $H_2O_2$ to ensure quantitative conversion to $SO_3$, was used for the S determination. The sample weighed 5.206 mg and the dissolved $SO_3$ ($SO_3 + H_2O \rightarrow H_2SO_4$) was titrated with 1.67 ml of $Ba(ClO_4)_2$ in 80% alcohol using the adsorption indicator Alizarin Red S. If a 9.00-ml aliquot of 0.01020 $F$ $H_2SO_4$ required 9.05 ml of the $Ba(ClO_4)_2$ for a stoichiometric titration under the same conditions, calculate the %S in the sample

$$Ba(ClO_4)_2 + H_2SO_4 \rightarrow BaSO_{4(s)} + 2HClO_4$$

$$F_{Ba(ClO_4)_2} \times ml_{Ba(ClO_4)_2} = F_{H_2SO_4} \times ml_{H_2SO_4} \times \text{reaction ratio}$$

$$F_{Ba(ClO_4)_2} \times 9.05 \text{ ml} = 0.01020 \, F \times 9.00 \text{ ml} \times 1$$

$$F_{Ba(ClO_4)_2} = 0.01014 \, F$$

$$\%S = \frac{F_{Ba(ClO_4)_2} \times ml_{Ba(ClO_4)_2} \times \text{reaction ratio} \times S \times 100}{\text{wt sample}}$$

$$\%S = \frac{0.01014 \, F \times 1.67 \text{ ml} \times 1 \times 32.06 \text{ mg/mmole} \times 100}{5.206 \text{ mg}} = 10.43\%$$

# Questions

1. What advantages does a precipitation titration offer?
2. What limitations does a precipitation titration offer?
3. What are the requirements that must be satisfied for a successful titration?
4. Show that

$$pK_{sp} = 3pCa + 2pPO_4$$

5. Explain how the common ion effect influences the titration break in a precipitation titration.
6. Which precipitation titration gives a bigger titration break. The second component is the titrant.

    a. $I^-$–$Ag^+$ or $Cl^-$–$Ag^+$
    b. $F^-$–$Ca^{2+}$ or $F^-$–$Pb^{2+}$

c. $CrO_4^--Pb^{2+}$ or $CrO_4^{2-}-Ag^+$

d. $F^--Ag^+$ or $F^--Cd^{2+}$

7. Using solubility products explain why the Mohr method is suitable for the determination of chloride ion but not iodide ion.

8. Explain how an adsorption indicator works.

9. What experimental techniques are used to improve the quality of adsorption indicators in end-point detection?

10. Explain why the pH of the solution can be lower when using dichlorofluorescein in comparison to fluorescein in the Ag–Cl titration.

11. What is a differential precipitation titration? List an example.

12. Lead carbonate has a $K_{sp}$ of $5.6 \times 10^{-14}$. Is $Na_2CO_3$ a good titrant for the titration of lead? Explain your answer.

# *Problems*

1. Calculate the pAg for each of the following. Assume that volumes are additive.

   *a. 25.4 ml of 0.141 $F$ AgNO₃ and 60.0 ml of 0.0854 F NaBr are mixed.

   b. 35.1 ml of 0.215 $F$ AgNO₃ and 41.6 ml of 0.109 $F$ NaBr are mixed.

   c. 41.1 ml of 0.121 $F$ AgNO₃ and 41.1 ml of 0.121 $F$ Na₃PO₄ are mixed.

2.* Calculate the pPb²⁺ and pIO₃⁻ for a solution prepared by mixing 25.0 ml of 0.100 $F$ Pb(NO₃)₂ and 25.0 ml of 0.100 $F$ NaIO₃.

3.* Calculate the pBa²⁺ and pSO₄²⁻ for the addition of 0, 10.0, 20.0, 40.0, 50.0, 60.0, and 70.0 ml of 0.100 $F$ BaCl₂ is added to 50.0 ml of 0.100 $F$ Na₃PO₄. Plot the data.

4. Calculate the pCa²⁺ and pF⁻ for the addition of 0, 10.0, 20.0, 40.0, 50.0, 60.0, and 70.0 ml of 0.050 $F$ Ca(NO₃)₂ to 50.0 ml of 0.100 $F$ NaF. Plot the data.

5.* Calculate the mg of silver left in solution after 80.0 ml of 0.125 $F$ KCl are added to 45.0 ml of 0.100 $F$ AgNO₃.

6.* Twenty (20.00) ml of a KCl solution yield a 0.2311-g AgCl precipitate. Calculate the formality of the KCl solution.

7. What is the formality of a KSCN solution if 60.00 ml require 37.15 ml of 0.1155 $F$ AgNO₃?

8.* Bismuth can be separated from several interferences by precipitation as BiOCl. A bismuth sample weighed 2.405 g. After precipitation of the BiOCl it was redissolved in HNO₃. To this was added 30.00 ml of 0.09555 $F$ AgNO₃ and the remaining silver was titrated with 14.15 ml of 0.08511 $F$ KSCN by the Volhard method. Calculate the percent Bi₂O₃ in the sample.

---

* Answers are listed at the end of the book for problems marked with an asterisk.

9. Zinc can be titrated with potassium ferrocyanide according to the reaction

$$3Zn^{2+} + 2K_4Fe(CN)_6 \rightarrow Zn_3K_2[Fe(CN)_6]_{2(s)} + 6K^+$$

If a 50.0-ml aliquot of a $Zn(NO_3)_2$ solution was taken and 18.10 ml of a 0.1310 $F$ $K_4[Fe(CN)_6]$ solution was required for the titration, calculate the mg of Zn per ml of solution.

10. Serum chloride can be determined by the Volhard method. A 2.00-ml serum sample was treated with 3.525 ml 0.1110 $F$ $AgNO_3$ and the remaining silver ion was titrated with a microburet with 1.802 ml of 0.09521 $F$ NaSCN using $Fe^{3+}$ as indicator. Calculate the mg of Cl per ml of serum.

11.* A 15.00-ml aliquot of a sodium chloride brine solution was titrated with 53.62 ml of 0.6300 $F$ $AgNO_3$. Calculate the weight of NaCl in g/liter for the brine.

12. A procedure for the argentimetric titration of malathion (330.4), **I** (insecticide), in technical grade product has recently been described. The reactions are

$$CO_2CH_2CH_3$$
$$|$$
$$HCSP(S)(OCH_3)_2 + 3KOH \rightarrow (CH_3O)_2P(S)SK + \begin{matrix} CHCO_2K \\ || \\ CHCO_2K \end{matrix} + 2CH_3CH_2OH + H_2O$$
$$|$$
$$CH_2CO_2CH_2CH_3$$
$$\mathbf{I}$$

$$(CH_3O)_2P(S)S^- + Ag^+ \rightarrow (CH_3O)_2P(S)SAg_{(s)}$$

If the sample weighed 1.0000 g and 26.19 ml of 0.1050 $F$ $AgNO_3$ was required, calculate the percent malathion in the sample.

13.* A piece of bone weighing 6.4152 g was ashed and the residue was dissolved and diluted to exactly 100 ml. A 50.0-ml aliquot was taken and titrated with $Th(NO_3)_4$, using a fluoride ion-selective electrode and a saturated calomel electrode. The end point was determined from the titration curve to be 6.53 ml of the $Th(NO_3)_4$, which was 0.08411 $F$. Calculate the mg of F in the bone.

14. In the micro determination of organofluorine the sample is fused with either Na or K metal in a specially designed nickel bomb. Subsequent workup produces a solution of inorganic fluoride. If the sample weighed 6.428 mg and after fusion 4.118 ml of 0.004156 $F$ $Th(NO_3)_4$ was required in a microtitration using a fluoride ion-selective electrode and a saturated calomel electrode for end-point detection, calculate the %F in the organic compound.

# Chapter
# Fifteen
# Complexes in
# Analytical Chemistry:
# Complexometric Titrations

## INTRODUCTION

A complex ion is one in which part or all of the coordination positions are occupied. In general, only in a gas phase at a high temperature is it possible for a metal ion to exist in a simple uncoordinated state. The instant the metal ion is dissolved in a solvent, a solvent sheath (solvation) forms around the metal ion and its anion by occupying the coordination positions. The extent of solvation and number of coordinated solvent molecules will be determined by the type of metal ion and solvent.

If the colors of dilute solutions of $Cu(ClO_4)_2$, $CuSO_4$, and $CuCl_2$ are compared, it is observed that the shade of blue is different for the three $Cu^{2+}$ solutions. It must be concluded that the $Cu^{2+}$ in the three solutions is coordinated differently and coordination between $Cu^{2+}$ and the anions must occur to some degree.

A more definitive description of a complexation reaction in solution would, therefore, be one in which one or more of the solvent molecules in the coordination sphere are replaced by another group or

$$M(H_2O)_x^{n+} + L^{m-} \rightleftharpoons [M(H_2O)_{x-1}L]^{n-m} + H_2O \qquad (15\text{-}1)$$

where L can be either a molecule or a charged ion. The remaining aquo groups in reaction (15-1) may be successively replaced by L to produce a series of complexes, such as $[M(H_2O)_{x-2}L_2]^{n-2m}$, $[M(H_2O)_{x-3}L_3]^{n-3m}$, etc.

In reaction (15-1), the metal ion in the complex is called the central ion and the groups bound to the central ion are ligands (also referred to as com-

**Table 15–1.   Examples of Multidentate Ligands**[a]

$$H_2NCH_2CH_2NH_2$$
Ethylenediamine
Bidentate

$$H_2NCH_2CH_2NHCH_2CH_2NH_2$$
Diethylenetriamine
Tridentate

Triethylehetetraamine
Tetradenate

Ethylenediaminetetraacetic acid
Hexadentate

[a] Atoms in the ligand that are involved in coordination are indicated by arrows.

plexing agents or coordinating groups*). The maximum number of ligands that can coordinate to the central ion is given by the coordination number for the central ion. Each metal ion will have its own characteristic coordination number. Frequently, it will have more than one possible coordination number. (Coordination in crystallography represents the number of surrounding neighbors and may or may not be equal to the coordination number exhibited in complexes.)

Ligands which are attached to the central ion at only one point are called unidentate ligands. The vast majority of these are inorganic and typical examples are $H_2O$, $NH_3$, and the halides. If each ligand has two or more coordinating sites, the ligands are called multidentate ligands. These are usually organic molecules or ions. A molecule or ion with two coordinating sites is bidentate, with three it is tridentate, and with four it is tetradentate. For five and six the prefixes penta- and hexa- would be used, respectively. Several examples are shown in Table 15–1.

In the multidentate type systems, the coordination results in the formation of rings. This particular type of coordination compound is called a chelate and the ligands involved in the coordination are chelating agents.

It is possible for a coordination compound to contain two or more central ions. In these systems some of the ligands will act as coordinating bridges and the resulting coordination compound is called a polynuclear complex.

Each type of coordination exhibits a certain kind of geometry. In analytical chemistry coordination numbers 2, 4, and 6 are the most common and most

---

* These terms will be used interchangeably in this text.

**Table 15–2. Coordination and Geometry of Complexes of Interest to the Analytical Chemist**

| Coordination number | Type | Description | Configuration | Examples |
|---|---|---|---|---|
| 2 | Linear | Two ligands at opposite ends of an axis passing through the central ion | X—M—X | $[H_3N—Ag—NH_3]^+$ |
| 4 | Tetrahedral | Four ligands at the corners of a tetrahedron with the central atom located at the center | (tetrahedral diagram with M and X) | $[FeCl_4]^-$ |
| 4 | Square planar | Four ligands at the corners of a square whose plane contains the central atom | (square planar diagram with M and X) | Ni complex of di-methylglyoxime |
| 6 | Octahedral | Four ligands are at the corners of a square, one is above and one is below the plane with the central atom at the center of the octahedron | (octahedral diagram with M and X) | Metal complexes of ethylenediamine-tetraacetic acid |

useful. Table 15–2 lists complexes illustrating these three coordination numbers and the type of geometry possible for each coordination.

In general, <u>significant properties</u> of the metal ion and ligand which influence coordination are the following.

<u>1.</u> *Sizes and charges.* These factors are strong influences in electrostatic bonds and forces.

<u>2.</u> *Dipole.* The value of the dipole moment will indicate the extent of charge separation in a ligand which in turn influences its ability to act electrostatically or in sharing of electrons.

<u>3.</u> *Deformability of the central ion.* In the presence of an electron field the electron structure of the central ion will be modified. Generally, increased deformability occurs as the number of inner subshell electrons increases.

<u>4.</u> *Polarizability of the ligand.* The electron structure of the ligand will be affected by the presence of an electric field provided by the metal ion and the effect increases with increasing electric field.

<u>5.</u> *Miscellaneous.* A number of factors such as steric requirements, strength of hydration, dielectric properties and others must also be considered.

Understanding of the coordination process is very important to the analytical chemist for two reasons. First, this information can be used to predict ligand structures that have desirable properties for applications in analysis. Second, many key chemical and biochemical reactions (organic and inorganic synthesis and mechanisms, enzyme reactions, metabolism reactions, drug re-

**Table 15–3.   Classification of Ligands of Analytical Interest**

### Inorganic

Neutral molecule                                    Negative ions

Unidentate

| | | |
|---|---|---|
| $NH_3$ | $CN^-$   $SCN^-$ | $F^-$ |
| $NH_2OH$ | $NO_2^-$   $CH_3COO^-$ | $Cl^-$ |
| $H_2O$ | $OH^-$   $NO_3^-$ | $Br^-$ |
| | $HCO_3^-$   $BO_2^-$ | $I^-$ |

Bidentate

| | | |
|---|---|---|
| $S_2O_3^{2-}$ | $SO_4^{2-}$ | $BO_3^{3-}$ |
| $SO_3^{2-}$ | $PO_4^{3-}$ | $C_2O_4^{2-}$ |

### Organic

Bidentate

Loss of two hydrogen ions

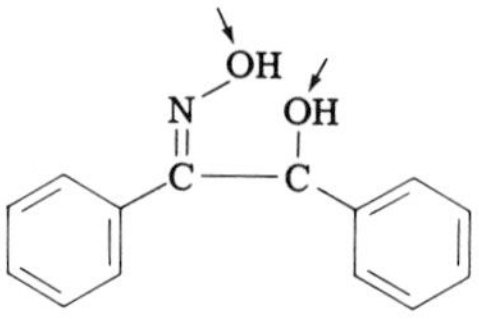

α-Benzoinoxime

$RAsO(OH)_2$

Alkyl- or aryl-
arsonic acids

Loss of one hydrogen ion

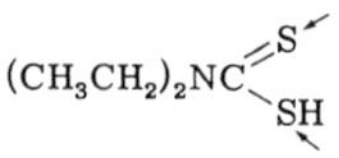

Diethyldithio-
carbamic acid

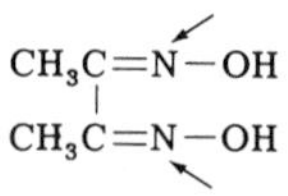

$CH_3C=N-OH$
$CH_3C=N-OH$

Dimethylglyoxime

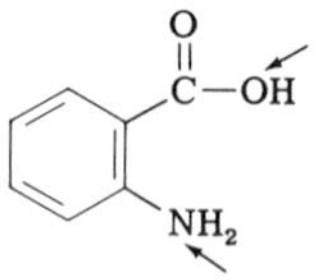

o-Aminobenzoic acid

Loss of no hydrogen ions

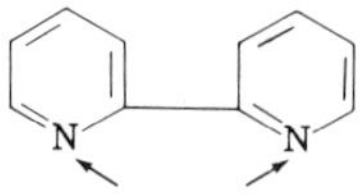

2,2′-Bipyridine

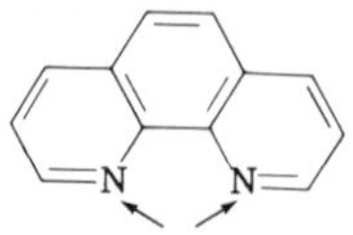

1,10-Phenanthroline

**Table 15–3.    (Continued)**

Multiringed

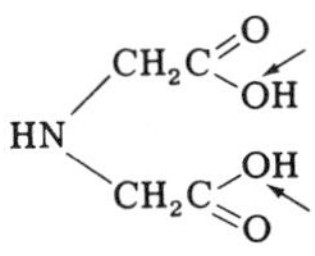

Iminodiacetic
acid

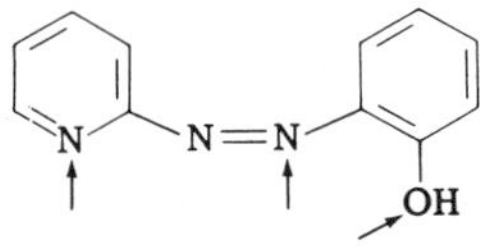

1-(2-Pyridylazo)-
2-naphthol

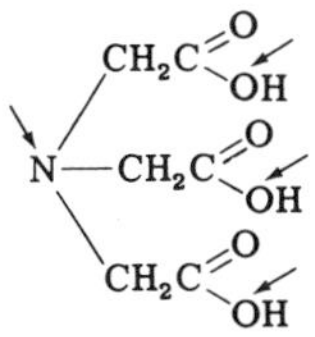

Nitrilotriacetic
acid

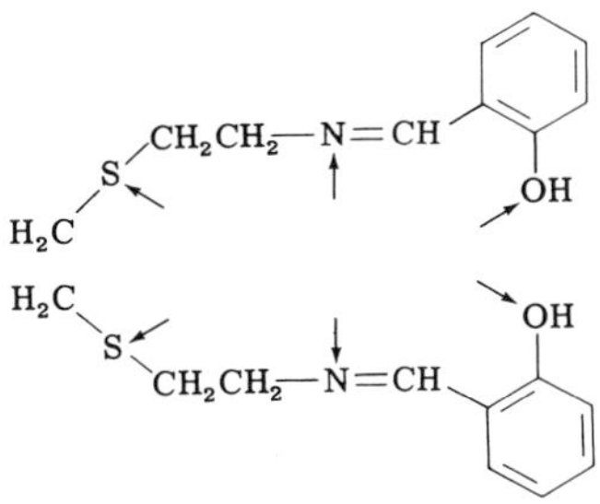

1,8-Bis(salicylidene-3,6-
amino)diethiaoctane

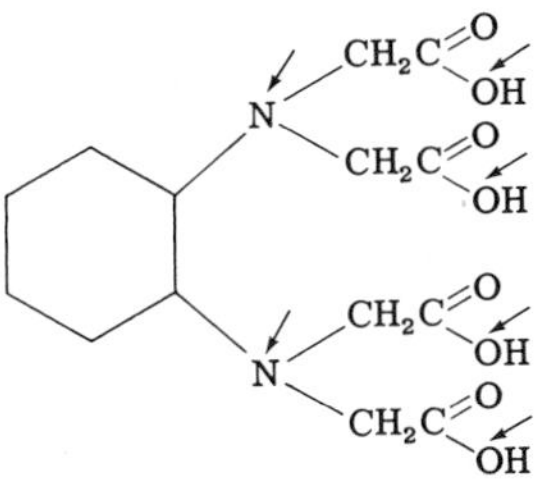

1,2-Diaminocyclohexane-
*N,N,N',N'*-tetraacetic acid

Ethylenediaminetetraacetic acid (see Table 15–1)

Salt formers[a]

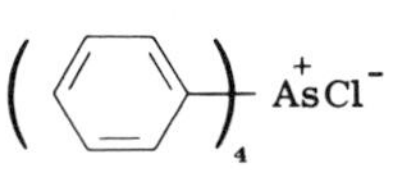

Tetraphenylarsonium
chloride

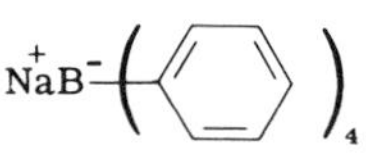

Sodium
tetraphenylboron

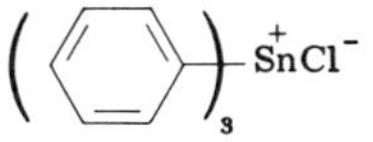

Triphenyltin
chloride

---

[a] See Table 15–1 for other examples. Coordination sites are marked by arrows. Many other ligands can be suggested. In many cases, however, the atoms involved in coordination have not been unequivocally established. It is also possible for solvent molecules to occupy coordination sites of the metal ion. Not shown are examples of dimer, trimer → oligimer formation.

lease, etc.) involve coordination and therefore, a basic knowledge in coordination chemistry is required to be able to investigate and understand these processes.

## CLASSIFICATION OF LIGANDS

There are several ways of classifying potential ligands. One obvious way is by whether the ligand is inorganic (see Table 15–3) or organic. In general, the organic ligands have a wider range of application in analytical chemistry.

Organic ligands can be divided according to whether they are salt formers or chelates. The chelates can be further divided into chelates involving the formation of one ring or multiple rings. In the one-ringed chelate group, there are ligands in which two hydrogen ions, one hydrogen ion, or no hydrogen ions are replaced in the coordination reaction. Table 15–3 lists examples of each of these classes.

## STABILITY

Understanding the formation and dissociation of complexes allows the prediction and/or calculation of the optimum experimental conditions in methods of analysis based on complexation. In addition, an accurate technique(s) must be available to determine the constants for these reactions in order to explain the behavior of new chemical systems that are influenced by complexation.

The constants, describing the formation and dissociation of complexes, are derived from the principles of equilibrium. Thus, for the formation of the complex $ML_2^{n+}$ successive equilibria and formation constant expressions can be written

$$M^{n+} + L \rightleftharpoons ML^{n+} \qquad K_{s_1} = \frac{[ML^{n+}]}{[M^{n+}][L]} \tag{15-2}$$

$$ML^{n+} + L \rightleftharpoons ML_2^{n+} \qquad K_{s_2} = \frac{[ML_2^{n+}]}{[ML^{n+}][L]} \tag{15-3}$$

where $K_{s_1}$ and $K_{s_2}$ are the stability constants (same as the formation constant in which the symbol $K_f$ is used) for the first and second step, respectively.

An overall equilibrium step and formation constant expression would be

$$M^{n+} + 2L \rightleftharpoons ML_2^{n+} \qquad K_s = \frac{[ML_2^{n+}]}{[M^{n+}][L]^2}$$

Alternatively, instability constants, $K_i$ (dissociation constants, $K_d$), which are the inverse of the stability constants, could be used.

$$ML^{n+} \rightleftharpoons M^{n+} + L \qquad K_{i_1} = \frac{1}{K_{s_1}}$$

It is possible to determine the "true" constants using activities and activity coefficients, however, in practice, this is often very inconvenient as well as very difficult. Usually the constants are determined at fixed ionic strength in moderately dilute solution which leads to the reasonable approximation of activity being equal to concentration. As pointed out previously, many complexes are of limited solubility in water and a mixed water–organic solvent or organic solvent must be used. These kinds of problems must be considered particularly when comparing stabilities of different complexes.

A number of factors related to the ligand and metal ion are major influences in the stability of complexes. These are as follows.

1. *Basicity of the ligand.* In general, stability of a series of complexes containing a basic site can be correlated to the ability of the ligand to accept a proton (Brønsted relationship); the more basic, the more stable the complex.

2. *Number of metal chelate rings per ligand.* Increasing the number of chelate rings between each ligand and metal ion leads to an increase in stability of the complex (known as the *chelate effect*).

3. *Size of chelate ring.* In general, formation of five- and six-membered rings provides the most stable complexes. Also, ligands that form chelates will be more stable than ligands that form complexes.

4. *Resonance effects.* Resonance can influence whether a five- or six-membered ring is formed and influence stability through conjugative effects.

5. *Steric effects.* This is a size factor and is therefore a combination of the spatial requirements of the ligand, the distances between the coordination sites, and the sizes of the central metal ions.

6. *Nature of the ligand.* The type of bond that is formed between the ligand and metal ion will influence stability.

7. *Nature of the metal ion.* The metal ion influences the type of binding between the ligand donor atom and the metal ion. The more the bonds tend toward electrostatic bonds, the more stable is the complex.

## KINETICS

The magnitude of the stability constant does not reveal how the system reaches equilibrium or how long it takes for equilibrium to be reached. Only after elucidating the kinetics of the reaction is it possible to discuss quantitatively these two important aspects of the complexation reaction.

The complexation reaction can involve a series of complicated steps and the time for the reaction to take place can vary over a wide range. For example, reaction (15-4) is virtually completed in the time it takes to mix the solutions, while reaction (15-5) is very slow.

$$[Cu(NH_3)_4]^{2+} + 4H^+ + nH_2O \rightleftharpoons [Cu(H_2O)_n]^{2+} + 4NH_4^+ \qquad (15\text{-}4)$$

$$[Co(NH_3)_6]^{3+} + H^+ + H_2O \rightleftharpoons [Co(NH_3)_5(H_2O)]^{3+} + NH_4^+ \quad (15\text{-}5)$$

To describe this type of rate behavior the terms *inert* and *labile* were suggested. The complex, $[Cu(NH_3)_4]^{2+}$, would be described as a labile complex while $[Co(NH_3)_6]^{3+}$ would be an inert complex. By definition a labile complex is one which undergoes a reaction in the time of mixing, assuming the experimental environment is at approximately "normal" conditions (mixing time of 1 minute, room temperature, and 0.1 $F$ solutions). An inert complex participates in reactions that are too slow to be measured or proceed at rates which can be followed by conventional methods, again assuming "normal" reaction conditions. The terms, inert and labile, should not be confused with the stability of the complex. There is no relationship between the stability constant for the formation of the complex and the complexes designation as being either labile or inert.

## ADVANTAGES AND PRACTICAL ASPECTS OF COORDINATION COMPOUNDS

Coordination compounds possess a variety of physical and chemical properties which are very useful to the analytical chemist. Several of the more significant ones are cited below.

**Production of a Characteristic Color.**    Formation of a colored coordination compound can provide evidence of the presence of a metal ion or anion. For example, if a drop of a neutral $Cu^{2+}$ solution is placed on a piece of filter paper, held over ammonia vapors, and treated with a drop of 1% alcoholic solution of dithiooxamide (rubeanic acid), **I**, a black or green coloration due to the formation of the polymeric coordination compound **II** is ob-

$$(15\text{-}6)$$

served and indicates the presence of copper. The limit of detection is 0.006 $\mu$g copper. Many other excellent spot tests based on the formation of coordination compounds are also known.

Many spot tests involving formation of a coordination compound are available for detecting the presence of an organic functional group. For example, a secondary amine can be converted to a dithiocarbamate derivative, III, which will coordinate with $Cu^{2+}$ to form a reddish-brown product that is soluble in $CHCl_3$ yielding a reddish solution.

$$R_2NH + CS_2 + NH_3 \longrightarrow R_2N-\overset{\overset{\displaystyle S}{\|}}{C}-S^- \ NH_4^+ \qquad (15\text{-}7)$$

**III**

$$2\,R_2NC\overset{\nearrow S}{\underset{\searrow S^-\ NH_4^+}{}} + Cu^{2+} \rightleftharpoons \left[ R_2NC\overset{\nearrow S\cdot}{\underset{\searrow S}{}} \cdot Cu \right]_2 + 2\,NH_4^+ \qquad (15\text{-}8)$$

Reaction (15-8) can also be used for the quantitative determination of copper since the color intensity of the solution is proportional to concentration of the coordination compound. For example, as little as 5 $\mu$g of $Cu^{2+}$ per 50 ml of $CHCl_3$ solvent can be determined using diethylammonium diethyldithiocarbamate. This particular method is typical of many coordination reactions that are routinely used for the spectrophotometric determination of inorganic cations and anions and for organic compounds. Development of color through a coordination reaction and subsequent spectrophotometric measurement accounts for one of the major applications of the coordination reaction in analytical chemistry (see Chapter 19).

Complex formation is often used for end-point detection in volumetric procedures. Several examples have already been cited. The Volhard method described in Chapter 14 is a typical example. Ferroin or related derivatives are used in oxidation–reduction titrations (see Chapter 11). This is a special case since a change in oxidation state removes the iron from the coordination compound. Coordination, to a certain degree, is probably involved in the functioning of adsorption indicators (Chapter 14). As will be described later, coordination compounds and their resulting color changes are an excellent method for detecting end points in chelometric titrations.

**Solubility.** The conversion of a substance into a coordination compound often results in a substantial change in solubility. This altered solubility is used advantageously in gravimetry (see Table 7–2) and separations.

There are numerous advantages in using organic precipitating agents in comparison to inorganic precipitating agents. Perhaps the main one is that organic precipitating agents generally offer a high degree of selectivity. In addition, the insoluble coordination compounds derived from organic precipitation agents are generally free of ionic coprecipitation and have a high molecular weight. Because of the latter property, a large precipitate weight represents a small amount of the substance being precipitated. Accordingly, weighing and filtering errors become less significant and the method can be applied to lesser amounts of sample.

If the complex is cationic, a large bulky organic anion can be used for charge neutralization. The type of anion chosen is determined by the kind of solubility property desired. Similarly, if the complex is anionic, the type of cation chosen will influence its solubility.

Converting the species of interest into a coordination compound will also alter solubility with respect to organic solvents. This is a very useful property in quantitative analysis. The coordination compound formed between $Cu^{2+}$ and diethyldithiocarbamate is insoluble. However, since this precipitate is soluble in $CHCl_3$, it is possible to determine the copper spectrophotometrically by taking advantage of the color of the complex in $CHCl_3$.

The differences in solubility can be used as a means of separating one species from another one. For example, one component of a mixture may form an insoluble coordination compound while the others form soluble coordination compounds or do not undergo a reaction with the organic precipitating agent. Similarly, the precipitating action can be used for concentrating a trace constituent.

Another useful separation technique is to allow the mixture to distribute itself between two immiscible phases. If the solvents are properly chosen, one phase will contain one coordination compound while the other phase contains the rest of the mixture in a coordinated or uncoordinated form. Often the experimental conditions such as pH, concentration levels, inert electrolyte concentration, and many others will play a significant role in determining the extent of the distribution. This technique of separation is known as solvent extraction (see Chapter 27).

**Chromatography.**      Like solvent extraction, chromatography is a very important separation method and many of its applications are based on complex formation. For example, $Fe^{3+}$ forms chloro complexes in hydrochloric acid solution.

$$Fe^{3+} + 4Cl^- \rightleftharpoons FeCl_4^- \tag{15-9}$$

Actually, five iron species, $FeCl^{2+}$, $FeCl_2^+$, $FeCl_3$, and $FeCl_4^-$, can be present in solution and the one that is dominant will be determined by the equilibrium constant for each step and by the HCl concentration. For simplicity assume reaction (15-9) to be a description of the equilibrium conditions. By increasing the hydrochloric acid concentration the iron complex is retained by the anion resin

$$RN(R')_3{}^+Cl^- + FeCl_4{}^- \rightleftharpoons RN(R')_3{}^+FeCl_4{}^- + Cl^- \qquad (15\text{-}10)$$

where R is a polymeric resin matrix. Upon reducing the hydrochloric acid concentration to below $1\,M$ HCl the iron complex dissociates (reverse of reaction (15-9) and the iron comes off the column. Different hydrochloric acid concentrations would be used for other metal ions. Consequently, by controlling the HCl concentration, metal ion mixtures can be separated.

Many other types of complexing agents can be used in chromatographic separations. Other supports besides ion exchange resins can be used and in some cases the property which accounts for the separation is a difference in solubility, while in other cases, it is a difference in volatility of the coordination compounds. These separation techniques will be discussed in Chapters 22–27.

**Selectivity.** Many ligands form stable complexes with only a few central ions. For example, dimethylglyoxime (DMG) (**IV**) forms stable, insoluble complexes with Ni(II) and Pd(II). Several other metal ions complex with DMG but they are not very stable or soluble. Thus, DMG is specific for Ni(II) and Pd(II).

Frequently, it is possible to improve the selectivity of a complexing agent by slight modification of its structure. The compound, 8-hydroxyquinoline (**V**) will precipitate Mg(II) and Al(III) as well as several other metal ions. Therefore, this reagent cannot be used for the separation of Mg(II) and Al(III). However, by putting a methyl group in the 2-position [2-methyl-8-hydroxyquinoline (**VI**)] the reagent will only complex with Mg(II). The methyl group sterically hinders the formation of the Al(III) complex.

IV          V          VI

A similar situation occurs with 1,10-phenanthroline (**VII**), which forms a colored complex with Fe(II) as well as with Cu(I). When methyl groups are placed in the 2- and 9-position (**VIII**) the reagent reacts only with Cu(I).

VII          VIII

These two reagents are used extensively for the spectrophotometric determination of iron and copper.

Another example is acetylacetone (**IX**), and its derivatives. Chelation

$$CH_3-\overset{\overset{O}{\|}}{C}-\overset{\overset{H}{|}}{CH}-\overset{\overset{O}{\|}}{C}-CH_3 \quad \rightleftharpoons \quad \overset{H\cdots}{\underset{H_3C}{\underset{}{}}} \qquad (15\text{-}11)$$

Keto form          **IX**          Enol form

occurs by replacement of the acidic hydrogen and coordination through the

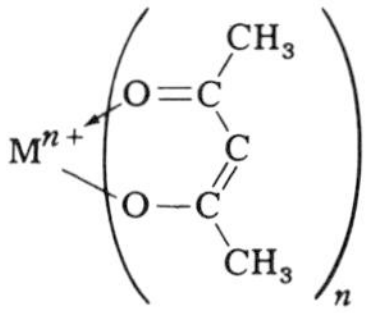

$n$ = charge on metal ion

two oxygens. This reagent has many applications and forms complexes with most transition elements.

The properties of acetylacetone can be varied extensively through structure modification. Complexes of $\alpha$-thenoyltrifluoroacetone, **X**, are similar to acetylacetone complexes except that the former are (1) more soluble in benzene and several other water-immiscible solvents and (2) form at low pH levels. The thiophene group promotes the solubility, while the highly electronegative properties of the thiophene and fluorine atoms result in the ligand being a stronger acid in comparison to acetylacetone.

If groups, such as isopropyl (**XI**) or *sec*-butyl (**XII**) are placed on carbon 3

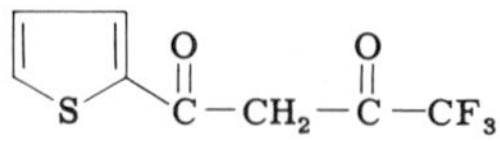

**X** 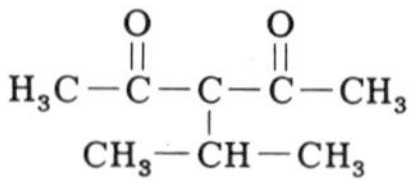          **XI** 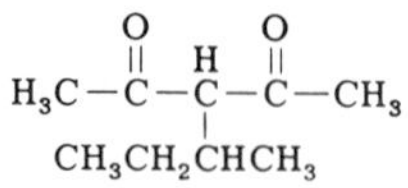          **XII**

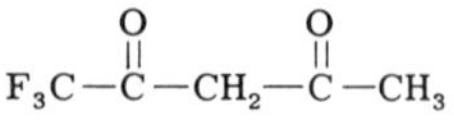

**XIII**

in acetylacetone, steric hindrance is introduced. This retards complex formation and formation of colored Fe(III) and Cu(II) complexes no longer occurs.

Insertion of several fluorine atoms into acetylacetone derivatives not only increases the acidity of the compound but also results, generally, in an increase in the vapor pressure of the complexes. This increase is volatility has been used advantageously for separations of metal ions by gas chromatography. Several reagents have been suggested for this purpose and perhaps the most useful one is trifluoroacetylacetone (**XIII**).

**Masking.**     The application of "masking agents" prevents a species other than the one being studied from participating in a reaction. In most cases masking involves the addition of a coordination compound which forms a complex with only the interferring species. The result is a reduction in concentration of the interferring ion to a level in which it no longer is subject to its usual reactions. Therefore, a separation to remove the interferences is not necessary.

The principal advantage of masking as a means of eliminating interferences is its simplicity. By adding excess masking reagent it is often possible to minimize or completely eliminate interfering effects of diverse ions. In doing this, there is no tedious expenditure of time or demand for special equipment or manipulative techniques.

It appears that the main disadvantage of applying masking is that it tends to be an art or trial and error process. Even though most masking agents are complexing agents, there are still insufficient formation constant data so that calculations of effectiveness of masking are minimal.

**Demasking.**     The process of demasking is the reverse of masking. Thus, the two processes can be represented as

$$M + L \underset{\text{demasking}}{\overset{\text{masking}}{\rightleftharpoons}} ML$$

In essence, the interference is freed so that it can be studied or analyzed.

There are two general methods of demasking. These are (1) to drastically change the hydrogen ion concentration, and (2) to effectively form a new complex or other un-ionized compound that is more stable than the masked species.

An example of the first type is the liberation of metal ions that are masked by cyanide ion. Most of the heavy metal ions form relatively stable complexes with cyanide ion. If a strong mineral acid is added, the complex breaks down in favor of the formation of hydrogen cyanide.

In some cases an increase in the pH can be used. For example, complexes of Al–oxalate, Zr–F, and Fe–SCN are decomposed if the solution is made basic. The hydrous metal oxide precipitate that is formed is filtered, which eliminates the masking agent, and then, the hydroxide salts are redissolved.

Examples of the second type are the following. Cyanide complexes can be broken by the addition of formaldehyde.

$$M(CN)_4^{2-} + 4HCHO + 4H_2O \rightarrow M^{2+} + 4HOCH_2CN + 4OH^-$$

For fluoride complexes, the metal can often be liberated by the addition of a borate salt. For example,

$$(SnF_6)^{2-} \rightleftharpoons Sn^{4+} + 6F^-$$

$$4F^- + BO_3^{3-} + 4H^+ \rightarrow BF_4^- + 3H_2O$$

Chemical destruction is a third method that can be applied. For this method to be successful, it is necessary that mild conditions be available for the reaction so that the masked species is not lost or changed.

In general, masking is potentially useful in any kind of measurement in which solutions are involved. Consequently, the scope of gravimetric, redox, chelometric, electrometric, and spectrophotometric determinations is greatly enhanced and simplified since separation procedures are often not necessary.

**Titrimetric Methods.**    Several different ligands can be used as titrants in quantitative analysis. This type of application of the coordination reaction is perhaps comparable to the importance of ligands in spectrophotometric analysis and for this reason it will be treated as a separate topic.


## COMPLEXOMETRIC  TITRATIONS *

A complexometric titration is one in which a soluble, undissociated, stoichiometric complex is formed during the addition of titrant to the sample solution. The techniques of carrying out this operation in the laboratory are typical of a volumetric titration procedure. Perhaps the key points in the general method are: (1) choosing a suitable chelating titrant; (2) choosing the experimental conditions that provides an optimum titration (this would include control of pH and the presence of competing ligands); and (3) selecting a suitable method of detecting the titration end point.

Complexometric titrations combine the advantages and limitations that complex formation and titration methods provide individually. For example, although the product of the reaction (a complex) is undissociated, it does not suffer from coprecipitation errors as in the case of precipitation titrations. The fact that a complexing agent coordinates with only certain metal ions provides selectivity. However, the stoichiometry is not always as clearly defined as in a redox, neutralization, or precipitation titration. If the complexing titrant is an organic compound, attention must be given to its solubility

---

* A complexometric titration refers to the use of any kind of ligand as a titrant while a chelometric titration implies that a chelating agent is used as a titrant.

properties. Many other advantages and disadvantages will become more clearly defined in the forthcoming discussion.

## TITRANTS

Most unidentate ligands form a series of complexes in a stepwise fashion with each step involving a formation constant. Usually, the values for the constants are small and are often very similar so that the titration reaction does not involve a clearly defined stoichiometry. Since a single stoichiometric complex is not formed, a stoichiometric point is not readily observed.

For example, ammonia might be suggested as a titrant for the determination of zinc(II) according to the reaction

$$Zn^{2+} + 4NH_3 \rightleftharpoons Zn(NH_3)_4^{2+} \qquad (15\text{-}12)$$

However, reaction (15-12) is not a true representation of what is happening in the solution, since a stepwise formation of the zinc–ammonia complexes takes place.

$$Zn^{2+} + NH_3 \rightleftharpoons Zn(NH_3)^{2+}, \quad K_{s_1} = 1.9 \times 10^2 = \frac{[Zn(NH_3)^{2+}]}{[Zn^{2+}][NH_3]}$$

$$Zn(NH_3)^{2+} + NH_3 \rightleftharpoons Zn(NH_3)_2^{2+}, \quad K_{s_2} = 2.1 \times 10^2 = \frac{[Zn(NH_3)_2^{2+}]}{[Zn(NH_3)^{2+}][NH_3]}$$

$$Zn(NH_3)_2^{2+} + NH_3 \rightleftharpoons Zn(NH_3)_3^{2+}, \quad K_{s_3} = 2.5 \times 10^2 = \frac{[Zn(NH_3)_3^{2+}]}{[Zn(NH_3)_2^{2+}][NH_3]}$$

$$Zn(NH_3)_3^{2+} + NH_3 \rightleftharpoons Zn(NH_3)_4^{2+}, \quad K_{s_4} = 1.1 \times 10^2 = \frac{[Zn(NH_3)_4^{2+}]}{[Zn(NH_3)_3^{2+}][NH_3]}$$

The formation constant for reaction (15-12) is represented by

$$K_s = K_{s_1}K_{s_2}K_{s_3}K_{s_4} = 4.25 \times 10^8$$

It would appear that a mixture of $NH_3$ and $Zn^{2+}$ in a 4 : 1 mole ratio would be far in the direction of the products. When stepwise formation is considered, a calculation at the mole ratio of four $NH_3$ to one $Zn^{2+}$ illustrates that the zinc ion is present in the solution in five forms, $Zn^{2+}$, $Zn(NH_3)^{2+}$, $Zn(NH_3)_2^{2+}$, $Zn(NH_3)_3^{2+}$, and $Zn(NH_3)_4^{2+}$. The concentration of each increases in the order listed. For example, the $Zn(NH_3)_4^{2+}$ is not quite 2 times the concentration of $Zn(NH_3)_3^+$. Consequently, a stoichiometric point is not observed at either a 1 : 1, 2 : 1, 3 : 1, or 4 : 1 ratio of ammonia to zinc(II). If the change in $Zn^{2+}$ concentration is followed as ammonia is added, a grad-

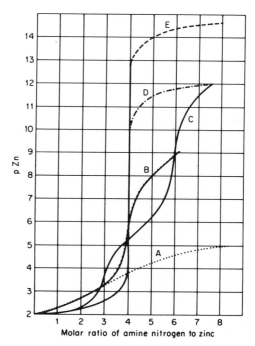

**Fig. 15–1.** Titration curves for the titration of zinc with polyamines. Titrant: (A) ammonia; (B) ethylenediamine; (C) diethylenetriamine; (D) triethylenetetramine; (E) triaminotriethylamine. [Taken from G. Schwarzenbach, *Analyst* **80**, 713 (1955).]

ual change would be experimentally observed. At no point is there a sharp, abrupt decrease in the $Zn^{2+}$ concentration (see Fig. 15–1). Most other systems employing unidentate complexing agents as titrants suffer from the same problem. In addition, many unidentate complexing agents coordinate at a rate that is too slow to be useful in a volumetric procedure.

Although organic multidentate ligands form more stable complexes than the unidentate ligands, their development as titrants was severely limited because of two factors. (1) The rates of reactions in most cases, although fast, are still not suitable for direct titration techniques. (2) Stepwise formation until the metal ion is fully coordinated will still occur. In general, the differences in stability for the individual steps are greater than for unidentate ligands but there is still difficulty in establishing an exact reaction stoichiometry.

In the mid-1940s, the polyaminocarboxylic acids and the polyamines were suggested to be useful as titrants. In general, these chelating agents are unique in that they satisfy the type of properties required for a successful titrant.

**Polyaminocarboxylic Acids.**     The polyaminocarboxylic acids combine the coordinating properties of the basic nitrogen and the carboxylate group.

The former group by itself shows a strong tendency to coordinate to Co, Ni, Cu, Zn, Cd, and other metal ions that form complexes with ammonia. The latter group, like the acetate ion, shows the tendency to coordinate to almost all metal ions. Individually, the ammonia or carboxylate group form weak complexes with metal ions. However, if the two groups are spaced properly in a single molecule in such a way that the molecule can act as a multidentate ligand with coordination to the metal ion from the nitrogen and carboxyl oxygen, a significant increase in stability is observed. Furthermore, multiple coordination through nitrogen and oxygen and the formation of five-membered rings will increase the chelate effect and improve complex stability. This was achieved by replacing the hydrogens on ammonia with acetate groups through a suitable synthetic route. Subsequent investigations with these reagents lead to the synthesis of ethylenediamine-$N,N,N',N'$-tetraacetic acid, EDTA (**XIV**).*

$$\begin{array}{c} HO_2CCH_2 \diagdown \qquad \diagup CH_2CO_2H \\ \qquad\qquad NCH_2CH_2N \\ HO_2CCH_2 \diagup \qquad \diagdown CH_2CO_2H \end{array}$$

**XIV**

$$H_4Y + H_2O \rightleftharpoons H_3O^+ + H_3Y^- \qquad K_1 = 1.00 \times 10^{-2} = \frac{[H_3O^+][H_3Y^-]}{[H_4Y]}$$

$$(15\text{-}13)$$

$$H_3Y^- + H_2O \rightleftharpoons H_3O^+ + H_2Y^{2-} \qquad K_2 = 2.16 \times 10^{-3} = \frac{[H_3O^+][H_2Y^{2-}]}{[H_3Y^-]}$$

$$(15\text{-}14)$$

$$H_2Y^{2-} + H_2O \rightleftharpoons H_3O^+ + HY^{3-} \qquad K_3 = 6.92 \times 10^{-7} = \frac{[H_3O^+][HY^{3-}]}{[H_2Y^{2-}]}$$

$$(15\text{-}15)$$

$$HY^{3-} + H_2O \rightleftharpoons H_3O^+ + Y^{4-} \qquad K_4 = 5.50 \times 10^{-11} = \frac{[H_3O^+][Y^{4-}]}{[HY^{3-}]}$$

$$(15\text{-}16)$$

---

* EDTA, also called (ethylenedinitrilo)tetraacetic acid, is commercially available under several trade names. These include Versene, Complexone-III, Nullapon, Sequestrene, Trilon B, and Idranal-III. Since EDTA has four replaceable hydrogens a convenient abbreviation representing the molecule is $H_4Y$. In this text $H_4Y$ or its various ionized forms ($H_3Y^-$, $H_2Y^{2-}$, $HY^{3-}$, $Y^{4-}$) are used in situations where importance is being assigned to the specific EDTA species in solution. On the other hand, the symbolism EDTA is used to represent the presence of the ligand where its form is implied by the necessary experimental conditions (generally, this will be the $H_2Y^{2-}$ species).

**Table 15–4.   Log Stability Constants of 1 : 1 Zinc : Ligand Complexes**

| Ligand | Structure | Log $K^e$ |
|---|---|---|
| Ammonia | $NH_3$ | 2.28 |
| Aminoacetic acid (glycine)[a] | $NH_2CH_2CO_2H$ | 5.33 |
| Iminodiacetic acid[b] | $NH(CH_2CO_2H)_2$ | 7.03 |
| Nitrilotriacetic acid (NTA)[c] | $N(CH_2CO_2H)_3$ | 10.45 |
| Ethylenediaminetetraacetic acid (EDTA)[d] | $CH_2N(CH_2CO_2H)_2$ $\mid$ $CH_2N(CH_2CO_2H)_2$ | 16.5 |

[a] $pK_{a_1} = 2.35$ ($H_2A^+$); $pK_{a_2} = 9.78$ (HA).
[b] $pK_{a_1} = 2.73$; $pK_{a_2} = 9.46$.
[c] $pK_{a_1} = 1.97$; $pK_{a_2} = 2.57$; $pK_{a_3} = 9.81$.
[d] $pK_{a_1} = 2.00$; $pK_{a_2} = 2.67$; $pK_{a_3} = 6.16$; $pK_{a_4} = 10.26$.
[e] Values for complexes with other metal ions are listed in Appendix V. (Taken from J. Bjerrum, G. Schwarzenbach, and L. G. Sillen, "Stability Constants: Organic Ligands," Part I, The Chemical Society, Burlington House, London, 1957.)

Table 15–4 lists the log stability constants for a series of 1 : 1 zinc : aminocarboxylic acid complexes. In some cases, additional coordination is possible. The chelate effect increases rapidly as the number of coordination sites in the ligand increases which results in a large increase in stability. For NTA and EDTA the rates of reaction are fast and a very stable, stoichiometric 1 : 1 zinc : ligand complex is formed. The stability constants for complexes between these ligands and other metal ions are listed in Appendix V.

Although NTA is suitable as a titrant, the more versatile reagent is EDTA. With this reagent as a titrant direct or indirect titration procedures have been devised for the determination of almost all metal ions and for many anions.

There are several reasons for the versatility of EDTA.

1.   EDTA forms stable, soluble, stoichiometric 1 : 1 complexes with metal ions.

2.   A certain amount of selectivity can be obtained because of differences in stability constants and through control of the pH of the solution. The metal ions can be divided according to their stability constants into three groups. This is shown in Table 15–5. In general, group I is titrated in basic conditions (pH 8–11), group II in acidic to slightly basic conditions (pH 4–7), and group III in acidic conditions (pH 1–4). (The actual pH for the titration will depend on the metal ion.) Furthermore, group III can be titrated in the presence of group II or I metal ions at the low pH. Group II in its pH range can be titrated in the presence of group I metals but not group III metals. However, a titration in the pH range 4–7 will represent a total of group II and III metals. At the higher pH 8–11 all three groups will be titrated. Thus, a group I metal can not be titrated in the presence of group II or group

**Table 15–5. Classification of Metal Ions in EDTA Titrations According to Log Stability Constants**[a]

|  |  |  |  |  |  |
|---|---|---|---|---|---|
| | | Group I | | | |
| | $Mg^{2+}$ | 8.69 | $Sr^{2+}$ | 8.63 | |
| | $Ca^{2+}$ | 10.70 | $Ba^{2+}$ | 7.76 | |
| | | Group II | | | |
| $Mn^{2+}$ | 13.58 | $Cu^{2+}$ | 18.79 | $TiO^{2+}$ | 17.3 |
| $Fe^{2+}$ | 14.33 | $Zn^{2+}$ | 16.5 | $V^{2+}$ | 12.70 |
| Rare earths | 15.3–19.8 | $Cd^{2+}$ | 16.59 | $VO^{2+}$ | 18.77 |
| $Co^{2+}$ | 16.21 | $Al^{3+}$ | 16.13 | | |
| $Ni^{2+}$ | 18.56 | $Pb^{2+}$ | 18.3 | | |
| | | Group III | | | |
| $Hg^{2+}$ | 21.8 | $Fe^{3+}$ | 25.1 | $Sn^{2+}$ | ~22 |
| $Bi^{3+}$ | ~23 | $Ga^{3+}$ | 20.27 | $Ti^{3+}$ | 17.7 |
| $Co^{3+}$ | ~36 | $In^{3+}$ | 24.95 | $Th^{4+}$ | 23.2 |
| $Cr^{3+}$ | ~23 | $Sc^{3+}$ | 23.1 | $V^{3+}$ | 25.9 |

[a] See Appendix V.

III metals as a result of pH control. It is assumed that the metal ions remain in solution at the suggested pH. In practice most metal ions will hydrolyze at some particular pH unless the experimental conditions are adjusted to prevent hydrolysis.

3. The disodium salt of EDTA as the dihydrate is acceptable as a primary standard. Since the free acid or monosodium salt of EDTA is insoluble in water, the disodium salt is used. A solution of the disodium salt, $Na_2H_2Y \cdot 2H_2O$, will be about pH 4–5. Plastic bottles are recommended for storage of solutions of $Na_2H_2Y$ since metal ions can be leached from glass bottles.

4. All the metal–EDTA complexes are soluble and most complexes form rapidly.

5. The stoichiometric point is readily detected by chemical or instrumental methods.

6. The titration is suitable for a semimicro to macro concentration range.

If $Na_2H_2Y$ is used, the reaction with a metal ion and the formation constant for the complex formed are represented by

$$M^{n+} + H_2Y^{2-} \rightleftharpoons [MY]^{n-4} + 2H^+ \qquad (15\text{-}17)$$

$$K_s = \frac{[H^+]^2[MY]^{n-4}}{[M^{n+}][H_2Y^{2-}]} \qquad (15\text{-}18)$$

Hence, the reaction (and titration) will be very sensitive to pH and all procedures in which EDTA is used as a titrant must include a buffer with suffi-

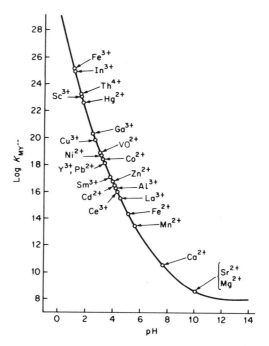

**Fig. 15–2.** Minimum pH for successful EDTA titrations of several metal ions. [Taken from C. N. Reilley and R. W. Schmid, *Anal. Chem.* **30,** 947 (1958).]

cient capacity to take care of the hydrogen ion produced during the titration. Hydrolysis of the metal ion and coordination of the metal ion with other ligands in the solution (see Chapter 16 for a detailed discussion of these effects) will influence the titration. Considering these effects Fig. 15–2 indicates the minimum pH that can be used for an effective titration of metal ions with EDTA.

**Polyamines.**     Polyamines are a class of chelating agents derived by substitution of the hydrogen atoms on $NH_3$ and $RNH_2$ by $-CH_2CH_2NH_2$. In this way both tertiary substituted polyamines and straight chain polyamines are possible. An example would be $\beta,\beta',\beta''$-triaminoethylamine, $N(CH_2CH_2\cdot NH_2)_3$, and triethylenetetramine, $H_2NCH_2CH_2NHCH_2CH_2NHCH_2CH_2NH_2$, respectively.

The simplest member of this group is ammonia itself. By building the molecule as outlined, the chelating effect becomes a deciding factor in the utility of these reagents as a titrant. This was illustrated in Fig. 15–1 where titration curves for ions with several different amines are shown. The stability constants for these complexes as well as the Ni(II) and Cu(II) complexes are listed in Table 15–6.

| Complexing agent[b] | Structure[a] | Nickel | | | | | | Zinc | | | Copper | | | |
|---|---|---|---|---|---|---|---|---|---|---|---|---|---|---|
| 1. Ammonia | $NH_3$ | 2.8 | 2.2 | 1.7 | 1.2 | 0.7 | −0.01 | 2.3 | 2.3 | 2.4 | 4.1 | 3.5 | 2.9 | 2.1 |
| 2. Ethylenediamine | $H_2NCH_2CH_2NH_2$ | 7.7 | 6.5 | 5.1 | | | | 5.9 | 5.2 | | 10.7 | 9.3 | | |
| 3. 1:3-Diaminopropane | $H_2NCH_2CH_2CH_2NH_2$ | 6.4 | 4.3 | 1.2 | | | | | | | 9.8 | | | |
| 4. Diethylenetriamine | $H_2NCH_2CH_2NHCH_2CH_2NH_2$ | 10.7 | 8.3 | | | | | 8.9 | 5.5 | | 16.0 | 5.3 | | |
| 5. 1:2:3-Triaminopropane | $H_2NCH_2CHNH_2CH_2NH_2$ | 9.3 | 6.5 | | | | | 6.8 | 4.3 | | 11.0 | 9.0 | | |
| 6. Triethylenetetramine (Trien) | $H_2NCH_2CH_2NHCH_2CH_2NHCH_2CH_2NH_2$ | 14.0 | | | | | | 12.1 | | | 20.4 | | | |
| 7. $\beta,\beta',\beta''$-Triaminotriethyl-amine | $\begin{matrix} & CH_2CH_2NH_2 \\ & / \\ N- & CH_2CH_2NH_2 \\ & \backslash \\ & CH_2CH_2NH_2 \end{matrix}$ | 14.8 | | | | | | 14.7 | | | 18.8 | | | |
| 8. Tetraethylenepentamine (Tetren) | $H_2NCH_2CH_2NHCH_2CH_2NHCH_2CH_2NHCH_2CH_2NH_2$ | 19.3 | | | | | | 16.2 | | | 22.4 | | | |

[a] Experimental conditions are for 20°C and ionic concentration of 0.1 M. Taken in part from G. Schwarzenbach, Analyst **80**, 713 (1955).

[b] Acidity constants are (1) $pK_a = 9.37$; (2) $pK_{a_1} = 7.30$, $pK_{a_2} = 10.11$; (3) $pK_{a_1} = 8.96$, $pK_{a_2} = 10.72$; (4) $pK_{a_1} = 4.42$, $pK_{a_2} = 9.21$, $pK_{a_3} = 10.02$; (5) $pK_{a_1} = 3.80$, $pK_{a_2} = 8.03$, $pK_{a_3} = 9.67$; (6) $pK_{a_1} = 3.32$, $pK_{a_2} = 6.67$, $pK_{a_3} = 9.20$, $pK_{a_4} = 9.92$; (7) $pK_{a_1} = 8.64$, $pK_{a_2} = 9.67$, $pK_{a_3} = 10.37$; (8) $pK_{a_1} = 2.6$, $pK_{a_2} = 4.1$, $pK_{a_3} = 8.2$, $pK_{a_4} = 9.2$, $pK_{a_5} = 10.0$.

As shown in Fig. 15–1, ammonia is not useful as a titrant since no true identifiable stoichiometric point is observed. Ethylenediamine and diethylenetriamine form 1 : 2 zinc to ligand complexes. Even though the chelating effect is partially present neither of these reagents are suitable titrants because of the complication of a two-step coordination. The full chelate effect is exhibited by triethylenetetramine and triaminotriethylamine (also tetraethylpentamine which is not shown in Fig. 15–1). For these titrants sharp, stiochiometric 1 : 1 zinc to ligand ratios are established and readily detected and therefore, are useful titrants.

The polyamines are excellent complexing agents like EDTA, but are not universal titrants since they form stable complexes with only those metal ions that normally coordinate with $NH_3$. The two polyamines that are the most versatile as titrants are triethylenetetramine (trien) and tetraethylenepentamine (tetren). Stability constants for their metal complexes are listed in Appendix V.

In general, commercially available trien and tetren are technical grade liquid reagents and will be contaminated with lower amines. The compounds are purified by recrystallization as the sulfate salt from sulfuric acid solution. If the sulfate ion interferes, precipitation and recrystallization as the nitrate salt is also possible. Solutions of these salts provide good stability and are usually used as titrants.

**Experimental Conditions.**     An EDTA titrant at 0.1–0.05 $F$ is usually used. However, since the end point detection (metallochromic indicators) is very sensitive, titrations in the semimicro and micro range are also easily done.

The EDTA titration is very sensitive to pH and a successful titration requires the selection of a suitable pH (see Fig. 15–2). Since hydronium ions are produced in the reaction [see reaction (15-17)], the buffer that is used to adjust the pH must also provide sufficient buffer capacity.

In some cases, the rate of reaction between the metal ion and complexometric titrant is too slow. Either of two techniques can be used to overcome this problem. A direct titration can be carried out at an elevated temperature or, a back-titration technique can be used.

Several substances are acceptable primary standards for the standardization of polyaminocarboxylic acid titrants. These are $CaCO_3$, pure metals such as Zn, Cd, and Cu, and the cadmium salt of 2-hydroxyethylethylenediaminetriacetic acid (**XV**). All of these except $CaCO_3$ can be used as standards for polyamine titrants.

$$\begin{array}{c} HOCH_2CH_2 \diagdown \qquad \diagup CH_2CO_2H \\ \qquad\qquad NCH_2CH_2N \\ HO_2CCH_2 \diagup \qquad \diagdown CH_2CO_2H \end{array}$$

**XV**

# END-POINT DETECTION

End-point detection techniques for complexometric titrations involve either a visual or instrumental response. By the use of visual indicators, the end point is detected by a pH, color, fluorescence, or phase change. Instrumental detection is based on measuring a change in an electrochemical property (potential, current, or resistance), an optical property (absorption or fluorescence), or a thermal property (heat of reaction).

**Visual End-Point Detection.**     The most practical and versatile method of visual end-point detection is with metallochromic indicators. A metallochromic indicator is a compound which is capable of acting as a complexing agent toward the metal ion. Under suitable conditions the metal–indicator complex that is formed has an intense color which is sharply different than the uncomplexed indicator. Since the indicator is added at a trace level, generally, no titration error is observed. An optimum stability constant for the metal–indicator complex is necessary. If it is too large, the sample will be overtitrated, while if it is too small, an undertitration is obtained. A precise treatment of the effect of the metal–indicator complex stability is possible by considering all the equilibria present.

At present, several different indicators can be used for each metal ion. Often it is possible to choose an indicator purely on the basis of ease in observing a color change.

Most visual metallochromic indicators, in addition to being complexing agents, are also acid–base indicators. Therefore, they are capable of undergoing a color change with a corresponding change in pH of the solution. Those indicators that do not function as acid–base indicators still undergo coordination with metal ions. A typical example of this latter class is thiocyanate ion.

Iron (III) forms a red complex with thiocyanate ion. As EDTA titrant is added at pH 3, the Fe(III) is removed from the Fe(III)–SCN$^-$ complex and since SCN$^-$ and Fe(III)–EDTA complex are colorless in solution, a color change from red to colorless takes place upon passing the stoichiometric point. The reactions describing this indicator action are

Indicator reaction:     $\underbrace{Fe^{3+}_{excess}}_{colorless} + SCN^- \rightleftharpoons \underbrace{Fe^{3+}-SCN^-}_{intense\ red}$

Titration:     $Fe^{3+}-\underbrace{SCN^-}_{intense\ red} + Fe^{3+} + EDTA \rightleftharpoons Fe^{3+}-EDTA + \underbrace{SCN^-}_{colorless}$

Eriochrome Black T (**XVI**), which was one of the initial indicators investigated, is useful for the titration of water hardness (Ca–Mg). The indicator has three acidic sites, two of which are involved in color changes. The color transitions for the indicator are

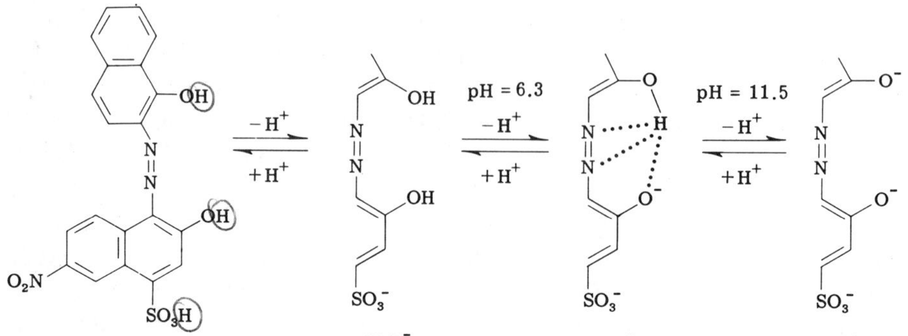


XVI $H_2E^-$ (Red) $HE^{2-}$ (Blue) $E^{3-}$ (Orange)

(15-19)

$$K_{a_1} = \text{Strong acid}, \quad K_{a_2} = 5.00 \times 10^{-7}, \quad K_{a_3} = 2.82 \times 10^{-12}$$

These series of equilibrium steps state that below pH 6.3 a solution of Eriochrome Black T will be a reddish color, between pH 6.3 to 11.5 it will be bluish, and above pH 11.5 it will be orange colored.

In the presence of Mg(II) the following can be written:

$$\underset{\text{Red}}{H_2E^-} \underset{}{\overset{\text{pH} = 6.3,\ -H^+}{\rightleftharpoons}} \underset{\text{Blue}}{HE^{2-}} \overset{\text{pH} = 11.5,\ -H^+}{\rightleftharpoons} \underset{\text{Orange}}{E^{3-}} \qquad (15\text{-}20)$$

$$HE^{2-} \xrightarrow{+Mg^{2+}} \underset{\text{Red}}{MgE^- + H^+} \qquad K_{MgE^-} \approx 10^7$$

At pH $=$ 10, Mg(II) forms a complex with the indicator. As the pH of the solution is lowered, the color of the complex begins to match the color of the indicator at the lower pH. Similarly, at a higher pH it is difficult to distinguish between the free indicator and metal–indicator complex. Consequently, an optimum pH for the titration of Mg(II) is in the range of 9–11. At higher pH values the hydrolysis of Mg(II) becomes a determining factor. A $NH_3$–$NH_4Cl$ buffer (pH 9.5–10) is convenient. As the stoichiometric point is reached, the color changes from the wine-red to the sky-blue color when the last traces of Mg(II) are removed from the Mg–$E^-$ complex by the EDTA.

It should be noted that the metal in the metal–indicator complex acts like a hydrogen ion with respect to indicator color. This property, in general, is characteristic of all metallochromic indicators possessing acid–base properties. For this reason, the indicator is used at a pH higher than a color transition pH. The color change for the titration, therefore, can be expected to change from a color closely resembling the indicator at a pH lower than the transition pH to the indicator color that is observed for the indicator at a pH above the transition.

Very careful control of the conditions is necessary for the indicator to function properly. For example, as the pH is lowered below 10, the Mg–E⁻ complex becomes less stable and will form only in the presence of a large excess of Mg(II). If the constant is too large, the end point occurs too late, while if it is not stable enough, a premature end point is found. Generally, the metal-indicator complex should be about one to two powers of 10 lower in stability than the metal–chelometric titrant complex.

Eriochrome Black T is not suitable for the EDTA titration of Ca(II) by itself, since the complex between the indicator and Ca(II) is not very stable. Thus, a small amount of Mg(II) is introduced into the Ca(II) solution by adding a measured aliquot of a dilute Mg(II) solution or by preparing the EDTA titrant with a small amount of Mg(II) dissolved in it. If the former method is used, the Mg(II) blank must be established. If the latter technique is used the Mg(II) blank is eliminated in the standardization procedure. However, the EDTA titrant containing small amounts of Mg(II) should not be used as a titrant for other metal ions.

In applying this titration to water hardness (Mg plus Ca) it is necessary to mask or remove interferences that are usually found in the water samples. Addition of sodium cyanide will prevent interference of Cu(II), Fe(III), and several other lesser important interferences through formation of metal-cyanide complexes.

Eriochrome Black T can be used for the titration of several other metal ions. Table 15–7 briefly describes these applications.

Several other azo dyes possessing both a chelating linkage and acid–base properties have been shown to be useful. No attempt will be made to review all useful metallochromic indicators. Instead, the discussion will be directed toward illustrating typical indicators for group I, II, and III metals in the EDTA titration scheme while emphasizing the types of structures that have been particularly useful as indicators.

In general, the trend in the synthesis of new metallochromic indicators is the modification of known chelating agents or introduction of a chelating group into a colored molecule. Usually, one or more sulfonic acid groups are included to increase water solubility. Also, functional groups can be introduced in positions which influence the color and the acid–base properties of the compound.

**Table 15–7. Applications of Eriochrome Black T as an Indicator in EDTA Titrations**

| Metal | Conditions |
|---|---|
| *Direct Titration* | |
| Mg, Zn, Cd, Pb, Mn | pH = 10 |
| Ca | pH = 10, presence of trace Mg |
| In, rare earths | pH = 8–9, tartrate |
| Sc | pH = 7–8, malate |
| *Indirect Titration* | |
| Al, $Fe^{3+}$, Co, Ni, Cu, rare earths, Ag, Platinum metals, $Hg^{2+}$, Ga | Usually pH = 10 using Mg for back titration |

One functional group on which many indicators are based is the azo group. Three typical indicators, naphthyl azoxine (**XVII**), NAS (**XVIII**), and arsenazo I (**XIX**) are shown below.

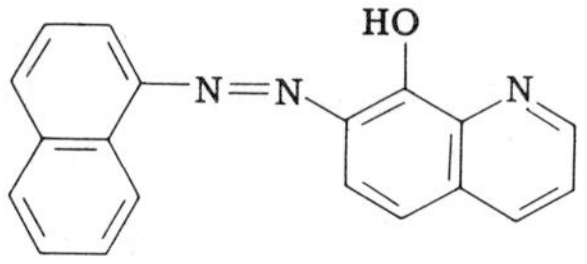

XVII 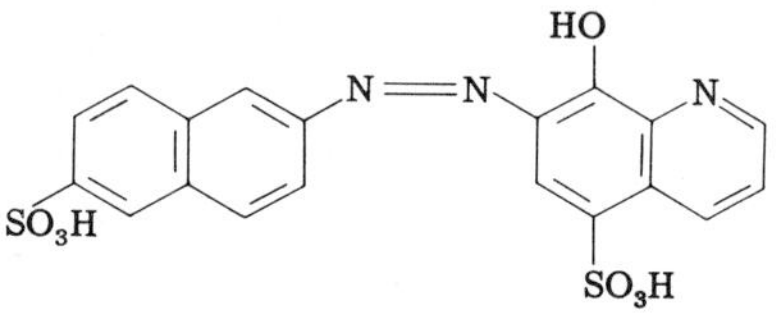                                    XVIII

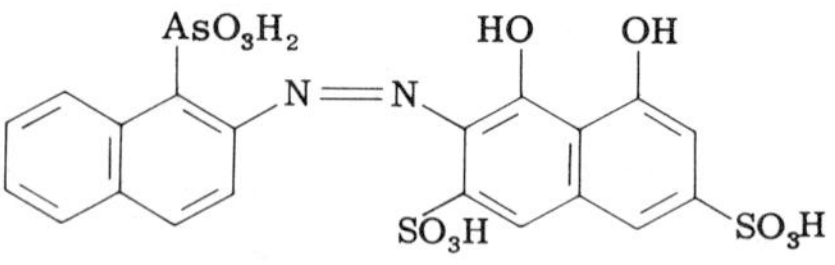

XIX

Indicators **XVII** and **XVIII** are similar in that they contain the same coordinating linkage found in 8-hydroxyquinoline. These two indicators find their greatest utility in the EDTA titration of group II metals. The color change for both compounds is from a pale yellow (M–Indicator) to a pinkish red (M–EDTA) with the intensity of the color very dependent on indicator concentration. Naphthyl azoxine is useful in the pH range 3–7, while NAS is useful at pH 3–9. It would appear that coordination occurs through the hydroxy and prydine nitrogen with potential coordination through the azo nitrogen . In general, NAS, because of a sharper color change and greater pH range, is a better indicator.

Arsenazo I is useful for group II and III metals particularly the rare earths and Th. The color change for the rare earth titration (pH = 4–6) is a reddish-violet to peach while the Th titration (pH = 3) color change is a deeper violet to peach. The coordination sites of the indicator have not been established. The applications of NAS and arsenazo I are summarized in Table 15-8.

Modification of existing acid–base indicators by the introduction of a chelating group is illustrated by the following two examples.

Methylthymol blue                      Calcein

**XX**                                   **XXI**

Methylthymol blue (**XX**) is prepared from the acid–base indicator thymol blue and is used for group III (also group II) metal ions at pH values below 7.2. The color change for this complexometric indicator is from a blue (M–Ind) to a lemon yellow color with EDTA as titrant. It is particularly useful for $Bi^{3+}$, $Th^{4+}$, and $Zr^{4+}$ in the pH range 2–3 while at a higher pH of 4–6 it is used for $Sc^{3+}$, rare earths, and several divalent group II metals.

Calcein (**XXI**), which is prepared from the acid–base indicator fluorescein, retains the fluorescent properties of fluorescein. Hence, end points in EDTA titrations using this indicator are detected by a change in fluorescence. At acidic pH values the indicator is used for back-titration detection. Excess EDTA is added to a metal ion solution and the remaining EDTA titrated with a standard $Cu(NO_3)_2$ solution. The end point is marked by quenching of fluorescence presumably through the coordination of the phenolic hydroxy, nitrogen, and acetate groups with Cu(II). At a high pH (pH = 12) the free indicator does not fluoresce but the Ca–Ind, Ba–Ind, and Sr–Ind complexes do. A direct titration of these metal ions with EDTA is possible and the end point is marked by a disappearance of fluorescence. This titration is particularly useful for the determination of calcium.

**Potentiometry.**     End points of metal–polyaminocarboxylic acid and polyamine titrations can be detected by potentiometry. Since there is no change in oxidation state during the titration, a redox couple and a suitable indicator electrode or an ion selective electrode must be introduced into the system.

It is possible in some cases to titrate a metal ion with a complexing titrant

**Table 15-8. Applications of NAS and Arsenazo I Indicators in EDTA Titrations**

| NAS[a] | | | Arsenazo I[b,c] | | |
|---|---|---|---|---|---|
| Ion titrated | Ion added | Masking conditions | Ion titrated | Ion added[d] | Masking agent |
| $Cd^{2+}$ | $Al^{3+}$ | 1 g NaF, pH 6.5 | $Ca^{2+}$ | (pH = 10) | |
| $Cu^{2+}$ | $Al^{3+}$ | 25 ml NaF, pH 5–6 | $Mg^{2+}$ | (pH = 10) | |
| $Pb^{2+}$ | $Al^{3+}$ | 50 ml NaF, pH 6 | $Dy^{2+}$ | $Ca^{2+}$, $Mg^{2+}$ | |
| $VO^{2+}$ | $Al^{3+}$ | 25 ml NaF, (B) Zn, pH 6 | $Sm^{3+}$ | $Ca^{2+}$(4:3) | |
| $Y^{3+}$ | $Al^{3+}$ | 10 ml 2,4-pentanedione, pH 6.5 | $Y^{3+}$ | $Ca^{2+}$ | |
| $Yb^{3+}$ | $Al^{3+}$ | 10 ml 2,4-pentanedione, pH 8 | $Er^{3+}$ | $Cu^{3+}$ | Cyanide |
| $Co^{2+}$ | $Cr^{3+}$ | Sodium acetate, pH 5.5–6.5 | $Y^{+3}$ | $Cu^{2+}$, $Co^{2+}$, $Ni^{2+}$ | Cyanide |
| $Cu^{2+}$ | $Cr^{3+}$ | Sodium acetate, pH 6.8 | $Y^{3+}$ | $Hg^{2+}$ (1:3) | Iodide |
| $Ni^{2+}$ | $Cr^{3+}$ | Sodium acetate, pH 6.5 | $Nd^{2+}$ | $Zn^{2+}$, $Cd^{2+}$, $Pd^{2+}$, $Hg^{2+}$, | Dithiocarbamate |
| $Cd^{2+}$ | $Fe^{3+}$ | 10 ml citrate, 50% acetone, pH 8.5 | $Y^{3+}$ | $Cd^{2+}$, $Hg^{2+}$, $Pb^{2+}$, $UO_2^{2+}$ | Dithiocarbamate |
| $Cu^{2+}$ | $Fe^{3+}$ | 25 ml citrate, 50% acetone, pH 8.5 | $Y^{3+}$ | $Zn^{2+}$ (2:3) | Dithiocarbamate |
| $Mn^{2+}$ | $Fe^{3+}$ | 10 ml citrate, 50% acetone, pH 8.5 | $Dy^{3+}$ | $Al^{3+}$ | Sulfosalicylate |
| $Pb^{2+}$ | $Fe^{3+}$ | 10 ml citrate, 50% acetone, pH 8.5 | $Er^{3+}$ | $Al^{3+}$ (2:5) | Sulfosalicylate |
| $Al^{3+}$ | $Mg^{2+}$ | (B)ᵉ Cu, pH 3–5 | $La^{3+}$ | $Al^{3+}$ | Sulfosalicylate |
| $Cu^{2+}$ | $MoO_4^{2-}$ | 10 ml citrate, pH 9 | $Pr^{3+}$ | $Al^{3+}$ | Sulfosalicylate |
| $Cu^{2+}$ | $Sn^4$ | 2 g NaCl, 25 ml NaF, pH 4 | $Sm^{3+}$ | $Al^{3+}$ (2:1) | Sulfosalicylate |
| $Ni^{2+}$ | $Sn^{4+}$ | 2 g NaCl, 25 ml NaF, pH 6 | $Y^{3+}$ | $Al^{3+}$ (2:1) | Sulfosalicylate |
| $Zn^{2+}$ | $Sn^{4+}$ | 2 g NaCl, 25 ml NaF, pH 6 | $Y^{3+}$ | $Cu$(1:3) | Thiourea |
| $Ni^{2+}$ | $Th^{4+}$ | 10 ml citrate, pH 8 | | | |
| $Zn^{2+}$ | $Th^{4+}$ | 10 ml citrate, pH 6.5 | | | |
| $Cu^{2+}$ | $WO_4^{2-}$ | 10 ml tartrate, pH 5–6 | | | |
| $Ti^{4+}$ | $WO_4^{2-}$ | (B) Cu, $H_2O_2$, tartrate, pH 4.5 | | | |
| $Cd^{2+}$ | $Zr^{4+}$ | 10 ml citrate, pH 6.5 | | | |
| $Cu^{2+}$ | $Zr^{4+}$ | (B) Zn, 10 ml citrate, pH 9 | | | |
| $Co^{2+}$ | $Zr^{4+}$ | 10 ml citrate, pH 6.5 | | | |

[a] Taken from J. S. Fritz, J. E. Abbink, and M. A. Payne. *Anal. Chem.* **33**, 1381 (1961).
[b] Taken from J. S. Fritz, R. T. Oliver, and D. J. Pietrzyk, *Anal. Chem.* **30**, 1111 (1958).
[c] pH = 5.5–6.5 except where indicated otherwise.
[d] Ratio of ion added to rare earth is 1:1 except where noted.

using an indicator electrode constructed from the same metal. The potential at the indicator electrode would be determined by the half-reaction

$$M^{n+} + ne \rightarrow M$$

and

$$E = E^\circ_{M^{n+},M} - \frac{0.0592}{n} \log \frac{1}{[M^{n+}]}$$

Since the metal ion concentration decreases during the titration, the potential at the indicator electrode also changes. In the vicinity of the equivalence point the pM changes abruptly producing a similarly abrupt change in potential. Usually, a saturated calomel electrode is used to complete the cell.

In theory the system described in the previous paragraph should provide excellent end-point detection. However, in practice only a few systems provide the kind of reproducibility, speed of response, and reversibility required for a simple, direct potentiometric titration.

These limitations are solved by using a "mercury electrode" or ion-selective electrode (Chapter 13) as the indicator electrode. With these types of electrodes and the adjustment of experimental conditions it is possible to follow any of the EDTA or polyamine tirations potentiometrically.

The mercury electrode (Fig. 15–3) consists of a drop of mercury in contact with the solution containing the metal ion and a drop or two of a solution of Hg(II)–chelate. If the chelate is assumed to be EDTA, a half-cell potential is determined by the $Hg/HgY^{2-}$ couple. However, this couple is also influenced by the $M^{2+}$ that is being titrated since it also forms a complex with EDTA.

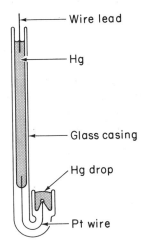

**Fig. 15–3.** Typical J-tube mercury electrode.

Therefore, prior to the addition of any EDTA titrant an equilibrium between the added $HgY^{2-}$ and $M^{n+}$ is established:

$$M^{n+} + HgY^{2-} \rightleftharpoons MY^{n-4} + Hg^{2+} \tag{15-21}$$

and the equilibrium constant is given by

$$K_{eq} = \frac{K_{MY^{n-4}}}{K_{HgY^{2-}}} = \frac{[Hg^{2+}][MY^{n-4}]}{[M^{n+}][HgY^{2-}]} \tag{15-22}$$

The Nernst equation for the $Hg^{2+}/Hg$ half-cell is given by

$$E = E^\circ_{Hg^{2+},Hg} - \frac{0.0592}{2} \log \frac{1}{[Hg^{2+}]} \tag{15-23}$$

Combination of Eq. (15-22) and the Nernst equation for the $Hg^{2+}/Hg$ half-cell defines the potential at the initial point and any other point during the titration:

$$E = E^\circ_{Hg^{2+},Hg} - \frac{0.0592}{2} \log \frac{[MY^{n-4}]K_{HgY^{2-}}}{[M^{n+}][HgY^{2-}]K_{MY^{n-4}}} \tag{15-24}$$

If Eq. (15-24) is rearranged into the form

$$E = E^\circ_{Hg^{2+},Hg} - \frac{0.0592}{2} \log \frac{[MY^{n-4}]K_{HgY^{2-}}}{[HgY^{2-}]K_{MY^{n-4}}} + \frac{0.0592}{2} \log \frac{1}{[M^{n+}]} \tag{15-25}$$

it can be seen that the potential at the stoichiometric point is determined primarily by the concentration of metal ion being titrated. This is apparent since the first log term contains two constants and the concentration of $HgY^{2-}$ and $MY^{n-4}$ approaches a constant value. Thus, the metal ion concentration and electrode potential changes sharply at the stoichiometric point.

The mercury electrode in combination with the reference electrode has been used in EDTA titrations and polyamine titrations. In general, potential changes are as large as 250 mV, very abrupt, and easily and accurately detected. The more stable the complex, the larger is the potential break for the titration. Table 15–9 summarizes the applications of the mercury electrode in EDTA titrations.

The potential response of the mercury electrode is affected by dissolved oxygen, particularly in micro titrations. This effect is most prominent in alkaline solution. If the solution is deaerated by passage of nitrogen before the titration, sharp, well-defined, and larger potential break curves are obtained. Halide ions will also interfere by formation of insoluble mercurous halides. Although very low halide concentrations can be tolerated, it is best if they are absent.

Ion-selective electrodes can be used to follow complexometric titrations. Since these electrodes respond to the free metal ion in solution a titration curve of pM vs volume of complexometric titrant is obtained. Details of the properties of these electrodes are described in Chapter 13.

## APPLICATIONS

In addition to water hardness, the clinical determination of calcium and magnesium in urine and serum is also possible by EDTA titration. Cal-red is usually used as the indicator for calcium, while Eriochrome Black T is used for magnesium.

Numerous other practical examples can be cited in which chelometric titrants are used. For example, each metal in a Mg, Cu, and Zn mixture can be titrated with EDTA at pH 10 ($NH_4^+$–$NH_3$) using Eriochrome Black T without a prior separation. First, an aliquot of the metal mixture is taken and the total metal ion is titrated. Cyanide ion is added to a second sample to mask Zn and Cu by formation of cyanide complexes. Titration with EDTA will then correspond to the amount of Mg in the sample. After reaching this end point, formaldehyde is added which destroys the Zn–cyanide complex and liberates the zinc ion. A continued titration with EDTA corresponds to the zinc content. The amount of Cu is then found by difference. Other examples of increasing the scope of EDTA titrations through masking are illustrated in Table 15–8.

The EDTA titration is ideally suited for the final measurement of metal ions after their mixtures have been separated. Thus, the EDTA titration is frequently combined with ion-exchange separations (see Chapter 26) and solvent extraction procedures (see Chapter 27).

**Table 15–9. Metal Ions That Have Been Quantitatively Titrated with EDTA Using the Mercury Electrode**[a]

| H | | | | | | | | | | | | | | | | | He |
|---|---|---|---|---|---|---|---|---|---|---|---|---|---|---|---|---|---|
| | | | | | | Metals titrated | | | | | | | | | | | |
| Li | Be | | | | | | | | | | | B | C | N | O | F | Ne |
| Na | Mg | | | | | | | | | | | Al | Si | P | S | Cl | A |
| K | Ca | Sc | Ti | V | Cr | Mn | Fe | Co | Ni | Cu | Zn | Ga | Ge | As | Se | Br | Kr |
| Rb | Sr | Y | Zr | Nb | Mo | Tc | Ru | Rh | Pd | Ag | Cd | In | Sn | Sb | Te | I | Xe |
| Cs | Ba | La | Hf | Ta | W | Re | Os | Ir | Pt | Au | Hg | Tl | Pb | Bi | Po | At | Ra |
| Fr | Ra | Ac | | | | | | | | | | | | | | | |
| | | | Ce | Pr | Nd | Pm | Sm | Eu | Gd | Tb | Dy | Ho | Er | Tm | Yb | Lu | |
| | | | Th | Pa | U | Np | Pu | Am | Cm | Bk | Cf | | | | | | |

[a] Taken from C. N. Reilley, R. W. Schmid, and D. W. Lamson, *Anal. Chem.* **30,** 953 (1958).

## *Questions*

1. Define the following terms: central ion, ligand, coordination, multidentate, chelate, coordination number, and polynuclear complex.
2. Differentiate between complexing agent and chelating agent.
3. Suggest how spot tests can be useful in practical situations.
4. Suggest the experiments that should be performed to demonstrate that reaction (15-6) is a suitable reaction for the spectrophotometric determination of Cu(II).
5. List several advantages of organic precipitating agents over inorganic precipitating agents.
6. Tetraphenylarsonium chloride can be used to precipitate Zn(II) from HCl solution. Write the reaction for this precipitation.
7. Suggest the coordination site(s) for picrolinic acid.
8. Write the reaction that takes place between Al(III) and 8-hydroxyquinoline.
9. What is the chelate effect?
10. Differentiate between an inert and a labile complex.
11. Explain why AgCl is more soluble in ammonical solution than in water.
12. Explain why Fe(III) is more easily reduced in the presence of 1,10-phenanthroline than in its absence.
13. Write a series of reactions illustrating the series of complexes that are formed between Co(II) and $NH_3$.
14. Differentiate between a dissociation constant and a formation constant.
15. Account for the high stability of EDTA–metal complexes.
16. Explain why unidentate ligands are usually poor complexing titrants.
17. Suggest the properties that a complexing agent should have if it is to be a suitable titrant.
18. Explain how a metallochromic indicator works.
19. List several different types of indicator electrodes that can be used to follow EDTA titrations potentiometrically.
20. Copper(II) will form 1:1 and 1:2 complexes with amino acids ($RHCNH_2CO_2H$). Write the stepwise formation of these complexes and indicate the points of coordination.
21. The ferrous porphyrin molecule, called heme, and its close relatives are capable of catalyzing many different biological reactions (for example, $O_2$ transport in blood). Suggest where the $Fe^{2+}$ is located in the porphyrin system (**I**).

**I**

22. The Mg–porphyrin system, chlorophyll, is responsible for the conversion of light energy to chemical energy. Suggest the possible location of $Mg^{2+}$ in the porphyrin system (**I**).

# Problems

1.* Perhaps the oldest titrimetric analysis based on a complex reaction is the "Liebig titration" for cyanide ion.

Titration: $\quad Ag^+ + 2CN^- \rightarrow Ag(CN)_2^-$

End point: $\quad Ag(CN)_2^- + Ag^+ \rightarrow Ag[Ag(CN)_2]_{(s)}$

If 24.10 ml of 0.05015 $F$ $AgNO_3$ solution was required to titrate a 0.6548-g sample containing cyanide ion, calculate the percent CN in the sample.

2.* Calculate the weight of $Na_2H_2Y_2 \cdot 2H_2O$ required to prepare 1500 ml of a 0.0250 $F$ EDTA solution.

3.* A pure 0.1184-g sample of pure $MgCO_3$ was dissolved, pH adjusted, and titrated with EDTA to the Eriochrome Black T end point with 17.61 ml of EDTA solution. Calculate the formality of the EDTA solution.

4. A 6.5115-g sample of a $ZnO$–$ZnSO_4$ ointment was dissolved and diluted to 250 ml. A 50-ml aliquot was taken, the pH adjusted, and titrated with 15.44 ml of 0.04918 $F$ EDTA. Calculate the percent Zn in the ointment.

5.* Several different brands of antiacid tablets are commercially available and usually contain a combination of $CaCO_3$, $MgCO_3$, $MgO$, and filler–binders. Ten tablets with a total weight of 6.6144 g were taken, dissolved and diluted to 500 ml. A 25.00-ml aliquot was taken, the pH adjusted, and titrated with 25.41 ml of 0.1041 $F$ EDTA with Eriochrome Black T. Calculate the percent alkaline earth as percent Mg in the sample and the mg of alkaline earth as mg of Mg/tablet.

6. Calculate the percent $Zn(NO_3)_2$ in a 0.6511-g sample that requires 21.42 ml of a 0.0756 $F$ EDTA titrant.

7. Calcium in serum can be determined by a micro EDTA titration. In the clinical procedure 100 $\mu l$ of serum is taken, 2 drops of 2 $F$ KOH is added, Cal-Red indicator is added, and the mixture is titrated with 0.001015 $F$ EDTA using a microburet. If 0.246 ml of titrant was required calculate the mg of Ca per 100 ml of serum. For healthy adults the Ca content of serum is from 9 to 11 mg/100 ml.

8.* The same method as described in Question 7 can be used for Ca in urine. Urine

---

* Answers are listed at the end of the book for problems marked with an asterisk.

also contains Mg which can be masked by the addition of 0.1 ml of 0.05 $M$ citrate solution. Because calcium in urine is greatly dependent on diet and pathological variations, an average healthy adult excretes 100–300 mg Ca/24 hours. A 24-hour sample is taken and diluted to exactly 1000 ml. A 10-ml aliquot was taken and treated as described with 5.12 ml of 0.01100 $F$ EDTA required for the micro titration. Calculate the mg of Ca excreted per 24 hours.

9. A common procedure for serum chloride or urine chloride is to titrate the sample with $Hg(NO_3)_2$ titrant with a microburet and diphenylcarbozene indicator according to the reaction

$$2Cl^- + Hg^{2+} \rightarrow Hg_2Cl_2$$

If 0.200 ml of serum is taken, 2 ml $H_2O$, 1 drop 1 $F$ $HNO_3$, and 3 drops of indicator are added and titrated with 0.995 ml of 0.01061 $F$ $Hg(NO_3)_2$, calculate the mg of $Cl^-$ per 100 ml serum.

10. Sulfate can be determined by an indirect EDTA titration procedure. A 0.5512-g sample of a soluble sulfate was dissolved, acidified with $HNO_3$, and to this was added excess $Pb(NO_3)_2$ solution. The $PbSO_4$ precipitate was removed by filtration, washed, and then dissolved in $NH_3$ solution by the addition of 50.00 ml of 0.1051 $F$ EDTA solution. After an appropriate period of time the excess EDTA was titrated (pH = 10) with 8.41 ml of 0.1181 $F$ $Zn(NO_3)_2$ using Eriochrome Black T as indicator. Calculate the $\%SO_4$ in the sample.

11. Wood's metal is a Bi–Pb–Cd–Sn alloy. A 2.318-g sample of the alloy was dissolved in hot $HNO_3$. Upon partial evaporation a precipitate of hydrated stannic oxide formed. This was filtered and ignited to $SnO_2$ and was found to weigh 0.3661 g. The filtrate and washings from above were diluted to 500-ml volume. A 50-ml aliquot was taken, pH adjusted to 2, and titrated with 11.20 ml of 0.05000 $F$ EDTA with xylenol orange indicator (titration of Bi). Small amounts of hexamine were added and the EDTA titration was continued to another end point (Pb + Cd) which required an additional 10.81 ml. o-Phenanthroline was added to the solution which freed the Cd from the EDTA complex. The liberated EDTA required 6.15 ml of a 0.04634 $F$ $Pb(NO_3)_2$ solution to affect a color change (Cd). Calculate the $\%Sn$, $\%Bi$, $\%Pb$, and $\%Cd$ in the Wood's metal.

# Chapter
# Sixteen
# Introduction to
# Multiple Equilibrium
# Systems

## INTRODUCTION

The concept of equilibrium and the measurement of the equilibrium constant has found its way into all areas of chemistry and related scientific fields. For example, biological processes in animals and plants can involve equilibrium steps which often include pH and associated buffer systems. Equilibria across an interface, such as cell walls and other membranes, are extremely important. Frequently, when the biological process undergoes unusual or undesirable changes, a disruption or alteration of an equilibrium process is responsible. Drug availability and its metabolism will involve equilibrium steps. Many other examples can also be cited.

Frequently, these kinds of systems will include complicated and multiple equilibrium steps. Therefore, the intention of this chapter is to provide an introduction to the writing and evaluation of complicated equilibrium systems by treating the metal–EDTA titration in detail. This particular system includes ionization of a weak polyprotic acid $H_4Y$ ($K_a$), hydrolysis of a metal ion ($K_{sp}$), and complex formation ($K_s$). Consequently, the formulism developed in this chapter can be applied to a more exact understanding of neutralization, precipitation, and complex formation all of which are part of the aforementioned equilibrium systems.

## EFFECT OF EXPERIMENTAL VARIABLES IN EDTA TITRATIONS

For successful chelometric titrations, selection of pH, buffer type and concentration, and masking agents must be controlled. These choices will depend on the metal ion and complexing titrant. As indicated previously, the stability constants cited are for conditions in which the titrant is in its more powerful coordinating form. The conditions used in the laboratory for the titration are not the same as the conditions used to determine the stability constant. Thus, it is possible to draw erroneous conclusions about a titration by simply looking at the stability constant for the reaction and ignoring the experimental conditions used for the titration. For this reason calculations leading to a predicted titration curve, if they are to be useful, must also account for the experimental conditions.

In the titration several competing equilibria are present. These are summarized in the following:

$$
\begin{array}{ccccc}
\mathrm{M}^{n+} & + & \mathrm{Y}^{x-} & \rightleftharpoons & \mathrm{MY}^{n-x}
\end{array}
$$

| $+$ | $+$ | $+$ | $+$ | $+$ | $+$ |
|-----|-----|-----|-----|-----|-----|
| $y\mathrm{OH}^-$ | $y\mathrm{L}$ | $x\mathrm{H}^+$ | $\mathrm{H}^+$ | $\mathrm{OH}^-$ | $\mathrm{L}$ |
| $\updownarrow$ | $\updownarrow$ | $\updownarrow$ | $\updownarrow$ | $\updownarrow$ | $\updownarrow$ |
| $\mathrm{M(OH)}_y^{n-y}$ | $\mathrm{ML}_y{}^n$ | $\mathrm{H}_y\mathrm{Y}$ | $\mathrm{MHY}^{1+n-x}$ | $\mathrm{MOHY}^{n-x-1}$ | $\mathrm{MLY}^{n-x}$ |
| Hydrolysis | Ligand effect | pH effect | Metal chelate derivatives | | |

$$(16\text{-}1)$$

where $\mathrm{M}^{n+}$ is a metal ion and $\mathrm{Y}^{x-}$ is the titrant. The importance of each effect will vary according to the experimental conditions. In general, the formation of the metal chelate derivatives is a minor effect and will not be considered in detail.

**pH Effect.**      It can be seen from Eq. (16-1) that the hydrogen ion is acting like a metal ion in that it competes directly with the metal ion for the EDTA titrant. For this reason metal ions that form weak EDTA complexes with the titrant (group I metals in Table 15–5) must be titrated in alkaline solutions, while for the more stable complexes titration is possible in acidic solution.

If $[\mathrm{Y}]'$ is designated as the total concentration of uncomplexed EDTA in all forms, the following can be written:

$$[\mathrm{Y}]' = [\mathrm{H}_4\mathrm{Y}] + [\mathrm{H}_3\mathrm{Y}^-] + [\mathrm{H}_2\mathrm{Y}^{2-}] + [\mathrm{HY}^{3-}] + [\mathrm{Y}^{4-}] \quad (16\text{-}2)$$

Rearrangement of the equilibrium constants expressions, Eqs. (15-13) to (15-16), in terms of $[\mathrm{H}_4\mathrm{Y}]$, $[\mathrm{H}_3\mathrm{Y}^-]$, $[\mathrm{H}_2\mathrm{Y}^{2-}]$, $[\mathrm{HY}^{3-}]$, and $[\mathrm{Y}^{4-}]$ (for example, $[\mathrm{H}_4\mathrm{Y}] = [\mathrm{H}^+][\mathrm{H}_3\mathrm{Y}^-]/K_1$), substituting into (16-2), and eliminating all forms except $[\mathrm{Y}^{4-}]$, yields the expression

$$[\mathrm{Y}]' = \frac{[\mathrm{H}^+]^4[\mathrm{Y}^{4-}]}{K_1 K_2 K_3 K_4} + \frac{[\mathrm{H}^+]^3[\mathrm{Y}^{4-}]}{K_2 K_3 K_4} + \frac{[\mathrm{H}^+]^2[\mathrm{Y}^{4-}]}{K_3 K_4} + \frac{[\mathrm{H}^+][\mathrm{Y}^{4-}]}{K_4} + [\mathrm{Y}^{4-}]$$

$$(16\text{-}3)$$

Dividing Eq. (16-3) by $[Y^{4-}]$ gives

$$\frac{[Y]'}{[Y^{4-}]} = \frac{[H^+]^4}{K_1 K_2 K_3 K_4} + \frac{[H^+]^3}{K_2 K_3 K_4} + \frac{[H^+]^2}{K_3 K_4} + \frac{[H^+]}{K_4} + 1 \qquad (16\text{-}4)$$

A new term can be defined relating $[Y]'$ to $[Y^{4-}]$. By definition

$$[Y]'\alpha_0 = [Y^{4-}] \qquad (16\text{-}5)$$

where $\alpha_0$ represents the fraction of the total EDTA species that exists as the $[Y^{4-}]$ form. Thus, to calculate the fraction of $[Y^{4-}]$ in solution at any pH,

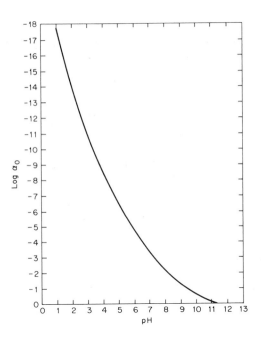

**Fig. 16-1.**   Values of $\alpha_0$ as a function of pH for EDTA.

| pH | $\alpha_0$ | pH | $\alpha_1$ |
|---|---|---|---|
| 2.0 | $3.7 \times 10^{-14}$ | 7.0 | $4.8 \times 10^{-4}$ |
| 3.0 | $2.5 \times 10^{-11}$ | 8.0 | $5.4 \times 10^{-3}$ |
| 4.0 | $3.6 \times 10^{-9}$ | 9.0 | $5.2 \times 10^{-2}$ |
| 5.0 | $3.5 \times 10^{-7}$ | 10.0 | $3.5 \times 10^{-1}$ |
| 6.0 | $2.2 \times 10^{-5}$ | 11.0 | $8.5 \times 10^{-1}$ |
|  |  | 12.0 | $9.8 \times 10^{-1}$ |

Eq. (16-4) is rewritten as

$$\frac{1}{\alpha_0} = \frac{[H^+]^4}{K_1 K_2 K_3 K_4} + \frac{[H^+]^3}{K_2 K_3 K_4} + \frac{[H^+]^2}{K_3 K_4} + \frac{[H^+]}{K_4} + 1 \qquad (16\text{-}6)$$

Using different hydrogen ion concentrations and inserting the equilibrium constants allows the calculation of $\alpha_0$, or the fraction of $Y^{4-}$ in solution, as a function of pH. This is shown in Fig. 16–1.

Since EDTA will exist in other forms as the pH is decreased, it is of interest to know their concentration as a function of pH. An equation relating the fraction of each form as a function of hydrogen ion concentrations and equilibrium constants can be derived in an analogous way as done for $\alpha_0$. These equations are summarized in Table 16–1. It should be noted that the symbolism defines the number of associated hydrogen ions.

**Table 16-1.  Summary of $\alpha$ Equations for EDTA**

$\alpha_0$ (implies fraction of $[Y^{4-}]$) $= \dfrac{[Y^{4-}]}{[Y]}$

$$\frac{1}{\alpha_0} = \frac{[H^+]^4}{K_1 K_2 K_3 K_4} + \frac{[H^+]^3}{K_2 K_3 K_4} + \frac{[H^+]^2}{K_3 K_4} + \frac{[H^+]}{K_4} + 1$$

$\alpha_1$ (implies fraction of $[HY^{3-}]$) $= \dfrac{[HY^{3-}]}{[Y]}$

$$\frac{1}{\alpha_1} = \frac{[H^+]^3}{K_1 K_2 K_3} + \frac{[H^+]^2}{K_2 K_3} + \frac{[H^+]}{K_3} + \frac{K_4}{[H^+]} + 1$$

$\alpha_2$ (implies fraction of $[H_2 Y^{2-}]$) $= \dfrac{[H_2 Y^{2-}]}{[Y]}$

$$\frac{1}{\alpha_2} = \frac{[H^+]^2}{K_1 K_2} + \frac{[H^+]}{K_2} + \frac{K_3}{[H^+]} + \frac{K_3 K_4}{[H^+]^2} + 1$$

$\alpha_3$ (implies fraction of $[H_3 Y^-]$) $= \dfrac{[H_3 Y^-]}{[Y]}$

$$\frac{1}{\alpha_3} = \frac{[H^+]}{K_1} + \frac{K_2}{[H^+]} + \frac{K_2 K_3}{[H^+]^2} + \frac{K_2 K_3 K_4}{[H^+]^3} + 1$$

$\alpha_4$ (implies fraction of $[H_4 Y]$) $= \dfrac{[H_4 Y]}{[Y]}$

$$\frac{1}{\alpha_4} = \frac{K_1}{[H^+]} + \frac{K_1 K_2}{[H^+]^2} + \frac{K_1 K_2 K_3}{[H^+]^3} + \frac{K_1 K_2 K_3 K_4}{[H^+]^4} + 1$$

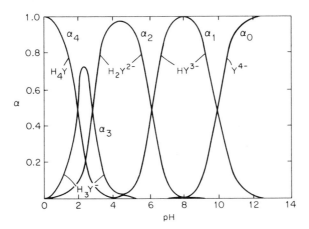

**Fig. 16-2.** Values of $\alpha_4$, $\alpha_3$, $\alpha_2$, $\alpha_1$, and $\alpha_0$ as a function of pH for EDTA.

Since the $\alpha$'s express the fraction of each possible form of EDTA at a fixed pH, their sum should equal 1 or

$$\alpha_0 + \alpha_1 + \alpha_2 + \alpha_3 + \alpha_4 = 1 \qquad (16\text{-}7)$$

A plot of $\alpha$ vs pH is shown in Fig. 16–2. From this figure it is apparent that above pH $= 10$ the major form of EDTA is $[Y^{4-}]$, from pH 6 to 10 it exists primarily as $[HY^{3-}]$, from pH 3 to 6 as $[H_2Y^{2-}]$, from pH 2 to 3 as $H_3Y^-$, and below pH $= 2$ as $[H_4Y]$.

The use of $\alpha$ is not restricted to EDTA and its derivatives. Similar equations and graphs can be derived for all polyprotic acids and bases and can be used for calculations of pH and composition of their solutions and their salt solutions. Several of these graphs, including ones for $H_3PO_4$, are shown in Appendix VI. Also, see the footnote on page 166 in Chapter 8.

The formation constant for a metal–EDTA complex is based on the concentration of $Y^{4-}$ in solution.

$$M^{2+} + Y^{4-} \rightleftharpoons MY^{2-}$$

$$K_s = \frac{[MY^{2-}]}{[M^{2+}][Y^{4-}]} \qquad (16\text{-}8)$$

However,

$$[Y^{4-}] = \alpha_0[Y]'$$

and therefore Eq. (16-8) becomes

$$K_s = \frac{[MY^{2-}]}{[M^{2+}]\alpha_0[Y]'} \qquad (16\text{-}9)$$

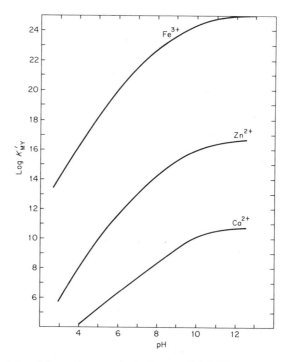

**Fig. 16–3.** Conditional formation constants for metal–EDTA complexes as a function of pH: effect of pH on the ligand.

Rearrangement of Eq. (16-9) leads to

$$K_s \alpha_0 = K_{Y'} = \frac{[MY^{2-}]}{[M^{2+}][Y]'} \tag{16-10}$$

where $K_{Y'}$ is a conditional constant that varies with $\alpha_0$ (and therefore, with pH) and $[Y]'$ stands for the formal concentration of EDTA in the mixture. This equation allows the calculation of the effective tendency for the reaction to take place at any chosen pH value providing the pH effect is the only competing step.

Consider the titration of a $1 : 1 : 1$ mixture of $Ca^{2+}$, $Zn^{2+}$, and $Fe^{3+}$. The log stability constants for their EDTA complexes (Table 15–5) are 10.7, 16.5, and 25.1, respectively. The conditional constants, as a function of pH, can be calculated by using the $\alpha_0$ values from Fig. 16–1, the log stability constants, and Eq. (16–10). A plot of this is shown in Fig. 16–3. It should be emphasized that in this calculation only the pH effect is considered and for other metal ratios the effect of mass action would have to be considered.

In Fig. 16–3 at pH 10 or higher, $Ca^{2+}$, $Zn^{2+}$, and $Fe^{3+}$ would be titrated as a group. The possibility of differentiation exists, since the order of stability is $Fe(III)-EDTA \gg Zn(II)-EDTA \gg Ca(II)-EDTA$, providing a technique is used to follow the titration continuously as the titrant is added. As the pH of the solution is reduced, the stability of the Ca–EDTA complex falls off rapidly and it should be possible to titrate $Zn^{2+}$ and $Fe^{3+}$ in the presence of $Ca^{2+}$, pH 4–6. Reducing the pH to about pH 2–3 causes a sharp change in the stability of the $Zn(II)-EDTA$ complex so that only $Fe^{3+}$ is titrated.

If a mixture of $Ca(NO_3)_2$, $Zn(NO_3)_2$, and $Fe(NO_3)_3$ is used and the pH raised to 10 with NaOH, the $Fe(III)$ would precipitate quantitatively as the hydrous oxide. Being amphoteric, part of the zinc would be precipitated as the hydrous oxide. The $Ca(II)$ ion would remain in solution. Consequently, for this mixture, more is involved than just the pH effect and the conditional constant must be further modified for the hydrolysis effect. Experimentally, to prevent precipitation another ligand is added to the solution. To keep $Zn(II)$ in solution at pH 10, a buffer composed of $NH_3–NH_4Cl$ would be added. Hydrolysis would be prevented since the zinc ion would be present in solution as zinc–ammonia complexes. To keep $Fe(III)$ in solution at pH 10 would require the presence of a powerful ligand. For this reason $Fe(III)$ would never or only rarely be titrated at these basic conditions.

**Hydrolysis Effect.**    Increasing the pH of the solution will lead to conditions at which metal ions will hydrolyze; this hydrolysis lowers the conditional constant. Therefore, an optimum pH exists for the titration of a given metal ion and is determined by the pH at which hydrolysis occurs, the stability of the metal–titrant complex, and the $K_a$ value(s) for the chelating titrant.

Hydrolysis for a divalent metal is represented by

$$M^{2+} + 4H_2O \rightleftharpoons M(OH)_{2(s)} + 2H_3O^+$$

and the solid in equilibrium with its ions by

$$M(OH)_{2(s)} \rightleftharpoons M^{2+} + 2OH^-$$

where the exact form of the hydrolyzed species is assumed to be a simple hydroxide; this is not always the case. The solubility product expression is

$$K_{sp} = [M^{2+}][OH^-]^2 \quad \text{and} \quad K_{sp} = \frac{[M^{2+}][K_w]^2}{[H_3O^+]^2}$$

Rearranging and converting to $-\log$ gives

$$-\log[M^{2+}] = -\log K_{sp} - 2\log[H_3O^+] + 2\log K_w$$

which can be written as

$$pM^{2+} = pK_{sp} + 2\,pH - 28 \tag{16-11}$$

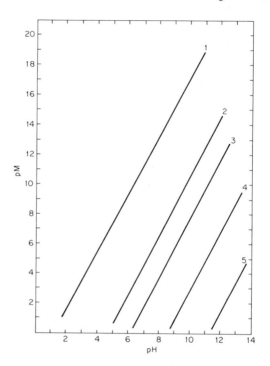

**Fig. 16-4.** Concentration of free metal ion (divalent metal ion) as a function of pH: effect of precipitation as the hydroxide. (1) $Hg^{2+}$ $pK_{sp} = 25.4$[Hg(OH)$_2$]; (2) Cu $pK_{sp} = 18.59$ [Cu(OH)$_2$]; (3) Zn $pK_{sp} = 15.68$ [Zn(OH$_2$]; (4) Mg $pK_{sp} = 10.74$ [Mg(OH)$_2$]; (5) Ca $pK_{sp} = 5.26$ [Ca(OH)$_2$].

With Eq. (16–11) it is possible to plot the free metal ion concentration as a function of pH. A plot of this type for several metal ions is illustrated in Fig. 16–4.

From Eq. (16–10) the following can be written:

$$K_s\alpha_0 = K'[M^{2+}] \tag{16-12}$$

Combining (16-11) and (16-12) and converting to log functions gives

$$\log K' = \log K_s + \log \alpha_0 + pK_{sp} + 2\,pH - 28 \tag{16-13}$$

Equation (16-13) describes the relationship between the conditional constant and the pH of the solution considering the opposing pH effects on hydrolysis and on chelating power of the titrant.

Figure 16–5 illustrates the calculated $K' - pH$ curve using Eq. (16-13) for four different metal ions. [Equation (16-13) must be modified for the calculation of the Sc(III) curve since it was derived specifically for a divalent metal ion.] In Fig. 16–5, the dashed lines signify the relationship if hydrolysis

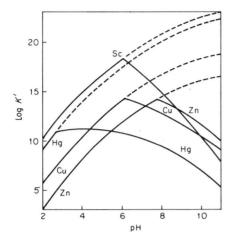

**Fig. 16-5.**   Log conditional stability constant as a function of pH in the presence of EDTA: effect of hydrolysis. [Taken from C. N. Reilley, R. W. Schmid, and F. S. Sadek, *J. Chem. Ed.* **36**, 555 (1959).]

were not taking place. From these data the optimum pH for the titration of Sc(III), Hg(II), Cu(II), and Zn(II) would be predicted to be 6, 3, 6, and 7.5, respectively.

The conditional constants at an optimum pH are compared to the thermodynamic constants in Table 16–2. It should be noted that, due to the effect of hydrolysis, the order of stability is altered upon titration at the optimum conditions.

Exact expressions describing hydrolysis are difficult to derive, because the hydrolysis reactions are not clearly understood. Even for cases where the exact hydrolysis species are known, often the equilibrium constants are not known. However, reasonable qualitative agreement between calculated and observed hydrolysis effects is usually observed.

**Table 16-2.   Comparison of the Thermodynamic and Conditional Stability Constant at Optimum pH Illustrating the Effect of Hydrolysis**

| Metal ion | log $K$ (thermodynamic) | log $K$ (conditional) |
|-----------|-------------------------|------------------------|
| $Sc^{3+}$ | 23.1 | 18.4 |
| $Hg^{2+}$ | 21.8 | 10.9 |
| $Cu^{2+}$ | 18.79 | 14.1 |
| $Zn^{2+}$ | 16.5 | 14.0 |

**Ligand Effect.**     A particular pH is achieved by the addition of a buffer to the solution. However, buffers are usually composed of species that can act as competing ligands. A typical example is ammonia–ammonium type buffers where the possibility of the formation of metal–ammonia complexes exists. The addition of a competing ligand can also prevent the hydrolysis of the metal ion.

It can be concluded from Eq. (16-1) that the presence of a competing ligand reduces the free metal ion concentration. This in turn leads to a lower conditional constant.

Calculation of the ligand effect is similar to calculation of the pH effect. Assume metal $M^{n+}$ is being titrated and ammonia is the competing ligand. If the metal shows maximum coordination of four toward ammonia, stepwise coordination to

$$M^{n+} + 4NH_3 \rightleftharpoons [M(NH_3)_4]^{n+}$$

with the constants $K_1$, $K_2$, $K_3$, and $K_4$ describing each step can be written.

If $[M]$ is the concentration of metal ion, $M^{n+}$, in solution, the total concentration of metal ion $[M]'$ in solution is given by

$$[M]'\beta_0 = [M] \tag{16-14}$$

where $\beta_0$ represents the fraction of the total metal ion in solution that is present as the free metal ion.*

The mathematical procedure for the derivation of $\beta$ is the same as the one used for $\alpha$. Therefore, from the equilibrium constant expressions for the metal–ammonia complexes and from

$$[M]' = [M] + [M(NH_3)] + [M(NH_3)_2] + [M(NH_3)_3] + [M(NH_3)_4]$$

it can be shown that

$$\frac{M'}{M} = \frac{1}{\beta_0} = 1 + K_1[NH_3] + K_1K_2[NH_3]^2$$
$$+ K_1K_2K_3[NH_3]^3 + K_1K_2K_3K_4[NH_3]^4 \tag{16-15}$$

If the stepwise constants are known, $\beta_0$ can be calculated as a function of ligand concentration (see Fig. 16–6). As the stability of the metal–ammonia complex increases or as the ammonia concentration increases, the concentration of free metal ion decreases. This in turn leads to a smaller conditional constant for the metal complex derived from the metal and titrant.

The effect of only the competing ligand on the equilibrium constant is seen

---

* $\beta$ is used rather than $\alpha$ to avoid confusion. Also, $\beta_0$ signifies fraction of uncoordinated metal ion, while $\beta_4$ would represent the fraction of metal ion that is coordinated with four ligands.

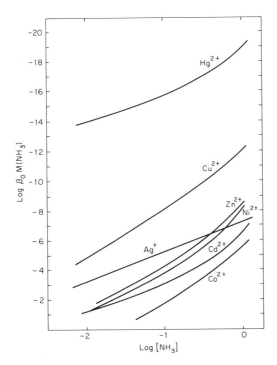

**Fig. 16–6.**   Effect of the concentration of the ligand ammonia on $\beta_0$.

by combining Eqs. (16-15) and (16-8),

$$K_s\beta_0 = K_{M'} = \frac{[MY^{2-}]}{[M]'[Y^{4-}]} \tag{16-16}$$

where $K_{M'}$ is the conditional constant accounting for the effect of a competing ligand in solution and $[M]'$ is the formal concentration of the metal ion in solution. Any competing ligand, providing the stepwise equilibrium constants are known, can be accounted for by this type of calculation. Thus, formation of hydroxy complexes which is preliminary to the hydrolysis effect can be handled.

**Conditional Constant: Combined Effect.**   The combined effect of pH and competing ligand on the stability constant for the metal complex formed in the titration is obtained by substitution of $[M]$ and $[Y^{4-}]$ in expression (16-8) by (16-5) and (16-14). Thus,

$$K_s = \frac{[MY]}{[M]'\beta_0[Y]'\alpha_0}$$

and

$$K_{M'Y'} = K_s \alpha_0 \beta_0 = \frac{[MY]}{[M]'[Y]'} \tag{16-17}$$

where $K_{M'Y'}$ is the conditional constant accounting for the pH and competing ligand effect. If the competing ligand is $OH^-$, depending on the pH and the system, the alteration of the metal ion may involve hydrolysis rather than the formation of competing soluble complexes.

Equation (16-17) allows the calculation of the conditional constant and the titration curve under a given set of experimental conditions. From practice it has been found that for a reasonably accurate titration the conditional formation constant must be at least $10^8$ or greater.

*Example 16-1.* Calculate the conditional formation constant for the titration of 50.0 ml of 0.100 $F$ $Zn^{2+}$ with 0.100 $F$ EDTA in a pH = 10 ammonical buffer (assume that in the buffer the concentration of the ammonia ligand, $NH_3$ = 0.100 $F$).

*Calculation of $\alpha_0$.* Since pH = 10, the $[H_3O^+]$ = 1 $\times$ $10^{-10}$. Substituting into Eq. (16-6) gives

$$\frac{1}{\alpha_0} = \frac{[10^{-10}]^4}{1.00 \times 10^{-2} \times 2.16 \times 10^{-3} \times 6.92 \times 10^{-7} \times 5.50 \times 10^{-11}}$$

$$+ \frac{[10^{-10}]^3}{2.16 \times 10^{-3} \times 6.92 \times 10^{-7} \times 5.50 \times 10^{-11}}$$

$$+ \frac{[10^{-10}]^2}{6.92 \times 10^{-7} \times 5.50 \times 10^{-11}} + \frac{[10^{-10}]}{5.50 \times 10^{-11}} + 1$$

$$\alpha_0 = 0.355$$

The same result can be obtained from Fig. 16–1.

*Calculation of $\beta_0$.* Since zinc (II) shows four coordination toward ammonia, four stepwise formation constants are needed. Substitution into Eq. (16-15) where $[NH_3]$ = 0.100 $F$ gives

$$\frac{1}{\beta_0} = 1 + 1.9 \times 10^2[0.100] + 1.9 \times 10^2 \times 2.1 \times 10^2[0.100]^2$$

$$+ 1.9 \times 10^2 \times 2.1 \times 10^2 \times 2.5 \times 10^2[0.100]^3$$

$$+ 1.9 \times 10^2 \times 2.1 \times 10^2 \times 2.5 \times 10^2 \times 1.1 \times 10^2[0.100]^4$$

$$\beta_0 = 4.68 \times 10^{-5}$$

The same result can be obtained from Fig. 16–6.

*Calculation of $K_{M'Y'}$.* The thermodynamic formation constant, $K_s$, for Zn–EDTA is given in Appendix V. Using this value and the calculated $\alpha_0$ and $\beta_0$ for the given experimental conditions in Eq. (16-16) gives the conditional formation constant, $K_{M'Y'}$.

$$K_{M'Y'} = K_s\alpha_0\beta_0$$

$$K_{M'Y'} = 3.2 \times 10^{16} \times 3.55 \times 10^{-1} \times 4.68 \times 10^{-5}$$

$$K_{M'Y'} = 5.32 \times 10^{11}$$

# CALCULATION OF A TITRATION CURVE

Calculation of the titration curve for a given set of experimental conditions is completed by considering the stoichiometry of the reaction and the conditional formation constant. Generally, the titration curve is a plot of pM vs volume of chelating titrant added.

*Example 16–2.*    Calculate the titration curve for the experimental conditions described in Example 16–1. The values for $\alpha_0$ and $\beta_0$ were calculated previously.
    The shape of the curve can be shown by calculation of only a few points.

*0 ml of Added EDTA.*    At the initial point of the titration the formal concentration of zinc ion is 0.100 *F*. However, this is not the concentration of the free zinc ion because of the formation of zinc–ammonia complexes. The uncomplexed zinc ion concentration in solution is calculated by Eq. (16-14) with a $\beta_0$ value calculated by Eq. (16-15) or interpreted from Fig. 16–6 (see Example 16–1).

$$\beta_0 = 4.68 \times 10^{-5} \ (\text{NH}_3 = 0.100 \ M)$$

$$[\text{Zn}^{2+}]' = 0.100 \ F$$

$$[\text{Zn}^{2+}] = [\text{Zn}^{2+}]'\beta_0 = 0.100 \times 4.68 \times 10^{-5}$$

$$[\text{Zn}^{2+}] = 4.68 \times 10^{-6} \ M; \quad \text{pZn} = 5.33$$

*25.0 ml of Added EDTA.*

$$50.0 \ \text{ml} \times 0.100 \ F = 5.00 \ \text{mmole of Zn}^{2+} \text{ started}$$

$$\frac{25.0 \ \text{ml}}{75.0 \ \text{ml}} \times 0.100 \ F = \frac{2.50 \ \text{mmole of EDTA added}}{2.50 \ \text{mmole Zn}^{2+} \text{ in excess}}$$

$$C_{\text{Zn}^{2+}} = [\text{Zn}^{2+}]' \cong \frac{2.50 \ \text{mmole}}{75.0 \ \text{ml}} = 0.0333 \ F$$

It is assumed in the above calculation that the Zn–EDTA complex is completely undissociated. Since the formation constant is very large, this assumption is reasonable. Using Eq. (16-14)

$$[Zn^{2+}] = 4.68 \times 10^{-5} \times 3.3 \times 10^{-2}$$

$$[Zn^{2+}] = 1.54 \times 10^{-6} \; M; \quad pZn = 5.81$$

The pZn at other points prior to reaching the stoichiometric point can be calculated in the same way. However, as the stoichiometric point is approached and the zinc ion concentration begins to decrease rapidly the above approximation should not be made.

*50.0 ml of Added EDTA.*   At 50.0 ml of added EDTA the stoichiometric point of the titration has been reached. At this point the total volume will be 100 ml. Since the formation constant of the complex is large, the amount of complex that dissociates is negligible in comparison to the concentration of the complex. Therefore,

$$C_{ZnY^{2-}} = [ZnY^{2-}] \cong \frac{0.100 \; F \times 50.0 \; \text{ml}}{50.0 + 50.0 \; \text{ml}} = 0.0500 \; M$$

Furthermore, it can be stated as an excellent approximation that the concentration of uncomplexed EDTA is equal to the concentration of uncomplexed zinc ion or

$$[Y^-]' = [Zn^{2+}]'$$

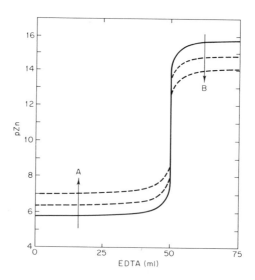

**Fig. 16-7.**  Zn–EDTA titration curve. Solid line: Calculated titration curve for the titration of 50.0 ml of 0.100 $F$ Zn (II) with 0.100 $F$ EDTA at pH 10. Dotted line: Effect of variables on the titration curve. (A) Increasing competing ligand concentration; (B) decreasing pH for the titration.

Substitution in Eq. (16-17) (see Example 16–1 for the calculation of $\alpha_0$):

$$3.2 \times 10^{16} \times 3.55 \times 10^{-1} \times 4.68 \times 10^{-5} = \frac{0.0500}{[Zn^{2+}]'[Zn^{2+}]'}$$

$$[Zn^{2+}]' = 3.07 \times 10^{-7} \ M$$

Calculation of the free $[Zn^{2+}]$ in solution is with Eq. (16-14):

$$[Zn^{2+}] = 4.68 \times 10^{-5} \times 3.07 \times 10^{-7}$$

$$[Zn^{2+}] = 1.44 \times 10^{-11}; \quad pZn = 10.8$$

*75.0 ml of Added EDTA.*

$$75.0 \ ml \times 0.100 \ F = 7.50 \text{ mmole of EDTA added}$$

$$\frac{50.0 \ ml \times 0.100 \ F = 5.00 \text{ mmole of } Zn^{2+} \text{ started}}{125 \quad ml \qquad\qquad 2.50 \text{ mmole EDTA in excess}}$$

$$C_{Y^{4-}} = [Y^{4-}]' \cong \frac{2.50 \text{ mmole}}{125 \text{ ml}} = 0.0200 \ M$$

Also,

$$C_{ZnY^{2-}} = [ZnY^{2-}]' \cong \frac{5.00 \text{ mmole}}{125 \text{ ml}} = 0.0400 \ M$$

Substitution into Eq. (16-17) gives

$$3.2 \times 10^{16} \times 3.55 \times 10^{-1} \times 4.68 \times 10^{-5} = \frac{0.0400}{[Zn^{2+}]' \ 0.0200}$$

$$[Zn^{2+}]' = 3.76 \times 10^{-12} \ M$$

and from Eq. (16-14)

$$[Zn^{2+}] = 4.68 \times 10^{-5} \times 3.75 \times 10^{-12}$$

$$[Zn^{2+}] = 1.76 \times 10^{-16}; \quad pZn = 15.75$$

Figure 16–7 illustrates the calculated titration curve for the titration of $Zn^{2+}$ with EDTA at pH 10 using an ammonical buffer. Qualitatively, the effect of a competing ligand and pH on the titration are also illustrated in Fig. 16–7. As the concentration of the competing ligand ($NH_3$) increases, that part of the titration curve preceding the stoichiometric point rises. In contrast, an increasing pH lowers the portion of the curve past the stoichiometric point. Consequently, if the titration is performed at a pH considerably lower than the optimum pH in the presence of a powerful competing ligand the pM break is greatly reduced. This reduction can in fact be large enough so that little or no break in pM occurs. Although Fig. 16–7 shows the effect qualitatively, a quantitative description is easily obtained by simply altering the experimental conditions in Examples 16–1 and 16–2. The minimum pH suggested in

Fig. 15–2 for an effective titration of metal ions with EDTA is based on a combination of these effects.

## OTHER EXAMPLES

The conditional constant concept and related calculations are not restricted to the metal ion–EDTA titration. In general, these can be extended to other types of complex equilibrium systems. The following two examples are only a partial illustration of this extention.

*Example 16–3.*     Calculate the solubility of $CaC_2O_4$ in solution maintained at a pH of 2; $K_{sp} = 4.00 \times 10^{-9}$, $K_{a_1} = 5.90 \times 10^{-2}$, $K_{a_2} = 6.40 \times 10^{-5}$.

This is not a simple solubility product type of system since $CaC_2O_4$ is the salt of a weak polyprotic acid. Therefore, the equilibria in this system are described by

$$CaC_2O_{4(s)} \rightleftharpoons Ca^{2+} + C_2O_4^{2-}$$

$$H_2C_2O_4 + H_2O \rightleftharpoons H_3O^+ + HC_2O_4^-$$

$$HC_2O_4^- + H_2O \rightleftharpoons H_3O^+ + C_2O_4^{2-}$$

and the equilibrium constant expressions are

$$K_{sp} = 4.00 \times 10^{-9} = [Ca^{2+}][C_2O_4^{2-}]$$

$$K_{a_1} = 5.90 \times 10^{-2} = [H_3O^+][HC_2O_4^-]/[H_2C_2O_4]$$

$$K_{a_2} = 6.40 \times 10^{-5} = [H_3O^+][C_2O_4^{2-}]/[HC_2O_4^-]$$

Since

$$[Ca^{2+}] = S \text{ (solubility)}$$

$$[C_2O_4^{2-}] = S\alpha_0$$

where the value $\alpha_0$ is the fraction of the total oxalate containing species present as $C_2O_4^{2-}$, or

$$C_{total} = H_2C_2O_4 + HC_2O_4^- + C_2O_4^{2-}$$

it can be shown [see Eqs. (16-2) to (16-6)] that

$$\frac{1}{\alpha_0} = \frac{[H_3O^+]^2}{K_{a_1}K_{a_2}} + \frac{[H_3O^+]}{K_{a_1}} + 1 \quad \text{or} \quad \alpha_0 = \left[ \frac{[H_3O^+]^2}{K_{a_1}K_{a_2}} + \frac{[H_3O^+]}{K_{a_1}} + 1 \right]^{-1}$$

Therefore,

$$4.00 \times 10^{-9} = (S)\,(S) \left[ \frac{[H_3O^+]^2}{K_{a_1}K_{a_2}} + \frac{[H_3O^+]}{K_{a_1}} + 1 \right]^{-1}$$

$$4.00 \times 10^{-9} = S^2 \left[ \frac{(1.00 \times 10^{-2})^2}{5.90 \times 6.40 \times 10^{-7}} + \frac{(1.00 \times 10^{-2})}{5.90 \times 10^{-2}} + 1 \right]^{-1}$$

$$S = 3.27 \times 10^{-4} \; M$$

f the pH is ignored, the solubility calculated form the $K_{sp}$ expression would be $4.00 \times 10^{-9})^{1/2} = 6.33 \times 10^{-5} \; M$

*Example 16-4.*    Calculate the molar concentration of the species in a phosphate solution at pH = 6.80 and a total phosphate concentration of 1.00 $F$.

The distribution diagram ($\alpha$ vs pH) for $H_3PO_4$ is given in Appendix VI. It can be concluded from this figure that the concentration of $H_3PO_4$ and $PO_4^{3-}$ are negligible and that $\alpha_1 = 0.28$ and $\alpha_2 = 0.72$. Since $C_{total} = 1.00 \; M = [H_2PO_4^-] + [HPO_4^{2-}]$ $H_3PO_4$ and $PO_4^{3-}$ are negligible), the concentration of the principle species are

$$[H_2PO_4^-] = 0.72 \; M$$

$$[HPO_4^{2-}] = 0.28 \; M$$

These would also be the concentrations of $NaHPO_4$ and $NaH_2PO_4$ that would be required to prepare a buffer of pH 6.80. Therefore, the distribution diagram for weak acids (or bases) can be used to predict concentrations for buffers prepared from these acids and their salts.

## Questions

1. What experimental factors can be adjusted which improves the selectivity of EDTA titrations?
2. Explain why the free $Ca^{2+}$ concentration in the presence of EDTA is greater in acid solution than in ammonical solution.
3. Explain why $Fe^{2+}$ is more readily oxidized in the presence of EDTA than in its absence.
4. What is the hydrolysis effect?
5. What is the significance of $\alpha_0$?
6. Derive an expression of $\alpha_0$ for $H_3PO_4$.
7. Derive an expression of $\alpha_1$ for $H_3PO_4$.
8. Derive an expression of $\alpha_2$ for $H_3PO_4$.
9. Derive an expression of $\alpha_3$ for $H_3PO_4$.
10. Using the equations derived in Questions 6 to 9 plot $\alpha$ vs pH. What is the significance of this plot?
11. What is the significance of $\beta_0$?
12. Draw a typical titration curve of pH vs ml of EDTA titrant.
    a. Illustrate the effect of pH on the shape of the titration curve.

b. Illustrate the effect of $NH_3$ (assume a metal : $NH_3$ complex can form) on th titration curve at different levels of $NH_3$ concentration.

c. Illustrate the effect of dilution on the titration curve.

d. Assume that the metal system is a mixture of two different metal ions. Illus trate the effect of a difference in $K$ for the metal–EDTA complexes.

## Problems

1.* Calculate the conditional formation constant for $Cu^{2+}$–EDTA as a function of pH (3–11) ignoring the hydrolysis and competing ligand effect.

2.* Calculate the conditional formation constant for $Cu^{2+}$–EDTA in a pH = 10 ammonical buffer (assume that $[NH_3] = 0.100 \ M$).

3.* Show by calculation of the conditional formation constants that $Cd^{2+}$ can be titrated in the presence of $Mg^{2+}$ at pH 5.5 with EDTA. (Ignore hydrolysis and competing ligand effect.)

4. Show by calculation of the conditional formation constants that $Th^{4+}$ can be titrated in the presence of $Nd^{3+}$ at pH = 3.0. (Ignore hydrolysis and competing ligand effect.)

5.* Calculate the conditional formation constant for $Hg^{2+}$–EDTA from pH 2–8 considering the pH and hydrolysis effect. What is the optimum pH for the titration of $Hg^{2+}$ with EDTA?

$$K_{sp,Hg(OH)_2} = 6 \times 10^{-26}$$

6. Determine the titration curve for the titration of $Cd^{2+}$ with EDTA at pH = 10, where $[NH_3] = 0.01 \ M$, by plotting pCd vs ml of titrant. Assume that 40.0 ml of $0.100 \ F \ Cd^{2+}$ is being titrated with $0.100 \ F$ EDTA. Calculate the titration curve again except using the condition $[NH_3] = 1.0 \ M$. Plot the two curves together and describe the effect of $NH_3$ concentration on the titration curve.

---

* Answers are listed at the end of the book for problems marked with an asterisk.

# Chapter
# Seventeen
# Fundamental
# Principles of
# Optical Spectroscopy

## INTRODUCTION

Absorption and emission of radiant energy by molecules and atoms is the basis for many methods in analytical chemistry. By interpretation of these data both qualitative and quantitative information can be obtained. Qualitatively, the positions of the absorption and emission lines or bands which occur in the electromagnetic spectrum indicate the presence of a specific substance. Quantitatively, the intensities of the same absorption and emission lines or bands for the unknown and standards are measured. The concentration of the unknown is then determined from these data.

The data obtained from a spectroscopic measurement are in the form of a plot of radiant energy absorbed or emitted as a function of position in the electromagnetic spectrum. This is known as a spectrum and the position of absorption or emission is measured in units of energy, wavelength, or frequency.

**Regions of the Electromagnetic Spectrum.**     Optical spectroscopy includes the region in the electromagnetic spectrum between 100 Å (124 eV) and 400 $\mu$m (3.1 × 10$^{-3}$ eV). The regions of the electromagnetic spectrum are tabulated in Table 17–1 along with the type of spectra obtained. At wavelengths shorter than the vacuum ultraviolet, nuclear interactions occur, and these waves are known as X-rays and $\gamma$-rays. At the other end of the electromagnetic spectrum, the longer wavelength regions are known as the microwave and radiowave regions (includes electron and nuclear magnetic resonance region) where precession of unpaired electrons and certain nucleii can be observed.

**Table 17–1. Regions of the Electromagnetic Spectrum**

| Region | Limits (common units)[a] | Wavenumber limits ($cm^{-1}$) | Frequency[b] limits (Hz) |
|---|---|---|---|
| X-rays | $10^{-2}$–$10^2$ Å | | $10^{20}$–$10^{16}$ |
| Far-ultraviolet (vacuum UV) | 10–200 nm | | $10^{16}$–$10^{15}$ |
| Near-ultraviolet | 200–400 nm | | $10^{15}$–$7.5 \times 10^{14}$ |
| Visible | 400–750 nm | 25,000–13,000 | $7.5 \times 10^{14}$–$4.0 \times 10^{14}$ |
| Near-infrared | 0.75–2.5 $\mu$m | 13,000–4,000 | $4.0 \times 10^{14}$–$1.2 \times 10^{14}$ |
| Mid-infrared | 2.5–50 $\mu$m | 4000–200 | $1.2 \times 10^{14}$–$6.0 \times 10^{12}$ |
| Far-infrared | 50–1000 $\mu$m | 200–10 | $6 \times 10^{12}$–$10^{11}$ |
| Microwaves | 0.1–100 cm | 10–$10^{-2}$ | $10^{11}$–$10^8$ |
| Radiowaves | 1–1000 m | | $10^8$–$10^5$ |

[a] Regions are most frequently expressed in these units.
[b] Calculated from $\nu = c/\lambda$.

The spectrum is divided into a series of regions corresponding to the type of absorption or emission obtained. For example, in the ultraviolet and visible region, electronic transitions of atoms and molecules are observed, whereas in the infrared region molecular vibration is observed.

**Units of Measurement.** Position of absorption or emission can be expressed by three different units; units of wavelength, frequency, and energy. The unit for wavelength, $\lambda$, is the centimeter with subdivisions in millimicrons (m$\mu$, $10^{-7}$ cm), nanometers* (nm, $10^{-7}$ cm), angstroms (Å, $10^{-8}$ cm), and micrometers ($\mu$m, $10^{-4}$ cm). The unit for frequency is cycles per second (Hz), and the units of energy are given in electron volts (eV, keV, meV), calories (cal, kcal), wavenumbers ($cm^{-1}$), and ergs. Conversion factors between different units of energy are shown in Table 17–2.

Since absorption and emission are quantized (each species has discrete molecular or atomic energy levels), relationships between energy, frequency, and wavelength can be established. Energy is related to frequency by

$$E = h\nu \tag{17-1}$$

where $E$ is the energy of the photon emitted or absorbed in ergs, $h$ is Planck's constant, $6.626 \times 10^{-27}$ erg-sec, and $\nu$ is the frequency in Hz. Frequency is related to wavelength through $c$, the velocity of light, by

$$\lambda(cm) \times \nu(Hz) = c(3 \times 10^{10} \text{ cm/sec}) \tag{17-2}$$

Substituting Eq. (17-2) into (17-1) gives the relationship

$$E = hc/\lambda \tag{17-3}$$

* Nanometers is preferred over millimicrons.

**Table 17–2. Conversion Factors**

| Unit | ergs/molecule | $cm^{-1}$ | cal/mole | eV/molecule |
|---|---|---|---|---|
| 1 eV | $1.602 \times 10^{-12}$ | 8068.3 | 23,063 | 1 |
| 1 cal/mole | $6.946 \times 10^{-17}$ | 0.3498 | 1 | $4.336 \times 10^{-5}$ |
| 1 $cm^{-1}$ | $1.985 \times 10^{-16}$ | 1 | 2.858 | $1.239 \times 10^{-4}$ |
| 1 erg/molecule | 1 | $5.036 \times 10^{-15}$ | $1.439 \times 10^{16}$ | $6.242 \times 10^{11}$ |

If $\lambda$ is measured in centimeters, then

$$\frac{1}{\lambda} = \bar{\nu} \tag{17-4}$$

where $\bar{\nu}$ is expressed in units of $cm^{-1}$ (reciprocal centimeters or kaysers). Thus,

$$E = h\bar{\nu}c \tag{17-5}$$

gives the final relationship needed for the conversion of units.

It should be noted that as energy increases the wavelength decreases, while the frequency is directly proportional to energy. The units are chosen so that the numbers are easily used in both communication and calculation. It is much easier to present position in the visible region using angstroms or nanometers as opposed to units of frequency. For example, 2000 Å or 200 nm is much more convenient to use than $1.5 \times 10^{15}$ Hz.

*Example 17–1.* Convert 2000 Å to cm, $cm^{-1}$, eV, erg, cal, and Hz.

| Conversion | | Factor* |
|---|---|---|
| Å → cm | 2000 Å = $2.000 \times 10^3$ Å | $10^{-8}$ cm/Å |
| | $2.000 \times 10^3$ Å $\times 10^8$ cm/Å = $2.00 \times 10^{-5}$ cm | |
| cm → $cm^{-1}$ | $\dfrac{1}{2.000 \times 10^{-5}} = 5.000 \times 10^4$ $cm^{-1}$ | $\dfrac{1}{cm} = cm^{-1}$ |
| $cm^{-1}$ → eV | $\dfrac{5.000 \times 10^4 \ cm^{-1}}{8068.3 \ cm^{-1}/eV} = 6.200$ eV | 8068.3 $cm^{-1}$/eV |
| cm → erg | $E_{ergs} = \dfrac{6.626 \times 10^{-27} \ \text{erg-sec} \times 3.00 \times 10^{10} \ \text{cm/sec}}{2.000 \times 10^{-5} \ \text{cm}}$ | $E = \dfrac{hc}{\lambda}$ |
| | $E_{ergs} = 9.940 \times 10^{-12}$ erg/molecule | |

---

\* Not all significant figures are included in the factor.

eV → cal      6.200 eV × 23,063 cal mole$^{-1}$ eV$^{-1}$      23,063 cal mole$^{-1}$ eV$^{-1}$

$$= 1.430 \times 10^5 \text{ cal mole}^{-1}$$

cm → Hz      2.000 × 10$^{-5}$ cm × $\nu$ = 3.00 × 10$^{10}$ cm/sec      $\lambda \times \nu = 3 \times 10^{10}$

$$\nu = \frac{3.00 \times 10^{10} \text{ cm/sec}}{2.00 \times 10^{-5} \text{ cm}} = 1.5 \times 10^{15} \text{ Hz}$$

**Absorption and Emission of Electromagnetic Radiation.** When an atom or molecule absorbs energy, the atom or molecule will move to a higher energy state, or excited state. A definite energy level can be assigned to each excited state, and the many possible levels will be characteristic of the particular atom or molecule. A simple energy level diagram for either atoms or molecules is given in Fig. 17–1. The two horizontal lines represent two energy levels within the species, $E°$ being the ground electronic state and $E^*$ being the excited electronic state. An electron is capable of undergoing a transition from the $E°$ to the $E^*$ state if energy in the form of light or heat is added. This absorption of energy causes the molecule or atom to be in an excited state.

Once in the excited state, the species can eliminate the excess energy by a number of processes. First, the energetic particle may collide with solvent molecules or other molecules and transfer its energy to its environment. Second, the species may deactivate by releasing a photon (emission) equivalent to the difference in energy of the two levels, $E^*$ and $E°$. In both cases, the molecule or atom terminates in the ground electronic state.

Absorption of white light by a potassium permanganate solution results in a purple coloration. This color is due to absorption of the green component in the white light. The combination of transmission of the red and blue components results in the production of a purple coloration. Since the green is being absorbed, the difference in energy levels ($E°$ and $E^*$) corresponds to about 5000 Å or about 3.5 eV. Similarly, when sodium is heated in a flame, the atoms are excited ($E°$ to $E^*$). These excited atoms emit yellow light corresponding to a wavelength of about 6000 Å (~3.0 eV). Thus, the fundamental difference between emission and absorption is that deactivation of

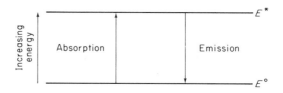

**Fig. 17–1.** A simple energy level diagram. $E°$, lower energy level; $E^*$, higher energy level.

an electron leads to emission of energy, while promotion of an electron leads to absorption of energy.

The movement of an electron from $E°$ to $E^*$ (absorption) requires the addition of energy and the energy for this transition is equal to the difference between the two energy levels. Emission is the reverse case in which the electron is deactivated from $E^*$ to $E°$ with emission of a photon. The energy of radiation emitted will be equivalent to the difference between $E°$ and $E^*$.

The questions that should be asked at this particular point are what is the experimental result of such transitions, what types of spectra are obtained, and what are their physical appearances?

**Differences in the Origin and Appearance of Molecular and Atomic Spectra.**     The atoms in the molecule can vibrate and rotate with respect to one another. These vibrations and rotations also have discrete energy states, but are of lower energy difference between the upper and lower levels when compared to electron transitions. A number of vibrational and rotational energy levels exist above each electronic level (see Fig. 17–2). Hence, it can be concluded that electronic transitions require the most energy, while rotational transitions require the least energy.

It is possible to have transitions from many vibrational and rotational energy levels within the ground electronic state to corresponding vibrational and rotational levels within the excited electronic state. Because of this, the spectrum obtained from molecular species appears to be broad. This type

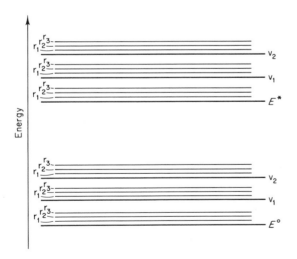

**Fig. 17–2.**   Molecular energy diagram. Electronic ($E$), vibrational (v), and rotational (r) energy levels are shown.

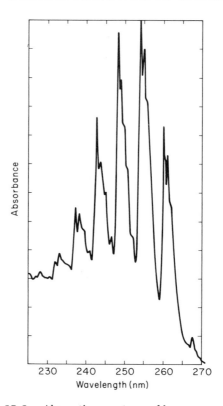

**Fig. 17–3.**   Absorption spectrum of benzene vapor.

of spectra is termed band spectra because the composite of the transitions is a series of lines which form the band. To illustrate this, the absorption spectrum of benzene vapor is given in Fig. 17–3. Each peak in the spectrum corresponds to a transition from different vibrational levels between the two electronic states $E°$ and $E^*$. Furthermore, there are many more transitions than peaks because the instrument used to observe this band, the spectrometer, is not capable of resolving each particular transition.

In the case of an atom, it is obvious that the species cannot vibrate or rotate as a molecule. Thus, the atom does not have vibrational or rotational energy levels. A transition between energy levels of an atomic species results in very sharp lines, as shown in Fig. 17–4. In this figure the absorption spectrum of sodium vapor is presented along with its partial energy level diagram. These spectra, which originate from atoms, are known as line spectra.

It becomes quite easy to determine whether the radiation absorbed or emitted by a species is from a molecule or atom. A molecule, as stated previously, produces a band spectrum, whereas the atom produces a line spectrum. Furthermore, since the energy difference between electronic

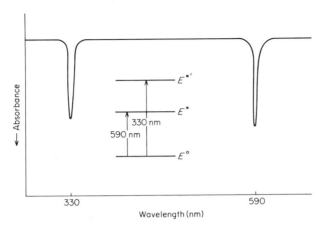

**Fig. 17–4.** Absorption spectrum of sodium vapor.

levels of molecules and atoms is different, the position of these bands or lines may be used to qualitatively determine the species being observed.

## BEER'S LAW: LAW OF PHOTOMETRY

Quantitative analytical spectroscopy is based on two fundamental laws. These laws apply to the change in radiant power of a monochromatic light beam with changing sample pathlength, $b$, and changing concentration, $c$.

The first of the two laws, generally attributed to Bouger, states that as the thickness of an absorbing sample increases, the amount of light transmitted through the sample at an absorbing wavelength decreases.

The incident power of monochromatic radiation is defined as $P_0$ and the transmitted power, or the amount of radiation which is not absorbed by the sample, is defined as $P$. If $P/P_0$ is plotted vs $b$, the result would be a curve similar to that depicted in Fig. 17–5. This nonlinear plot gives the relationship between the ratio of $P/P_0$ and $b$.

The equation describing the curve in Fig. 17–6 is given by

$$\ln P/P_0 = -kb \qquad (17\text{-}6)$$

where $k$ is a proportionality constant relating to the amount of absorption of radiation.

Another useful term is transmittance and is defined as

$$T = P/P_0 \qquad (17\text{-}7)$$

and it follows that percent transmittance ($\%T$) is the product of $T \times 100\%$

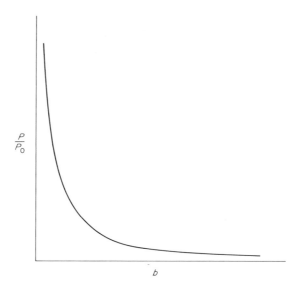

**Fig. 17–5.** Plot of $P/P_0$ vs pathlength.

The recommended term used is absorbance $(A)$ and may be defined in equation form as

$$-\log T = -\log P/P_0 = A \qquad (17\text{-}8)$$

The second law states that as the concentration of an absorbing solution increases, the transmitted power at an absorbing wavelength decreases in a logarithmic manner equivalent to the pathlength relationship.

$$\ln (P/P_0) = -k'c \qquad (17\text{-}9)$$

Hence, Eq. (17-9) can be used to relate concentration to the ratio of $P$ and $P_0$ where the constant $k'$ is related to the intensity of absorption.

By combining Eqs. (17-6) and (17-9) and converting to base 10 logarithms, the following is obtained:

$$\log(P/P_0) = -kbc \qquad (17\text{-}10)$$

Inserting absorbance [Eq. (17-8)] gives the Beer's Law relationship

$$A = kbc \qquad (17\text{-}11)$$

where $k$ is the absorptivity for the species being observed. This constant is a relative measure of the absorption intensity of a compound and is dependent upon the solvent, wavelength, and total environment, but independent of pathlength and concentration.

Since percent transmission and absorbance are used extensively, the

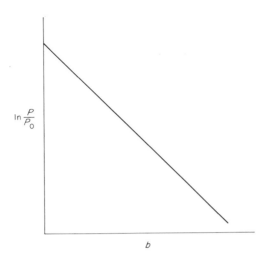

**Fig. 17–6.** Plot of ln $(P/P_0)$ vs pathlength.

relationship between these two terms can be written as

$$A = \log \frac{1}{\%T/100} = 2 - \log \%T \tag{17-12}$$

For a particular molecular species the absorbance, percent transmission, and transmission are a function of concentration. The pathlength and concentration are defined in any units. Since absorbance is a dimensionless number, absorptivity has the inverse units of pathlength and concentration. If the pathlength and concentration are expressed in centimeters and moles/liter, the units for the absorptivity are liters mole$^{-1}$ centimeter$^{-1}$. For this case the absorptivity is the molar absorptivity, $\epsilon$, and Eq. (17-11) becomes

$$A = \epsilon bc \tag{17-13}$$

*Example 17–2.*    A water solution of a colored compound has a molar absorptivity ($\epsilon$) of 3200 at 525 nm. Calculate the absorbance and percent transmission of a $3.40 \times 10^{-4}$ $F$ solution if a 1.00 cm cell is used.

$$A = \epsilon bc = 3200 \text{ liters/mole-cm} \times 1.00 \text{ cm} \times 3.40 \times 10^{-4} \text{ mole/liter}$$
$$A = 1.09$$
$$A = 2 - \log \%T$$
$$1.09 = 2 - \log \%T$$
$$-0.91 = -\log \%T; \quad \%T = 8.1\%$$

*Example 17–3.*    Antimycin, an experimental fungicide, has an absorption maximum at 320 nm. If a $1.1 \times 10^{-4}$ $M$ solution of this compound produces an absorbance of 0.52, calculate the molar absorptivity of this compound (assuming a 1-cm pathlength).

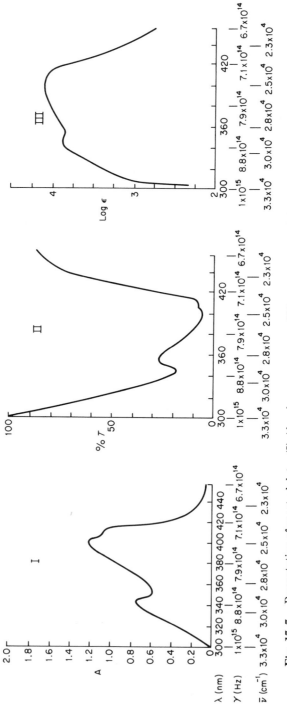

**Fig. 17-7.** Presentation of spectral data. (I) Absorbance vs energy; (II) percent transmission vs energy; (III) log ε vs energy.

$$A = \epsilon bc$$

$$\epsilon = \frac{A}{bc}$$

$$\epsilon = \frac{0.52}{1.1 \times 10^{-4} \text{ mole liter}^{-1} \times 1.0 \text{ cm}} = 4.8 \times 10^{-3} \text{ liter mole}^{-1} \text{ cm}^{-1}$$

**Presentation of Data: The Spectrum and Concentration**     The criterion for presentation of a spectrum must follow a plot of energy (emitted or absorbed) vs a function of absorption of radiation. The most common methods are to plot either absorbance, percent transmission, or log $\epsilon$ vs wavelength, energy or frequency (Fig. 17–7).

In Fig. 17–8 absorption curves a, b, c, and d represent increasing concentrations of the sample material. As the concentration increases, the absorbance $(A)$ increases while the percent transmission $(\%T)$ decreases. An example of a Beer's law plot is also shown in Fig. 17–8. A linear function as predicted by Beer's law is found when $A$ is plotted vs concentrations a, b, c, and d at a wavelength near the absorption maxima. The slope of the line is the molar absorptivity, $\epsilon$, provided the pathlength is 1 cm. A calibration curve of $\%T$ vs concentration can also be used, but has the disadvantage of not providing a linear relationship.

**Deviations from Beer's Law.**     A deviation from Beer's law is observed when a plot of concentration vs absorbance is nonlinear as shown in Fig. 17–8. Deviation toward the ordinate is known as a positive deviation, while

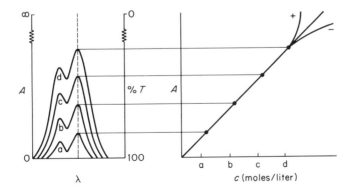

**Fig. 17–8.**   Correlation of spectra with concentration. This figure represents the conversion of spectral data into a Beer's law plot. Note that deviations can occur in a positive $(+)$ or negative $(-)$ direction.

the deviation toward the abcissa is known as a negative deviation. In either case, strict adherence to Beer's law (absorbance being proportional to concentration) does not exist in all regions. However, in most systems there is a region of concentration in which a linear relationship exists.

The more important factors that produce deviation are:

1. Environment, such as temperature, pressure, and solvent
2. Instrumental errors, such as stray radiation, stability of the radiation source, detector, wavelength selector, slit control, electronics, and reliability of the optical parts
3. Chemical deviations, including changes in chemical equilibrium such as pH, presence of complexing agents, competitive metal ion reactions, and concentration dependence
4. Refractive index changes in the sample
5. Nonmonochromaticity of radiation


## INSTRUMENTATION

In general, the basic design of the instrumentation used in the optical regions of the electromagnetic spectrum is the same. The individual components of the instrument, however, might be different according to the optical region being studied. For example, an infrared detector which responds to heat change is more efficient than a photocell. The latter detector is more useful in the ultraviolet and visible regions. All the optical components must be transparent toward the region being studied. Thus, different substances are used for the optics in the various regions.

The instrument can be divided into a series of components. These are the (a) radiation source, (b) monochromator, (c) sample container, and (d) detector. For an absorption instrument the source and sample portions are separated as shown in Fig. 17–9a. In comparison, an emission instrument combines the source and sample into one unit as shown in Fig. 17–9b.

The signal in an absorption measurement is the ratio of $P$, the transmitted

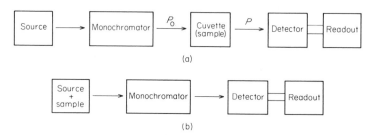

(a)

(b)

**Fig. 17–9.** Block diagram of a spectrometer. (a) Absorption spectrometer; (b) emission spectrometer.

monochromatic radiation and $P_0$, the incident radiation, while for emission the intensity of the emitted radiation is measured.

**Sources.**    The source units for absorption spectrometers must meet the following requirements. First, the emitted signal ($P_0$), in most cases, must be continuous radiation in the region being studied, and, second, it must be stable. Finally, the source must emit a measurable signal throughout the region. Ideally, a source, which gives a uniform intensity over the region, is most desirable. Unfortunately, this type of source is not available, and precautions must be taken to allow for changes in intensities by placing a wedge or an iris diaphragm in the radiation path. The purpose of these devices is to control liner response of the detector throughout the spectral region. Another method of varying the source intensity is to change the power to the source itself as the region is scanned.

**Monochromator.**    The monochromator is used to separate polychromatic radiation into a suitable monochromatic form. Several important advantages are gained by the use of monochromatic radiation. First, Beer's law is based on monochromatic radiation. Therefore, a closer adherence to Beer's law should be expected if the instrumental requirement of monochromatic radiation is fulfilled. Other advantages are that the sensitivity of the measurement is increased and the interference due to adverse compounds is decreased.

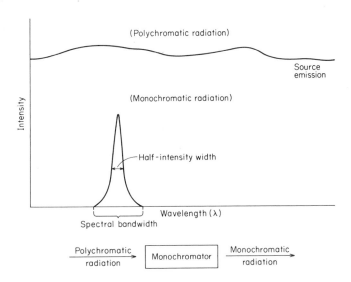

**Fig. 17–10.**  Comparison of monochromatic and polychromatic radiation. The polychromatic radiation is the radiation incident to the monochromator, while the monochromatic radiation is the energy transmitted by the monochromator (see block diagram).

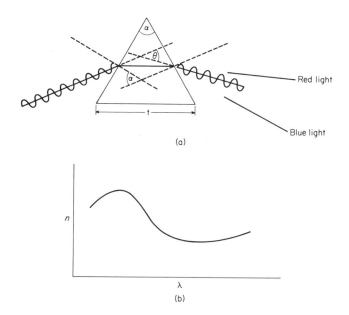

**Fig. 17–11.** Interaction of prism with radiation. (a) Refraction of radiation by a prism is shown where $\alpha$ is the refracting angle of the prism and $\theta$ is the angle of refraction; (b) change in refractive index with wavelength.

The monochromator unit consists of the following: (1) focusing lens, (2) an entrance slit, (3) a dispersing device, and (4) an exit slit.

The dispersing device controls the monochromatic character of the radiation which is incident to the sample. Figure 17–10 illustrates the conversion of polychromatic radiation to monchromatic radiation. The useful dispersing devices are prisms and gratings. There are, however, monochromators which use filters, such as colored glass, interference filters, and solution filters. The latter devices normally produce a transmitted wavelength of relatively large spectral bandwidths which can be reduced by using a combination of filters. A disadvantage of this arrangement is that some energy is always absorbed and reflected by the filter. Thus, the intensity of the radiant beam is reduced.

The most popular methods of producing monochromatic radiation are by prisms and gratings. Prisms separate white light into its components by means of refraction. This is illustrated in Fig. 17–11a. The separation of the wavelength is based upon the change in refractive index of the prism material as a function of wavelength. Thus, the dispersion of the individual wavelengths is nonlinear. That is, the separation between two wavelengths at higher energy is less than for two wavelengths of lower energy (see Fig. 17–11b).

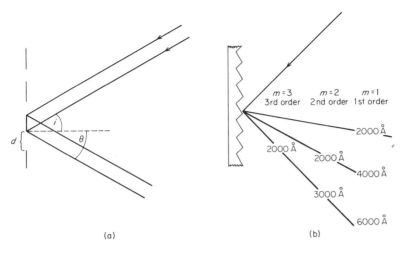

**Fig. 17–12.** Diffraction of radiation by a grating. (a) The diffraction of radiation where $d$ is the groove spacing, $i$ is the incident angle, and $\theta$ is the angle of diffraction; (b) the different orders observed in diffraction.

Gratings are prepared by cutting grooves in transmitting or reflecting plates. A typical surface would have between 2500 and 60,000 lines/inch, and its physical appearance is essentially an indistinguishable set of fine parallel straight lines. The grating operates on the basis of constructive and destructive interference of radiation (see Fig. 17–12). One of the main advantages of the grating is that it provides a linear dispersion with $\lambda$.

A prism or grating type instrument has two main advantages over a filter type instrument. First, the former can be used to scan an absorption spectrum, while the latter is used to measure absorption at one wavelength. The second advantage is that prism or grating instruments, because of higher resolution, will provide more detail in the spectrum and a linear absorbance response over a larger concentration range.

**Sample Containers.** Sample containers must satisfy two main requirements. They must be made of substances which are transparent in the wavelength region of interest, and they must be reproducible in pathlength or be designed in such a way that their pathlength may be easily determined. Table 17–3 lists some of the more commonly used substances for cells and their region of transparency. Also, other optical parts of the instrument, such as lenses, must be made from transmitting substances.

Designs for a variety of sample containers, which are available for observing absorption spectra in the ultraviolet, visible, and infrared regions, are shown in Fig. 17–13. Each sample cell is designed for a specific use. For example,

**Table 17–3.  Cell and Prism Material**

| Material | Wavelength region |
|---|---|
| *Ultraviolet–Visible* | |
| $SiO_2$ | 200 nm–4 $\mu$m |
| Soft glass | 350 nm–2.5 $\mu$m |
| Pyrex | 300 nm–2.5 $\mu$m |
| Vycor | 280 nm–2.5 $\mu$m |
| *Infrared[a]* | |
| NaCl (rock salt) | 15 $\mu$m |
| KBr | 27 $\mu$m |
| Irtran-2[b] | 14 $\mu$m |
| Crystal quartz | 4 $\mu$m |
| KCl | 20 $\mu$m |
| TlBr-TlCl | 30 $\mu$m |

[a] Wavelength listed is the maximum wavelength which is transmitted.

[b] A series of synthetic commercial materials several of which are resistant to aqueous solutions.

the cell in Figure 17–13a is used in the ultraviolet–visible region for obtaining the spectra of solutions. This cell has a 1-cm pathlength and the material from which it is constructed determines the wavelength region that is transmitted. For the ultraviolet–visible region the cell is usually made of quartz. If glass is used, the cell is useful only in the visible region.

Cells in simpler spectrometers generally use "test tube-type" cuvets. Because the surface is curved and inhomogeneous with respect to wall thickness, refraction of a certain amount of incident radiation occurs. Thus, different cuvette positions in the cell holder will produce different absorbances. Care must be taken to place the cuvet in the instrument in the same orientation for each measurement.

Gas cells generally are characterized by long pathlengths. A useful infrared cell for gaseous samples is shown in Fig. 17–13b, where the cell faces are sodium chloride.

Solid sampling in the infrared region is possible by one of several techniques. For example, the solid can be homogeneously mixed with KBr and pressed carefully into a pellet. This is illustrated in Fig. 17–13c.

Unfortunately, sampling for quantitative infrared measurements is more difficult than in the ultraviolet and visible regions because of difficulty in

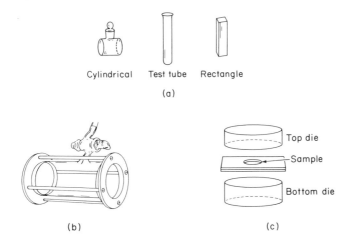

Cylindrical    Test tube    Rectangle

(a)

Top die

Sample

Bottom die

(b)                              (c)

Fig. 7–13.  Several cells useful in absorption spectroscopy. (a) Ultraviolet–visible cells; (b) infrared gas cell; (c) KBr pellet.

controlling the pathlength of the sample and its physical state. According to Beer's law, the absorbance is directly proportional to the pathlength. This means that the pathlength must be accurately known or reproducible to apply Beer's law to infrared measurements. The latter is most important if calibration curves are used. Therefore, as the several techniques are described below, it is important to consider the question, is the pathlength reproducible or measurable?

The sample will exist in one of three physical states, solid, liquid, or gas. The latter two are the easiest to handle. A liquid can be used neat or made into a solution. When this is placed into a cell, for example, a cavity cell, where a spacer separates two NaCl windows, the problem of pathlength appears to be minimal. The width of the spacer can be measured, or alternatively, the distance between the windows can be measured. Unfortunately, the cell arrangement requires the NaCl windows to be optically flat, and this is not always achieved in making routine measurements. In addition, water vapor from the air, sample, or from handling or cleaning can cause etches in the NaCl plates and affect the pathlength. If one is very careful in handling the liquid or solution sample in such a cell, the pathlength is fairly reproducible.

Gases are handled in a cell of the variety shown in Fig. 17–13. Pathlengths in these arrangements are easily reproduced and controllable.

The solid presents more of a problem. If it can be dissolved, the resulting solution can be handled in the cavity-type cell. The only difficulty is that the solvent should not have absorbance peaks in those regions in which the

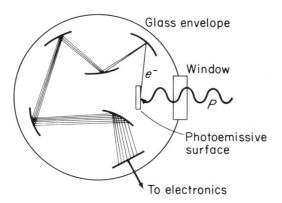

**Fig. 17–14.** The photomultiplier detector.

solid absorbs. If solutions can not be used, several other techniques are available. The pellet technique, previously mentioned, can be used. In this case the KBr–solid intimate mixture is subjected to 20,000–30,000 psi in a stainless steel hydraulic press. The pellet that is obtained should be transparent. Often anomalous spectra are found and are usually traced to a poorly cleaned press or to a change in the structure of the sample which is brought about by the pressure or grinding process. The dimensions of the pellet are easily measured, but the reproducing of pellet thickness is difficult.

In the "mull technique" a few milligrams of the solid is placed on a NaCl plate, and a drop or two of mineral oil is added. The mixture or mull is smeared by placing another NaCl plate on top of it in sort of a sandwich-type arrangement. Usually, the two plates are rubbed together until a clear mull is obtained. In this case, it is very difficult to reproduce or measure the film thickness. To obtain the best spectrum, a complete breakdown of the crystal into individual unit cells is desired, and, therefore, this should also be one of the goals of any sampling technique.

In the previous discussion reference is made only to NaCl as the material for cell windows. Other materials can also be used (see Table 17–3).

**Detectors.** The common detection devices are photomultipliers, photographic plates, thermocouples, and photoconductive cells. Each device is suited to a particular region of the electromagnetic spectrum.

Photomultipliers operate on the principle of photon amplification. A photon strikes a photocathode causing emission of electrons which are multiplied by striking a series of anodes resulting in electron multiplication (Fig. 17–14). In this way radiant energy is converted to electrical energy which can be easily measured through electronics. The detector is very sensitive and rapid in its response to radiation in the spectral range from 1000 to 12,000 Å.

**Table 17–4. Detectors for Electromagnetic Radiation**

| Region | Detector | Comments |
|--------|----------|----------|
| UV–visible | Photoemissive cell (or photocell) | Useful in 200 nm to 1 $\mu$m, sensitivity depends on alkali metal used, can be damaged by high–intensity radiation |
| | Photomultiplier cell | Same as photocell except large amplification brought about in tube through anode–dynode arrangement |
| | Photovoltaic cell (or Barrier layer cell) | Useful in 400–800 nm, rugged but suffers from fatigue |
| | Photographic plate and film | Chemical development needed, properties depend on the emulsion |
| | Eye | Not too sensitive, region of response varies but usually about 400–750 nm |
| Near IR | Photoconductive cell | PbS or PbSe or other semiconductor, used in region of 0.7–3.3 $\mu$m |
| IR | Thermocouple | High nonselective sensitivity in region 0.8–40 $\mu$m |
| | Bolometer | Resistance wire or thermistor which is part of a Wheatstone Bridge setup, high nonselective sensitivity in region of 0.8–40 $\mu$m |
| | Pneumatic cell (or Golay cell) | Based on total energy falling on detector, flexible and very sensitive, useful in range of 0.8–1000 $\mu$m |

The photographic plate or film has a particular advantage in that it integrates radiant energy over a period of time. Developing facilities, however, are needed to effectively use this type of detector.

The thermocouple is used mainly in the infrared region and operates on the basis of heat detection. Since neither the photographic plate or film or the photomultiplier are efficient in this region, the thermocouple is preferred in most infrared instruments.

Photoconductive cells, such as the barrier layer cell and PbS cell operate on the principle of an increase in conduction due to the impinging radiant energy. This type of detector is used in the visible, near infrared, and infrared regions.

Actually, detectors may be classed into either heat detectors (thermocouple, thermistor, bolometer, thermopile, Golay detector) or photon detectors (photomultiplier, barrier layer cell, PbS cell, photographic plate, or film). The main requirements are that the detectors respond to the region of application and that they are stable. Table 17–4 presents common detectors, their ranges, and restrictions.

A number of commercial instruments are readily available at a wide range of costs. The price range is directly proportional to the sophistication of the instrument. For example, useful features such as double beam, recording, automatic scanning, increased wavelength range, variable or controlled slit width, interchangeable optics, stability and linearity of electronics, and reliability of all instrumental parameters can add substantially to the overall cost of an instrument. In addition, the type of instrument to be used should reflect the nature of the problem.

## *Questions*

1. What are the different regions of the electromagnetic spectrum and for what are they used?
2. Describe the three methods of expressing position in the electromagnetic spectrum.
3. What is the difference between emission and absorption of radiation?
4. How do atomic and molecular spectra differ?
5. Describe Beer's law and define all of the terms.
6. What are the relationships between $P/P_o$, $T$, $\%T$, and $A$?
7. What are the different methods of presenting a spectrum and how are they related?
8. What causes deviations from Beer's law?
9. Describe the components of a spectrophotometer and discuss the use of each component.
10. How do a prism and a grating differ?

## *Problems*

1. Convert the following to units of wavelength ($\lambda$) in nm.
   *a.  6000 Å        b.  6.32 $\mu$m        c.  70.1 kcal/mole        d.  1 cm
   e.  3.2 eV/molecule  f.  73.0 ergs/molecule  g.  90.0 cal/mole  *h.  30,000 Hz
   i.  7200 Å      j.  0.3 keV/molecule      k.  60,000 cm$^{-1}$

2. Convert the following to units of energy in calories.
   *a.  7.9 eV/molecule    b.  80,000 Hz    c.  105 ergs/mole  *d.  0.70 kcal/mole
   e.  1.5 cm    f.  5200 Å    g.  9.0 keV/molecule    h.  300 cm$^{-1}$

3. Sodium emits lines at 590 nm and 3300 Å. Calculate the energy of each transition in eV/molecule.

4.* Convert the units in question 1 to units of frequency in Hz.

5. Convert the following absorbance values to $\%T$.
   *a.  0.21      b.  0.65      c.  0.78      d.  0.04      *e  1.21      f.  1.75
   g.  0.001      h.  1.9

---

* Answers are listed at the end of the book for problems marked with an asterisk.

6. Convert the following $\%T$ values to $A$.

  *a.  32%     b.  5.4%     c.  72%     *d.  52%     e.  0.01%

7. Calculate the absorbance of the following solutions.

  *a.  $1.03 \times 10^{-3}$ $M$ solution; $\epsilon = 720$; pathlength $= 1.0$ cm

  b.  16.0 $M$ solution; $\epsilon = 2.0$; pathlength $= 0.1$ cm

  c.  $3.2 \times 10^{-5}$ $M$ solution; $\epsilon = 30{,}000$; pathlength $= 1.0$ cm

8. Calculate the molar absorptivity of the following solutions.

  *a.  $A = 0.71$; concentration $= 1 \times 10^{-4}$ $M$; pathlength $= 1.0$ cm

  b.  $A = 0.53$; concentration $= 4.5 \times 10^{-1}$ $M$; pathlength $= 10.0$ cm

  c.  $A = 1.2$; concentration $= 4.7$ g/liter; pathlength $= 1.0$ cm; molecular weight of solute $= 120$

  d.  $A = 0.45$; concentration $= 3.1$ mg/ml; pathlength $= 2.0$ cm; molecular weight of solute $= 73$

9. A solution containing 0.701 mg of solute per 100 ml of solvent gives a percent transmittance of 40% in a 1.00 cm cell.

  *a.  What is the absorbance of the solution?

  b.  What would the absorbance and $\%T$ be if a 2.00-cm cell were used?

  c.  What would the absorbance and $\%T$ be if the solution were 0.420 mg of solute per 100.0 ml of solvent?

10*. From the series of spectral data in Fig. 17–7 calculate the concentration of the solute in the solvent (*Hint*: use I and III) assuming a 1.0-cm pathlength.

# Chapter Eighteen
# Qualitative Analysis: Ultraviolet, Visible, and Infrared

## INTRODUCTION

In general, spectroscopy is routinely used to examine unknown materials, both in the crude and pure states. These measurements can establish the presence, as well as the absence, of different elements, of functional groups, or provide other structural information. Fortunately, low cost spectrometers have become readily available in recent years. Thus, these instruments are now commonplace in most laboratories. Although spectroscopic methods are invaluable in routine practice, additional techniques should be utilized in confirming the structural characteristics of the molecule.

The basis for the identification is that the absorption spectrum shows a number of absorption bands associated with structural units within the molecule. For example, the absorption observed in the ultraviolet and infrared for the carbonyl group in acetone is at the same wavelength as the absorption observed for the carbonyl in diethyl ketone. Similarly, the location for hydroxyl group absorption in the infrared for methanol, ethanol, and propanol is essentially the same.

This constancy in the appearance of the bands, due to a particular structural grouping, led to the assignment of group frequencies. In some cases, these assignments were made on the basis of repeated experimental observations, while in others, assignments were arrived at from physical information or empirical relationships. More recently, calculations have been used to predict the frequencies.

In general, the interpretative process depends on comparing the position (wavelength) of absorption, the intensity ($\epsilon$) of the absorption, and the appearance of new bands in the spectrum (which are due to influence of one structural grouping on another one) to known assignments. This procedure is basically the same with complicated spectra.

# ELECTRONIC SPECTRA: ORGANIC COMPOUNDS

The spectrum obtained for a simple organic molecule in the gaseous state consists, as described in Chapter 17, of narrow peaks. Each peak represents a transition from the vibrational and rotational levels in the electronic ground state to a corresponding combination in the excited state. If the vibrational and rotational states were absent, a single discrete line corresponding to each transition within the possible electronic states would be found. As the compounds become more complex, an increase in the number of transitions between rotational and vibrational levels of two electronic states is possible. The net result is that the narrow peaks are more tightly packed and lead to broad absorption bands. The same result is observed when the vapor is dissolved in the solvent.

**Transitions.**    Electronic transitions in organic molecules in the vast majority of cases involve transitions of sigma ($\sigma$) electrons, $n$-electrons, and $\pi$-electrons. A simplified energy diagram illustrating the transitions is shown in Fig. 18–1.

Sigma electrons are located in $\sigma$-bonds. A typical example would be the single valence bond between two carbons atoms as found in saturated hydrocarbons. The $\sigma$-electrons are tightly held and the energy of the ultraviolet or visible region is not sufficient to overcome this attraction.

$n$-Electrons are nonbonding electrons found on atoms such as N, O, halogens, and S. These are less firmly held than $\sigma$-electrons. In this case ultraviolet and visible energy is sufficient to cause the excitation process to take place.

The nonbonding electron or $n$-electron can undergo two types of transitions

$$n \rightarrow \pi^* \qquad \text{and} \qquad n \rightarrow \sigma^*$$

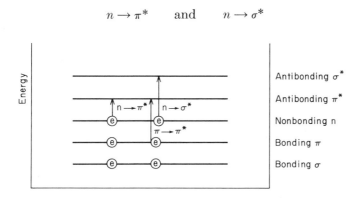

**Fig. 18–1.**   Simple energy level diagram for an organic molecule. Transitions occur when promotion of an electron is made from a filled orbital to an unfilled orbital.

where the nomenclature states that an $n$-electron is promoted to an excited $\pi^*$ and in the second case promoted to an excited $\sigma^*$ state. Absorption, although at a longer wavelength than for saturated hydrocarbons, occurs below 200 nm. Ethers, thioethers, disulfides, alkyl halides, and alkyl amines have $n$-electrons and are classified as being transparent in the ultraviolet.

An unsaturated bond contains four electrons, two of which are $\pi$-electrons and two are $\sigma$-electrons. Of the types of electrons existing in the molecule, the $\pi$-electrons are the easiest to excite. Generally, the transition of a $\pi$-electron results in absorption in the ultraviolet or visible region. Typical examples are benzene, ethylene, and the carbonyl group, where a $\pi$–$\pi^*$ notation is used to describe the transition.

**Chromophores.** When an organic compound absorbs ultraviolet or visible radiation, the two most important spectral characteristics are the position of the absorption band ($\lambda_{max}$) and its intensity ($\epsilon$). The $\lambda_{max}$, in addition to being qualitatively useful, is also a measure of the energy required for the transition. On the other hand, intensity, which is useful for quantitative purposes, is largely dependent on the polarity of the excited state and on the probability of the transition taking place after an interaction between the electronic system and the radiant energy.

To aid the chemist in transferring the above measurements into meaningful information, those organic groups which undergo $n \rightarrow \pi^*$ and $\pi \rightarrow \pi^*$ transitions are conveniently classified as chromophoric groups or chromophores. They are the color-exciting groups. The molecule containing a chromophore is called a chromogen.

Table 18–1 lists data for the simplest compounds containing single chromophoric groups. Not all chromophores absorb strongly nor do all the chromophores absorb in both the ultraviolet and visible regions. The type of solvent that is used can influence the wavelength and intensity of absorption. Table 18–2 lists the terminology which is useful in describing spectral changes.

Noticeably absent from Table 18–1 are groups such as the hydroxyl, amino, and halogen groups. In practice, it is found that these groups, which are called auxochromic groups or auxochromes, will cause shifts in the absorption maxima and changes in the intensity of a chromophore. The position of the chromophore in relation to the auxochrome is very important. For example, only if the auxochrome (the group contains an heteroatom possessing $n$-electrons) is directly attached to a chromophore, which results in $n$–$\pi$ conjugation, are the spectral changes observed.

**Multiple Chromophores.** A compound which has two ethylenic groups that are well separated from one another will behave chemically as though the unsaturation were completely independent of each other. An example

**Table 18–1.   Absorption Data for Chromophores[a]**

| Chromophoric group | System | Example | $\lambda_{max}$ (nm) | $\epsilon_{max}$[b] | Solvent |
|---|---|---|---|---|---|
| Ethylene | $RCH=CHR$ | Ethylene | 193 | 10,000 | Vapor |
| Acetylene | $RC\equiv CR$ | Acetylene | 173 | 6,000 | Vapor |
| Carbonyl | $RR_1C=O$ | Acetone | 188 | 900 | *n*-Hexane |
| Carbonyl | $RHC=O$ | Acetaldehyde | 293.4 | 11.8 | Alcohol |
| Carboxyl | $RCOOH$ | Acetic acid | 204 | 60 | Water |
| Amido | $RCONH_2$ | Acetamide | 208 | | |
| Nitrile | $RC\equiv N$ | Acetonitrile | 160 | | |
| Azo | $RN=NR$ | Azomethane | 347 | 4.5 | |
| Nitroso | $RN=O$ | Nitrosobutane | 300 | 100 | Ether |
| | | | 665 | 30 | |
| Nitro | $RNO_2$ | Nitromethane | 271 | 18.6 | Alcohol |
| Nitrate | $RONO_2$ | Ethyl nitrate | 270 | 12 | Dioxane |
| Nitrite | $RONO$ | Amyl nitrite | 218.5 | 1,120 | Petroleum ether |
| | | | 356.5 | 56 | |

[a] From E. A. Braude, *Ann. Repts. Chem. Soc.* **42**, 105 (1945).
[b] $\epsilon$ is defined as the molar absorptivity at maximum absorption, $\lambda_{max}$.

of this would be 1,5-hexadiene. If this diene ($CH_2=CHCH_2CH_2CH=CH_2$) is reacted with $Br_2$, the $Br_2$ will add across both double bonds. The reaction is much the same as if propene were used, the difference being that the hexadiene is capable of adding twice as much $Br_2$.

When compounds are examined optically, a similar observation is made. Absorption for the diene occurs at the same wavelength as for propene, while the absorption intensity for the diene is larger than for propene. In general, then, if two or more choromphores are in the same chromogen the spectra obtained will be, in general, a summation of the spectra for each chromophore. This will be the case when the chromophores are separated by two or more single bonds.

The chemical behavior of a compound in which the two unsaturated groups are separated by one single bond, such as 1,3-butadiene, $CH_2=CHCH=CH_2$,

**Table 18–2.   Definitions Describing Spectral Changes**

| Term | Definition |
|---|---|
| Bathochromic | The wavelength change is to longer wavelength or lower energy. *red shift* |
| Hypsochromic | The wavelength change is to shorter wavelength or higher energy. *blue shift* |
| Hyperchromic | The absorption intensity is increased. |
| Hypochromic | The absorption intensity is decreased. |

is very different than for 1,5-hexadiene. For the butadiene, $Br_2$ addition is at the 1,4 and 1,2 positions with the former case being the predominant product. In essence, in the conjugated system the $\pi$-electrons are spread over at least four atomic centers.

It is, therefore, not surprising that the optical properties are also affected. When two chromophoric groups are conjugated, as in the case of 1,3-butadiene, the $\pi-\pi^*$ transition undergoes a bathochromic shift of 15–45 nm. Ethylene has absorption at 193 nm, while 1,3-butadiene has a more intense peak at 217 nm. If more conjugation is introduced into the molecule, the absorption is shifted to even longer wavelengths.

Conjugation is possible with other alternating chromophores or with two

**Table 18–3.   Other Conjugated Chromophoric Systems**

| Chromophore | Example | $\lambda_{max}$ (nm) | $\epsilon_{max}$ | Solvent |
|---|---|---|---|---|
| $>C=C-C=C<$ | Butadiene | 217 | 20,900 | Hexane |
| $>C=C-C\equiv C-$ | Vinylacetylene | 219<br>228 | 7,600<br>7,800 | Hexane |
| $>C=C-C=O$ | Crotonaldehyde | 218<br>320 | 18,000<br>30 | Ethanol |
| $-C-C=C-\overset{O}{\overset{\|\|}{C}}-C-$ | 3-Penten-2-one | 224<br>314 | 9,750<br>38 | Ethanol |
| $-C\equiv C-C=O$ | 1-Hexyn-3-one | 214<br>308 | 4,500<br>20 | Ethanol |
| $>C=C-CO_2H$ | cis-Crotonic acid | 206<br>242 | 13,500<br>250 | Ethanol |
| $-C\equiv C-CO_2H$ | n-Butylpropiolic acid | 210 | 6,000 | Ethanol |
| $>C=C-C=N-$ | N-n-Butylcrotonaldimine | 219 | 25,000 | Hexane |
| $>C=C-C\equiv N$ | Methacrylonitrile | 215 | 680 | Ethanol |
| $>C=C-NO_2$ | 1-Nitro-1-propene | 229<br>235 | 9,400<br>9,800 | Ethanol |

different chromophores separated by a single bond. Examples of other conjugated chromophoric systems are given in Table 18–3.

**Aromatic Systems.**     At first glance the aromatic systems might be classified along with the other conjugated systems. The spectra that are observed, however, are slightly different in appearance. Benzene, for example, absorbs strongly at 198 nm ($\epsilon_{max}$ of 8,000) and rather weakly at 255 nm ($\epsilon_{max}$ of 230) in cyclohexane as solvent (see Fig. 17–3). The band at 255 nm is a very broad absorption band extending from 230 to 270 nm and consists of a series of multiple peaks or fine structure. These peaks are the result of the vibrational sublevels and their influence on electronic transitions ($\pi \rightarrow \pi^*$).

As benzene is substituted with a single functional group three effects on the spectra are generally observed. First, detail is lost in the fine structure of the bands. Second, the intensity of the absorption is increased. Third, bathochromic shifts take place.

In some cases, the overall effect of the substitution is very large, since the added group is involved in $n$–$\pi$ conjugation with the benzene ring. Examples of this are for groups such as —OH, —NH$_2$, —NO$_2$, and —CHO. Other groups, such as halogens and —CH$_3$, cause auxochromic effects.

Similar type absorption behavior can be expected for substituted naphthalene, anthracene, and other polyaromatics. However, as the number of rings increases, the conjugation increases and the absorption occurs at a longer wavelength.

Many heterocyclic compounds are also capable of absorbing in the ultraviolet. Addition of functional groups will cause shifts in $\lambda_{max}$ just as in the case of substituted benzene derivatives.

## INORGANIC SPECTRA

Many inorganic substances are transparent *- charge transfer* toward ultraviolet–visible radiation. However, through the use of appropriate ligands many of the inorganics can be converted to complexes which will not be transparent in the ultraviolet–visible region. This feature of certain complexes allows the analytical chemist to analyze for trace ions in a multitude of systems.

Complex formation can be generally described as the tendency for metal ions and ligands to bond in a manner such that the coordination sphere of the particular ion is filled (see Chapter 15). By the use of specific ligands and metal ions, highly colored solutions are obtained, as in the equilibrium described by

$$M^{n+} + mL^- \rightleftharpoons ML_m^{n-m} \tag{18-1}$$

colorless                   highly colored

The absorption spectrum of a complex generally originates from three types of transitions:

1. Excitation within the ligand
2. Excitation of the metal ion
3. Charge transfer excitation

Each type of transition will be discussed in the following paragraphs.

**Excitation within the Ligand.**   Since most ligands are organic molecules, the discussion pertaining to the spectra of organic molecules will also apply in this section. Transitions such as $n \rightarrow \pi^*$, $n \rightarrow \sigma^*$, $\pi \rightarrow \pi^*$, and $\sigma \rightarrow \sigma^*$ can be observed in the ultraviolet–visible region of the electromagnetic spectrum. Upon complexation, a change in the wavelength of maximum absorption and molar absorptivity can be expected to occur. Although slight in most cases, these changes can be observed and are similar to alterations due to protonation of the ligand.

**Excitation of Metal Ions.**   When a metal ion binds to a ligand, it can be expected that the energy levels of the metal ion will be altered. These transitions, which normally have molar absorptivities of the order of 1 to 100, are not used in quantitative analytical chemistry. These absorption bands can, however, still be used for structure elucidation.

**Charge Transfer Transition.**   A transition of this type may be described as the movement of an electron from the ligand to the metal or from the metal to ligand in the complex. Frequently, the intense color that a complex possesses is the result of a charge transfer transition. This intense color of the complex is of great value to the analytical chemist for trace ion determination. The following types of transitions are thought to be the origin of charge transfer bands:

1. Promotion of an electron from $\sigma$ bonding orbitals to the unoccupied orbitals of the metal ion
2. Promotion of an electron from the $\pi$ levels of the ligand to the unoccupied orbitals of the metal ion
3. Promotion of $\sigma$-bonded electrons to unoccupied $\pi$-orbitals on the ligand

Charge transfer transitions are very intense ($\epsilon = 10^4$–$10^5$) and are found in the ultraviolet and visible region. The position of the wavelength of maximum absorption is determined by the ease at which the electron can undergo its transition. In other words, the energy of absorption is a function of how easy the ligand and metal ion are oxidized or reduced. Normally the transition

**Table 18–4.   Spectral Characteristics of Selected Cu(I)–Phenanthroline Complexes**

| Compound | Neocuproine | Bathocuproine | 1,10-Phenanthroline |
|---|---|---|---|
| Structure | | | |
| Molar absorptivity | 7950 | 14,160 | 7,250 |
| Solvent | Isoamyl alcohol | n-Hexyl alcohol | n-Octyl alcohol |
| λ_max (nm) | 454 | 479 | 435 |

occurs in which the metal ion is reduced and the ligand is oxidized. However, the opposite may be observed when a metal ion of low oxidation state is complexed with a ligand of high electron affinity as in the Fe(II)–phenanthroline complex.

In order to increase the sensitivity of an organic complexing agent, it is necessary to increase the molar absorptivity. As suggested in the previous section, this is done by increasing the number of conjugation centers within the ligand. As the amount of conjugation is increased, the ability for a charge transfer transition to take place also is increased. This is illustrated in Table 18–4 where the spectral characteristics of the Cu(I)–neocuproine, Cu(I)–bathocuproine, and Cu(I)–1,10-phenanthroline complexes are compared. The Cu(I)–neocuproine and Cu(I)–1,10-phenanthroline complexes would be expected to have very similar spectral characteristics since the methyl groups substituted on the phenanthroline ring are not involved in conjugation. The opposite is the case for Cu(I)–bathocuproine where the benzene rings substituted on the phenanthroline ring are involved in the conjugation. This results in a significantly higher molar absorptivity.

# VIBRATIONAL SPECTRA: QUALITATIVE IDENTIFICATION

The infrared region from 2.5 to 15 $\mu$m is used to observe the vibrational movements of groups of atoms and molecules. The positions of the absorption bands in the infrared region, just as in the ultraviolet and visible regions, play an important role in the identification of functional groups and elucidation of molecular structure since all molecules will undergo this type of motion. As a practical tool, this region affords excellent opportunities for structural elucidation.

**Table 18–5.   Preliminary Check List Useful for Infrared Interpretation**

| Spectrum | Assignment |
|---|---|
| 1. Absorption at 2.5–3.2 $\mu$m | O—H, N—H Compounds |
|    Check: a.  5.7–6.1 | Acids |
|           b.  5.9–6.7 | Amides (usually two bands) |
|           c.  7.5–10.0 | —O— compounds |
|           d.  about 15.0 | Primary amines (broad) |
| 2. Sharp absorption at 3.2–3.33 $\mu$m | Olefins, aromatics |
|    Check: a.  5.0–6.0 | Benzenoid patterns (weak) |
|           b.  5.95–6.10 | Olefins |
|           c.  6.10–6.90 | Aromatics (two bands) |
|           d.  11.0–15.0 | Aromatics (several very strong bands) |
| 3. Sharp absorption at 3.35–3.55 $\mu$m | Aliphatics |
|    Check: a.  6.7–7.0 | —CH$_2$—, —CH$_3$ |
|           b.  7.1–7.4 | —CH$_3$ |
|           c.  13.3–13.9 | —(CH$_2$)$_4$— |
| 4. Two weak bands at 3.4–3.7 $\mu$m | Aldehydes |
|    Check: 5.7–6.1 | Aldehydes and ketones |
| 5. Absorption at 4.0–5.0 $\mu$m | Acetylenes, nitriles |
| 6. Strong sharp bands at 5.4–5.8 $\mu$m | Esters, acyl halides (1 peak) |
|    Check: 7.5–10.0 | Anhydrides (2 peaks) |
| | —O— compounds |
| 7. Strong sharp bands at 5.7–6.11 $\mu$m | Aldehydes, ketones and acids |
| 8. Strong bands at 7.5–10.0 $\mu$m | —O— compounds (note: may be confused with skeletal bands) |
| 9. Strong bands at 11.0–15.0 $\mu$m | Aromatics, chlorides |

In order to insure a correct and rapid interpretation of the spectrum a systematic approach is usually taken by comparing the spectrum against a preliminary check list such as the one shown in Table 18–5. Using these data as a guide a more intricate interpretation can be made by referring to the literature, to pamphlets and books containing known spectra, and to correlation charts as shown in Fig. 18–2.

Although it is often possible to make conclusions regarding structural properties of a molecule from infrared spectral data, it is best to include other measurements in the structural studies. Generally, the procedure is to usually reinforce the infrared interpretations with several of these other approaches. For example, physical properties such as boiling point, melting point, and refractive index are easily determined. In addition, elemental analysis, nuclear magnetic resonance (nmr) spectra, mass spectra, ultraviolet data, and electrochemical properties can be measured. It should also be emphasized that to routinely interpret infrared spectra, it is necessary to gain an ample amount of practice.

**Examples of Spectra.**    It will become clear that it is very difficult to use ultraviolet, visible, and infrared spectroscopy to interpret the total structure of a molecule. There are, however, certain structural characteristics which may be gleaned from this data. It is also important in this type of work to know the history of the sample because this will give an indication as to the type of molecule to be observed.

*Example 18–1.*    An unknown sample does not absorb in the ultraviolet or visible region. Its infrared spectrum is shown in Fig. 18–3. What conclusions about the structure of the unknown can be drawn from the information?

If the preliminary checklist is consulted, the data can be tabulated as in Table 18–6.

From the data obtained thus far, several functional groups, such as aldehydes, ketones, acids, nitriles, and halogens can be eliminated. Also the molecule is not aromatic or unsaturated because of the absence of absorption in the ultraviolet and infrared regions. The molecular species does contain CH, probably as $CH_3$, and an OH or $NH_2$ group which is hydrogen-bonded (see Fig. 18–2).

It is now necessary to distinguish between NH and OH. From Fig. 18–2, it can be seen that the NH and OH occur at the following wavelength ranges:

| NH | O–H (primary) |
|----|---------------|
| 2.6–3.3 | 2.6–3.3 |
| 5.7–6.3 | 6.8–7.3 |
| 7.7–11.4 | 7.4–7.8 |
| 11.4–14.5 | 9.5–10 |

The spectrum fits most closely with that of a primary alcohol rather than a primary amine. Thus, it is concluded that the molecule is a saturated primary alcohol.

**Table 18–6.   Preliminary Interpretation of the Infrared Spectrum in Fig. 18–3**

| Spectrum (see Fig. 18–3) | Check ($\mu$m) | Assignment |
|---|---|---|
| 1. Broad band at 3 $\mu$m |  | OH or NH (hydrogen-bonded) |
| No absorption | 5.7–6.1 | Not an acid |
| No absorption | 5.9–6.7 | Not an amide |
| 9.8 $\mu$m | 7.5–10.0 | possibly —O— |
| 16 $\mu$m |  | possibly primary, amine (—$NH_2$) |
| 2. No absorption | 3.2–3.33 | Not unsaturated |
|  |  | Not aromatic |
| 3. Bands 3.4 and 3.55 $\mu$m |  | Aliphatic C—H |
| Band ~7.0 $\mu$m |  | —$CH_2$— or $CH_3$ |
| 7.0 $\mu$m | 7.1–7.4 | $CH_3$ |
| No absorption | 13.3–13.9 | No —$(CH_2)_4$— |
| 4. No weak bands at 3.4–3.7 $\mu$m |  |  |
| No absorption | 5.7–6.1 | No aldehyde or ketone |
| 5. No absorption | 4.0–5.0 |  |
| 6. No absorption | 5.4–5.8 |  |
| 7. No absorption | 5.7–6.1 |  |
| 8. Band 9.8 $\mu$m |  | —O— |
|  |  | Alcohol, ether |
| 9. No absorption | 11.0–15.0 |  |

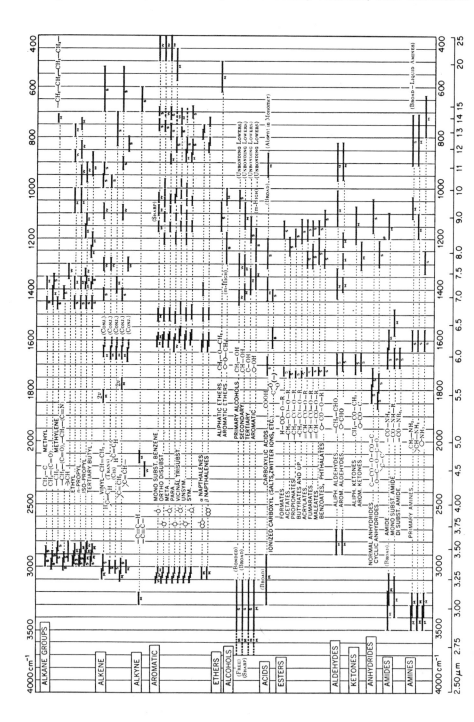

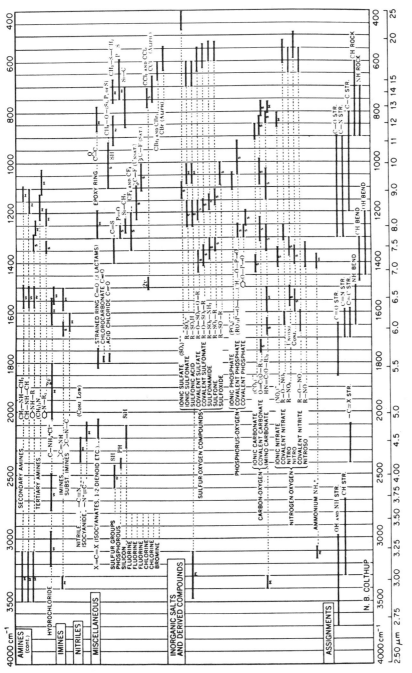

**Fig. 18–2.** Characteristic infrared group frequencies. Overtone bands are marked 2ν. (Courtesy of N. B. Colthup, Stamford Research Laboratories, American Cyanamid Co., and the editor of the *Journal of the Optical Society*.)

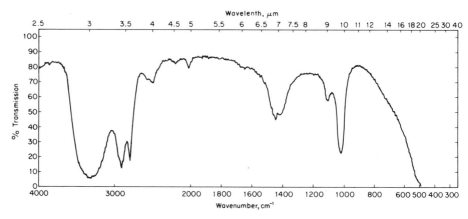

**Fig. 18–3.**  Infrared spectrum of unknown compound. (© Sadtler Research Laboratories, Inc., Philadelphia, Pa.)

Additional information is required to establish the exact structure of the molecule. Therefore, other physical properties, such as molecular weight, boiling point, melting point, physical state, and elemental analysis, are given below:

Mol wt, 32.03
Boiling point, 64.7°C
Physical state, clear liquid at room temperature

%C    12.6
%H    37.5
%O    49.9
      ‾‾‾‾‾
      100.0%
Empirical formula   $CH_4O$

From the additional data, it becomes obvious that the molecule in question is methanol.

# Questions

1.  Explain how a spectrum of a compound is used for qualitative analysis.
2.  Explain and give examples of the types of transitions which occur in organic molecules.
3.  Describe the terms bathochromic, hypsochromic, hyperchromic, and hypochromic shifts and give an example of each.
4.  Describe the spectral effects of adding chromophores to a molecule.
5.  What are the types of spectral transitions which are associated with complexes?
6.  Which type of spectral transition would be used for the maximum sensitivity for a quantitative determination?

7. How is the infrared region of the electromagnetic spectrum used for qualitative analysis?

## Problems

1. From the data given in Table 18–1, predict the $\lambda_{max}$ and $\epsilon_{max}$ of the following compounds:

   a. Nitroethane
   b. Proprionic acid
   c. Methyl isobutyl ketone
   d. Formaldehyde

   e. Nitrobenzene
   f. Hydrogen cyanide
   g. Propene

2. From the data given in Tables 18–1 and 18–3, predict the $\lambda_{max}$ and $\epsilon_{max}$ of the following compounds.

   a. Hexatriene
   b. OCHCHCHO
   c. $NCCH_2CH_2CH_2NO_2$

   d. $CH_2=CHCO_2H$
   e. $HO_2CCH_2CO_2H$

3. A compound, $C_6H_{14}$, has a boiling point of 68.8°C. Its spectrum is given below. What is the structure of the compound?

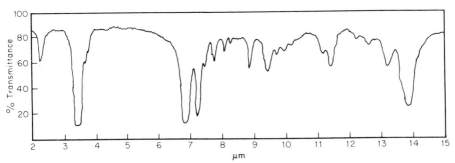

4. This compound has a molecular weight of 84.15 and an empirical formula of $CH_2$. What is its structure?

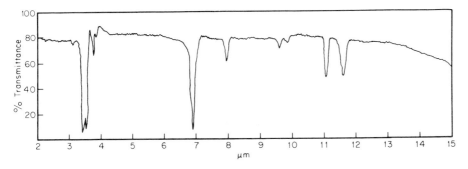

5.  A compound of molecular weight 119.39 has a boiling point of 61°C. What is its structure?

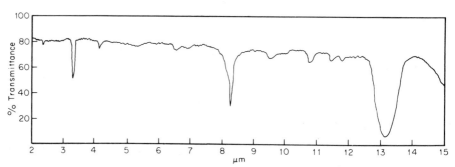

# Chapter Nineteen
# Quantitative Analysis in Absorption Spectroscopy

## INTRODUCTION

Many articles dealing with the use of absorption spectroscopy in quantitative analysis can be found in the primary literature. Typical applications would be for drug assay, determination of purity of organic compounds, clinical analysis, and trace metal analysis. In this chapter the important steps for quantitative measurements in absorption spectroscopy are discussed and illustrated by several examples.

In evaluating spectrophotometric procedures it is necessary to look for certain characteristics. The following checklist contains the more important parameters that should be a part of the procedure or included in the discussion of the procedure.

1. Molar absorptivity.
2. Stability and sensitivity with respect to time and temperature.
3. Effect of pH.
4. Absorption spectra of reactants and products.
5. Nature of reaction which includes establishing the stoichiometry and other experimental details.
6. Beer's law plot and concentration range in which the linear relationship is followed.
7. Interferences and how they are eliminated.

Many simple inorganic and organic compounds cannot be determined by absorption because of their low molar absorptivities. Thus, sensitivity and accuracy can be improved by employing one of several experimental techniques.

1. Precision colorimetry. This is an instrumental modification which

expands the absorbance or percent transmission scale. When properly used, it permits the determination of species with low molar absorptivities.

2. Complex formation. The inorganic ion is converted into a complex. Since coordination, ring formation, and often spatial configurations are altered, the absorption becomes greatly enhanced as well as shifted to other wavelengths (see Chapters 15 and 18).

3. Photometric titration. This is an ordinary volumetric measurement. The only difference is that the eye is replaced by an absorption instrument with the advantage being that the change in absorption is more easily and accurately detected by the instrument. Any reaction in which a change in absorption takes place can be followed by this technique.

4. Quenching. This technique, which is probably the least used of the four, relies on the fact that the substance being analyzed will react with another substance of high absorbance. In the course of the reaction the absorption is drastically reduced. Thus, two measurements are necessary. First, the absorption of the reacting reagent must be obtained. Second, the absorption after addition of the sample is determined and the difference between the two is correlated to concentration of the sample. This procedure requires a calibration curve.

## QUANTITATIVE ANALYSIS: ULTRAVIOLET–VISIBLE

The basis of quantitative spectrophotometry is the adherence of a system to Beer's law. When radiation impinges on a sample in a cuvette, each quantum of radiation can affected as described in Fig. 19–1. If the amount of reflection (A), scattering (B), and refraction are minimal, the ratio of $P/P_0$ will follow Beer's law, provided no chemical problems are encountered (see Chapter 17). Thus, a quantitative relationship between transmission or absorption and concentration is possible. If, however, the number of re-

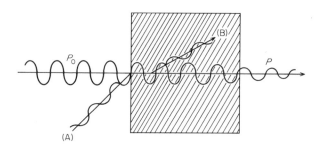

**Fig. 19–1.** Interaction of radiation with matter. Radiation can be absorbed, reflected (A), scattered (B), and transmitted when passing through a medium.

flections or the amount of light scattering is large, these properties will curtail the use of Beer's law. Fortunately, reflection or scattering is not always undesirable and, in fact, both can be used in quantitative relationships.

There are several useful ways of handling the absorption data in analysis. Although the absorbance scale covers the range of zero to infinity, the best accuracy is obtained in the absorbance range of 0.1 to 1.0. Therefore, the experimental conditions should be designed to give absorbance data in this range. If the solutions provide too high of an absorbance, they should be diluted. Similarly, if the absorbance is too low, the solutions should be concentrated. Usually, these decisions are based on a preliminary spectrophotometric measurement. Under favorable experimental and instrumenta-conditions the error in a quantitative determination can be expected to be 2% or better.

As will be illustrated in the following example, Beer's law can be applied to quantitative determinations with many variations. All of the calculations, with the exception of one, depend upon a linear correlation between absorbance and concentration of the absorbing species. Only the calibration method can be used when a deviation from Beer's law occurs.

**Standard Comparison.** The absorbances for the unknown, $A_1$, and the standard, $A_2$, are described through Beer's law by

$$A_1 = \epsilon_1 b_1 c_1$$

$$A_2 = \epsilon_2 b_2 c_2$$

Thus,

$$\frac{A_1}{A_2} = \frac{\epsilon_1 b_1 c_1}{\epsilon_2 b_2 c_2}$$

However, $\epsilon_1 = \epsilon_2$ (same compound) and $b_1 = b_2$ (same cell used for both measurements; often 1 cm) and

$$\frac{A_1}{A_2} = \frac{c_1}{c_2}$$

*Example 19–1.* A 1.000-g sample of steel is dissolved in $HNO_3$. The Mn in the sample is oxidized with $KIO_3$ to $KMnO_4$ and diluted to 100 ml. The absorbance reading for this solution in a 1.00 cm cell at the prescribed wavelength is 0.700. A $1.52 \times 10^{-4} F$ solution of $KMnO_4$ served as a standard and under the same conditions its absorbance was 0.350. What is the percent Mn in the steel?

$$\frac{0.700}{0.350} = \frac{c_1 \text{ (moles/liter)}}{1.52 \times 10^{-4} \text{ (moles/liter)}}$$

$$c_1 = 3.04 \times 10^{-4} M = [KMnO_4] = [Mn]$$

$$\%Mn = \frac{c_1 \times At\ wt\ Mn \times dilution\ factor}{sample\ weight} \times 100$$

$$\%Mn = \frac{3.04 \times 10^{-4}\ moles/liter \times 54.94\ g/mole \times 1.00\ liter \times 100\ ml/1000\ ml \times 100}{1.000\ g}$$

$$\%Mn = 0.17\%$$

**Standard Addition.**    In this case, two solutions are prepared: solution A containing only the unknown, and solution B containing a measured volume of solution A plus a measured portion of a standard. Solution A is then compared to solution B. Thus, the absorbance of solution A and B is given by

$$A_{unk} = \epsilon b c_{unk}$$

$$A_{unk+std} = \frac{\epsilon b (V_1 c_{unk} + V_2 c_{std})}{V_T}$$

where $V_1$ is the volume of the unknown in solution A and $V_2$ is the volume of the standard, $V_T$ is the total volume or, $V_1 + V_2$, if no further dilution is taken. Taking a ratio gives

$$\frac{A_{unk}}{A_{unk+std}} = \frac{c_{unk}}{(V_1 c_{unk} + V_2 c_{std})/V_T}$$

$$c_{unk} = \frac{A_{unk} V_2 c_{std}}{V_T A_{unk+std} - A_{unk} V_1} \tag{19-1}$$

*Example 19-2.*    Two 5.00-ml solutions of a sample prepared as previously described in Example 19–1 are taken. To one solution is added 5.00 ml of standard $KMnO_4$ solution $(1.00 \times 10^{-4}\ M)$. The data are summarized below:

|  | Unknown solution | Unknown + st'd solution |
|---|---|---|
| Original solution | 5.00-ml sample | 5.00-ml sample |
| Solution added | none | 5.00 ml of $1.00 \times 10^{-4}\ M$ st'd |
| Total volume | 5.00 ml | 10.00 ml |

If the absorbance of the first solution is 0.700 and the absorbance of the second is 0.465, what is the concentration of the unknown solution?

$$c_{unk} = \frac{A_{unk}\ V_2\ c_{std}}{V_T A_{unk+std} - A_{unk} V_1}$$

$$c_{unk} = \frac{(0.700)\ (5\ ml)(1.00 \times 10^{-4}\ M)}{(10\ ml)(0.465) - (0.700)(5\ ml)}$$

$$c_{unk} = 3.04 \times 10^{-4}\ M$$

**Beer's Law.**    The concentration of an absorbing solution can be calculated with Beer's law providing the absorptivity of the absorbing species is known.

*Example 19–3.*    From the data given in Example 19–1 calculate the percent Mn in the steel sample.

$$\epsilon = \frac{A}{bc} = \frac{0.350}{1.00 \text{ cm} \times 1.52 \times 10^{-4} \, M} = 2300 \text{ liter/mole cm}$$

With this constant and the data for the unknown, the concentration of the unknown can then be calculated.

$$c = \frac{A}{\epsilon b} = \frac{0.700}{2300 \text{ liter/mole cm} \times 1.00 \text{ cm}} = 3.04 \times 10^{-4} \, M$$

The weight and percent manganese in the sample would be calculated as shown in Example 19–1.

**Calibration.**    In this method, a series of standard solutions containing known concentrations of the absorbing species are prepared. Their absorbances are measured and plotted against concentration. Subsequently, an unknown is treated similarly and its absorbance is used to read the concentration directly from the calibration curve.

The calibration method is used more than any other method for a quantitative determination. This method offers the advantage of averaging a number of values to obtain a line which best fits the data (see Chapter 3). Thus, the determination will be more accurate than that obtained by using only one of the data points to calculate the molar absorptivity.

Samples which are this easily handled are not often encountered in practice since a sample generally has other elements which will interfere with the analysis in one of several ways. For example, other species in the solution may absorb at the same wavelength. Other typical problems encountered are solubility of reagents, stability of color, purity, availability of reagents, and instrumental parameters.

*Example 19–4.*    From the data in Example 19–1 and the calibration curve in Fig. 19–2 calculate the percent Mn in the sample.

$$A_{\text{unknown}} = 0.700$$

From the calibration curve a concentration of $3.00 \times 10^{-4} \, M$ is obtained. The percent Mn is then calculated as shown in Example 19–1.

*Example 19–5.*    Alcuronium, a muscle relaxant, has an absorption maximum at 292 nm. A series of standard solution were prepared and the absorbances determined

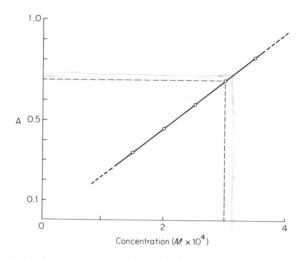

**Fig. 19–2.** Beer's law calibration plot for potassium permanganate.

as shown below:

| Concentration $(M)$ | Absorbance |
|---|---|
| $5.00 \times 10^{-6}$ | 0.22 |
| $1.00 \times 10^{-5}$ | 0.43 |
| $2.00 \times 10^{-5}$ | 0.85 |
| Unknown | 0.73 |

If the same cell (1.00 cm) was used for all determinations, calculate the concentration of the unknown solution.

Method 1:

$$\frac{A_{std}}{A_{unk}} = \frac{c_{std}}{c_{unk}}$$

$$\frac{0.85}{0.73} = \frac{2.00 \times 10^{-5}\ M}{c_{unk}}$$

$$c_{unk} = 1.72 \times 10^{-5}\ M$$

Method 2:

$$A = \epsilon bc$$

$$\epsilon = A/bc = 0.85/(1.00\ \text{cm})(2.00 \times 10^{-5}\ \text{mole liter}^{-1})$$

$$\epsilon = 4.3 \times 10^{4}\ \text{liter mole}^{-1}\ \text{cm}^{-1}$$

$$c = A/\epsilon b = 0.73/4.3 \times 10^{4}\ \text{liter mole}^{-1}\ \text{cm}^{-1} \times 1.00\ \text{cm}$$

$$c = 1.70 \times 10^{-5}\ M$$

# ANALYSIS OF ORGANIC COMPOUNDS

As previously stated in Chapter 18, a large number of organic molecules absorb radiation in the ultraviolet and visible region. Those that have high molar absorptivities can be determined directly. Those that do not can be converted chemically into derivatives which have high molar absorptivities.

The determination of a single absorbing component is generally simple, if it is assumed that the component to be analyzed is the only species that absorbs in the sample mixture or, if it is the only species in the sample that absorbs at the wavelength chosen for the analysis. Experimentally, the spectrum of the sample is obtained and the wavelength for the measurement is chosen (generally the wavelength of maximum absorbance).

Tetracycline hydrochloride, whose structure is given below, can be deter-

mined spectrophotometrically. This drug is used as an antimicrobial medicinal and as a broad spectrum antibiotic in animals. The absorption spectrum of this compound is given in Fig. 19–3 and the absorption maxima are at 220, 268, and 355 nm.

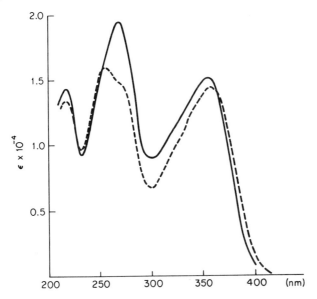

**Fig. 19–3.** Absorption spectra for tetracycline (—) and epitetracycline (- - -) in 0.1 $N$ $H_2SO_4$. [Reprinted with permission from A. P. Doerschuk, B. A. Butler, J. R. D. McCormick, *JACS*, **77**, 4687 (1955). Copyright by the American Chemical Society.]

If the percent tetracycline in a tablet is to be determined, the following experimental techniques would be carried out. A series of standards (between $10^{-4}$ and $10^{-5}$ $M$) are prepared in 0.10 $F$ HCl solution and the absorbance is determined at 355 nm for each standard.

To prepare the sample, 10 tablets are crushed and homogeneously mixed. A portion of this is accurately weighed, diluted to the appropriate volume, and the absorbance determined. The concentration is then calculated by comparison with the standards.

*Example 19–6.* Ten tablets of tetracycline hydrochloride (mol wt 480.9) are processed as described previously and a portion weighing exactly 0.4500 g is dissolved in 1.000 liter of 0.10 $F$ HCl. Exactly 10.0 ml of the solution is subsequently diluted to 100.0 ml in a volumetric flask. The absorbance of this solution using a 1.00-cm cell is 0.940. From the data below calculate the percent tetracycline hydrochloride in the tablet.

| Standard | Concentration (moles/liter) | Absorbance | $b$ (cm) |
|:---:|:---:|:---:|:---:|
| 1 | $2.50 \times 10^{-5}$ | 0.46 | 1.00 |
| 2 | $4.20 \times 10^{-5}$ | 0.76 | 1.00 |
| 3 | $5.00 \times 10^{-5}$ | 0.90 | 1.00 |
| 4 | $6.40 \times 10^{-5}$ | 1.15 | 1.00 |

If the Beer's law plot is linear in the region where the unknown is to be determined, a standard comparison method can be used. Hence,

$$c_{sample} = \left(\frac{A_{sample}}{A_{standard}}\right)(c_{standard}) = \frac{(0.940)(2.50 \times 10^{-5})}{0.460}$$

$$c_{sample} = 5.11 \times 10^{-5} M$$

In order to insure that the calculation is correct, the sample can be compared with the other standards and the values averaged.

| Standard | Concentration of unknown ($M$) |
|:---:|:---:|
| 1 | $5.11 \times 10^{-5}$ |
| 2 | $5.19 \times 10^{-5}$ |
| 3 | $5.22 \times 10^{-5}$ |
| 4 | $5.23 \times 10^{-5}$ |

$$c_{sample} = 5.18 \times 10^{-5} M \pm 0.04 \times 10^{-5} \text{ (av. deviation, } \bar{d})$$

The percent tetracycline hydrochloride in the tablet is calculated by

$$\%\text{TC} \cdot \text{HCl} = \frac{\text{g TC} \cdot \text{HCl}}{\text{g sample}} \times 100$$

$$\%\text{TC} \cdot \text{HCl} = \frac{1.000 \text{ liter} \times 0.100 \text{ liter}/0.010 \text{ liter} \times 5.19 \times 10^{-5} M \times 480.9 \text{ g/mole} \times 100}{0.450 \text{ g}} = 55.6\%$$

## INORGANIC ANALYSIS

A common way of converting a nonabsorbing species into an absorbing species is through a complexation reaction. If the chelating agent is properly chosen, large molar absorptivities are obtained. Consequently, spectrophotometry employing ligands is often applied to trace metal ion determinations.

Several complexing agents which are used in spectrophotometric analysis of metal ions are listed in Table 19–1. In general, the requirements, which are essential for success in a spectrophotometric determination using a complexing agent, will be determined by the following:

1. The complexation reaction must be complete and stoichiometric.
2. The complex must be stable.
3. The complex must absorb in the ultraviolet or visible region.
4. The absorption spectrum of the complex should not overlap with the absorption spectra of the ligand of metal ion.

Two additional advantages are gained by converting the metal ion to a complex. For example, a chelating agent will often react only with a few metal ions, thus, providing selectivity. Second, even when several metal ions form complexes with the same reagent, the absorption characteristics may differ enough to allow the determination of one metal ion in the presence of the others.

The absorbance of solutions of complexes are influenced by several variables. Probably, the most important is the effect of equilibrium. This is illustrated in the following example.

Iron(III) and thiocyanate form a soluble red complex:

$$Fe^{3+} + SCN^- \rightleftharpoons FeSCN^{2+}; \quad K = \frac{[FeSCN^{2+}]}{[Fe^{3+}][SCN^-]} = 154 \quad (19\text{-}2)$$

The question that should be considered at this point is how much thiocyanate must be added to the iron to insure 100% complex formation? Since the equilibrium constant is small, the amount of thiocyanate is going to have to be in large excess because the absorbance is dependent upon the concentration of thiocyanate. Beer's law for this system relates the absorbance to the concentration of complex since it is the only absorbing species

$$A = \epsilon b [FeSCN^{2+}]$$

Assume that the concentration of iron(III) is held constant. Since the molar absorptivity of $FeSCN^{2+}$ and the pathlength are constant, as the thiocyanate concentration is increased, the equilibrium in Eq. (19-2) shifts to the right forming more complex. Thus, an excess of thiocyanate is necessary for determining the concentration of iron(III) accurately.

If a 99.99 mole% of the iron(III) is needed in the complex form, the relative

**Table 19–1. Some Typical Reagents Used in Spectrophotometric Analysis[a]**

| Reagent | Structure | Ion analyzed[b] |
|---|---|---|
| 1,10-Phenanthroline | | Fe(II) |
| 2,9-Dimethyl-1,10-phenantholine | | Cu(I) |
| Sulfosalicylic acid | | Al(III),Ti(IV) |
| Thiourea | | Bi(III), Os |
| Nitroso R salt | | Co(II) |
| 8-Hydroxyquinoline | | Zn(II), Al(III), Ce(III), Ga(III), In(III), Mg(II), Sc(III), others |
| Dithizone | | Pb(II), Hg(II), Zn(II), Bi(III) |
| Benzoin $\alpha$-oxime | | Cu(II), Mo(V) |
| Dithiooxamide | | Ni(II), Co(II), Cu(II), Bi(III) |
| 1-Nitroso-2-naphthol | | Co(II) |

**Table 19–1.  (Continued)**

| Reagent | Structure | Ion analyzed[b] |
|---|---|---|
| Rhodamine B | $(C_2H_5)_2N$ ... $\overset{(+)}{N}(C_2H_5)_2$  $Cl^{(-)}$ ... $CO_2H$ | Sb(V) |
| Sodium diethyldithiocarbamate | $\begin{matrix} C_2H_5 \\ C_2H_5 \end{matrix} N - C \overset{S}{\underset{S^{(-)}}{\diagup}}$  $Na^{(+)} \cdot 3\,H_2O$ | Cu(II) |
| Quinalizarin | HO  O  OH  OH  HO  O | B |
| Toluene-3,4-dithiol | SH  SH  $CH_3$ | Sn(IV), Mo(V), W(VI) |
| 2,4-Xylenol | OH  $CH_3$  $CH_3$ | $NO_3^-$ |
| Thoron | OH  HO  $SO_3^-Na^+$  $-N=N-$  $SO_3^-Na^+$ | Th(IV), Zr(IV) |
| Dimethylglyoxime | HO  OH  N  N  $H_3C-C-C-CH_3$ | Ni(II) |

**Table 19–1. (Continued)**

| Reagent | Structure | Ion analyzed[b] |
|---|---|---|
| Arsenazo | | Hf(IV), Zr(IV) |
| Acetylacetone | $CH_3CCH_2CCH_3$ | Be(III) |
| 2,2', 2''-Terpyridine | | Co(II) |
| Dibenzoylmethane | | UO$_2$(II) |

[a] In some cases a precipitate is formed which is dissolved in an organic solvent.

[b] The ions listed are the ones that are most frequently analyzed by the reagent. If only one ion is listed it should not be assumed that the reagent is specific for just that ion.

ratio of the concentration of $Fe(SCN)^{2+}$ to $Fe^{3+}$ will be 99.99/0.01. Substituting this value into the equilibrium constant expression, a thiocyanate concentration which is 6490 times greater than the iron concentration is calculated.

*Example 19–7.* From the data below determine the concentration of iron in solution. Assume that the thiocyanate is present in a large excess.

| Standard | $b$ (cm) | $A$ at 465 nm | Concentration of complex ($M$) | $\epsilon$ |
|---|---|---|---|---|
| 1 | 1.00 | 2.00 | $5.00 \times 10^{-4}$ | $4.00 \times 10^3$ |
| 2 | 1.00 | 1.20 | $1.00 \times 10^{-4}$ | $1.20 \times 10^4$ |
| 3 | 1.00 | 0.60 | $5.00 \times 10^{-5}$ | $1.20 \times 10^4$ |
| 4 | 1.00 | 0.13 | $1.00 \times 10^{-1}$ | $1.30 \times 10^4$ |
| Unknown | 1.00 | 0.54 | | |

At a first glance, it becomes obvious that Beer's law is not linear for the concentration range selected. Standard 1 is too concentrated while standard 4 is too dilute because the absorbance is close to zero where a high error is expected. Another standard could be prepared in the concentration range between $1.00 \times 10^{-4}$ and $5.00 \times 10^{-5}$ $M$ to check the values of the molar absorptivity for these two concentrations. Assume that the third concentration gives the same molar absorptivity ($1.20 \times 10^4$).

The unknown can now be calculated directly through Beer's law:

$$A = \epsilon bc$$

$$0.54 = (1.2 \times 10^4 \text{ liter/mole cm}) (1.00 \text{ cm}) (c_{\text{complex}})$$

$$c_{\text{complex}} = 4.50 \times 10^{-5} \text{ moles/liter}$$

Although thiocyanate has been used as a complexing agent for the determination of iron(III), the system does not adhere to all of the previously mentioned requirements. The complex is unstable for long periods of time resulting in a decrease in absorbance. Complexation with 1,10-phenanthroline is a more acceptable method for the spectrophotometric determination of iron. This complex has the advantages of long term stability and a high molar absorptivity. The disadvantage, however, is that the iron(III) must be reduced quantitatively to iron(II) after dissolution of the sample.

*Example 19–8.* Ten (10.00) milliliters of a water sample containing a trace amount of iron is transferred to a separatory funnel. A series of standards are prepared simultaneously. Hydroxylamine hydrochloride is added to reduce the ferric ion to the ferrous state, the solutions are buffered, and bathophenthroline is added to each solution. To the solutions 6.00 ml of isoamyl alcohol (immiscible with water) is added to extract the complex (Fe[bathophen]$_3^{2+}$). Subsequently, the absorbance of each extract is determined at 533 nm using a 1.00-cm cell. From the data below calculate the concentration of iron in ppm in the original sample.

| Iron conc. in water ($\mu$g/ml) | Volume of solution taken (ml) | Iron conc. in isoamyl alcohol ($\mu$g/ml) | Absorbance |
|---|---|---|---|
| 0.100 | 10.00 | 0.167 | 0.08 |
| 0.100 | 20.00 | 0.333 | 0.16 |
| 1.000 | 5.00 | 0.833 | 0.41 |
| 1.000 | 10.00 | 1.667 | 0.83 |
| 1.000 | 20.00 | 3.333 | 1.61 |
| Unknown | 10.00 | | 0.54 |

From the data Beer's law for this system is linear in the absorbance range of the unknown. Thus, the relationship

$$\frac{A_{\text{sample}}}{A_{\text{standard}}} = \frac{c_{\text{sample}}}{c_{\text{standard}}}$$

is valid, if a standard that is approximately the same absorbance is used for comparison. Therefore,

$$c_{\text{sample}} = (0.54/0.41) \times 0.833 \ \mu\text{g/ml} = 1.10 \ \mu\text{g/ml}$$

The total weight of the iron in solution is, therefore, 11.0 $\mu$g and the concentration of iron in the water sample is 11.0 $\mu$g/10.00 ml or 1.10 ppm.

## MULTICOMPONENT ANALYSIS

It is often possible to determine the amount of each component in a mixture by spectrophotometry even though their absorption spectra overlap. The reason for this is that absorbances are additive. In Fig. 19–4 the spectra of two components are shown (a and b). If the two are mixed, spectrum c is obtained. Careful examination would reveal that the same result would be obtained by adding spectra a and b.

To handle the mathematics of the problem two wavelengths, $\lambda_1$ and $\lambda_2$, are selected. Since the total absorbance at $\lambda_1$ and $\lambda_2$ are due to the sum of components I and II at both wavelengths, the following equations can be written:

$$A_{\lambda_1} = \epsilon_{\lambda_1}^{I} b c_{\lambda_1}^{I} + \epsilon_{\lambda_1}^{II} b c_{\lambda_1}^{II} \tag{19-3}$$

$$A_{\lambda_2} = \epsilon_{\lambda_2}^{I} b c_{\lambda_2}^{I} + \epsilon_{\lambda_2}^{II} b c_{\lambda_2}^{II} \tag{19-4}$$

The superscripts and subscripts refer to the component and wavelength, respectively.

Several conditions will simplify the problem. For example, the same cell can be used for all measurements and if a 1.00-cm cell is used, the $b$ term drops out of the two expressions. In either case, $b$ is readily determined. On the other

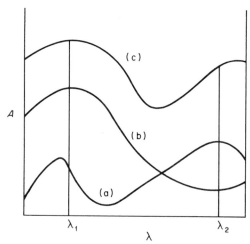

**Fig. 19–4.** Absorption curves. (a) Spectrum of component I; (b) spectrum of component II; (c) spectrum of sum of components I and II.

hand, since molar absorptivities are dependent on wavelength,

$$\epsilon_{\lambda_1}^{I} \neq \epsilon_{\lambda_2}^{I} \quad \text{and} \quad \epsilon_{\lambda_1}^{II} \neq \epsilon_{\lambda_2}^{II}$$

However,

$$c_{\lambda_1}^{I} = c_{\lambda_2}^{I} = c^{I} \quad \text{and} \quad c_{\lambda_1}^{II} = c_{\lambda_2}^{II} = c^{II}$$

and therefore Eqs. (19-3) and (19-4) simplify to

$$A_{\lambda_1} = \epsilon_{\lambda_1}^{I} c^{I} + \epsilon_{\lambda_1}^{II} c^{II} \tag{19-5}$$

$$A_{\lambda_2} = \epsilon_{\lambda_2}^{I} c^{I} + \epsilon_{\lambda_2}^{II} c^{II} \tag{19-6}$$

Thus, two equations with six unknowns are obtained. Four of the unknowns are experimentally found. For example, the total absorbances at $\lambda_1$ and $\lambda_2$ are measured, and the molar absorptivities are determined for each wavelength from spectra of the pure compounds (Fig. 19–4a and b). Consequently, two equations with two unknowns, $c^{I}$ and $c^{II}$, remain which can be solved simultaneously.

For more complicated systems it is only necessary to write the appropriate number of equations and have available all the molar absorptivities at the selected wavelengths. The fact that $n$ equations are written, where $n$ is the number of components and wavelengths chosen, and must be solved simultaneously is not as difficult as it first appears. For complicated cases, it is relatively easy to carry out the computation with a computer. Also, the method can be applied to ultraviolet, visible, and infrared regions with about equal ease.

When tetracycline is allowed to stand in acid solution, a reversible reaction known as epimerization occurs as shown below.

Tetracycline                                    Epitetracycline

Although this slight change in structure seems insignificant, the drug is inactive when it is in the epi form. Thus, it is important to determine the ratio of tetracycline to epitetracycline. Spectrophotometry can be used effectively to determine this ratio. Figure 19–3 presents the spectrum of both tetracycline and epitetracycline. Even though the molecules are similar, there is a change in the absorption spectrum due to the difference in structure. The determination is based on the difference in the molar absorptivities of tetracycline and epitetracycline at 267 and 254 nm.

From the observed absorbances at these two wavelengths, the concentration of each component can be calculated. First, however, the molar absorptivity for each component at both wavelengths must be determined. These are 16,000 and 19,000 for tetracycline at 254 and 267 nm, respectively, and 16,000 and 15,000 for epitetracycline at 254 and 267 nm, respectively. If the measurements are made at the two prescribed wavelengths, the following can be written using the molar absorptivities and assuming a pathlength of 1.00 cm.

At 254 nm: 
$$A_{254} = \epsilon_{tet}bc_{tet} + \epsilon_{epi}bc_{epi} \tag{19-7}$$

$$A_{254} = 16,000c_{tet} + 16,000c_{epi} \tag{19-8}$$

At 267 nm: 
$$A_{267} = \epsilon_{tet}bc_{tet} + \epsilon_{epi}bc_{epi} \tag{19-9}$$

$$A_{267} = 19,000c_{tet} + 15,000c_{epi} \tag{19-10}$$

*Example 19–9.* Consider a set of typical analytical data for the tetracycline problem. Figure 19–5 gives the spectrum of a solution of a mixture of tetracycline and epitetracycline. How much of each is present?

The absorbances obtained from the spectrum for the two prescribed wavelengths are 0.750 and 0.790, respectively. Substituting into Eqs. (19-8) and (19-10) the following is obtained:

$$0.750 = 16,000\,c_{tet} + 16,000\,c_{epi}$$

$$0.790 = 19,000\,c_{tet} + 15,000\,c_{epi}$$

and solving simultaneously gives

$$c_{tet} = 2.17 \times 10^{-5} \text{ mole/liter}$$

$$c_{epi} = 2.37 \times 10^{-6} \text{ mole/liter}$$

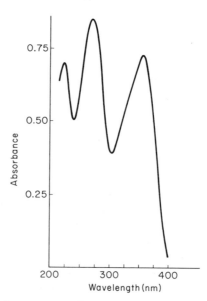

**Fig. 19–5.** Absorption spectrum of a mixture of tetracycline and epitetracycline.

# AUTOMATION IN SPECTROPHOTOMETRIC ANALYSIS

Two principal reasons have contributed to the development of instruments to perform absorption measurements automatically. First, many absorption methods are available for the determination of organic and inorganic samples. Second, many routine analyses that must be done repeatedly are often done by absorption methods. For example, many clinical, environmental, and industrial and pharmaceutical quality control analyses are now routinely performed by automated spectrophotometric instruments. This is particularly true in the clinical laboratory.

In the development of automated absorption instrumentation several major problems had to be overcome. For example, since the instrumentation must allow the introduction of samples automatically, it is necessary that each sample be distinguishable from the next sample, the reagents must be introduced in appropriate amounts, chambers must be provided to allow the reactions that produce the color to occur, and finally, a flow-through absorption cell must be available. The following example illustrates a typical automated instrument.

There is a continuing interest in the determination of chloride ion (as HCl) in the upper atmosphere. In this region several reactions originate from the diffusion of halocarbons into the ozone layer and ultimately degrade photochemically:

$$\text{halocarbons} + \text{sunlight(UV radiation)} \rightarrow \text{Cl atoms}$$

Subsequently, the Cl atoms undergo a reaction with hydrocarbons to produce HCl levels approaching 1 ppb:

$$\text{Cl atoms} + CH_4 \rightarrow HCl + CH_3 \text{ radicals}$$

Since a wide region of the upper atmosphere is to be continuously examined, many Cl samples will be collected, and because of this large number of samples an automated analytical procedure is desirable.

The procedure for obtaining the sample is to lower a paper saturated with a quaternary ammonium hydroxide ($R_4N^+OH^-$) from an airplane flying at high altitudes. The sample is collected by pumping air through the paper at a controlled time and flow rate with the chloride being trapped on the paper by the acid–base reaction

$$R_4H^+OH^- + HCl \rightarrow R_4N^+Cl^- + H_2O$$

Approximately 250 samples can be taken on each flight. In the laboratory, the $R_4N^+Cl^-$ is dissolved in a fixed volume of water and introduced into the automated system shown in Fig. 19–6. The tray, which holds many samples and standards, rotates at a fixed time; thus, each sample is introduced automatically. The $Fe(NO_3)_3$ solution, sample, and air are mixed at point A with

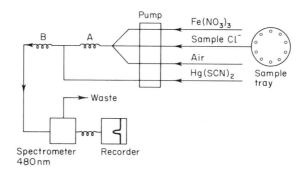

**Fig. 19–6.** Automated system for the determination of chloride.

the air not only facilitating mixing but more importantly, providing a bubble that separates adajacent samples. The $Hg(SCN)_2$ is introduced at point B and the reaction

$$2Cl^- + Hg(SCN)_2 + 2Fe^{3+} \rightarrow Hg(Cl)_2 + 2[Fe(SCN)]^{2+}$$

takes place. Flow rates of the solutions are controlled by the diameter of the tubes used in the pump. Eventually, the red color of the $Fe(SCN)^{2+}$ complex is determined in the spectrometer. In the example shown in Fig. 19–6 40 samples can be analyzed per hour at Cl levels as low as 0.03 ppb in the original sample.

In automated procedures parameters such as flow rates, solution concentrations, temperature, mixing times, tube lengths, tube diameters, and sample size must be carefully controlled and reproduced. Once this is achieved, the operator is primarily responsible for maintaining a continuous supply of reagents and the work-up of the samples and their loading into the sample tray. Another important feature of automation is that the instrumentation is readily interfaced with a computer.

## QUANTITATIVE INFRARED

The quantitative interpretation of infrared spectra is also based on Beer's law. As was pointed out earlier, the two most difficult problems in applying Beer's law in the infrared region are: (1) knowing the pathlength accurately or being able to reproduce it, and (2) knowing the molar absorptivity. Unfortunately, both are largely affected by the sampling technique and are often difficult to control.

The thickness of a sample in the infrared, if a cell technique is used, varies between 0.1 and 0.01 mm. If the absorbance of a solution is to be related to the concentration, the pathlength of the cell or a number of cells must be known

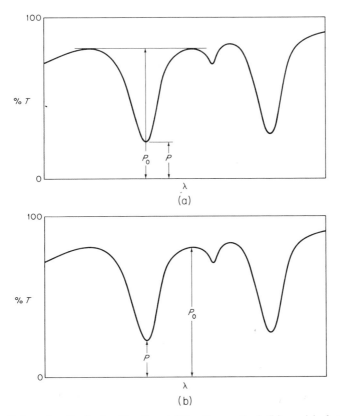

**Fig. 19–7.** Infrared methods of calibration. (a) Baseline method; (b) empirical ratio method.

to within $\pm 1\%$. This problem can be solved by using the same cell for all measurements.

The second problem, that of the determination of $\epsilon$, is not solved as readily. The molar absorptivity is dependent upon the instrumental parameters which are used for each particular measurement. Reproducibility of $\epsilon$ from day to day and instrument to instrument is questionable.

Furthermore, in order to calculate the molar absorptivity of a species at a particular wavelength the ratio of $P$ to $P_0$ must be determined. Since the windows of the cell are not as smooth as quartz or glass windows, the amount of scattering of radiation varies significantly from cell to cell. This problem may be circumvented to some extent by making measurements as illustrated in Fig. 19–7.

Fortunately, $\epsilon$ and pathlength need not be calculated since generally standard comparison is used in quantitative infrared spectrometry. Thus, by using the same cell, the pathlength, scattering, and molar absorptivity remain the same for a series of measurements.

## PHOTOMETRIC TITRATIONS

In a photometric titration the end point of the titration is determined with a spectrophotometer. In the titration of iron(II) with permanganate

$$5Fe^{2+} + MnO_4^- + 8H^+ \rightarrow Mn^{2+} + 5Fe^{3+} + 4H_2O$$

the wavelength of observation would be set at 520 nm (the wavelength at which permanganate absorbs). The cell containing the iron(II) sample is placed in the instrument and the permanganate titrant added in small increments. The absorbance readings would remain relatively constant until permanganate is in excess. At this condition the absorbance would rise linearly with added permanganate. The intersection of the lines marks the end point.

Many different reactions can be followed in the same way. All that is needed is for the reactants or products to undergo some change in absorbance as the titration is carried out. Figure 19–8 shows two possible titration curves.

The advantages of a photometric titration are the following:

1. Slight changes in color are readily detected by the spectrophotometer.
2. The method can be applied to solutions that are highly colored which would interfere with visual indicators.
3. A series of points are used to determine the end point.
4. Since an extrapolation method is used to arrive at the end point, reactions in which equilibrium constants are not favorable can often still be used. In these cases the data appears to be curved in the end point region.
5. Indicator color changes can easily be detected.

The instrumentation does not have to be complicated in order to obtain good results. A simple experimental setup is shown in Fig. 19–9. For visible measurements it can be imagined that the beaker is resting in the cell com-

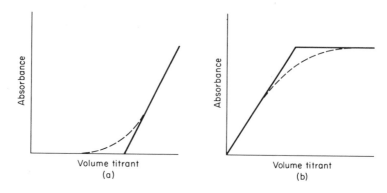

**Fig. 19–8.** Typical shapes of photometric titration curves. (a) Titrant alone absorbs; (b) product of reaction absorbs.

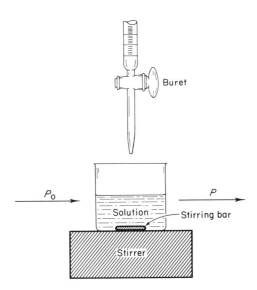

**Fig. 19–9.** Experimental setup for photometric titrations.

partment of one of several commercial instruments. Similarly, ultraviolet instruments can be used in much the same way.

Table 19–2 contains a list of several practical photometric titrations that have been reported in the literature. In many cases a high degree of selectivity

**Table 19–2. Some Typical Methods of Analysis by Photometric Titrations**

*Acid–Base Methods*

Phenols titrated with NaOH; absorbance due to the formation of phenolate ion is followed.

*Oxidation–Reduction Titrations*

Ce(III) titrated with Co(III); formation of Ce(IV) is followed.

*Complexometric Titrations*

Bi(III) titrated with EDTA; Cu(II) added and the appearance of Cu–EDTA complex followed or thiourea is added and the disappearance of the Bi–thiourea complex is followed.

Fe(III) titrated with EDTA; disappearance of Fe–sulfosalicylic acid complex is followed.

Cu(II) titrated with EDTA or Trien; formation of Cu–EDTA or Cu–Trien complex is followed.

*Precipitation Titrations*

$SO_4^{2-}$ titrated with Ba(II); appearance of turbidity is followed.

$F^-$ titrated with Th(IV); reaction of Th(IV) with the indicator SPADNS followed in the presence of precipitate formation.

is imparted to the method by virtue of following the reaction spectrophotometrically.

## Questions

1. What factors must be considered in evaluation of a spectrophotometric procedure?
2. Write Beer's law and state what each term is dependent upon.
3. What effect does a colloidal suspension have on an absorbance reading?
4. What is the best absorbance range for quantitative absorption spectrophotometry?
5. If an absorbance measurement for a solution is too low, how can this be corrected?
6. If an absorbance measurement for a solution is too high, how can this be corrected?
7. State the different ways of using Beer's law for a quantitative determination.
8. Why are complexes used to quantitatively determine metal ions?
9. State the requirements for use of a complex in a spectrophotometric determination.
10. State how an analysis of a multicomponent system is performed.
11. What are the problems of using the infrared region for quantitative analysis? How are these problems circumvented?
12. What are the requirements for a spectrophotometric titration?
13. What are the advantages of using a spectrophotometric titration?
14. How is the end point determined for a spectrophotometric titration?

## Problems

1.* A $1.2 \times 10^{-5}$ $M$ solution of a compound has an absorbance of 0.21 at its wavelength of maximum absorption. If the pathlength of the cell is 1.0 cm, calculate the molar absorptivity.

2. A $2 \times 10^{-3}$ $M$ solution of a compound has an absorbance of 0.52 at its wavelength of maximum absorption. If the pathlength of the cell is 0.1 cm, calculate the molar absorptivity of the compound.

3.* A compound has a molar absorptivity of 13,200 liters/mole cm at its wavelength of absorption. The absorbance for a solution of this compound in water is 0.41 using a 1.0-cm cell. Calculate the concentration of the solution.

---

\* Answers are listed at the end of the book for problems marked with an asterisk.

4.* Using the calibration curve in Fig. 19–2, determine the concentration in moles/ liter and grams/liter of a solution which has an absorbance of 0.72.

5. From Fig. 19–2 determine the molar absorptivity of potassium permanganate assuming a 1.0-cm pathlength.

6. From Fig. 19–3 determine which wavelength is most sensitive to changes in trace concentrations of tetracycline.

7. A solution containing 3.00 ppm has a transmittance of 65.0% in a 1.0 cm-cell.
   *a. Calculate the absorbance of the solution.
   *b. Calculate the transmittance and absorbance for a solution containing 5.2 ppm of the solute.
   *c. What is the molar absorptivity of the solute if its molecular weight is 155?

8. The transmittance (at 520 nm) of a 5.00-ppm solution of potassium permanganate using a 1.00-cm cell is 27.0%.
   a. Calculate the absorbance of the solution.
   b. Calculate the absorbance and % transmission of a solution containing 3.20 ppm $KMnO_4$.
   c. If a 0.100-g sample of a steel sample is dissolved, oxidized to $MnO_4^-$ and diluted to 100.0 ml and the absorbance of the solution is 0.52, calculate the milligrams of manganese in the original sample.
   d. What is the molar absorptivity of potassium permanganate?
   e. Calculate the %T and absorbance for a 0.1-ppm solution using a 10.0-cm pathlength.

9. What is the molar absorptivity of a compound having a molecular weight of 192 if a 0.0150% solution by weight has a transmittance of 27.0% through a 1.0-cm pathlength?

10. A 1.0000-g sample of a drug excipient material containing a trace amount of iron is dissolved in nitric acid, boiled, and diluted to 100 ml. A 10-ml aliquot was taken and the solution is treated as described in Example 19–8. An absorbance of 0.27 in a 1-cm cell at 533 nm was obtained. Using the data in the example determine the concentration of iron in the original sample in % and ppm Fe.

11. A $1.00 \times 10^{-3}$ $M$ solution of a drug shows an absorbance of 0.400 at 270 nm and 0.010 at 345 nm. A $1.00 \times 10^{-4}$ $M$ solution of a metabolite of the drug has 0.000 absorbance at 270 nm and 0.460 at 345 nm. The drug and its metabolite were extracted from a urine sample and diluted to 100 ml. The absorbance of this solution was 0.325 and 0.720 at 270 and 345 nm, respectively. Calculate the nmoles of the drug and its metabolite in the 100-ml sample.

12.* Ammonia can be determined spectrophotometrically with Nessler's reagent (alkaline solution of KI and $HgCl_2$) according to the reaction.

$$2K_2[HgI_4] + 2NH_3 \rightarrow NH_2Hg_2I_3 + 4KI + NH_4I$$

A 500-ml sample of drinking water was made alkaline and the ammonia steam distilled. This was collected, Nessler's reagent added, and diluted to 250 ml. The absorbance at 425 nm was found to be 0.461. A standard was prepared by dissolving 3.1410 g of $NH_4Cl$ per liter of solution and 10 ml of this was diluted to a liter. A 25-ml aliquot of this was taken, Nessler's reagent added, and diluted to 100 ml. The absorbance of this solution was found to be 0.515. Calculate the mg of $NH_3$ in the water sample. Express the result as ppm $NH_3$.

13. Dithizone is a very sensitive reagent for the determination of Pb, Hg, Cu, and Bi. (See Experiment 26). A 50.0-ml sample of water was treated as described in Experiment 26 and the Hg–dithizone complex was extracted into 25 ml of $CHCl_3$. Its absorbance was 0.515 at 510 nm in a 1-cm cell. A 10-ml aliquot of $Hg(NO_3)_2$ (0.0075 g/liter) was treated the same way and had an absorbance of 0.611. Calculate the concentration of Hg in the water sample in mg/ml and in ppm.

14.* A soluble complex $MX^+$ dissociates according to the reaction

$$MX^+ \rightleftharpoons M^+ + X$$

The metal ion, $M^+$, and the ligand, X, do not absorb at 560 nm but the complex $MX^+$ does. A solution that was known to be $2.10 \times 10^{-4}$ $M$ in $MX^+$ had an absorbance of 0.481 at 560 nm in a 1-cm cell. Another solution was prepared by taking 10 ml of $1.28 \times 10^{-3}$ $M$ $M^+$ and 10 ml of $1.31 \times 10^{-3}$ $M$ X and diluting to exactly 100 ml. If the absorbance of this solution was 0.278, calculate the formation constant for the complex $MX^+$.

15.* Cholesterol in blood is determined by isolation of the cholesterol from the blood with $CHCl_3$. This $CHCl_3$ extract is treated with acetic anhydride and conc. $H_2SO_4$ and the color that is produced is measured at 630 nm. A sample of blood (0.050 ml) was treated as described and an absorbance of 0.518 was determined (1-cm cell). If the volume of the final $CHCl_3$–acetic anhydride–$H_2SO_4$ extract was 10.0 ml and a 1.00-ml aliquot of a cholesterol standard (50 mg/liter) treated in the same way had an absorbance of 0.462, calculate the mg cholesterol/100 ml of blood.

16. Phenylbutazone tablets NF were purported to contain 100 mg of drug per tablet. The analyst weighed out 30 tablets (6.3020 g) and reduced them to a fine powder. A 0.2026-g portion was extracted with alcohol, filtered, and the filtrate diluted to 100 ml with alcohol. A 10.0-ml aliquot of this solution was diluted to 1 liter with 0.1 $F$ NaOH. Its absorbance value in a 1-cm cell was 0.622. A literature value for the absorptivity was reported to be 66 liter $g^{-1}$ $cm^{-1}$.
a. Calculate the mg of drug in a tablet of average weight.
b. Does this brand comply with NF limits on phenylbutazone tablets?
c. Criticize the analyst's procedure.

17. The acidic form of a monobasic acid absorbs at 475 nm ($\epsilon = 3.4 \times 10^4$ liter $mole^{-1}cm^{-1}$) while the basic form does not absorb at this wavelength. In a buffered solution of pH 3.90, the absorption of a $2.72 \times 10^{-5}$ $M$ solution of the acid was 0.261 at 475 nm. Calculate the $K_a$ for the weak acid. A 1-cm cell was used for all the measurements.

18.* A solution containing a mixture of tetracycline and epitetracycline is found to have absorbances of 0.67 and 0.72 at 254 and 267 nm, respectively. From the molar absorptivities given in the text, calculate the ratio of tetracycline to epitetracycline.

19. Convert the spectrum of tetracycline to absorbance vs. wavelength assuming the solution is $1 \times 10^{-4}\ M$ and a 1.00-cm pathlength (see Fig. 19–3).

20.* A 10.00-ml aliquot of a $KMnO_4$ solution is titrated with $0.01000\ F\ H_2C_2O_4$. The end-point is detected photometrically

$$2MnO_4^- + 5C_2O_4^{2-} + 16H^+ \rightarrow 2Mn^{2+} + 5CO_2 + 8H_2O$$

by monitoring the disappearance of the permanganate at 520 nm. From the data below calculate the concentration of the $KMnO_4$.

| ml $H_2C_2O_4$ | $A$ | ml $H_2C_2O_4$ | $A$ | ml $H_2C_2O_4$ | $A$ |
| --- | --- | --- | --- | --- | --- |
| 0.00 | 1.43 | 2.00 | 0.60 | 3.50 | 0.09 |
| 0.50 | 1.21 | 2.25 | 0.51 | 3.75 | 0.04 |
| 0.75 | 1.11 | 2.50 | 0.43 | 4.00 | 0.02 |
| 1.00 | 1.03 | 2.75 | 0.35 | 4.25 | 0.01 |
| 1.50 | 0.82 | 3.00 | 0.27 | 4.50 | 0.00 |
| 1.75 | 0.71 | 3.25 | 0.18 | 4.75 | 0.00 |

21. Ten tablets of sulfanilamide (2.510 g) are powdered, extracted with 100 ml of alcohol, filtered, and the filtrate is diluted to exactly 1.00 liter with $0.10\ F$ NaOH solution. The absorbance of the resulting solution is 0.99 at 250 nm. What is the weight of sulfanilamide in each tablet if the absorptivity is 12 liter $g^{-1}$ $cm^{-1}$ at the prescribed wavelength? (Assume a 1.0-cm pathlength.)

22.* A 10.0-mg/liter solution of procaine hydrochloride has an absorbance of 0.65 at 290 nm. What is the concentration of a solution which has an absorbance of 0.93? Calculate the molar absorptivity of procaine hydrochloride at 290 nm if its molecular weight is 272.8.

# Chapter
# Twenty
# Spectroscopy of
# Atoms

## INTRODUCTION

Atomic spectroscopy is primarily used for the determination of trace metals in many types of samples composed of organic or inorganic matrices. The techniques used for this purpose are atomic emission spectroscopy and atomic absorption spectroscopy. The basis for the observation of atomic emission and atomic absorption has been presented in Chapter 17 and is summarized in Fig. 20–1.

In emission, the atoms of interest are vaporized by input of thermal energy by either combustion or electrical discharge. The emission intensity, which is observed in the form of line spectra, is proportional to concentration and is dependent on the temperature of the system.

In absorption radiation incident on the metal vapor causes electronic transitions from the ground state to selected excited states. The measurement of the ratio of the transmitted power to the incident power is proportional to concentration.

Atomic fluorescence spectroscopy is the newest technique to be developed. In this method, radiation impinging on a vapor metal sample causes the promotion of electrons into excited states. The atoms then return to the ground state with emission of radiation. The measurement of this emitted radiation requires that the detector be placed at an angle to the incident radiation.

## ATOMIC EMISSION

**Flame Emission.**     Flame emission spectrometry is a special area of emission spectroscopy in which a flame is used to excite the atoms. When a

*450*

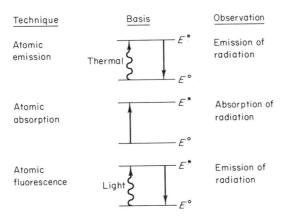

**Fig. 20–1.** Basic processes of atomic emission, atomic absorption, and atomic fluorescence.

solution containing an ion is nebulized through a flame, a series of processes occur:

1. The solvent is vaporized leaving particles of salt.
2. The salt is subsequently vaporized and dissociated into atoms.
3. Some of the atoms are excited by the flame.
4. The excited atoms emit radiation characteristic of their species.

The efficiency of forming excited atoms in a flame is low. Other processes, such as formation of molecular species, incomplete vaporization, and incomplete excitation, decrease the emission intensity of the atoms.

Because of the relatively low energy of the flame, not all elements can be excited as in arc excitation to a usable extent. The main application of flame emission is the quantitative determination of the alkali and alkaline earth elements at concentrations as low as 0.1 $\mu$g/ml solution (0.1 ppm).

The instrument that is used to observe emission from flames is shown in Fig. 20–2. The basic components are the flame, monochromator, and detector readout system. The flame is produced with a burner–nebulizer assembly as

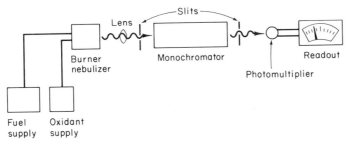

**Fig. 20–2.** A block diagram of a flame photometer.

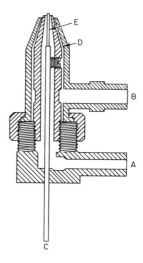

**Fig. 20–3.** Beckman burner-atomizer unit. (A) Oxygen, or air duct; (B) acetylene or hydrogen duct; (C) palladium capillary for the analytical solution; (D) acetylene or hydrogen inlet; (E) oxygen or air inlet.

shown in Fig. 20–3. The fuel and oxidant are fed into two separate chambers within the burner and mix outside the exit orifices. Thus, a turbulent flame is produced. As the oxidant flows past the sample capillary a vacuum is produced which draws the solution into the flame. The more common flame gases (fuel and oxidant) are listed in Table 20–1.

The monochromator is similar in optical design to those mentioned in chapter 17. It consists of entrance and exit slits, lenses, and a light dispersing device (prism or grating). A photomultiplier is generally used as a detector, and is coupled to an amplifier and meter readout.

**Interferences.** An interference in a flame is observed when the number of excited species is caused to increase or decrease. Interferences can be classed

**Table 20–1. Some Common Flame Gas Mixtures**

| Fuel | Oxidant | Temperature (°C) |
|------|---------|------------------|
| Hydrogen | Air | 2000 |
| Hydrogen | Oxygen | 2700 |
| Acetylene | Air | 2000 |
| Acetylene | Oxygen | 2800 |

into two catagories, chemical and spectral. Chemical interference occurs when a species in the flame reacts with the atoms, thus decreasing the emission. An example of this is the reaction between calcium and phosphorus containing molecules. If a solution of calcium and a soluble phosphorus compound is nebulized into a flame, the concentration of calcium atoms would be decreased due to the formation of calcium-containing molecules in the flame. Thus, the emission intensity of calcium decreases as the phosphorous concentration is increased. Spectral interferences can be observed when the emission of a second species in the flame occurs at the same wavelength as the compound being measured. As an example, consider a solution of calcium and sodium nebulized into the flame where the sodium is to be determined. The sodium emission is measured at 5889 Å. From the intensity of emission, it appears that there is more sodium than was actually placed in the solution. The reason for this is that another species, CaO, which is produced by the flame, is also emitting at this wavelength. Combustion products from the fuel and oxidant also have a tendency to interfere with the formation of metal atoms in the flame by converting the atoms into metal oxides and hydroxides. These are often very stable molecular species and thus, reduce the metal atom concentration appreciably.

In order to determine the concentration of a metal ion in solution, it is necessary to determine the extent of both spectral and chemical interferences. For most samples the effect is minimized by the addition of the interference to the standards or by a standard addition technique.

**Quantitative Determination.** In quantitative analysis, the emission intensity is correlated to the concentration of the emitting species through a calibration curve (intensity vs concentration). The method is very sensitive for certain elements such that solution concentrations of less than 1 ppm can be analyzed with an accuracy of greater than $\pm 5\%$. A list of elements that have been determined by flame photometry is shown in Table 20–2.

The use of flame photometry for the determination of certain metal ions has replaced tedious and time consuming methods. In clinical chemistry, the method is used to rapidly determine the concentrations of sodium and potassium in serum and urine. For example, a urine sample to be analyzed for sodium is diluted 1:1000 and the emission intensity at 589 nm is compared with a series of standard sodium ion solutions through a calibration plot. In order to determine potassium the urine is diluted 1:250. The average excretion of these metal ions in urine by a healthy person is approximately 75–200 mEq of sodium per 24 hours and 40–80 mEq of potassium per 24 hours.

**Plasma Emission.** Plasma emission spectrometry uses a special type of high temperature source which has been developed over the last ten years.

Table 20–2.  Elements Excitable in the Air-
Acetylene Flame[a]

| Element | Wavelength | Detectability limit (moles/liter) |
|---------|-----------|-----------------------------------|
| Ag | 3280.7 | 0.000005 |
| Ba | 5535.5 | 0.001 |
| Ca | 4226.7 | 0.00001 |
| Cs | 4555.3 | 0.0005 |
| K | 4044.2 | 0.0002 |
| Li | 6707.9 | 0.000001 |
| Mg | 2852.1 | 0.0002 |
| Na | 5890.0 | 0.00001 |
| Rb | 4215.6 | 0.0001 |
| Zn | 3072.1 | 0.5 |

Others: Au, Cd, Co, Cr, Cu, Dy, Fe, Ga, Gd, Hg,
In, La, Mn, Nd, Ni, Pb, Pr, Rh, Ru, Sc,
Tl, Y

[a] From R. Mavrodineau, "Flame Spectroscopy," Wiley, New York, 1965.

The plasma is produced by inductively or capacitatively coupling an ionizable gas with the magnetic field of a radiofrequency source, or with the
electric field of a microwave source. The production of the plasma is dependent on the ability of high velocity electrons to ionize the confined gas,
and thus sustain the plasma. The plasma can assume a flamelike configuration
or be confined in a quartz tube.

The basic instrument for observation of plasma emission is shown in Fig.
20–2, the main difference being that the burner–nebulizer system is replaced
by the plasma as the source. In a typical experiment, a gaseous or partially
desolvated analyte is injected into the plasma source. The sample is vaporized,
the atoms are excited via the high temperature, and subsequently emit characteristic radiation. The intensity of the radiation is then related to the concentration in the original sample by means of standards.

The plasma emission source has advantages over the flame in that it is not
dependent on a combustion process and the resulting temperatures are much
higher than that of the flame. Because of the increased temperatures the
efficiency of the excitation process is increased, thus increasing the sensitivity
for most elements. In addition, the number of chemical interferences are
decreased due to both the temperature effect and the simplicity of the flowing
gas, which is usually argon.

The plasma source has one major disadvantage in that difficulty has been

**Table 20–3.  Applications of Plasma Emission Spectrometry**

| Element | Sample type | Detection limit |
|---------|-------------|-----------------|
| As | Gas chromatographic effluent, pesticides | 20 pg |
| As | Direct solution | 0.03 $\mu$g/ml |
| C | Gas chromatographic effluent, organics | 10 ng |
| Hg | Gas chromatographic effluent, organics | 0.5 pg |
| Hg | Direct solution | 3.0 pg/ml |
| S | Gas chromatographic effluent, organics | 0.2 pg |
| Se | Direct solution | 0.04 $\mu$g/ml |
| Zn | Direct solution | 0.6 pg/ml |

experienced in sustaining the plasma when large amounts of sample are injected into the source unit. For this reason samples are usually limited to gaseous materials and desolvated solutions.

This technique has been applied to the determination of elements through direct solution analysis and to the determination of compounds as effluents from gas chromatographic columns (see Chapter 24). Table 20–3 presents a partial summary of applications and sensitivities.

**Arc Emission.**    The emission spectra of atoms can be observed by arc emission spectroscopy. A typical emission spectrograph is presented in Fig. 20–4. It is composed of a source unit, a monochromator, and a photographic film or plate detector. The source unit contains two carbon electrodes in a configuration in which the lower electrode is shaped in the form of a cup and the upper electrode is pointed. A solid sample is placed in the cup and an arc is generated between the two electrodes from a high wattage power supply.

The emitted radiation passes through the appropriate optics and is dispersed

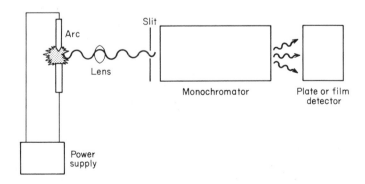

**Fig. 20–4.**  An emission spectrograph.

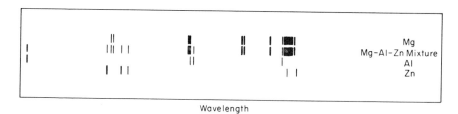

Fig. 20–5.   Emission spectrograms for several metals.

into its individual wavelength components by the monochromator. The dispersed radiation is then detected by the photographic plate. After development of the plate, a series of lines and bands appear as shown in Fig. 20–5.

**Qualitative Analysis.**     Since an atom emits a line spectrum which is characteristic of that species, the position where these lines occur in the electromagnetic spectrum can be used for qualitative analysis. Two approaches to plate interpretation can be taken. In the first method, the wavelength of the sample lines are determined and compared to standard tables of wavelength of emission lines for the individual elements. The second method involves direct comparison with the spectrum of a single element or a standard mixture of metals on the same photographic plate. In either case, at least three predominant lines should match to confirm the presence of an element.

In most cases the intensities of the lines are also important. Certain lines in the spectrum of an element are very intense and are the first to appear at very low concentrations. Thus, these lines should be present for an identification to be positive. Tables which list the wavelength of lines also cite the relative intensities of each line.

The emission spectrum of an atomic species is recorded on an uncalibrated plate. Thus, in order to determine the wavelength of each line it is necessary to calibrate the photographic plate. This is done by exciting a metal (usually copper) in the arc and recording its spectrum. The plate is then moved vertically to a different position and the sample containing the unknown metals is excited. Hence, the emission is recorded below the spectrum of the standard metal. Approximately 12 plate positions are available such that the emission specta of a number of samples can be obtained on the same photographic plate. After the plate is developed, it is calibrated by using the lines of the standard metal as a wavelength reference to convert any position across the plate to wavelength units.

If the dispersing device in the instrument is a grating, the dispersion of radiation is linear with respect to wavelength. This allows the plate to be calibrated in units of Å/mm (reciprocal linear dispersion). Thus, the wavelength of any line can be determined by accurately measuring the distance between a line of the standard and line of a sample, multiplying this value by the reciprocal linear dispersion, and adding or subtracting the product to the wavelength of the standard. For example, if the reciprocal linear dispersion of an instrument is 5.706 Å/mm and a sample line lies 7.931 mm in a lower energy direction from a 3247.54 Å copper reference line, the wavelength of the sample line is 3292.79 Å.

In order to determine the reciprocal linear dispersion of an instrument, it is necessary to use two or more reference lines of known wavelength. The distance (in mm) between these lines is determined and the difference in wavelength of the two lines is divided by the measured length between them. The result of this calculation gives the reciprocal linear dispersion in Å/mm. A number of sets of lines should be measured to insure a high accuracy for this number.

If the instrument employs a prism, the dispersion is nonlinear with respect to wavelength or energy and plate calibration is more difficult. The plate is calibrated by measuring the position of a number of standard lines and solving a set of simultaneous equations considering a series of constants and the measured data. The wavelength of the sample lines can be determined by use of the measured positions and calculated constants, or by comparison with standards.

**Quantitative Analysis.**    The intensities of the emitted lines and thus the densities of the lines on the photographic plate are related to the quantity of material present in the original sample. The densities of the lines are measured with a densitometer (Fig. 20–6) which has many of the features of the spectrophotometer. The main differences are that the densitometer does not have a monochromator and the cuvette (cell) is replaced by a plate or film. The instrument correlates the incident power ($P_0$) and transmitted power ($P$) to the density of a particular line. Hence, the density is related

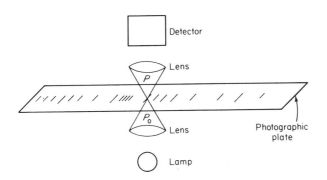

**Fig. 20–6.**   The densitometer.

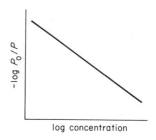

**Fig. 20–7.** Calibration curve for relating density of a line to concentration.

to the concentration through a calibration plot of log density $(P_0/P)$ vs the logarithm of the concentration.

In a typical experiment, a calibration curve is prepared with the aid of an internal standard. A series of exposures are taken in which the density of the line of the sample is compared with the density of the line of another element. In doing this the effect of change in excitation conditions is eliminated. After a series of measurements is taken the log of the ratio of the line intensities of the sample and the internal standard is plotted as a function of log sample concentration. An example is shown in Fig. 20–7 where an unknown sample concentration is determined by the densities of lines of the sample and standard. The concentration of the unknown is then determined by comparison with the calibration curve.

The internal standard method is usable if the lines adhere to the following requirements:

1. Both lines must respond in intensity in a similar manner to a change in excitation conditions.

2. The lines of the two elements must both originate from atoms or ions.

Emission spectroscopy is primarily used as a qualitative or semiquantitative tool because the accuracy is not as good as for other methods. It is generally used for concentrations of metals in solids ranging from a few percent down to less than 1 ppm with an accuracy of better than 15%. The greatest advantages of this technique are:

1. The photographic plate integrates the light intensity. Thus, trace metal determination is possible.

2. Solid samples can be vaporized and excited by the process. Gases and solutions can also be handled with ease.

The disadvantages of this technique are:

1. The instrumentation and facilities are costly.
2. The technique is time consuming if only one analysis is desired.
3. Standards must often be synthesized.

## ATOMIC ABSORPTION

Atomic absorption spectrophotometry is an absorption method where radiation is absorbed by nonexcited atoms in the vapor state. This method has advantages over flame emission because:

1. More elements can be quantitatively determined.
2. The spectral interferences are decreased.
3. The sensitivity is higher for most elements.

The instrument is composed of a light source, a cell, monochromator, and detector system. A diagram of the system is presented in Fig. 20–8. The light source (hollow cathode) emits a line radiation, which is of the exact wavelength of the element being determined since the source is made of the sample element. Thus, if iron is to be determined the source element should be made of iron.

The sample is nebulized into a premixed gas–air burner designed for a long path length (see Fig. 20–9). The radiation is passed into the mono-

**Fig. 20–8.** Block diagram of an atomic absorption spectrophotometer.

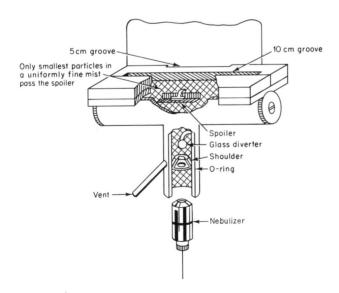

**Fig. 20–9.** A laminar flow burner used for atomic absorption spectrophotometry.

Table 20–4. Some Elements That Can Be Determined by Atomic Absorption Spectrophotometry[a]

| Chemical species | Material analyzed | Concentration range, in ppm in solution and standard deviations ($\pm$) | Analytical line (Å) |
|---|---|---|---|
| Li | Test solutions | 0.03–4 | 6707.8 |
| Na | Plants | 0.2–2000 | 3232.6 |
| Na | Soil extracts | 0.1–0.5(0.08)–5(0.2) | 5890.9 |
| | | | 5895.9 |
| K | Soil extracts | 0.1–1(0.06)–10(0.1) | 7664.9 |
| | | | 7699.0 |
| Cu | Copper-based alloys | 25(0.24)–50(0.12) | 3247.5 |
| Cu | Test solutions | 2–200 | 2227.8 |
| Rb | Test solutions | 0.1–20 | 7800.2 |
| Ag | Test solutions | 0.1(0.04)–10(0.05) | 3280.7 |
| Cs | Test solutions | 0.2–20 | 8521.1 |
| Au | Test solutions | 1(0.15)–50(0.1) | 2428.0 |
| Au | Test solutions | 2–200 | 2676.0 |
| Mg | Plants, soils, lysimeter, and drainage waters, blood sera, milks | 0.3(0.02)–3(0.06)–10(0.3) | 2852.1 |
| Mg | Blood sera | 0.3(0.003)–2(0.02) | 2852.1 |
| Mg | Plants, soil extracts | 0.5(0.08)–5(0.06) | 2852.1 |
| Ca | Blood sera | 4–10(0.1)–15 | 4226.7 |
| Ca | Plants | 2.5–50 | 4226.7 |
| Ca | Soil extracts | 2.5–50 | 4226.7 |
| Sr | Test solutions | 0.2–20 | 4607.3 |
| Ba | Test solutions | 8–1000 | 5535.6 |
| Zn | Leaded brass | 5(0.03)–25(0.05) | 2138.6 |
| Zn | Plants | 1(0.01)–10(0.3) | 2138.6 |
| Cd | Test solutions | 0.03–4 | 2288.0 |
| Hg | Test solutions | 10–1000 | 2536.5 |
| Ga | Test solutions | 3–500 | 2874.2 |
| Tl | Test solutions | 1–100 | 2767.9 |
| Sn | Test solutions | 5–350 | 2863.3 |
| Pb | Leaded brass | 100–200 | 2833.1 |
| Cr | Test solutions | 0.2–20 | 3578.7 |
| Sb | Test solutions | 2–200 | 2311.5 |
| Bi | Test solutions | 2–300 | 3067.7 |
| Mo | Test solutions | 0.5–80 | 3132.6 |
| Mn | Leaded brass | 10–75 | 4030.7 |
| Mn | Soils, soil extracts, plants | 0.5(0.04)–25(1.0) | 2794.9 |
| Fe | Plants | 2.5(0.14)–125(4.1) | 2483.3 |
| Co | Test solutions | 0.2–20 | 2407.2 |
| Ni | Leaded brass | 10–50(1) | 3414.8 |
| Rh | Test solutions | 2(0.4)–100(0.6) | 3434.9 |
| Pd | Test solutions | 2(0.1)–100(0.5) | 2476.4 |
| Pt | Test solutions | 10(3.0)–100(2.0) | 2659.4 |

[a] From R. Mavrodineau, "Flame Spectroscopy," Wiley, New York, 1965.

chromator and measured at the detector. The amount of radiation absorbed is proportional to the concentration of the element in the sample. A calibration curve is prepared by measuring the absorbance of a series of standard solutions.

Atomic absorption can be used for measuring very low concentrations of metal ions in solutions as illustrated by the sensitivities shown in Table 20–4. This method has been widely applied to biological, agricultural, metallurgical, geological, and pollution samples.

For example, in the area of pollution, the Council of British Archeology (1964) prepared a list of historic towns in England, Scotland, and Wales for which a comprehensive study of the environment in these areas was needed. One study dealt with the atmospheric lead contamination by automobiles in Warwick, England. Samples of air were passed through Watman No. 1 filter paper to remove the particulate material from the air. The volume of air was recorded on a gas meter. The filter paper was subsequently treated

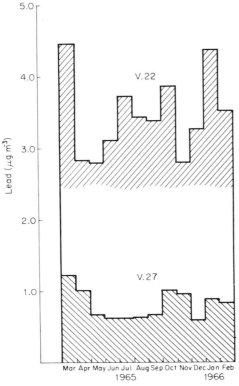

**Fig. 20–10.** The influence of traffic on atmospheric lead concentrations. Listed are the monthly mean values. V.22, heavily traveled; V.27 lightly traveled. (From J. Bullock and W. Lewis, *Atmospheric Environment* **2**, 517, 1968.)

with 10% nitric acid, filtered to remove the insoluble material, and diluted to volume. The lead concentration was then determined by atomic absorption spectrophotometry using standards which were treated similarly. The results indicated that the heavily traveled roads maintained a much higher concen-centration of lead than the lesser traveled roads (Fig. 20–10).

*Example 20–1.*    From the data below calculate the concentration ($\mu g/m^3$) of lead in the air sample. The amount of air passed through the filter was 1000 $m^3$. The filter was dissolved in 10% $HNO_3$, filtered, and diluted to 100.0 ml in a volumetric flask. A series of standards were also prepared in 10% $HNO_3$. The data obtained on these samples are as follows:

| Conc. of Pb ($\mu g/ml$) | Absorbance |
|---|---|
| 2.00 | 0.15 |
| 4.00 | 0.31 |
| 6.00 | 0.47 |
| 8.00 | 0.60 |
| 10.00 | 0.77 |
| Unknown sample | 0.58 |

By plotting the data ($A$ vs $c$) the concentration of the unknown is determined to be 7.50 $\mu g/ml$. The concentration of the sample can be calculated as follows:

$$7.50 \ \mu g/ml \times 100.0 \ ml = 750 \ \mu g \ \text{Pb total}$$

$$\frac{\text{Total Pb}}{\text{Volume of air}} = \frac{750 \ \mu g}{1000 \ cm^3} = 0.750 \ \mu g \ \text{Pb}/cm^3$$

# ATOMIC  FLUORESCENCE

Atomic fluorescence spectroscopy is the newest of the techniques used for the determination of metals. In comparison with atomic absorption where the absorpton of radiation from a hollow cathode is measured, atomic fluores-cence is the observation of emission after the atomic species is excited by a selected wavelength. The schematic diagram of an atomic fluorescence spectrometer is presented in Fig. 20–11. It should be noted that the source is placed orthogonal to the optical axis of the system and the radiation is modulated to detect only the resonance radiation of the sample caused by the source.

The most successful source for atomic fluorescence is the electrodeless discharge lamp. The lamps are sealed quartz tubes containing argon and the metal of interest. These tubes are driven by a microwave generator for

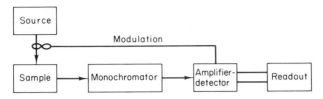

**Fig. 20–11.**   Schematic diagram of an atomic fluorescence spectrometer.

vaporization and excitation of the metal and produce very intense atomic lines of usually long lifetimes. A long warmup time for stabilization is required.

Although a number of papers have appeared in the primary literature on the utility of this technique, few articles reflect application to real samples. The determination of wear metals in oils and determination of selected metals in metabolic fluids have been reported. Since the observation for atomic fluorescence and emission is similar, methods for quantitative determination are the same.

## Questions

1.  Describe the methods used to excite atoms.
2.  Describe the basic components of an arc spectrograph.
3.  How is an arc spectrogram used for qualitative identification?
4.  What is the purpose of a densitometer and how is it used in quantitative determinations?
5.  What are the differences between atomic absorption and flame emission?
6.  Describe the basic components of an atomic absorption unit.

## Problems

1.* A sample (1.2456 g) containing sodium is dissolved, diluted to 100 ml, and analyzed using the 590-nm Na line. From the data given below, determine the percent sodium in the original sample.

| Concentration of Na (mg/liter) | Emission reading |
|:---:|:---:|
| 0.5 | 24 |
| 1.0 | 49 |
| 2.0 | 103 |
| 2.5 | 120 |
| 4.0 | 190 |
| Unknown | 121 |

---

* Answers are listed at the end of the book for problems marked with an asterisk.

2. A sample (2.9674 g) containing zinc is dissolved in dilute acid, diluted to 100.0 ml, and analyzed by absorption. From the data below, determine the ppm zinc in the original sample.

| Concentration of Zn ($\mu$g/ml) | Absorbance |
|---|---|
| 1.0 | 0.11 |
| 3.0 | 0.30 |
| 5.0 | 0.54 |
| 6.0 | 0.67 |
| 7.0 | 0.79 |
| 20.0 | 1.13 |
| Unknown | 0.37 |

3. An organic compound containing calcium is to be analyzed quantitatively for that element. The compound, 0.7350 g, is dissolved, diluted to exactly 100.0 ml, and its emission intensity compared to that of a series of standards. Determine the percent composition of calcium in the sample.

| Concentration Ca (mg/ml) | Emission intensity |
|---|---|
| 0.50 | 23 |
| 0.75 | 35 |
| 1.00 | 45 |
| 1.25 | 58 |
| 1.50 | 70 |
| Unknown | 62 |

4.* An iron sample (0.9421 g) containing a trace amount of copper is determined by atomic absorption by standard addition. The iron is dissolved in acid and diluted to exactly 100.0 ml using a volumetric flask. To 25 ml of this sample solution, 25 ml of 4.50 ppm copper solution is added. From the following data, calculate the concentration (in ppm) of the copper in the original sample.

Absorbance of sample = 0.22

Absorbance of sample with standard added = 0.31

5. A metallic sample is excited using an arc spectrograph. The predominent lines which are found after development of the photographic plate are 2427.92, 2675.96, 2802.20, 3247.51, 3273.99, 5105.50, 5153.24, and 5218.22 Å. What is the composition of the alloy?

6. A sample is excited in an arc and the emission is recorded by photographic plate. From the following lines, determine qualitatively the composition of the sample.

| | | |
|---|---|---|
| 2288.02 Å | 3610.50 Å | 3748.28 Å |
| 2816.19 Å | 3719.96 Å | 6231.76 Å |
| 3261.06 Å | 3737.133 Å | 6243.36 Å |
| 3403.61 Å | 3745.57 Å | 6438.47 Å |
| 3466.20 Å | | |

# Chapter
# Twenty-One
# Luminescence

## FLUORESCENCE

Fluorescence is a form of luminescence in which light is emitted from an irradiated sample. This is illustrated in Fig. 21–1. A sample containing a fluorescent compound is irradiated with light of energy $\Delta E_1$ which promotes an electron from energy level $E$ to energy level $E^*$. The compound now can lose its energy by two modes: (1) by collision with solvent molecules (collisional deactivation, illustrated by $\Delta E_2$ and $\Delta E_4$ in Fig. 21–1); and (2) by radiation of energy $\Delta E_3$ after a partial energy loss of $\Delta E_2$.

As shown in Fig. 21–2, the activation wavelength is of higher energy than the emission or fluorescence wavelength. However, some compounds fluoresce at a wavelength equal to the activation energy (resonance fluorescence). The total process of excitation ($\Delta E$), vibrational relaxation ($\Delta E_2$), and emission ($\Delta E_3$) takes between $10^{-4}$ and $10^{-8}$ seconds.

**Intensity of Emission.**   The intensity of fluorescence ($F$) is given by

$$F = 2.303\phi P_0\epsilon bc \qquad (21\text{-}1)$$

where $\phi$ is the quantum yield, $P_0$ is the incident intensity, $\epsilon$ is the molar absorptivity, $b$ is the pathlength, and $c$ is the concentration. This equation only holds for solutions of very low concentrations.

The quantum yield, $\phi$, is a measure of the efficiency of production of fluorescent radiation

$$\phi = \frac{\text{number of photons emitted}}{\text{number of photons absorbed}} \qquad (21\text{-}2)$$

If, for example, $\phi = 1$, then every photon which is absorbed would be emitted

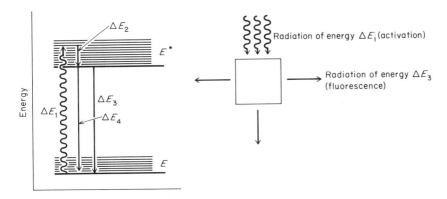

**Fig. 21-1.** Energy level diagram depicting fluorescence.

as fluorescence radiation. The quantum yield, however, is always less than 1, and only in a few cases does the yield approach unity. In the majority of systems, the quantum yield is very low. For this reason, many compounds are often classified as nonfluorescent even though their structures are conducive to fluorescence.

In Eq. (21-1) the fluorescence intensity is directly proportional to the incident intensity, $P_0$, and concentration, $c$. As $P_0$ is increased (or the power of the source increased), the amount of fluorescence observed will also increase. As the concentration of the fluorescing component increases, its fluorescence increases.

If the fluorescence spectra of a series of solutions of increasing concentration $(c_1 < c_2 < c_3 < c_4)$ are measured, plots similar to those presented in Fig. 21-3a will be obtained. The maxima at a given wavelength, $\lambda_1$, are then plotted versus concentration and the calibration curve as shown in Fig. 21-3b

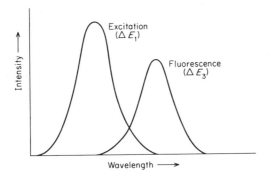

**Fig. 21-2.** Activation and fluorescence spectra.

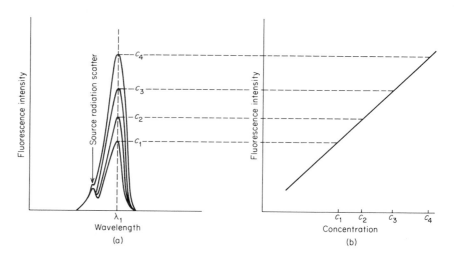

**Fig. 21–3.** Change in fluorescence intensity as a function of concentration. (a) Fluorescence spectrum at different concentrations; (b) calibration plot.

is obtained. Normally, the concentration, in which the method is useful, is in the part per million range or less.

**Factors Affecting the Quantum Yield.**    As previously inferred, the fluorescence of all molecules cannot be observed. Molecules that fluoresce with a high quantum efficiency generally have one or more of the following structural components:

1. A high molar absorptivity
2. A number of conjugated double bonds or high resonance stability
3. An electron-donating group, such as $NH_2$ and OH
4. A relatively rigid structure, such as metal complexes, in the molecule

Not all of these structural features are required to produce fluorescence in a molecule.

Certain structural features also tend to inhibit fluorescence (quenching). Functional groups such as iodide, bromide, nitro ($-NO_2$), and carboxylate ($-CO_2H$) tend to decrease the fluorescence of the molecule. A decrease in fluorescence may also occur when the molecule complexes with heavy metal ions such as mercury.

The term quenching is applied to all processes which tend to decrease the quantum yield. Examples of these are:

1. Collisional deactivation by solvent
2. Energy consumed by bond breakage
3. Absorption of the fluorescence by another component of the solution

Generally, fluorescence has been applied to the greatest extent in the life sciences where trace concentrations of metabolites, drugs, etc. must be determined quantitatively. For example, thiamine and riboflavin and their metabolites can be determined in trace amounts in biological tissues and fluids. Fluorescence is also used in quality control of drug dosage.

**Instrumentation.** In order to measure fluorescence, it is necessary to have an emission source in the ultraviolet region, a detector (ultraviolet–visible), and a method of separating the activation and fluorescent radiation. A block diagram for a simple instrument is illustrated in Fig. 21–4. Radiation from a source is passed through a filter and imposed upon the sample in the cuvette. The fluorescent radiation emitted by the sample is then passed through another filter and the amount of energy is measured by the detector.

The wavelength scans in Fig. 21–4 indicate the distribution of radiation at any point along the light path. Radiation is initially emitted by a high-pressure xenon lamp giving light at all wavelengths. The radiation is then passed through a filter, prism, or grating to give the proper activation energy and impinges upon the sample. After the sample fluoresces (in all directions), the emitted radiation is measured at right angles by a detector. The purpose of the secondary filter is to remove unwanted scattered radiation. In less sophisticated instruments, a fixed wavelength source, such as the

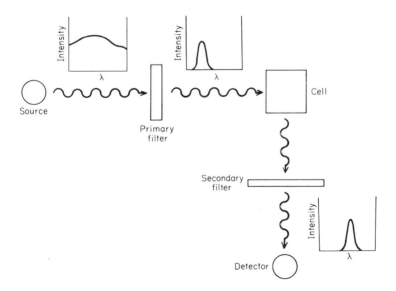

**Fig. 21–4.** Block diagram of a fluorometer. Scans of the spectra are given at each point of change.

mercury lamp, is used. By combination with different filters it is possible to select certain wavelengths for excitation.

The fluorescent intensity is measured as percent transmission. Therefore, as the fluorescent intensity increases, the percent transmission increases. This requires standardization of the 100% T level on the fluorometer. Either of two techniques can be used: (1) an aqueous $H_2SO_4$ solution of quinine (quantum yield of about 1) or a piece of glass containing uranyl ion ($UO_2^{2+}$) (high quantum yield) is inserted into the sample chamber and the instrument set at 100% T; or (2) the most concentrated standard of the sample being determined is used to set 100% T. After the fluorometer is standardized (0 and 100% T), the calibration curve is prepared by measuring %T for a series of standards. These data are plotted against concentration yielding a calibration curve which is often nonlinear.

Tetracycline (**I**) is used as a general antibiotic and can be determined fluorometrically after conversion to anhydrotetracycline (**II**) and subsequent

I                                           II

complexation with aluminum. The following is a typical procedure for the determination of tetracycline in serum. After extraction of 0.2 ml of serum from the metabolic system and addition of 9.0 ml of water and 1.0 ml of 30% trichloroacetic acid, the solution is mixed and centrifuged. Then 8 ml of the filtrate is taken, 1 ml of 5 $M$ hydrochloric acid is added, and the solution boiled for 30 minutes. After the solution is neutralized with sodium hydroxide solution, the pH is adjusted with a buffer to 4.5 and 2 ml of chloroform are added. Then 1 ml of the solution is mixed with 1 ml of 0.1% $AlCl_3 \cdot 6H_2O$ in absolute ethyl alcohol and the solution is allowed to stand at least 1 hour for development of the complex. Subsequently, fluorescence is observed at 550 nm using an activation wavelength of 475 nm. A plot of fluorescence intensity versus concentration for a series of standards gives a straight line over a concentration range of 0.1–20 μg tetracycline/ml of sample. Table 21–1 presents the results of the study of tetracycline concentration in dog serum as a function of time.

It is interesting to note that anhydrotetracycline also fluoresces; however, the quantum yield of the compound is about 30 times smaller than the aluminum complex. Thus, formation of the complex provides increased sensitivity.

Polycyclic aromatic hydrocarbons, which are identified by the symbol PAH, are known carcinogens. Typical PAHs are benz(a)anthracene, benz(a)pyrene,

Table 21–1. Results of the Study of Tetracycline Concentration in Dog Serum as a
Function of Time

| Intravenous dose | Time after dose (hours) | Concentration ($\mu$g/ml serum) |
|---|---|---|
| 7.5 mg/kg | 0.5 | 10.3 |
| | 1.5 | 7.0 |
| | 3.0 | 4.3 |
| | 5.0 | 3.5 |
| | 8.0 | 2.7 |
| | 24.0 | 0.5 |

benz(e)pyrene, and chrysene. They are found in the atmosphere as the result
of burning organic materials at temperatures that are sufficient to break
C—C and C—H bonds. Radicals that form under these conditions then react
to form the PAHs. It is very important to be able to determine PAH levels
in the atmosphere, and since these compounds are highly conjugated and
fluoresce in the visible region at high fluorescent efficiencies, procedures
based on fluorescent measurements have been devised.

The polycyclic aromatic hydrocarbons, which occur as suspended particulates in air, are collected on a fiberglass filter. The PAHs are extracted from
the filter with benzene and the benzene is then evaporated under nitrogen.
The residue that remains is separated into the individual PAHs by thin-layer
chromatography (see Chapter 25) using a pentane–ether (19:1) eluting
mixture. After development, the layer is allowed to dry and the spots are
detected as fluorescent spots under ultraviolet light.

If, for example, benz(a)pyrene (BaP) is to be determined quantitatively,
the spot corresponding to that compound is removed and placed in a flask
containing 2.0 ml of $H_2SO_4$. After the compound dissolves, the fluorescent intensity of the solution is measured at 545 nm using an excitation wavelength
of 525 nm. Comparison of this fluorescent intensity with the fluorescence
of standards allows the calculation of the BaP concentration in the air sample. As little as 0.01 $\mu$g of BaP can be detected by this method.

## PHOSPHORESCENCE

Several different molecules have the capability of undergoing delayed
emission or phosphorescence. This emission is observed at right angles to the
activation beam similar to that in fluorescence. The range of times for emission

of phosphorescent radiation from a molecule is $10^{-4}$ to 1 second. Due to this fact, phosphorescence is usually observed only in rigid solvent structures where collisional deactivation is minimized. Thus, phosphorescence of a solution is generally observed at liquid nitrogen temperatures.

Since the same structural features are needed for a molecule to fluoresce and phosphoresce, compounds that phosphoresce will often fluoresce. The two are distinguished by use of a rotating shutter which introduces a delay in the emission after activation.

Phosphorimetry is not generally used as a routine method of analysis. Only when other methods fail does this method become invaluable in the trace concentration range (ppb to ppm).


## CHEMILUMINESCENCE

Chemiluminescence is a subdivision of luminescence in which the energy is supplied to the molecule by a chemical reaction. In other words, the product of a reaction is formed in an excited electronic state. The "firefly" reaction is an excellent example. Luciferin ($LH_2$) reacts with adenosine triphosphate (ATP) in the presence of the enzyme luciferase (E) to form a complex

$$E + LH_2 + ATP \overset{Mg^{2+}}{\rightleftharpoons} E \cdot LH_2 \cdot AMP$$

$$E \cdot LH_2 \cdot AMP + O_2 \rightarrow E + \text{products} + AMP + \text{light} \qquad (21\text{-}3)$$

This complex then reacts with oxygen to form products with the emission of radiation. The amount of light emitted is proportional to the concentration of ATP if the concentrations of the other reactants are held constant for a series of runs.

A schematic diagram of an instrument useful for this measurement is shown in Fig. 21–5, where the essential components consist of an injection system for the reactants, a cell cavity, and a detector. By injecting a constant concentration of luciferase, magnesium sulfate, and luciferin into standard ATP samples, a calibration curve of luminescence intensity vs concentration can be made. This technique is especially valuable because all living microorganisms contain ATP. Thus, the method may be used to monitor the effectiveness of biocides, effect of drugs, and detection of microorganisms in foods.

One application of chemiluminescence is in the monitoring of ppb levels of NO in air. The chemiluminescence measurement is based on the gas phase reaction of NO with ozone:

$$NO + O_3 \rightarrow NO_2^* + O_2$$

$$NO_2^* \rightarrow NO_2 + \text{light}$$

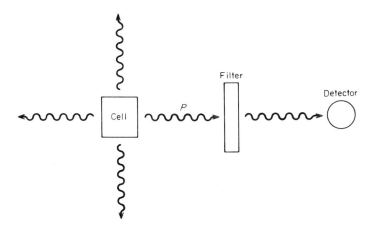

**Fig. 21–5.**   Diagram of an instrument used for observing chemiluminescence.

The electronically excited nitrogen dioxide, $NO_2^*$, formed in the reaction decays to the ground state with the emission of energy in the form of light. In a typical observation, air containing the NO is pumped at a constant rate through a chamber containing $O_3$. The resulting emission is continuously monitored by a light-sensitive cell such as a photomultiplier.

## LIGHT SCATTERING

If light is passed through a cell containing suspended particles, radiation can be observed at all angles. This process, as illustrated in Fig. 21–6, is called scattering. The incident radiation, $P_0$, impinges upon the sample, resulting in a transmitted intensity, $P_t$, and a scattered intensity $P_s$. The amount of scattering is dependent upon the number of particles in the suspension, the size of the particles, and the pathlength. The relationship

$$\log \frac{P_0}{P_t} = Kbc \qquad (21\text{-}4)$$

expresses the dependence of the scattering $[\log (P_0/P_t)]$ upon the concentration, $c$, and the pathlength, $b$. The turbidity constant, $K$, is a proportionality term which relates to the size and shape of the particles.

The relationship between the scattered radiation, $P_s$, and concentration, $c$, is

$$\frac{P_s}{P_0} = bk'c$$

where $k'$ is a constant depending upon the system.

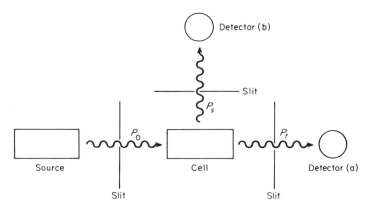

**Fig. 21–6.** Block diagram of a scattering photometer. The position of detector (a) measures the transmitted radiation. The orientation of detector (b) allows measurement of the scattered radiation.

A block diagram of a simple instrument is presented in Fig. 21–6. Two types of measurements can be made. First the amount of light passed through the solution $(P_0/P_t)$ can be observed. This measurement is known as turbidimetry. Nephelometry applies to measurements where the amount of scattered radiation, which is measured at an angle (usually 90°), is monitored $P_0/P_s$. It should be noted that as the transmitted radiation decreases the scattered radiation increases.

Light scattering has been used for molecular weight determinations of macromolecules and detection of end points in titrations where the product forms a precipitate. Turbidimetric measurements have been applied to turbidity measurements of brackish solutions and the determination of the level of smog. For example, the suspended particulate material in city water is generally monitored by the use of turbidimeters.

## Questions

1. What is the difference between phosphorescence and fluorescence?
2. How does an increase in quantum yield affect the intensity of fluorescence?
3. What parameters determine fluorescence intensity?
4. What are the basic components of the fluorimeter?
5. What are the basic components of the phosphorimeter?
6. What is chemiluminescence?
7. What are the requirements for causing light to be scattered?
8. What is the difference between turbidimetry and nephelometry?

## *Problems*

1.* Solutions (50 ml) containing various concentrations of zinc ion were complexed with an excess of 8-hydroxyquinoline and extracted into 100 ml of chloroform. The fluorescence of each solution was determined and is given below.

| Zinc (mg/100 ml) | Fluorescence reading (%T) | Zinc (mg/100 ml) | Fluorescence reading (%T) |
|------------------|---------------------------|------------------|---------------------------|
| 3                | 15.2                      | 12               | 61.0                      |
| 4                | 20.1                      | 14               | 72.3                      |
| 6                | 31.0                      | 16               | 85.7                      |
| 8                | 39.8                      | 18               | 95.3                      |
| 10               | 51.1                      | 20               | 100%                      |

An unknown solution was treated the same as the standards and a fluorescence reading of 37% T was obtained. Determine the concentration of zinc in mg/100 ml and moles/liter.

2. Assume that the fluorescence reading of 10.3 $\mu$g tetracycline/ml serum is 53.7% T. Calculate the fluorescence reading for 7.0, 4.3, and 3.5 $\mu$g/ml serum.

3. Using the data in Problem 2, what is the concentration of tetracycline in serum if the fluorescence reading is 37.2% T?

4.* Using the data in Problem 2, if a dog has 2 quarts of blood and the fluorescence reading of 1 ml of serum is 25.7% T, what is the total amount of tetracycline in the blood?

5. An unknown amount of riboflavin (10 tablets) was dissolved in exactly 1 liter of water. One milliliter (1.00) of the solution was diluted to 1.00 liter and its fluorescence measured (42.0% T). A standard containing 9.05 mg/liter riboflavin had a fluorescence intensity of 32% T. What is the average amount of riboflavin in each tablet?

6.* A solution of thiamine hydrochloride gave a fluorescence reading of 52.3% T. Five milliliters (5.00) of a standard (0.2 mg/liter thiamine HCl) were added to 5.00 ml of sample. What is the concentration of thiamine in the unknown if the fluorescence intensity of the latter solution is 67.0% T?

---

* Answers are listed at the end of the book for problems marked with an asterisk.

# Chapter
# Twenty-Two
# Separations:
# Introduction to
# Chromatography

## SCOPE

The goal in separations, as applied to chemical systems, is to part or divide a heterogeneous or homogeneous mixture or mass into its individual units, components, or even into elements. There are many different methods and techniques for performing separations which are based on sound fundamental chemical and physical principles. Some of the separation procedures are very complex, while others are relatively simple and require little expenditure in money, attention, and effort.

The area of separations has experienced a rapid growth in recent years. During this period not only have underlying principles been extended and more clearly defined, but newer techniques have been developed. Thus, many mixtures which were difficult or impossible to separate in the past can now be separated, some even in a routine manner. Examples of these are the separation of amino acids, rare earths, isolation of 10 to 40 atoms of newly prepared transuranium elements, isotopes, homologous series of organic molecules, minerals from water, and many others.

A separation can be considered to be a form of sample pretreatment where those components of the sample, which will interfere in the determination, are removed. Or, if each component of the sample is to be determined, separation is performed such that each component is isolated. Consequently, a method of analysis can be chosen which is based on the amount of the component present rather than on the kind of interferences present. The net result is that the separation method provides additional selectivity for the ordinary instrumental and chemical methods.

A separation procedure can be used for purification, qualitative identification, or quantitative determination. There are many different separation procedures, some of which have been known and used for a long time [gravimetry (Chapter 7) and distillation], while others have been recently developed (gas chromatography and thin-layer chromatography). It is not possible to discuss all of these techniques in this book. In this chapter a general introduction to separation methods and chromatography is given. In the next five chapters several of the more important and versatile separation methods in analytical chemistry and related health sciences are discussed in more detail.

# CLASSIFICATION OF SEPARATION METHODS

Several different approaches have been used to classify separation methods. These include division according to whether they involve batch discrete equilibrium steps or continuous nonequilibrium steps, type of force involved,

Table 22-1. Methods of Separation

| Method | Basis |
|---|---|
| Precipitation | Differences in solubility |
| Distillation | Differences in volatility |
| Sublimation | Differences in vapor pressure |
| Extraction | Differences in solubility between two phases |
| Crystallization | Property of solubility usually at lower temperature |
| Zone refining | Crystallization usually at elevated temperature |
| Flotation | Differences in density between substance and liquid |
| Ultrafiltration | Size of substance in comparison to filtering device |
| Dialysis | Osmosis; flow of a system through a membrane |
| Electrodeposition | Electrolysis at inert electrodes |
| Chromatographic Methods | |
| Adsorption column chromatography | Distribution of solute between a solid and liquid phase on a column |
| Partition column chromatography | Distribution of a solute between two liquids on a column |
| Thin-layer chromatography | Adsorption or partition on a open thin sheet |
| Paper chromatography | Partition on a paper sheet |
| High pressure liquid chromatography | Column liquid chromatography under high inlet pressure |
| Ion exchange chromatography | Exchange of ions |
| Molecular sieves | Size of solute |
| Gel permeation (filtration) | Size of solute |
| Gas chromatography | Distribution of a gaseous solute between a gas and liquid or solid phase |
| Zone electrophoresis | Separation on a sheet in the presence of a electrical field |

whether they are mechanical, physical, or chemical in nature, and the type of heterogeneous equilibria involved. All of these suffer from one kind of limitation or another. Because of the widely diverse properties on which separations are based, it is not possible to assemble all separation techniques into one classification scheme.

Table 22–1 lists many separation methods according to commonly used descriptive titles. A brief description of each is given. Those that will be discussed in this book are chromatographic methods (column chromatography, sheet methods, gas chromatography, and ion exchange), solvent extraction, and gravimetry (Chapter 7).

## CHROMATOGRAPHY

Of all the different types of separation methods, chromatography has the unique position of being applicable to all types of problems in all areas of science, and has undergone explosive growth in the last 25 years. Several facets of chromatography are common to all chromatographic techniques. These will be discussed in the remainder of this chapter. Subsequently, several of the more specific chromatographic techniques will be discussed in the following chapters.

**History.**     Two investigators at the turn of the century, David Day, a geologist and mining engineer, and Mikhail Tswett, a botanist and physical chemist, independently carried out the initial chromatographic experiments. Day's work was primarily with the separation of crude oil on Fuller's earth (he called the process "fractional diffusion" because of its similarity to distillation), while Tswett's experiments were with separation of components in dissolved leaf extracts.

Tswett's investigations are, perhaps, more clearly identifiable with current techniques. Much of the current terminology in adsorption chromatography was coined by Tswett. In his experiments, a leaf extract sample in petroleum ether was passed through a column of $CaCO_3$. Passage of pure ether was continued through the column and various chlorophyll pigments, etc. were separated into a series of differently colored and easily distinguished zones. Unfortunately, the potential of these early experiments, even though Tswett's experiments were reported in great detail (about 50 papers), were not recognized and further development in chromatography did not occur until the late 1920s and early 1930s.

**Classification in Chromatography.**     Chromatography can be divided into the following general areas: adsorption chromatography; partition chromatography; exclusion chromatography; and ion-exchange chroma-

**Table 22–2.  Phase Combinations in Chromatography**

| Stationary phase | Mobile phase | Type |
|---|---|---|
| Solid | Liquid[a] | Absorption, ion exchange,[b] exclusion[b] |
| Solid | Gas | Adsorption |
| Liquid | Liquid | Partition |
| Liquid | Gas | Partition |

[a] Adsorption is also possible between the mobile liquid phase and stationary phase.

[b] Ion exchange and exclusion methods fit in this category from a description point of view; the actual process involves more than is implied here.

tography. These should not be confused with laboratory operations. For example, partition can be done on paper (paper chromatography), in a column (column partition, reversed phase, and gas chromatography), or on a thin layer (thin-layer chromatography). The fundamental principle, one of partition, is the same. The difference is the manner in which the partition effect is experimentally carried out.

Furthermore, it is significant to consider the type of phases that are present, All chromatographic processes involve a mobile phase which passes over a stationary phase. Therefore, the solute is distributed between the two phases and the particular reason for the type of distribution is the heart of the chromatographic system.

**Adsorption and Partition Chromatography.**    Table 22–2 summarizes the type of phase combinations. The solute distribution between the two phases, in general, can be described as shown in Fig. 22–1. Cases (a) and (b) in Fig. 22–1 are similar in that they both involve surface interactions. This type of interaction (adsorption) is the result of intermolecular forces between surface atoms of the solid and molecules of the external solute and involve one or some combination of the following.

1. London forces between all surfaces and adsorbed molecules
2. Electrostatic forces between polar surfaces and any adsorbed molecule or between nonpolar surfaces and polar adsorbed molecules
3. Charge transfer forces between strong electron donors and acceptors
4. Formation of hydrogen bonds

These forces, which lead to physical adsorption, are weaker than those found for covalently or ionically adsorbed species. These latter forces lead to chemisorption and are not as ideally suited to chromatographic applications as physical adsorption.

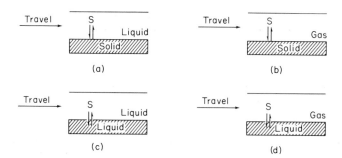

**Fig. 22–1.** Reversible transfer of a solute between mobile and stationary phases. S = solute.

Two important characteristics of adsorbents are that they have large surface areas and high porosity. In some cases these are influenced by the ability of the absorbent to swell in the presence of a solvent.

Partition is illustrated in Fig. 22–1c and d. In both cases solute molecules are in equilibrium across an interfacial boundary that occurs between the mobile and stationary phase. The solute is distributed throughout the stationary phase in partitioning and not just at the surface as in adsorption. It follows, therefore, that solubility in the two phases is an important part of partitioning. If the system were static, rather than dynamic, the process, in general, would be identical to solvent extraction.

Specific and general properties which influence solubility are important parameters in the partitioning of the solute between the two phases. Thus, forces which control adsorption are also those which can influence solubility. For example, hydrogen bonds, dipole–dipole interactions, and charge transfer processes between solvent and solute molecules (intermolecular) affect solubility. In addition, solute–solute and solvent–solvent interactions (intramolecular) must be considered.

In partition methods, the stationary liquid phase is held in a column or sheet by an inert support. It is reasonable to expect that the inert support may, in some cases, influence the action of the solute and act as an adsorbent. Similarly, in adsorption, the adsorbent may retain some of the mobile liquid phase as a stationary phase and produce a partitioning effect. Therefore, the chromatographic process is not very often purely partition or adsorption.

**Exclusion Chromatography.**      Exclusion chromatography includes those chromatographic processes in which separation of the sample components takes place according to molecular size. Several of these techniques are recent in development and at present, it is convenient to divide them into two types: gel permeation and sieving separations.

In gel permeation, separation takes place according to size of the solute molecules. Mixtures of low molecular weight compounds, as well as large molecular weight compounds and polymers, can be separated. This technique has been used with great success in separation of sugars, polypeptides, proteins, lipids, asphalts, butyl rubbers, polyethylenes, polystyrenes, silicone polymers, and many others. It is also possible, after suitable calibration, to determine molecular weights of the different chain length polymers in the mixture.

The solid supports (gels) that are used are three-dimensional networks of crosslinked polymer chains. These gels are able to swell in a particular solvent and as the gel structure swells the spaces between the polymer chains increase in size. For a particular gel, there will be a critical size of a molecule that can just penetrate the interior. Larger molecules will pass unhindered through the column, while molecules smaller than the critical size will be retarded differently according to their size.

In addition to size factors (molecular weight, linear molecule, coiled molecule, etc.), the role of adsorption, partition, and electrical factors must also be considered. Several typical separations are shown in Fig. 22-2.

The most useful materials for sieving separations are natural and synthetic zeolites (metal–aluminosilicates). A typical zeolite, also called molecular sieve, consists of a basic formula $M_{2/n}O \cdot Al_2O_3 \cdot xSiO_2 \cdot yH_2O$, where M is a cation of $n$ valence, and contains permanent cavities and channels in their structure. It is the size of these cavities which determines the sieve properties of the zeolite while the large surface area and availability of M, Al, Si, and O in the zeolite structure determine its adsorptive properties.

**Ion-Exchange Chromatography.** In ion-exchange chromatography a reversible exchange of ions is possible between ions in a liquid phase (mobile phase) and a solid, insoluble substance containing ionic sites (solid phase). The zeolites can act as an ion exchanger. Usually, M in the zeolite is a sodium ion and can be exchanged with other cations. (This is the principle of water softening. Zeolites are often used to exchange $Na^+$ for $Ca^{2+}$, $Mg^{2+}$, $Fe^{3+}$, and other multivalent cations found in hard water.) Several types of natural and synthetic cation and anion exchangers are available (see Table 22-3).

In general, the solid phase can be described as a polymeric matrix, a crystalline lattice, or a modified naturally occurring substance containing fixed ionic groups and mobile counterions of opposite charge. These solid phases are insoluble, permeable, exchange ions on a stoichiometric basis, and they tend to show a well-defined selectivity of one ion over another. Cation exchangers exchange cations while anion exchangers exchange anions.

$$R_C^-H^+ + M^+ \rightleftharpoons R_C^-M^+ + H^+ \qquad (22\text{-}1)$$

$$R_A^+OH^- + Cl^- \rightleftharpoons R_A^+Cl^- + OH^- \qquad (22\text{-}2)$$

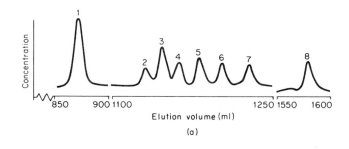

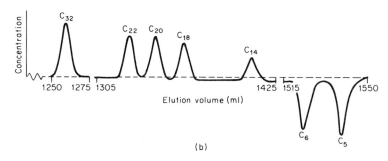

**Fig. 22–2.** Examples of gel permeation chromatography. (a) Separation of triglycerides. Column: 160 ft × ⅜ in., flow rate: 0.4 ml/min; packing: 500 Å polystrene–divinylbenzene gel; peaks: 1, polystyrene; 2, triarachidin; 3, tristearin; 4, tripalmitin; 5, trimyristin; 6, trilaurin; 7, tricaprin; 8, *o*-dichlorobenzene. (b) Separation of hydrocarbons. Column conditions are the same as (a). Reversal in peaks is found because the detection of the peaks is by refractive index. (Taken from J. J. Kirkland, ed., "Modern Practice of Liquid Chromatography," Wiley-Interscience, New York, 1971.)

**Table 22-3.   Ion-Exchange Materials**

| Inorganic materials | Organic materials |
|---|---|
| **Cation Exchangers** ||
| Natural: | Natural: |
|   zeolites, clays |   peat, lignite |
| Synthetic: | Modified: |
|   $MgO$, $SiO_2$, $Al_2O_3$, $SiO_2–Al_2O_3$ |   natural sulfonated coal and wood |
| | Synthetic: |
| |   polymeric resin matrix containing acidic exchange site |
| **Anion Exchangers** ||
| Natural: | Synthetic: |
|   dolomite |   polymeric resin matrix containing basic exchange sites |
| Synthetic: | |
|   heavy metal silicates | |

where $R_C$ and $R_A$ are the cation and anion exchangers containing negative and positive ionic sites, respectively.

The introduction of the ion-exchange concept occurred in 1850 when the exchange of ions in soil was first described. However, from the standpoint of analytical chemistry, it was not until 1935 that ion exchange became important to the analytical chemist. At that time a series of polymeric ion exchangers capable of exchanging ions were first synthesized. These exchangers had the advantage of high exchange capacities, were readily penetrated by solvents, possessed stability, and provided reproducible response.

The most widely used ion exchangers are based on a polystyrene–divinylbenzene copolymer structure. Divinylbenzene is a crosslinking agent and imparts strength to the polymer by joining the chains together at various positions. The structure for a typical cation and anion resin is

where R is $-SO_3^-H^+$ or $-CH_2N(CH_3)_3^+Cl.^-$

The sulfonated polymer is a strongly acidic cation exchanger while the quaternary amine resin is a strongly basic anion exchanger. Both are typical insoluble, strong electrolytes (100% ionized). The exchange can be represented as

$$ResSO_3^-H^+ + Na^+ \rightleftharpoons ResSO_3^-Na^+ + H^+ \qquad (22\text{-}3)$$

and

$$ResNR_3^+Cl^- + OH^- \rightleftharpoons ResNR_3^+OH^- + Cl^- \qquad (22\text{-}4)$$

where Res is the resin matrix.

Crosslinking is very important because it affects two properties of the ion exchanger: strength and swelling. As the crosslinking in the ion exchanger decreases, the swelling of the ion-exchange bead increases. Accompanying this is an increase in the tendency for fracture of the polymer chains.

Most of the exchange sites are located within the ion exchange bead (see Fig. 22–3). If these interior sites are to be accessible, the exchanger must swell so that the solvent can enter the interior of the exchanger. In general, 8–12% crosslinking is used in commercial resins.

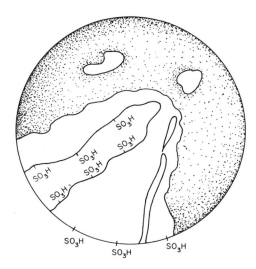

**Fig. 22–3.**   A typical cation exchanger illustrating interior and exterior exchange sites.

Other types of functional groups  (Table 22–4) can be introduced onto a polymeric matrix. For some of these, the polystyrene–divinylbenzene ploymer matrix is not suitable. All the cation exchangers in Table 22–4 are weakly acidic ion exchangers, except the sulfonic acid ion exchanger and will act like weak electrolytes in that they will dissociate only over a limited pH range. For example, the carboxylic acid ion exchanger is useful as an ion exchanger only above pH 4. Similarly, the weak base ion exchangers (all except $R_4N^+OH^-$ in Table 22–4) can not be used as ion exchangers in basic solution because of incomplete dissociation. It is also possible to introduce chelating and oxidation–reduction groups onto a polymer matrix.

**Table 22–4. Common Func-
tional Groups on Ion Ex-
changers**

| Cation | Anion |
|---|---|
| $-SO_3H$ | $-NH_2$ |
| $-CO_2H$ | $-NHR$ |
| $-OH$ | $-NR_2$ |
| $-SH$ | $-NR_3^+$ |
| $-PO_3H_2$ | |

## GENERAL CONCEPTS

There are many similarities in experimental techniques, operating conditions, and basic principles in chromatographic methods. This section discusses these common characteristics.

Perhaps, the overall main goal in chromatography is that it serve as a tool for resolving mixtures of known as well as unknown molecules. This is done by percolating a mobile phase containing the mixture over a large bulky stationary phase, which is usually of large surface area. The separation, if conditions are properly chosen, occurs at the two phases because of differences in the retention of the different kinds of molecules in the mixture toward one of phases.

The main experimental procedure includes preparation of the stationary phase, introduction of the sample, passing the sample over the stationary phase, and collecting the individual components for quantitative or qualitative purposes. If the stationary phase is packed into a tube, the method is termed *column chromatography*. In contrast, thin, flat systems, such as paper or thin layers, are *sheet techniques*. The solid material acting as the stationary phase in adsorption chromatography is the *adsorbent*, whereas the inert solid used to retain a liquid film in partition is called the *support*. Passing of the mobile phase over the adsorbent or support is known as *development* and during this process, the sample components isolate themselves into *zones* or *bands*. The process of development is *elution* and the mobile phase used for the development is the *eluting agent*. Eventually, the zones are *detected* or *visualized*. A *chromatogram* represents this entire process and can be described as a plot of concentration of the sample components as a function of *elution time* or *elution volume* (*retention time* or *retention volume*).

**The Heart of Chromatography.**     A physical picture of what occurs in a chromatographic system is conveniently illustrated by considering column chromatography. This also applies, at least in principle, to sheet methods as well.

In column chromatography, the stationary phase is placed as a slurry in a cylindrical tube that is plugged at the bottom by a piece of glass wool or an inert porous disk. The sample is dissolved in a minimum of solvent, applied to the column, and passed into the column with the liquid mobile phases. (In gas chromatography, the column is dry-packed.) Two typical columns are shown in Fig. 22–4. If the appropriate eluting conditions are chosen, the individual solutes of the mixture emerge at the bottom of the column as a function of elution volume or elution time.

The sequence of events that takes place in the column between the time the sample enters and leaves the column is shown in Fig. 22–5 as a function

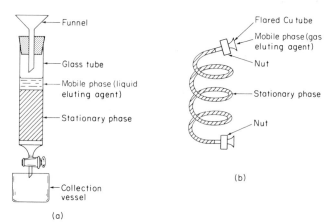

**Fig. 22–4.** Typical column designs. (a) Liquid column chromatography; (b) gas chromatography.

of time. A typical column (a), is prepared and a sample mixture, represented as X is transferred to the column (b). In (c) the sample passes into the column, and as the eluting agent is passed, the separation of X begins to take place (d). A continued flow of eluting agent results in a complete separation of X (e). If the effluent is analyzed as it emerges from the column, a chromatogram as shown in (f) is obtained. Most often the chromatogram appears as either a well-defined, gaussian-shaped elution peak (solid line in Fig. 22–5f) or one with tailing (dotted line); the gaussian-shaped chromatogram is preferred.

Both solutes, $S_1$ and $S_2$, travel the same length of column (d). However, the speed, $v$, at which they travel is given by

$$v_{S_1} = \frac{d}{t_{S_1}} \quad \text{and} \quad v_{S_2} = \frac{d}{t_{S_2}} \tag{22-5}$$

where $t$ is the time of appearance, such that $v_{S_1} \gg v_{S_2}$. In general, if there is any difference in $v_{S_1}$ and $v_{S_2}$, separation should be feasible.

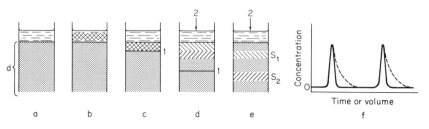

**Fig. 22–5.** Steps in the separation of a two-component mixture. (1) Sample solvent front; (2) addition of eluting agent.

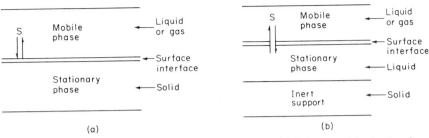

**Fig. 22–6.** Mass transfer of a solute between two phases. (a) Solute participates in adsorption; (b) solute participates in partition. S = solute.

During the separation process the solute is participating in mass transfer between the two phase conditions. Since the system is dynamic, it is not always possible for the system to be exactly at the point of equilibrium throughout the column. However, by adjustment of experiment conditions, generally, the chromatographic experiment is carried out as close to equilibrium as possible. Figure 22–6 illustrates the two types of transfer that can occur between the phases. In (a) transfer by adsorption (also ion exchange and exclusion) between a liquid or a gas and a solid phase is illustrated while (b) represents mass transfer through partition between a gas or liquid phase and a liquid phase. As previously stated, factors which influence the mass transfer differences are H-bonding, ion exchange, solubility, and various types of polar forces and interactions.

**Chromatographic Development.** Chromatographic development is possible by four different techniques: elution; gradient elution; frontal analysis; and displacement. Although it is theoretically possible to apply each of these to all chromatographic techniques, in practice, only those that are useful and offer distinct advantages are used in a specific technique. In general, elution is more widely used than the others.

**Elution.** In this method of development, the sample contained in a very small volume of solvent is added to the column. Subsequently, an eluting mixture is passed through the column which causes the solutes to travel at different speeds. For adsorption, separation is due to differences in adsorption affinity of the solutes, while in partition it is the difference in the solute's ability to distribute itself. This method of development is very versatile and is used in all chromatographic techniques.

**Gradient Elution.** Gradient elution development is characterized by an intentional variation of eluting conditions. Initially, the chromatographic experiment is performed with poor eluting conditions and it is changed to excellent eluting conditions gradually.

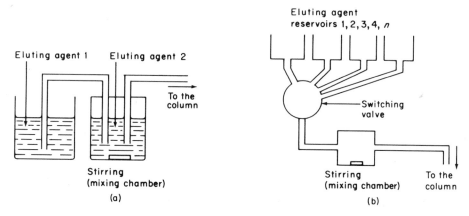

**Fig. 22–7.**   Two devices for preparing a gradient.

Two experimental methods for changing the eluting power are shown in Fig. 22–7. In (a), which is the simpler of the two, the eluting agent is mixed in the mixing chamber and its eluting power is controlled by the delivery of eluting components into the mixing chamber. Figure 22–7b, which represents a more complex way of controlling the gradient, employs an automatic switching valve.

The nature of the mixing chamber and rates at which the eluting components are mixed will determine whether the change in the gradient is linear or exponential. In general, the more similar the chromatographic behavior of the components of the mixture, the more gradual the gradient should be in order to produce a separation. The gradient usually involves a change in pH, in solvent composition, or in ionic concentration.

Gradients are used to sharpen tailing zones into well-defined gaussian zones and to separate closely related solutes. It is also possible to use a temperature gradient for development.

**Frontal Analysis.**   This method of development is characterized by a continuous passing of the sample solution through the column. During passage the various solutes accumulate in an elution order. For example, in absorption the solute that is least adsorbed concentrates at the front, while the most adsorbed is last. Since the sample is continuously fed into the column only the first component can be isolated free of the others.

Frontal analysis is primarily used for purification of the least retained solute. Recycling is also usually part of the procedure. For these reasons, this technique of development is more often used in pilot plant and commercial operations rather than in analytical applications.

**Displacement.**     In this technique, a sample is introduced into the column and subsequently, the displacing agent is passed through the column. A useful displacing agent is one that is preferentially retained by the column. As the flow of the displacing agent is continued, its retention zone increases causing all other solutes to be forced ahead of the displacer. This results in the solutes aligning themselves into a retention sequence with the least retained appearing at the front. Eventually, the solutes emerge from the column in this order followed by the displacer. It is difficult to obtain complete separation of one solute from the other, since the trailing edge of one zone overlaps with the leading edge of the next. The main advantage of this method is that large sample quantities can be used while the main disadvantage is that the chromatographic system is completely charged in the form of the displacing agent after completion of the separation. For these reasons displacement elution is primarily used for purification rather than for quantitative separations.

**Quantitative Analysis.**     Two general techniques are used to determine the components as they are chromatographically separated. In the first, each component is collected in a separate container, and subsequently determined by a suitable chemical or instrumental procedure. The second is based on the fact that the area under the chromatographic peak is directly proportional to concentration of the component within that peak. The following describes briefly the methods available for determining the areas. These apply to all column chromatographic techniques; their application to sheet methods are described in Chapter 25.

There are several methods for determining the peak area and relating this to concentration. These are peak height, triangulation, planimetry, cutting and weighing the peak, and disc and electronic digital integration. All require a proper introduction of the sample, suitable chromatographic equilibria within the column, a known, reproducible flow rate for the mobile phase, a fast, efficient, and reliable detector response, and an accurate, sensitive recorder.

Once the areas are determined, calibration curves can be prepared whereby the areas of a series of standards are plotted versus the concentration of the standard. The unknown is chromatographed under the same conditions and with the area of its peak its concentration is determined from the calibration curve. An internal standard procedure can also be used. In general, this is used routinely only in gas chromatography (Chapter 24).

**Peak Height.**     Peak height is measured as the distance from the baseline to the peak maxima as shown in Fig. 22–8a. For best results, the peak should be a well-defined, gaussian shape, and completely resolved from other peaks. Thus, the height, if the peak is considered to be a triangle, is proportional

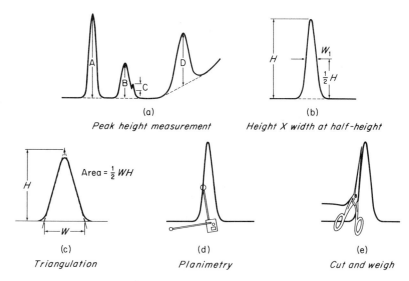

(a)                               (b)

*Peak height measurement*      *Height X width at half-height*

(c)                        (d)                        (e)

*Triangulation*           *Planimetry*          *Cut and weigh*

**Fig. 22–8.** Techniques for the measurement of peak area in chromatography.

to the area, which is proportional to concentration. Of all the methods, this one is the fastest and generally the easiest to carry out.

If the peaks are not completely resolved the baseline is identified as shown in Fig. 22–8a. A baseline drift can contribute a substantial error. Another limitation is that the linear range for the calibration curve of peak height versus concentration is small.

**Triangulation.** The methods in this group are based on the fact that chromatographic peaks approximate a triangle. These are illustrated in Fig. 22–8b and c. In the first the area is computed by multiplying the peak height by the peak width at $\frac{1}{2}$ the height of the peak. The width $w_1$, at $\frac{1}{2}$ height is used to eliminate errors due to tailing or other factors that would tend to broaden the base of the peak. In the second method the width at the base of the peak is used, see Fig. 22–8c, and the area is calculated by Area = $\frac{1}{2}wH$. The latter method suffers not only from the limitations of the first method, but also from broadening at the base and the difficulty in drawing tangents to the peak to identify the base. Because of these errors the first method is recommended over the second even though both are relatively rapid and simple. In general, triangulation methods require a peak to base (or $\frac{1}{2}H$) ratio of at least 5 to 10.

**Planimetry.** A planimeter is a mechanical device that is used to trace the perimeter of the peak. The result is an integration of the peak with its area being recorded directly on a dial. This device and its use is illustrated in Fig. 22–8d.

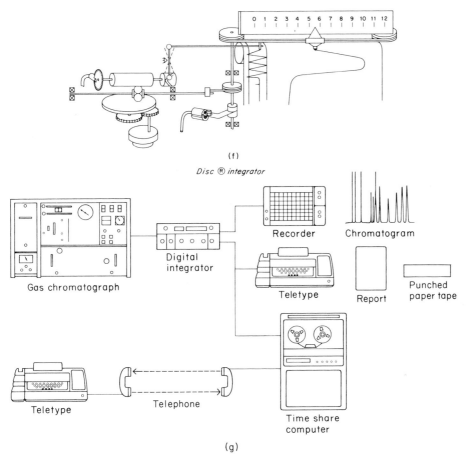

(f)

*Disc ® integrator*

(g)

*Chromatograph integrator-time share system*

**Fig. 22–8f, g**

Using a planimeter is tedious and requires not only skill but also patience by the operator. Unlike the triangulation methods, the true peak area is measured. Consequently, non-Gaussian type peaks are handled with a greater degree of accuracy than would be obtained if triangulation were used.

**Cut and Weigh.**     The cut and weigh technique is a perimeter technique whereby the peak is cut out and weighed on the analytical balance. This is illustrated in Fig. 22–8e. Thus, if a calibration curve was being prepared, the peak for each standard would be cut out, weighed, and the weight plotted vs weight of the standard. Since the entire peak is being weighed, this method is perferred over triangulation if peak shapes are non-Gaussian. However, it is more time consuming than the triangulation method. Also, the accuracy and precision of the method is very dependent on the quality and uniformity of the recorder paper. Since cutting will destory the chromatogram, often the chromatogram is first Xeroxed and then either the Xerox copy or the original recording trace is cut up.

**Disc and Electronic Integration.** The disc integrator is a mechanical device which essentially does automatically what the planimeter does manually. The recorder is equipped with the disc integrator and operates a second pen. Thus, a permanent record of the integration is provided along with the chromatogram.

Electronic integration of two general types are used. These are real-time digital integrators and digital computers. Both are very precise and eliminate the use of recorders and their inherent limitations. It is beyond the scope of this book to go into the details of the electronic integrators. Basically, they automatically convert the chromatographic signal into a numerical form. With modern computer technology, not only is the peak area determined but also all data calculations and interpretations can be handled. The disc and electronic integration systems are illustrated in Fig. 22–8f and g.

**Accuracy–Precision.** It is difficult to generalize concerning the accuracy and precision of the area determination methods since the limiting factor in an actual chromatographic separation may be in the chromatography or in the control of the instrumental parameters. However, under ideal circumstances the precision and accuracy of planimetry and $A = \frac{1}{2}wH$ triangulation is the poorest while it is the best for disc and electronic integration. Intermediate precision and accuracy is provided by triangulation using peak height $\times$ width at $\frac{1}{2}H$, peak height, and cut and weigh methods with the latter being the best of the three.

# SOME FUNDAMENTAL CONCEPTS

**Distribution Value.** One of the most fundamental concepts, as well as one which is of great practical significance, is the relationship describing the amount of solute retained by one phase in relation to another phase. In its general form, the relationship can be expressed as a distribution coefficient, $K$,

$$K = \frac{C_1}{C_2} \tag{22-6}$$

where $C_1$ and $C_2$ are the concentrations of the solute in phase 1 and phase 2, respectively. This basic equation may be modified for a particular chromatographic technique. The value of $K$ is dependent on temperature and pressure and independent of concentration. However, in practice, $K$ is independent of concentration only over a limited concentration range.

In partition methods, the distribution coefficient is called the partition coefficient (also applies to solvent extraction). Also, $C_1$ is the concentration of solute in the stationary phase, $C_S$, and $C_2$ is the concentration of solute in the mobile phase, $C_M$, or

$$K_p = \frac{C_S}{C_M} \tag{22-7}$$

The larger the $K_p$ value, the greater is the retention of the solute in the stationary phase.

The coefficient describing adsorption is defined like the partition coefficient except that $C_S$ is the amount of solute adsorbed by the stationary phase and $C_M$ is the amount of solute in the mobile phase where both are expressed in the same units. Therefore, a large adsorption coefficient indicates a high retention for the solute.

In ion exchange, the distribution coefficient is modified to account for the volume of solution and weight of resin. Thus,

$$K_D = \frac{\text{amount of ion on resin per gram of resin}}{\text{amount of ion in solution per milliliter of solution}} \tag{22-8}$$

and is referred to as the batch distribution coefficient.

In sheet methods retention of solutes, whether by partition or adsorption, are described by their migration relative to that of the eluting agent. This ratio has the symbol $R_f$ and $R_R$ where $R_f$ is for linear flow in one direction and $R_R$ is for linear flow in a radial direction. Both are defined as

$$R_f \text{ (or } R_R) = \frac{\text{distance traveled by solute}}{\text{distance traveled by mobile phase}} \tag{22-9}$$

As the $R_f$ increases, the retention of the solute decreases.

The $R$ values vary with the type of migration, sorbent, and solvent. Concentration effects are minor in partition while the $R$ value usually decreases with decreasing solute concentration in adsorption. In addition, experimental conditions, such as the solvent, sorbent, porosity, solute concentration, and temperature, should be controlled if $R$ values are to be reproducible.

**Resolution.** Two general factors will determine the effectiveness of the separation: distance between zone centers as they migrate and compactness of zones. As the distance between the zone centers increases, the degree of separation increases and a measure of this difference is termed resolution.

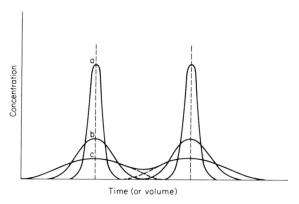

**Fig. 22–9.** Effect of zone broadening. It should be assumed that areas under the peaks for each component are equal.

As zones travel, however, they tend to spread and broaden and even though zone centers are well separated, overlapping of the bands can occur. This is illustrated in Fig. 22–9 where peak volumes are at the same value. In passing from a to c, zone broadening occurs, which leads to overlappage even though the peak maxima are still separated by the same distance.

In general, it can be shown that resolution is affected by the distribution coefficient for the second component of the two-component mixture, the selectivity, and the number of theoretical plates corresponding to the second component. As each of these increases, the resolution of the two components will increase.

Figure 22–10 illustrates how resolution can be improved by changes in the column experiment. In curve a, poor resolution is obtained. In curve b, the column efficiency, which is represented by the number of theoretical plates, is improved by altering column parameters such as flow rate, particle size, column diameter, and column temperature. In curve c, resolution is increased by improvement in selectivity, that is, a more favorable stationary–mobile phase selection was made. If the experimental conditions are such that the distribution coefficients are small, selectivity is also small. Thus, poor resolution will be obtained. This is illustrated in curve d.

Resolution can be calculated by the equation (see Fig. 22–10c)

$$R = \frac{t_{R_2} - t_{R_1}}{w_2 - w_1} = \frac{2\Delta t}{w_2 - w_1}$$

where $t_R$ is the retention time of compound 1 and 2 and $w$ is the width at the base of the peak in time units of peak 1 and 2. If retention volume is used in the chromatogram rather than retention time, the distance between the two peaks and their width are expressed in volume units.

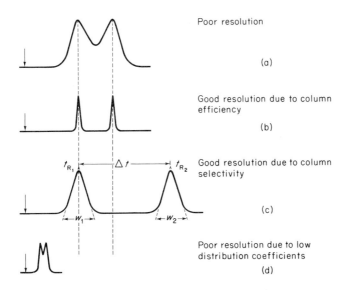

Poor resolution

(a)

Good resolution due to column efficiency

(b)

Good resolution due to column selectivity

(c)

Poor resolution due to low distribution coefficients

(d)

**Fig. 22–10.** Changes in resolution due to changes in column efficiency, selectivity, and distribution coefficients.

**Peak Shape.** If the distribution coefficient is independent of concentration a plot of the concentration of the component in the stationary phase vs the concentration in the mobile phase will be linear. A graph of this type is called an isotherm. Nonlinear isotherms are common, particularly the convex type. The three possible isotherms are illustrated in Fig. 22–1. Also illustrated are the shapes of the resultant chromatographic peak and how retention changes as a function of sample size for each of the isotherms. In general, linear convex isotherms are experimentally found in all areas of chromatography. In contrast, concave isotherms, when observed, usually occur only in partition chromatography. These are usually the result of overloading the column with excessive sample. As illustrated in Fig. 22–11, a concave isotherm produces an increase in retention time with increased sample size, while the reverse is observed for a convex isotherm.

**Selectivity.** Selectivity is a measure of the preference a stationary phase shows for one solute over another, and is expressed as a ratio, $\alpha$. Therefore, by definition

$$\alpha = \frac{K_1}{K_2} \qquad (22\text{-}10)$$

where $K_1$ and $K_2$ are distribution coefficients for two different solutes. The

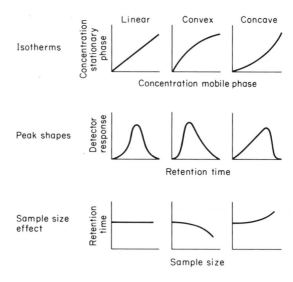

**Fig. 22–11.** Chromatographic isotherms and their effect on peak shape and retention times.

choice of which is $K_1$ and $K_2$ is arbitrary, however, for convenience $K_1$ is usually designated as the slower moving solute.

In essence, $\alpha$ describes the relative rates of migration of two solutes and, therefore, is a direct measure of zone to zone separation. The greater the difference between the distribution values, the larger the value of $\alpha$ and thus, the better the separation.

**Efficiency.**     Efficiency in chromatography can be expressed quantitatively by the number of theoretical plates, $N$, or the height equivalent to a theoretical plate, $H$ (or HETP). The two are related by

$$H = L/N \qquad (22\text{-}11)$$

where $L$ is the column length. In general, a theoretical plate can be defined as that length of column in which the solute undergoes one complete equilibration between the two phases.

The number of plates is given by

$$N = 16(V_R/W)^2 \qquad (22\text{-}12)$$

where $V_R$ is the retention volume* and $W$ is the width of the elution peak in volume units (see Fig. 22–12). [Retention volume in Eq. (22-12) can be

---

\* Retention time $\times$ flow rate of solvent = retention volume.

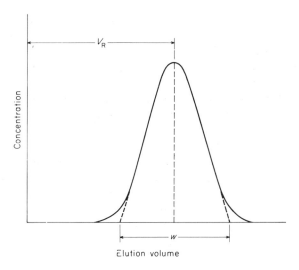

**Fig. 22-12.**   Calculation of the number of plates from an elution peak.

replaced by retention time which means that the width is measured in time units.]

As shown in Fig. 22–9 a decrease in column efficiency is indicated by zone spreading. Zone spreading within a column (or a sheet) originates from three main sources. These are multiple path flow, molecular diffusion, and nature of mass transport. These are briefly described in the following paragraph.

Multiple path flow means that the molecules of a sample will travel through the column at different rates. For example, size, shape, and packing of the particles will determine the kinds of pathways available for travel. Molecular diffusion describes the movement in a longitudinal fashion away from a compact zone as the band travels through the column. In general, this effect is minor in relation to the other two unless very, very slow flow rates are employed. Mass transfer effects are most often the major contributor to zone broadening. The solute molecule is continually passing into and out of the mobile and stationary phase (see Figs. 22–1 and 6). In doing so some molecules of a given zone at a given moment will have faster rates than others and therefore, because of this random back and forth motion (only the average rate can be measured) the zone will broaden as the zone moves down the column.

Zone broadening can also occur in other parts of the chromatograph, for example, in the injection system, in the connecting tubing, or in the detector. These effects are minimized in a well-designed chromatograph.

The value of $H$ is a measure of the efficiency of the chromatographic system; the smaller the value of $H$, the more efficient and the better the resolution is for the particular column at a fixed set of experimental conditions.

Since $N$ is inversely related to $H$, an increase in $N$ improves the efficiency and resolution. Assuming the system is at equilibrium, $H$ is independent of length and dependent on particle size of the stationary phase, diffusion coefficient of the solute in the mobile phase, velocity of the mobile phase, and the kinetics of the system. However, the value of $N$ is directly proportional to length of band travel. Thus, an increase in the length increases the number of plates which increases the resolution. For example, it can be shown that a quadrupled increase in the column length should increase the peak to peak distance by 4 and the resolution by a factor of 2.

The length of the column cannot be increased without optimum limits. Generally, the limits are the result of practical rather than theoretical considerations. Increasing the length excessively often leads to broad bands. Since the solute is diluted, it is more difficult to detect. Also, the separation can be very time consuming and very wasteful with respect to eluting agent.

## SUMMARY

In this chapter the concept of separation and the specific technique of chromatography have been introduced. The main emphasis has been to illustrate the underlying principles and concepts. Very little attention has been given to the actual performance, the type of equipment and instruments required, or to the many practical applications of these separation methods. These facets will be considered in the following five chapters.

## *Questions*

1. Suggest the experimental procedure employed in distillation and dialysis.
2. How would the two techniques suggested in Question 1 be used in analytical chemistry?
3. Differentiate between an adsorption and partition chromatographic process.
4. List the forces that influence adsorption and partition.
5. Suggest an experiment or series of experiments that would prove that molecular sieve materials separate mixtures on the basis of size.
6. How can you prove that ion-exchange resins exchange ions stoichiometrically?
7. Using equilibrium expressions show why ion-exchange resins containing carboxylic acid groups are not useful below pH of 4.
8. Differentiate between elution and gradient elution.
9. What is the largest and smallest value for $R_f$?
10. What is the largest and smallest value for $K_D$?

11. Does a large $K_D$ mean a high or low retention of the solute to an ion exchange resin?
12. Does a large $R_f$ mean a high or low retention of the solute to the stationary phase in a sheet method?
13. Define resolution.
14. What properties in the chromatographic system influence resolution?
15. If the chromatographic peak in Fig. 22–12 was obtained with a $75 \times 1$ cm column and $W$ and $V_R$ were 3.7 ml and 42.0 ml, respectively, calculate the number and height equivalent to a theoretical plate.
16. Suggest the effect on $H$ if each of the following changes were incorporated into the column chromatographic experiment.
    a. An increase in flow rate
    b. An increase in the diameter of the column
    c. An increase in the column length
    d. An increase in the temperature of the column

# Chapter
# Twenty-Three
# Column Methods

## INTRODUCTION

Either a liquid mobile phase or a gas mobile phase is used in column chromatography. This chapter will consider column adsorption and partition processes employing liquid mobile phases. Gas mobile phases will be considered in Chapter 24.

Ion exchange and exclusion chromatography are also carried out in columns. Because of the broad applications of ion exchange, this column technique is considered separately in Chapter 26. Although exclusion chromatography is equally important, particularly as a means for the separation of macromolecules, space does not permit a detailed discussion of this technique (see Chapter 22).

In all applications of liquid column chromatography, the mixture that will be separated is introduced as a small concentrated sample at the top of the column followed by the addition of the liquid mobile phase (eluting agent). If the separation is due to adsorption, it is dependent on the interactions between the solute and the absorbent surface and solvent. If the separation is due to partition, it is dependent on solute distribution between the mobile and stationary liquid phase. Therefore, successful separation in both cases requires a careful selection of column dimensions, mobile liquid phase, and stationary phase.

## COLUMNS

Chromatographic tubes come in different sizes, shapes, designs, and are made from glass or some other chemically inert material. The main goal for all of these is that they support the stationary phase (absorbent or sta-

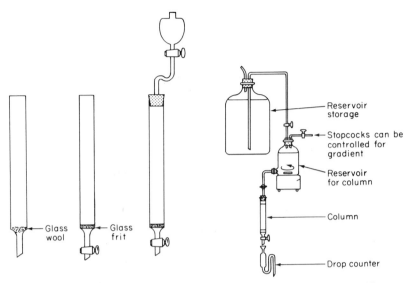

**Fig. 23-1.**  Several types of columns used in column chromatography.

tionary liquid on an inert support) in a column and permit control of solvent input and effluent collection. The length to diameter ratio should be at least 10 or larger. In general, if the mixture contains closely related components, long columns are needed while larger column diameters are used to accommodate larger amounts of samples.

In almost all cases, columns are designed for complete elution of the zones. Flow of the mobile phase is controlled by a stopcock or pinch clamp employing gravity or by pumps and the tubing carrying the mobile phase to the column and the effluent away from the column should be inert with respect to the mobile phase used. Common materials for the tubing are stainless steel, glass, Teflon, and polyethylene, and joints are made with connectors, sleeves, or "Swaglock" type connectors. Figure 23-1 illustrates several different conventional types of columns. In all cases, some sort of plug (glass wool or a glass, steel, or Teflon porous disk) is placed at the bottom of the column to prevent the packing from falling out.

Overdesigning a column can lead to waste and poor operation. If the column is too long, excessive volumes of mobile phase are required before the components of the mixture emerge from the column. Also, the zone in the column may broaden because of diffusion in all directions (see Fig. 23-2).

Although an excessively slow flow rate can sometimes lead to zone broadening, in general, a slow flow rate is preferred over a fast one, if optimum flow conditions have not been predetermined. Too fast of a flow rate often leads to a nonequilibrium system and extensive tailing.

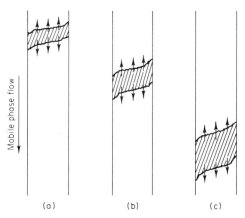

**Fig. 23–2.** Band broadening in column chromatography during elution. Time on column (c) > (b) > (a).

Packing of the column is very important. In most column applications, a slurry (column packing dispersed in the eluting agent or sample solvent) is used to introduce the packing into the column. A continued passage of the solvent aids the settling of the packing particles. Since the sample and mobile phase seeks out the path of least resistance, channeled or loosely packed columns provide poor separation. These undesirable properties will usually result if columns are allowed to run dry and replenishing the mobile phase will not eliminate the channeling. Hence, the column must be reslurried and allowed to settle again.

Packing the column too tightly hinders the flow of the mobile phase and produces a large pressure drop across the column. Packing of columns identical to previously used columns is one variable that is difficult to reproduce.

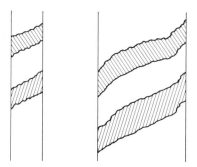

**Fig. 23–3.** Effects of nonuniform flow and poorly packed column on resolution for narrow and wide columns.

Recently, in certain chromatographic techniques column packings have been shown to be more reproducible if the stationary phase particles are dry-packed.

Column diameter and packing can influence the apparent resolution. This is illustrated in Fig. 23–3. The distance between the two zones is identical in the narrow and wide column. However, because of the nonuniform flow in the wide column, the zones appear to overlap as they emerge from the column. Poorly packed columns can also lead to the same type of experimental observation.

## ADSORPTION OR PARTITION

Both adsorption and partition column methods have advantages and limitations in their application. In general, column adsorption chromatography is less difficult to carry out than column partition chromatography. However, two other general factors, which must be considered before choosing between the two, are the types of compounds being separated and the reason or goal of the separation.

Although the adsorption and partition mechanism can involve complicated processes, a single factor, polarity, stands out. Consequently, a general discussion of column chromatography can be based on this factor and from this, reliable predictions of suitable supports, mobile phases, and stationary phases are possible.

**Advantages and Limitations in Adsorption and Partition.**    Awareness of the advantages (or limitations) of adsorption in comparison to partition and vice versa facilitate the selection of the appropriate technique for the particular separation problem. These are briefly itemized below in the form of comparisons of one technique to the other.

### ADSORPTION

1. Adsorption is experimentally easier than partition.

2. Adsorption provides a more uniform, reproducible behavior, since only a solid and a mobile eluting phase are involved.

3. Adsorption is very sensitive to steric differences for similar molecules and therefore, is applicable to the separation of mixtures of these kinds of molecules.

4. The polarity of the mobile phase can be varied widely since miscibility with a stationary liquid phase is not involved.

5. Adsorption chromatography is suitable for large amounts of sample.

6. Adsorption chromatography is preferred for the separation of mixtures whose components differ widely in polarity or structure.

## PARTITION

1. Partition provides a much larger resolving power than adsorption even though it is more difficult to reproduce the amount of stationary liquid phase and interaction between the stationary and mobile liquid phases.

2. Partition, in general, is more suitable to low concentrations of a mixture. (Recent advances have overcome this limitation in certain cases.)

3. Since partition is very dependent on solubility in two liquids, small differences in molecular weight will influence partitioning. Therefore, partitioning is preferred for the separation of homologous series.

4. The tendency is for the partition coefficient to be independent of concentration over a greater range than for the adsorption coefficient. (Partition provides a linear isotherm over a wider concentration range.)

5. There appears to be a more rational relationship between structure and substituent influence in partition in comparison to adsorption.

It would appear, in general, that adsorption is the more versatile method and would most often be the method of choice. However, as pointed out previously, adsorption and partition are based on polarity differences. Since adsorption increases with an increase in polarity, it becomes more difficult to elute very polar molecules from the absorbent. For this reason adsorption

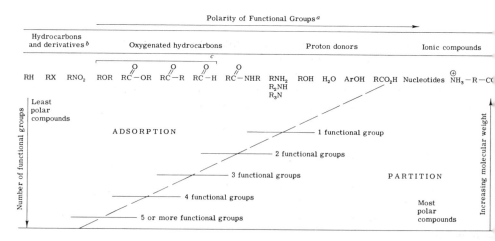

**Fig. 23–4.** Guideline for choosing between adsorption and partition. *Notes:* [a] In the case of polyfunctional compounds, the location on this diagram should be determined by the most polar of the groups. [b] Double bonds should be considered functional groups, although it should be noted that a double bond does not increase polarity as much as the functions. Aromatic rings also increase polarity, but more so than three double bonds. [c] The esters, ketones, and aldehydes have similar polarities. (Taken from J. M. Bobbitt, A. E. Schwarting, and R. J. Gritter, "Introduction to Chromatography," Reinhold, New York, 1968.)

is used for nonpolar or less polar molecules while partition is used for polar molecules. A convenient general guideline is shown in Fig. 23–4. Since polarity increases with an increase in number of functional groups and decreases with increasing carbon content and molecular weight, a diagonal line is used for the division.

## ADSORPTION

Many different adsorbents and eluting agents can be used in column adsorption chromatography. Only the more common type of adsorbents will be discussed. Similarly, elution will be discussed from the standpoint of a general elution order, rather than specifying specific eluting conditions.

**Adsorbent.** Many different adsorbents have been prepared, modified, and applied. In general, the more important characteristics of a good adsorbent are: large surface area, available polar sites, and reproducibility in the degree of activation. The latter property, which is a measure of the adsorbing power and often refers to the extent of removal of surface water, is the most difficult to control and reproduce.

The two most common ones, alumina and silica gel, and several other adsorbents are listed in Table 23–1 according to adsorbing power. Not listed are several synthetic organic polymers which have recently been shown to be good adsorbents.

The adsorption sequence for an adsorbent follows polarity. This is illustrated in Table 23–2 where the order of adsorption for functional groups is listed. For specific compounds or adsorbent the order might be slightly different.

Table 23–1. Common Adsorbents for Column Adsorption Chromatography

| Increasing adsorption power |
|---|
| Sucrose |
| Cellulose |
| Starch |
| Calcium caronate |
| Calcium sulfate |
| Calcium phosphate |
| Magnesium carbonate |
| Calcium oxide |
| Silicic acid (silica gel) |
| Charcoal |
| Magnesium oxide |
| Aluminum oxide |

Table 23–2.  General Adsorption Sequence
of Functional Groups

| | | |
|---|---|---|
| Increasing polarity → | Increasing adsorption → | Acids and bases<br>Hydroxy, amino, thio, and nitro groups<br>Aldehydes, ketones, and esters<br>Halogen compounds<br>Unsaturated hydrocarbons<br>Saturated hydrocarbons |

**Alumina.**     Alumina ($Al_2O_3$) is available in many modifications. The formula, $Al_2O_3$, is deceiving since, depending on the extent of drying and preparation, it will have $Al^{\delta+}$, Al—OH, AlO—H, and Al—O⁻ sites which are the sites responsible for adsorption. Alumina is activated by heating it in an oven at 200°C or at 400°C (3 hours). These drying procedures provide two different types of activated alumina, which can be used in chromatography. Neither of these grades is completely anhydrous. In fact, anhydrous alumina is a poor chromatographic adsorbent. Alumina can also be graded according to it being acid, base, or neutral washed.

**Silica Gel.**     Silica gel ($SiO_2$) is activated by heating it to 160°C (3 hours). Not all water is removed and adsorption sites similar to those in alumina, except containing silicon, are also part of the silica gel structure. Like alumina, silica gel is compatible with water and most common organic solvents. However, silica gel will swell and the amount of swelling is determined by the type of solvent used.

**Eluting Agent (Mobile Phase) in Adsorption.**     In choosing an eluting agent (solvent mobile phase), two considerations must be made. First, does the solvent satisfy the practical factors such as suitable viscosity, stability, compatibility with detection, solubility with respect to the sample, suitable purity, and easily removed? Second, does the solvent provide maximum resolution for the separation of the sample in a reasonable time? These two considerations also apply in choosing a suitable liquid mobile phase in partition chromatography.

In general, a polarity related factor can be used to correlate solvents as eluting phases in adsorption chromatography. Since the extent of adsorption can be qualitatively predicted according to functional group (Table 23–2) it should follow that solvents will follow a similar order. A grouping of solvents in order of chromatographic strength is called an eluotropic series. A typical

**Table 23-3. Eluotropic Series for Common Solvents**

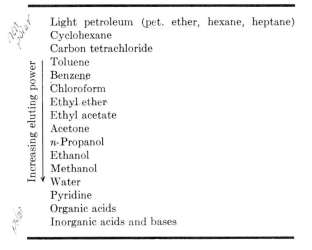

| | |
|---|---|
| *non-polar* | Light petroleum (pet. ether, hexane, heptane) |
| | Cyclohexane |
| | Carbon tetrachloride |
| ↑ | Toluene |
| Increasing eluting power | Benzene |
| | Chloroform |
| | Ethyl ether |
| | Ethyl acetate |
| | Acetone |
| | n-Propanol |
| | Ethanol |
| | Methanol |
| ↓ | Water |
| | Pyridine |
| | Organic acids |
| *polar* | Inorganic acids and bases |

series listing only common solvents is shown in Table 23–3 where the less polar solvents are at the top and the most polar at the bottom.

The more polar the adsorbed solute, the more polar the mobile phase must be if it is going to be a suitable eluting agent. This is the reason why adsorption is not generally useful for separation of polar compounds. That is, the binding is so strong that the eluting agents do not possess sufficient polarity to overcome this binding.

Mixed solvents can be used providing they are miscible. Thus, the eluotropic series in Table 23–3 is broadened considerably in polarity range. For example, ethanol–chloroform mixtures can be used and by a gradual change in the ratio of the two solvents a gradual change is incorporated into the eluting mixture.

## PARTITION

A suitable support in column partition chromatography must have a large capacity to hold the stationary liquid phase and must be inert toward the mobile phase and sample components. The support may be one which holds polar solvents or nonpolar solvents as the stationary phase. The latter support is used in the technique known as reversed phase chromatography. Table 23–4 lists several of the supports commonly used for the two types of partition column chromatography. Those that are useful for conventional partition have polar properties, while those that are useful for reversed phase are nonpolar. Kieselguhr, silica gel and cellulose are currently used the most.

Table 23–4.  Some Common Supports for
Partition Column Chromatography

| |
|---|
| Kieselguhr[a] |
| Silica gel |
| Cellulose |
| Starch |
| Glass powder[a] |
| Powdered rubber[a] |
| Acetylated cellulose[a] |
| Various organic polymers[a] |

[a] Can be used for reversed phase chromatography.

**Silica Gel.**     Silica gel that is used for adsorption can also be used for partition; however, it must first be deactivated by impregnating it with water or some other polar solvent. Although silica gel is widely used in partition, it often fails in completely satisfying the requirement of no interaction with the mobile phase or sample. That is, its adsorptive properties often contribute to the chromatographic behavior.

**Kieselguhr.**     Kieselguhr (diatomaceous earth) is available in several different grades and has the advantage over silica gel of not usually participating in a competing adsorptive interaction. Also, it can be used as a support in reversed phase chromatography.

**Cellulose.**     The advantage offered by powdered cellulose is that it is a column packing which duplicates the sheet method, paper chromatography. Thus, preliminary studies can be carried out on paper and from these data, conditions for column separations are predicted. Most often cellulose is used where large quantities of a sample (preparative scale) are being separated.

**Impregnating the Support.**     Columns in partition chromatography are more difficult to prepare because the support must retain a stationary liquid phase. Usually, the stationary liquid phase is introduced onto the support prior to packing of the column. Several different techniques are available for impregnating the support and the one that is chosen will depend on the inert support and the kind and amount of stationary liquid phase required.

The most common technique is to mix the support and the liquid intimately in a mortar and pestle or some other container. The impregnated power should be free flowing and not wet. It is also possible to prepare a column of the support and pass the stationary liquid through the column until the support is evenly coated with the liquid.

**Table 23–5. Suggested Solvent Systems for Column Partition Chromatography**[a]

| Stationary phase | Mobile phase |
|---|---|
| | Normal Partition |
| Water | Alcohols (n-butanol, isobutanol) |
| Water plus acid | Hydrocarbons (benzene, toluene, cyclohexane, |
| Water plus alkali | hexane) |
| Water plus buffer components | Chloroform |
| Aqueous alcohols (MeOH, EtOH) | Ethyl acetate |
| Alcohols (MeOH, EtOH) | Ethylene glycol monomethyl ether |
| Formamide | Methyl ethyl ketone |
| Glycols (Ethylene, propylene, glycerol) | Pyridine |
| | Reversed Phase Partition |
| n-Butanol | Water |
| Octanol | Water plus acid |
| Chloroform | Water plus alkali |
| Chlorosilanes and silicones | Water plus buffer components |
| Mineral oil | Aqueous alcohols (MeOH, EtOH) |
| Paraffin | Alcohols (MeOH, EtOH) |
| | Formamide |
| | Glycols (ethylene, propylene, glycerol) |

[a] Taken from J. M. Bobbitt, A. E. Schwarting, and R. J. Gritter, "Introduction to Chromatography," © 1968 by Litton Educational Publishing, Inc. Reprinted by permission of Van Nostrand Reinhold Company.

For reversed phase supports, a common technique is to dissolve the stationary liquid in a volatile solvent, the support is added to this solution, and the volatile solvent is removed by evaporation leaving the support uniformly coated. This procedure permits an accurate control of the amount of stationary phase on the support if the two are initially weighed quantities.

**Eluting Agent (Mobile Phase) in Partition.** In general, molecules that are more soluble in the mobile phase move faster than those which are less soluble. Also, the more soluble the molecule is in the stationary phase, the more slowly it will move down the column. Therefore, in partition two liquids must be chosen; one that acts as the stationary phase and one that acts as the mobile phase.

It is often difficult to predict the optimum combination of stationary and mobile liquid phase since the partitioning system may involve mixed solvents, solutions of salts, buffers, or complexing agents. A further complication is that the two phases must be in equilibrium during the chromatography and if a third or more components are introduced into the eluting mixture the equilibrium conditions are altered.

Table 23–5 lists several partition systems for conventional and reversed phase chromatography which serves as a starting point for selection of optimum conditions. Fortunately, more detailed discussions of solvent selection are readily available in many texts devoted to partition chromatography.

## BONDED PHASES

A bonded phase consists of a solid support to which an organic moiety is attached via a chemical bond. It is nonextractable, thermally stable, and often is hydrolytically stable. Although others are used, the principal solid supports modified by a bonded phase are silica and alumina.

In designing bonded phases the initial idea was that a stationary phase would be obtained that exhibited characteristics of liquid–liquid chromatography (partitioning between the bonded phase and the mobile liquid phase), while maintaining the stationary phase stability of the liquid–solid system. It is beyond the scope of this book to discuss the results in detail. However, in general, the bonded phase can be thought of as acting as a hybrid between adsorption and partition chromatography. Thus, as an approximation, a polar bonded phase exhibits properties similar to a composite of a polar solid support and a polar stationary liquid, while a nonpolar bonded phase has properties of a nonpolar solid support and a nonpolar stationary liquid.

A wide variety of functional groups ranging from very polar to nonpolar properties, and consequently, exhibiting widely diverse selectivities, can be introduced into the solid support as a bonded phase. Although there has been considerable success in using bonded phases in column chromatography, and advances have been made in their synthesis, they still suffer from two main limitations.

1. Bonded phases will tend to have lower efficiencies in comparison to the solid support that is modified.

2. Bonded phases have very low loading capacities and, thus, smaller sample sizes must be used.

## DETECTION OF COLUMN EFFLUENT

Historically, column effluent was collected in individual aliquots each of which was analyzed by some suitable chemical or instrumental method. Eventually, automatic fraction collectors were devised which replaced manual collection. Once the column experiment starts, the fraction collector will automatically collect a preset volume of effluent in a container and then automatically switch to the next empty container. Therefore, the column system can be operated unattended providing ample eluting agent is available.

In recent years, instrumentation has been improved to the point where column effluent can be followed continuously. The instrumental response, as a function of eluting agent flow rate, provides the chromatogram for the separation. The area under each curve is proportional to concentration, and analysis is possible, if the flow rate of the mobile phase is carefully controlled and the system is calibrated with standards.

Two general types of detection devices are available. These are bulk property detectors and solute property detectors. The bulk property detector measures a change in some overall physical property in the mobile phase as it emerges from the column. Two typical examples are the measurement of refractive index and conductance. The solute property detector is sensitive to changes in a physical property of the solute as it emerges from the column in the mobile phase. A typical example is the measurement of ultraviolet and/or visible absorption. In general, the solute property detectors are more sensitive particularly if the mobile phase does not contribute to the property being measured.

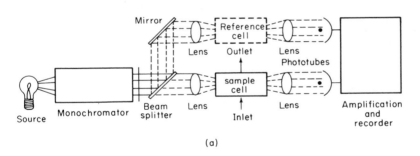

(a)

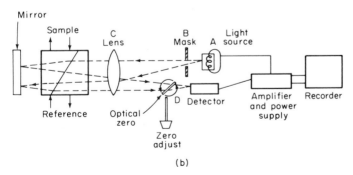

(b)

**Fig. 23-5.** Detectors in chromatography. (a) Ultraviolet—visible detector; (b) refractive index detector. (Courtesy of Waters Associates, Inc.)

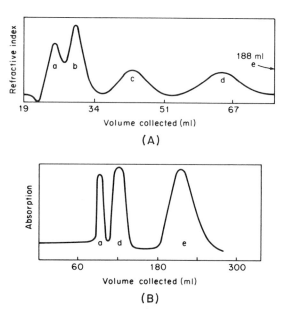

**Fig. 23–6.** Separations using refractive index and absorption detection. (A) Separation of an aniline mixture using 80% acetonitrile. (B) Separation of an aniline mixture using 70% acetonitrile. (a) 2,6-Dichloro-4-nitroaniline; (b) o-nitroaniline; (c) 4-methyl-2-nitroaniline; (d) N,N-dimethyl-p-nitroaniline; (e) p-nitroaniline. [Reprinted with permission from T. C. Gilmer and D. J. Pietrzyk, *Anal. Chem.* **43**, 1585 (1971). Copyright by the American Chemical Society.]

The two detectors which have the widest range of application are the absorption and refractive index detectors. Diagrams for these two instruments are illustrated in Fig. 23–5. In general, the effluent, as it emerges from the column, passes into the absorption or refractive index cell. If a solute is present, the absorption (must choose the correct wavelength) or refractive index changes and is plotted on a recorder to provide the chromatogram. The cells should be small in volume (less than 0.5 ml) to prevent diffusion of the band into a larger volume. In the better detection system a double beam principle is used. Hence, the eluting agent minus the sample is passed through a reference cell and the absorption or refractive index of the unknown cell (column effluent) is measured in comparison to the reference cell.

The absorption detectors are available with monochromators or with filters and are very sensitive, often detecting as little as nanogram quantities. Of course, the solute must absorb and the sensitivity is determined by the molar absorptivity for the solute and the degree of transparency of the eluting mixture. In general, detector response is not influenced by modest changes in temperature or flow rate.

All substances have a refractive index and therefore, a refractive index detector can be used for any solute. However, any slight change in solvent will produce a different refractive index and for this reason a double beam design must be employed. Furthermore, slight changes in flow rate, temperature, and mobile phase variations will affect detector sensitivity. Consequently, the refractive index detector is not as sensitive as the absorption detector and is used only in the microgram range.

These detectors are applicable to all liquid column techniques. Fig. 23–6 illustrates several different chromatograms obtained by refractive index or absorption detection.

## LIQUID CHROMATOGRAPHY APPARATUS

There are four basic parts to a liquid chromatographic instrument. Two of these, the column and detector, have already been described. The other two are the injection system and the mobile phase–pump system. Figure 23–7 illustrates a typical instrument and many variations of this general design are commercially available.

The multiport valve allows the selection of the appropriate mobile phase for a continuous flow or a gradient flow into a small mixing chamber which is placed between the pump and reservoirs to smooth out the changes in the

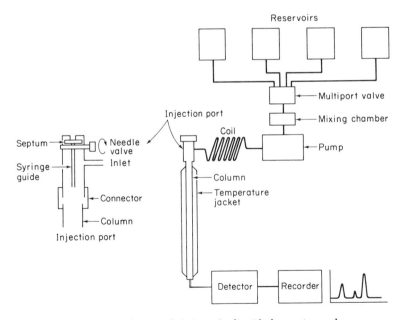

**Fig. 23–7.** A typical design of a liquid chromatograph.

mobile phase. The pump, a peristaltic pump which is incapable of developing a high inlet pressure, pulls the mobile phase from the reservoirs, pushes it through the column, and into the detector while maintaining a constant reproducible flow rate. A coil placed between the pump and column dampens the pulsating action of the pump. Inserted above the column is the injection system. Two general types are available, neither of which requires the opening or stopping of the chromatographic experiment. The one shown in Fig. 23–7 is a general design of a septum injection port where a syringe is used to inject the sample through an inert septum directly into the flowing mobile phase. Usually, some device is available to prevent the pressure of the mobile phase from being applied to the septum except when the injection is made. The second type of injection port involves a valve which can be rotated from the mobile phase tubing to a sample tubing. Thus, the sample is injected into the sample line, the valve is rotated, and the mobile phase washes the sample into the main line. After this, the valve is returned to its original position and the sample mixture passes into the column, separation takes place, and the effluent is detected and the chromatogram recorded.

## HIGH-PERFORMANCE LIQUID CHROMATOGRAPHY

In recent years, a technique known as high-performance liquid chromatography (HPLC) has been developed. The instrumentation needed for this is similar to that shown in Fig. 23–7 except for two changes: (1) The low pressure pump is replaced by a pump that is capable of producing pulse-free inlet pressures of 1000–6000 psi, and (2) the column is a narrow bore (less than 1/4 inch) metal (stainless steel) tube containing the stationary phase as micro particles. (The narrow column is used for analytical work. Recently, instrumentation has been designed that is capable of handling preparative level separations employing columns of 1–2 inches in diameter while still maintaining high inlet pressures.) Since the instrument is under a high pressure from the pump to the end of the column, all connections must be strong enough to prevent leaks.

Adsorption, partition, ion exchange, and exclusion chromatographic techniques are possible in HPLC. Furthermore, the theory which describes these chromatographic processes under low inlet pressure does not change just because a high inlet pressure is used.

In HPLC, stationary phase particles of uniform size ranging from 5 $\mu$m to 50 $\mu$m are packed into narrow-bore columns. With this arrangement high linear velocities (speed of travel of the sample through the column) is obtained at the high inlet pressure. It is the increase in linear velocity, and not the fact

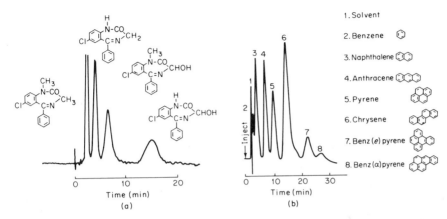

**Fig. 23-8.** Examples of high-performance liquid chromatography. (a) Separation of some benzodiazepines. Column: 1 m × 1 mm of 36–75 μm Durapak-OPN, hexane–isopropanol at 1.0 ml/min, 8 μg total sample. (b) Revised phase separation of bused-ring aromatics. Column: 1 m × 2.1 mm of hydrocarbon polymer on Zipax 37 packing, water–methanol. (Taken from J. J. Kirkland, "Modern Practice of Liquid Chromatography," Wiley-Interscience, New York, 1971.)

that a high pressure is used that leads to the advantages of better resolution, better efficiency, faster separation times, and better sensitivity (part of this is due to the use of low dead-volume detectors) in comparison to low inlet pressure chromatography.

Although HPLC is relatively new, it has already grown into a widely used and valuable technique in organic analysis. For example, it is routinely used for separations and subsequent analysis of environmental, pharmaceutical, and biological samples. Two typical HPLC separations are illustrated in Fig. 23-8. The time at which each peak appears is characteristic of the compound (qualitative analysis) and the area under each peak is proportional to its concentration (quantitative analysis). Details of the procedures for calibration of the chromatogram were described in Chapter 22.

In gas chromatography, which is described in Chapter 24, organic compounds must have an appreciable vapor pressure before this technique can be applied to their separation. This limits gas chromatography to approximately 60% of the known organic compounds. HPLC techniques, in contrast, do not have this kind of requirement and are for all practical purposes suitable for the separation of all organic molecules. Together, the two techniques provide the analytical chemist with methods for the determination of most simple as well as complex mixtures of organic compounds at not only macro levels but also at trace levels.

## Questions

1. List the important parameters that must be controlled in preparing a column for column chromatography.
2. Explain why solubility is more of a dominate factor in column partition chromatography than in column adsorption chromatography.
3. Suggest an experiment or series of experiments which would demonstrate the differences in adsorptive power between alumina and silica gel.
4. Alumina can be graded according to the amount of water that it contains (degree of activation). Suggest a procedure for grading alumina.
5. Suggest an adsorption sequence for a mixture of *o*-, *m*-, and *p*-nitroaniline.
6. Which of the eluting systems would be expected to provide better eluting conditions in column adsorption chromatography?
   a. Chloroform or ethanol
   b. (1:10) Chloroform–ethanol or (10:1) chloroform-ethanol
   c. Chloroform or heptane
   d. (1:10) Chloroform–heptane or (10:1) chloroform–heptane
   e. (1:1) Pyridine–water or (1:1) ethanol–water
7. Describe a procedure for impregnating an inert support with a stationary phase.
8. What is reversed phase chromatography?
9. In column chromatographic separation of a two component mixture where one component is at a trace level, is it better for the trace component to emerge first or second from the column? Why?
10. Suggest a general design of a fluorometer that could be used for detection of column effluent.
11. Why does the flow rate have to be controlled in a column separation?

# Chapter
# Twenty-Four
# Gas Chromatography

## INTRODUCTION

Gas chromatography is a technique whereby the components of a mixture in the gaseous state are separated as the sample passes over a stationary liquid or solid phase. Differences in the interactions between the sample components and the stationary phase account for the separation. Although gas phase separations through adsorption were reported in the early 1900s, it was not until 1952 that the true significance and potential of gas chromatography was realized. Since that time, gas chromatography has been developed to the point where it is now applicable to almost every area within the physical and chemical sciences. The method has many advantages in that it can be used for qualitative and quantitative analyses, the time of analysis is short, the instrument is simple, the sensitivity is high, and the method is applicable to about 60% of the organic compounds known to man.

Many examples can be cited to illustrate the advantages of this separation technique. Separations, which are very difficult or virtually impossible by other techniques, can be simple and straightforward with gas chromatography. The successful separation of *cis* and *trans* isomers, of oxygen isotopes, or of pesticides is a matter of choosing the proper column and conditions.

The description of gas chromatography in this chapter will be from an experimental approach. Hence, a discussion of the chromatogram, instrument, and factors affecting separation will be presented.

## THE CHROMATOGRAM AND ITS INTERPRETATION

As the components of a mixture are eluted from the column, they pass directly into the detector. The response of the detector is plotted as a function of time

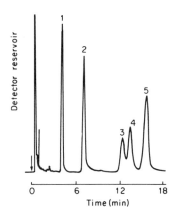

**Fig. 24-1.** Gas chromatogram of fatty acids. (1) Myristic; (2) palmitic; (3) stearic; (4) oleic; (5) linoleic. [From K. Hammarstrand, "Gas Chromatographic Analysis of Fatty Acids," Varian Aerograph Walnut Creek, California, 1966.]

or carrier gas volume on a strip chart recorder. This plot is a gas chromatogram and an example is presented in Fig. 24–1. Two important features should be noted. First, the components of the mixture are eluted from the column at different time intervals from injection. This time interval is the retention time, $t_R$, and is constant, if all of the separation conditions remain the same on repeated injections. Consequently, this is the basis for qualitative analysis since the retention times between a standard and an unknown are comparable. Alternatively, the $x$ axis of the chromatogram can be reported as retention volume, $V_R$, where $V_R$ is the product of the retention time and the carrier gas flow rate.

The second characteristic, which is the basis for quantitative analysis, is that the areas of the elution peaks are proportional to concentration. If a series of standards are injected into the instrument, the size of each elution peak will be proportional to the amount of material. Analysis of an unknown mixture is completed by injecting a known volume of the unknown and comparing the resulting peak area to the calibration curve. The methods for measuring peak areas were described in Chapter 22.

## INSTRUMENTATION

A block diagram showing the essential components of a gas chromatograph is presented in Fig. 24–2. These are a tank of carrier gas which acts as the mobile phase, an injection port for introducing the sample, the column, and a detector with appropriate readout. The carrier gas flows throughout the system carrying the sample (in the vapor state) which is introduced into the

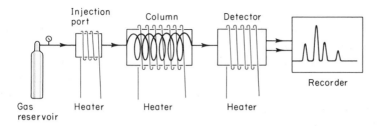

**Fig. 24–2.** The gas chromatograph.

system by a syringe through the injection port. Separation occurs in the column because of partition or adsorption of the sample components on the stationary liquid or solid phase, respectively. After separation each component is detected by the detector as it emerges from the column.

**The Carrier Gas.** The mobile phase (carrier gas) in gas chromatography is usually helium or nitrogen. These gases are used most often because they fulfill the following requirements.

1. The carrier gas should be inert.
2. The carrier gas should be inexpensive as large quantities are used.
3. The carrier gas should allow the detector to respond in an adequate manner.

A high-pressure gas cylinder is used as a carrier gas reservoir. Attached to the cylinder is a pressure regulator to reduce and control the gas flow through the column and a flow meter to control the rate of the gas flow.

**The Injection Port.** The injection port is placed so that the sample is introduced directly into the carrier gas. It is designed for instantaneous injection and vaporization of a sample so that the sample is introduced immediately into the column. The port is constructed of a heavy mass that is maintained at an elevated temperature and contains a pliable septum through which samples are injected.

Solid, liquid, and gas samples are conveniently injected into the sample port by a calibrated syringe. For gases, a gas tight syringe (0.5–10.0 ml) can be used. An alternative method is to use a sample loop whereby the gas sample is first introduced into a chamber of known volume and then, by the use of valves, passed into the carrier gas flow.

Liquids are introduced neat or as solutions with syringes of 0.1–100-$\mu$l capacity. The sample is drawn into the syringe a number of times to ensure exclusion of any gas bubbles and subsequently is injected very rapidly into the gas stream.

Solids can be treated in two ways. First, if a suitable solvent can be found, the material is dissolved and injected as a solution. Second, the solid may be injected directly by use of a special syringe. It is designed so that the solid is packed into the end of the syringe by "tapping." The sample is then injected directly into the instrument through a septum by a plunger. A major problem with this type of sampling procedure is that it suffers from a lack of reproducibility since the amount of sample taken up by the syringe cannot be measured accurately.

Samples which cannot be vaporized rapidly and completely at the operating temperature of the instrument should not be injected into the injection port since these samples will not move appreciably. Repeated injections of this type of sample will cause the port to clog, ruin the column, or change the response characteristics of the detector. Samples which have low vapor pressures, however, can be converted chemically to compounds which have higher vapor pressures, and these derivatives can be injected into the instrument. This is a very common and useful technique since it increases the number of compounds that can be separated by gas chromatography. Consequently, classes of compounds, such as amino acids, lipids, and high-molecular-weight polymers which are normally not very volatile can be separated after conversion to volatile derivatives. For example, low vapor pressure organic acids can be converted into low boiling acid chlorides, steroids can be silylated, or metal ions can be complexed with hexafluoroacetyl acetone. In all cases, the vapor pressure of the derivative is much higher than the vapor pressure of the original molecule.

**The Column.**     The choice of the packing in the gas chromatographic column provides the basis for the separation process. The column container is usually a stainless steel or copper ($\frac{1}{16}$ in., $\frac{1}{8}$ in., $\frac{1}{4}$ in., o.d.) tube packed with either a solid substrate [gas–solid chromatography (GSC)] or a liquid coating on an inert solid [gas–liquid chromatography (GLC)]. The column is placed in an oven such that the temperature can be controlled (25–400°C). Glass tubes are frequently used for biological samples or for compounds which undergo reactions with copper or stainless steel.

A GSC column is prepared by filling a straight tube of desired length (3–12 ft) and diameter with the substrate after plugging one end with glass wool. A funnel is placed on the other end and the packing material is vibrated until it fills the column. The other end is then plugged and is coiled such that it will fit into the instrument oven. Conditioning of the column is accomplished by baking at the prescribed temperature* (while carrier gas is flowing) to remove foreign materials. Usually, it takes a period of 6–12 hours to condition a column.

---

* A column is conditioned above the temperature of normal use, but below the temperature at which the column decomposes or vaporizes.

A column for GLC is prepared by a slightly different method, since a high boiling liquid must be uniformly deposited on the inert support. In general the stationary phase of known weight is dissolved in a volatile solvent and the desired amount of inert packing is then added to the solution. Subsequently, the solvent is removed while the mixture is continuously agitated. After all the solvent is evaporated, the uniformly coated substrate which can be described as percent liquid phase (g liquid/g support), is packed into the column and conditioned.

Capillary columns of diameters of $\frac{1}{16}$ inches or less and lengths of 200 ft or more can be used. These columns are not packed, but the inner walls are coated with a liquid layer. Consequently, they are used for partition chromatography. The advantage of this type of column is that it provides a high degree of efficiency (small plate heights). However, this is accomplished by a sacrifice in loading capabilities, which means that only small sample sizes can be efficiently used.

**Choice of Phase and Support.**     In choosing the correct system for separation of a mixture, particular attention should be given to the stability of the column toward the components of the mixture. If the column reacts with the compounds injected on to the column, erroneous elution peaks, background detector drift, or destruction of the instrument could result.

In general, the useful adsorbents for GSC will have very large surface areas or a high level of porosity. The mechanics of experimentally using GSC are less complicated than for GLC. Even so, GSC techniques are less often used because they tend to be limited to the separation of permanent gases and nonpolar low molecular weight compounds. For example, nitrogen and oxygen can be separated on a molecular sieve column at ambient temperatures. For higher molecular weight or polar compounds excessive tailing is observed.

The number of inert supports and liquid stationary phases available for GLC is almost unlimited. In general the liquid phase chosen should satisfy the following requirements for a satisfactory separation:

1. It should be a good solvent for the components of the sample.
2. The solvent power of the liquid should be different for each component of the sample.
3. The liquid phase should have a very low vapor pressure.
4. It should be thermally stable.
5. It should be chemically inert toward the sample.

Frequently, the selection of the proper column support and liquid phase is based on trial and error or from previous experience (consultation of the literature on gas chromatography). Recently, considerable effort has been in the direction of putting the selection on a more fundamental basis. The major consideration appears to be the polarity of the stationary phase and

**Table 24–1.   A Partial List of Stationary Phases and Their Application**

| Stationary phase (liquids) | Application |
|---|---|
| Adiponitrile | Hydrocarbons |
| Apiezon L | Alcohols, aldehydes, ketones, aromatics, fatty acids, pesticides |
| Asphalt | Aromatics |
| Beeswax | Essential oils |
| Carbowax 200 | Aldehydes and ketones |
| Carbowax 20M | Alcohols, aromatics, gases, halogenated compounds, pesticides |
| Silicone gum rubber SE30 | Alcohols, aromatics, bile and urinary compounds, drugs and alkaloids, fatty acids, gases, pesticides, sugars, vitamins |

of the mixture to be separated. As a general rule, the best separation is obtained when the liquid phase is structurally similar to the compounds being separated. For example, if a series of hydrocarbons are to be separated, a liquid phase which is highly nonpolar, such as a silicone oil, would give the best results. If, on the other hand, water and methanol are to be separated, a semipolar liquid phase is desirable. For mixtures containing molecules of widely differing polarity, more than one column would have to be used. Table 24–1 presents some of the more common liquid phases and their applications.

The main purpose of the solid phase (or support) is to provide support of the thin, uniform film of liquid phase. It should be porous, have a large surface area, be inert, provide mechanical strength, and be uniform in particle size. Probably the most used solid phase is diatomaceous earth (kieselguhr) and is available under several different commercial tradenames. These will differ in pretreatment (acid, base, neutral washed) and certain physical properties. Other supports and their suppliers are listed in Table 24–2.

**Detectors.**     Although there are many detectors available for monitoring the effluent from a gas chromatographic column, only three, the thermal

**Table 24–2.   Solid Supports for Gas Chromatography**

| | |
|---|---|
| Chromosorb W | (Johns Manville) |
| Chromosorb P | (Johns Manville) |
| Firebrick | |
| Anachrom U | (Analabs) |
| Anachrom A | (Analabs) |
| Porapak Q | (Waters Associates) |
| Glass beads | |

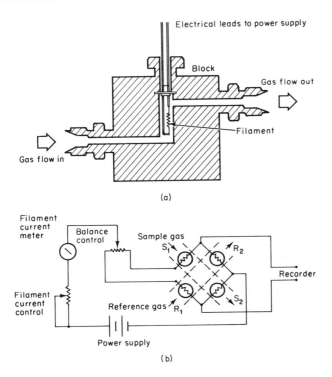

Fig. 24–3. Thermal conductivity detector and wheatstone bridge. (a) Detection cell; (b) Wheatstone bridge. [From H. M. NcNair and E. J. Bonelli, "Basic Gas Chromatography," Varian Aerograph, Walnut Creek, California, 1969.]

conductivity, flame ionization, and electron capture detectors, will be discussed. In general, detectors must meet certain requirements in order to be useful as monitoring devices. Among these are sensitivity, stability, a reasonable lifetime, and linear response characteristics over a range of sample concentrations. The thermal conductivity, flame ionization, and electron capture detectors satisfy these requirements.

The thermal conductivity detector is based on the principle that a hot object will lose heat at a rate that is determined by the composition of the surrounding gas. Furthermore, the rate of heat loss is a measure of the gas composition. Figure 24–3a illustrates a typical thermal conductivity detector.

The filament in the detector is made of a material whose electrical resistance varies greatly with temperature. This means that it has a high temperature coefficient of resistance. In operation the wire filament is heated by passing a constant current through it. The temperature of the wire (often over 100°C above the block temperature) will be determined by the applied current and by the surrounding gas. If pure carrier gas, such as helium, is flowing

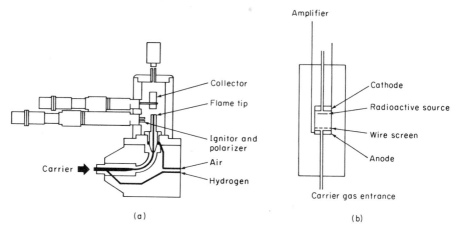

**Fig. 24–4.** (a) A flame ionization detector and (b) an electron affinity detector. [From H. M. McNair and E. J. Bonelli, "Basic Gas Chromatography," Varian Aerograph, Walnut Creek, California, 1969.]

over the wire at constant conditions, heat loss is constant and hence the filament temperature is also constant. A large heat change is favored by gases that have large thermal conductivities. Since thermal conductivity increases with a decrease in molecular weight, lower molecular weight gases, such as helium and hydrogen, are ideal as carrier gases when using a thermal conductivity detector.

If the gas composition around the filament changes, for example, when a sample peak emerges from the column, the filament's temperature changes and this causes a corresponding change in electrical resistance in the filament. This resistance change is what is measured and eventually put on the recorder.

The circuit used for measuring resistance is a Wheatstone bridge circuit* and a typical one used for a thermal conductivity cell is shown in Fig. 24–3b. In using the circuit, pure helium gas (prior to reaching the sample port) passes first through the reference cell (a thermal conductivity cell), through the sample port, through the column, and then through the sample cell (a second thermal conductivity cell). If pure gas is in both cells, the bridge is in balance. However, if a sample peak is in the sample cell, the bridge is out of balance, and the electronic circuit is designed so that an output signal is generated to return the cell to balance. It is this output signal that is finally recorded.

Sensitivity of the thermal conductivity detector is affected by several factors. To increase the sensitivity, the filament current can be increased,

---

* The Wheatstone bridge circuit is a frequently used circuit in scientific instruments and several variations are possible.

the temperature of the block holding the filament can be decreased, the carrier gas that is used should be one with the highest thermal conductivity (He), and a reduced flow rate of the carrier gas should be used.

The flame ionization detector consists of a small $H_2$–air flame in an electrostatic field. A typical design is shown in Fig. 24–4a. As the carrier gas containing a sample enters the burner it is mixed with the $H_2$ and air. If the sample is organic (and not fully oxidized), it is combusted in the flame, ion fragments and free electrons are produced, which in turn changes the electric current. Therefore, the change in current, which is measured by a set of electrodes, is proportional to the concentration of the sample. Since the detector responds to only oxidizable carbon atoms, an increase in number of carbon atoms per molecule results in a detection limit.

The electron capture (or electron affinity) detector operates on the basis of electron absorption by compounds that have an affinity for electrons. Thus, the compound must have an electronegative group or element. Since chlorine has a high efficiency for the capture of electrons, this detector is routinely used in monitoring pesticides in the environment. Under normal operating conditions nanogram quantities of chlorinated pesticide can be detected.

The basic components of the detector are presented in Fig. 24–4b. The

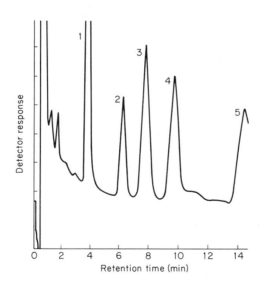

**Fig. 24–5.** Gas chromatogram of (1) lindane; (2) heptachlor; (3) aldrin; (4) heptachlor epoxide; and (5) dieldrin. The separation was made at 180°C using a SE-30 column. The weight of each component was 0.0824 ng. [From B. J. Gudzinowicz, "The Analysis of Pesticides, Herbicides, and Related Compounds Using the Electron Affinity Detector," Research Department, Jarrell-Ash Company, Waltham, Mass.]

detector is composed of a radioactive source which emits electrons, a cathode which repels the electrons, an anode and wire screen which collect electrons. As a compound enters the chamber, electrons are absorbed and a decrease in current is observed at the anode. If the concentration of the sample increases, a resultant decrease in current will be observed. Thus, the detector will respond to changes in the amount of sample being eluted from the column. A typical chromatogram for a series of pesticides in which an electron-capture detector was used is shown in Fig. 24–5. It should be noted that quantities less than 0.1 ng are easily detected.

Table 24–3 lists the detector parameters for the thermal conductivity, flame ionization, and electron-capture detectors. The first two will respond to all organic compounds (flame ionization will not respond to completely oxidized carbon compounds) and consequently both are widely used. For greater sensitivity the flame ionization detector is used over the thermal conductivity detector. Since the electron-capture detector responds only to compounds containing electronegative atoms, it is restricted to detection of these kind of compounds. One other limiting feature is that the response is essentially nonlinear. Several other specialty type detectors for gas chromatography are surveyed in Table 24–3.

**Table 24–3.  Some Detectors for Gas Chromatography**

| Detector | Specificity | Operation | Minimum detection[a] |
|---|---|---|---|
| Thermal conductivity | Responds to all compounds | Change in resistance of detectors | $10^{-5}$ g |
| Flame ionization | Responds to most compounds | Change in conductivity of flame | $10^{-11}$ g |
| $^{63}$Ni electron capture | Halogenated compounds | Change in ionization due to $^{63}$Ni radio-activity | $10^{-12}$ g |
| Flame photometric | Sulfur, phosphorous compounds | Flame emission of $S_2$*, HPO* | $10^{-12}$ g |
| Hall electrolytic conductivity | Nitrogen and halogen-containing compounds | Thermal degradation of compounds to $NH_3$ or HCl; dissolution in water; measurement of change in conductivity. | $10^{-12}$ g |
| Photoionization | Responds to most compounds | Irradiation with UV; change in conductivity | $10^{-12}$ g |

[a] The minimum detection quantity that can be detected will depend on the compound. The numbers cited indicate optimum conditions.

# FACTORS AFFECTING SEPARATION

The resolution of chromatographic peaks in gas chromatography, as in other column chromatographic techniques, is determined by both column and stationary phase efficiency.* The former is a measure of the broadening that the band undergoes as it passes through the column. In general, column design and operating conditions will determine the extent of broadening. Quantitatively, this is expressed by the height equivalent to a theoretical plate (HETP) for the column.

Stationary phase efficiency is a measure of the interaction between the sample components and the stationary phase, and determines the relative position of the sample components in the chromatogram. The importance of this is illustrated by the fact that compounds having the same vapor pressure can still be easily separated, provided the appropriate stationary phase is chosen.

A more detailed discussion of column and stationary phase efficiency was given in Chapter 22. In gas chromatography the following parameters are the ones of practical interest that can be altered to improve the efficiency of the separation.

1. *Partical size and surface area.* An increase in surface area or a decrease in particle size tends to increase the number of theoretical plates. Accompanying this, however, is a decrease in carrier gas flow rate (at the same applied pressure). In general a 60/80 mesh particle size is used in a $\frac{1}{4}$-inch column.

2. *Carrier gas flow rate.* There is an optimum flow rate which gives the maximum efficiency. This parameter, however, can only be experimentally determined. If the gas flow rate is too slow the eluted peaks will tend to be broad while if the flow is too fast, the peaks will not be resolved.

3. *Type and amount of stationary phases.* This variable is a key factor in determining the efficiency of the column. Therefore, considerable care must be exersized in choosing the correct stationary phase. If the wrong liquid is chosen, separation will not be obtained.

The amount of stationary phase will affect the column performance in several ways. As the concentration of the liquid phase increases, the number of theoretical plates for the column will also increase. However, this is not without limit since too much of the liquid tends to make the particles sticky resulting in undesirable packing qualities. Excessive liquid support can also cause tailing in the chromatogram. In general, light loading is preferred and most columns will contain from 1 to 15% liquid phase.

---

* The term *stationary phase efficiency* rather than *solvent efficiency* is used here since the mobile gas phase is usually not changed, unlike column liquid chromatography where it is frequently changed during the separation.

4. *Column length.* As the length of a column increases, the number of theoretical plates will increase. There is, however, a practical limit to the length because problems relating to the gas flow are encountered with long columns. Most columns are between 1 and 10 m (excluding capillary columns).

5. *Column diameter.* As the diameter of the column decreases, the number of theoretical plates will increase.

6. *Column temperature.* The maximum operable temperature for the column is determined by the vapor pressure of the liquid phase, the vapor pressure of the sample, and the efficiency of separation. The temperature should not be high enough to vaporize the stationary phase since this will destroy the uniformity of the column. On the other hand, the temperature has to be high enough to maintain the sample in the vapor state. Hence, adjustment of the temperature is made to give a high number of theoretical plates, which leads to optimum resolution, while maintaining reasonable elution times for the components of the sample.

With experience and the availability of tables of separations, liquid phases, and supports, it is possible to predict conditions for a separation and expect a reasonable degree of success.

## QUALITATIVE ANALYSIS

Qualitative identification by gas chromatography is achieved by two general approaches. The first method is based on the comparison of retention time (or retention volume) of the unknown to retention times (or volumes) for a series of standards. The second method is based on the collection of each peak as it emerges from the detector and subsequent characterization through the use of chemical and instrumental tests. For example, the effluent can be directed into an infrared spectrometer or a mass spectrometer.

If retention data is used for characterization it is necessary to carefully control the flow and temperature parameters, since it is these data for the standard and unknown that are being compared. Since many different compounds can have the same retention time (or volume) it is necessary to use several columns which differ in selectivity to be sure that the unknown and the standard are in fact the same.

Figure 24–6 illustrates a chromatogram for an unknown alcohol mixture and a chromatogram for a series of known alcohols obtained under identical conditions. By comparison, several of the peaks in the unknown are identified. Confirmation of the identification can be accomplished by collecting each of the identified peaks and examining each of these by other instrumental techniques.

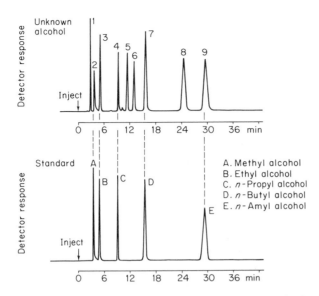

**Fig. 24–6.** Identification of an unknown by comparison to standards. [From H. M. McNair and E. J. Bonelli, "Basic Gas Chromatography," Varian Aerograph, Walnut Creek, California, 1969.]

## QUANTITATIVE ANALYSIS

The primary application of the more than 100,000 gas chromatographs which have been sold by instrument suppliers is for the quantitative analysis of gases, liquids, and solids, almost all of which are organic compounds. The procedures for calibrating the chromatogram have been discussed in Chapter 22. Because of the nature of the detector response the handling of the data must be modified. Also, because gas chromatography provides such a high level of efficiency and resolution it is possible to use an internal standard for calibration. These added modifications are briefly discussed in the following.

**Normalization.**    In the technique of normalization the percent composition is determined by measuring the area of each peak and dividing the individual areas by the total area. This assumes that all peaks from the sample have been eluted and that each compound has the same detector response. The first is verified by using several columns which differ in resolving power to examine the sample. The second is more difficult to evaluate since detector response is not the same for different compounds. If the compounds are similar, for example, a homologous series, it can be assumed that the detector response is the same. For other cases a detector-response correction must be applied. Also, the procedure for determining the detector response is different for the different detectors; details of this are provided elsewhere.*

**Absolute Calibration.**    To avoid having to determine detector response, exact amounts of the sample can be chromatographed. A calibration curve of either peak area or peak height vs concentration of the sample is made. The unknown is then chromatographed, its peak area or height determined, and the concentration of the unknown is obtained from the calibration curve.

In this type of calibration, it is assumed that all the experimental parameters are reproduced between the standards and the unknown. Of particular importance is the need to reproduce the flow rate and detector response.

**Internal Standardization.**    The procedure of using an internal standard in gas chromatography is similar to its use in other instrumental techniques. In the procedure an internal standard is added to a series of known sample weights. After injection of each, the ratio of the peak areas of the sample to standard is plotted vs the weight ratio. The calibration curve should be linear. To determine the weight of an unknown, a known weight of the internal standard is added to a known weight of the unknown and the mixture injected into the chromatograph. After the area of the two peaks and their ratio is determined, the weight of the unknown is obtained from the calibration curve.

The principle advantages of internal standardization as a calibration method is that minor changes in detector response, instrumental parameters, or in the quantities injected are compensated for since the area ratio will not be altered. The main limitation is the choice of the internal standard. In general, the requirements for a satisfactory internal standard are:

1. The chromatographic peak of the standard must be resolved from other peaks.

2. The retention times for the sample component and the standard must be close.

3. The concentration of the standard should be approximately the same as the unknown.

4. The structure of the standard should be similar to the unknown.

---

* H. M. McNair and E. J. Bonelli, "Basic Gas Chromatography," Varian Aerograph, Walnut Creek, California, 1969.

D. W. Grant, "Gas–Liquid Chromatography" Van Nostrand Reinhold, New York, 1971.

J. Novak, "Quantitative Analysis by Gas Chromatography," M. Dekker, Inc., New York, 1975.

## APPLICATIONS

Gas chromatography has been successfully applied to the separation of mixtures of numerous organic and inorganic compounds that have appreciable vapor pressure. Virtually every scientific discipline has benefited from this separation technique. The examples cited in the remaining part of this chapter illustrate not only the scope of gas chromatography but also several techniques which broaden its applications.

Many kinds of samples occur as or are converted into permanent gases. These are readily separated by gas chromatography, usually by using a solid stationary phase rather than a liquid stationary phase. For example, the determination of elemental carbon, hydrogen, and nitrogen in organic compounds is often required when characterizing organic molecules. This can be done by an automated gas chromatograph. In this technique, a sample is weighed into a sample boat (0.5–0.7 mg) and placed in a boat cavity. The sample boat is packed with oxidizing catalysts and placed in the sample port and sealed in position. The instrument automatically combusts the sample for a prescribed period of time after which the products ($CO_2$, $H_2O$, and $N_2$) are swept by carrier gas to the chromatographic column.

The combustion of the sample takes place in a helium atmosphere and the oxygen is supplied by the thermal decomposition of manganese dioxide. In order to insure that the products of the oxidation are water, carbon dioxide, and nitrogen, a series of reaction tubes are placed between the (sample) port and the column. These are shown in Fig. 24–7. The reactions which occur in the oxidation tube are

$$CO + CuO \rightarrow CO_2 + Cu$$

$$H_2 + CuO \rightarrow H_2O + Cu$$

Subsequently, the product gases are swept into the reduction furnace where the excess oxygen is removed and the nitrogen oxides are reduced to nitrogen:

$$2Cu + O_2 \rightarrow 2CuO$$

$$2NO_x + 2xCu \rightarrow N_2 + 2xCuO$$

Hence, the helium carrier gas, which contains only $CO_2$, $H_2O$, and $N_2$, is swept onto the gas chromatographic column, separated and detected. The percentage of each element is obtained by measuring the relative heights of

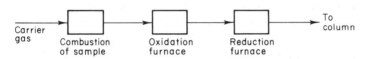

**Fig. 24–7.**  Block diagram of the carbon, hydrogen, and nitrogen autoanalyzer.

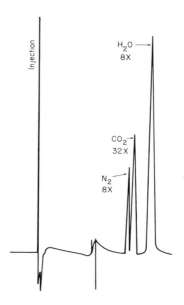

**Fig. 24–8.**   A typical C–H–N chromatogram.

each peak. Calibration of detector response to each of the three compounds is accomplished by combustion of a material of known formula. Figure 24–8 illustrates a combustion chromatogram.

A completely automated, computer operated C, H, and N instrument is commercially available. The samples are weighed by an automated micro-balance, loaded into the instrument, and the computer which is programmed to carry out the analysis takes over. About every tenth sample is a standard. Based on the results for the standard the computer adjusts the column temperature, gas flow, and combustion conditions, and provides a printed readout of %C, %H, and %N for each sample. About 15 minutes are required for each analysis. Since 40 samples can be loaded into the instrument, it can operate overnight unattended. In a testing laboratory, this instrument would be in operation on a 24-hour basis.

Many other gaseous samples are routinely determined by gas chromatography. For example, components of air, and natural gas can be separated and determined. In air pollution studies, lead alkyls, polynuclear hydrocarbons, CO, auto exhaust (over 75 compounds have already been identified), volatile aldehydes and ketones, and oxides of sulfur and nitrogen are routinely monitored by gas chromatography. By choosing the appropriate detector as little as 1 ng is readily detected.

Figure 24–9 illustrates a chromatogram for the separation of impurities in ethylene (permanent gases). The two chromatograms are the result of an

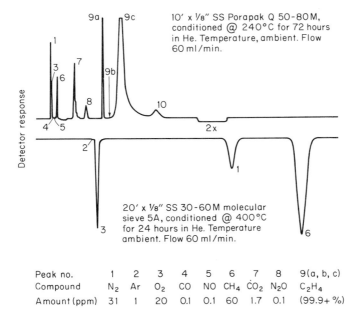

**Fig. 24–9.** Determination of impurities in ethylene ($C_2H_4$) using dual column gas chromatography. [From H. M. McNair and E. J. Bonelli, "Basic Gas Chromatography," Varian Aerograph, Walnut Creek, California, 1969.]

instrument modification. A stream splitter (1:1) is introduced into the gas flow after the sample port. Each arm of the splitter leads to a separate column. Thus, each injection is subjected to two different columns and complete separation is obtained. It should be noted that an ambient column temperature is used for the separation which is typical of separations of permanent

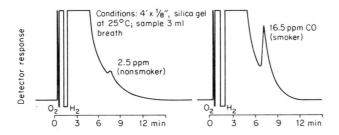

**Fig. 24–10.** Determination of trace amounts of carbon monoxide in breath. [From H. M. McNair and E. J. Bonelli, "Basic Gas Chromatography," Varian Aerograph, Walnut Creek, California, 1969.]

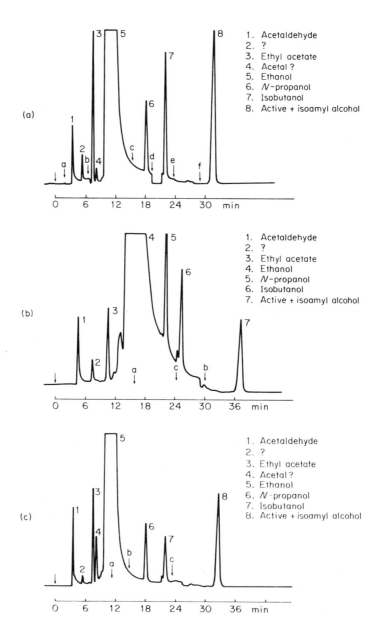

**Fig. 24–11.** Gas chromatograms of alcoholic beverages. (a) Bourbon; (b) vermouth; and (c) rum. [From C. Scott, N. Hadden, and E. Bonelli, "Analysis of Alcoholic Beverages," Varian Aerograph, Walnut Creek, California, 1966.]

gases. Figure 24–10 shows the chromatograms and subsequent determination of CO in the breath of smokers and nonsmokers.

Many complex liquid and solid samples are readily separated by gas chromatography. For example, pesticide and herbicide residues (see Fig. 24–5) petroleum distillates and products, carbohydrates, fatty acids, vitamins, resins, solvents, essential oils such as peppermint oil, foods, and adultered foods have been separated. Figure 24–11 illustrates chromatograms for several alcoholic beverages where the lower boiling, more common organic compounds present have been identified. In many alcoholic beverages, such as wines, over one hundred compounds have been identified.

The resolving power provided by a capillary column is often required to resolve very complex mixtures. For example, using this column technique crude oil distillation cuts have been carefully studied. Often hundreds of compounds are found, many of which have been identified. Figure 24–12 illustrates a separation of a crude oil distillation cut. In this example 66 compounds most of which are saturated hydrocarbons containing from 5 to 9 carbons have been identified.

Two other useful modifications are the conversion of the sample into a derivative and temperature programming. Some compounds have low vapor pressures and consequently are not easily eluted in the gas chromatograph. Others have even lower vapor pressures such that they will not even exist at an appreciable concentration level as a vapor within the temperature limits of the gas chromatograph. Some of these compounds can be converted into derivatives which have appreciable vapor pressures. Separations are then carried out with the derivatives. Ester formation and silylation are two reactions frequently used to prepare derivatives. Figure 24–13 illustrates an improvement in resolution after converting the sterols (free $-OH$ group) to the tetramethylsilyl derivatives [$-OSi(CH_3)_3$ group].

Temperature programming is a gradient technique whereby the sample is

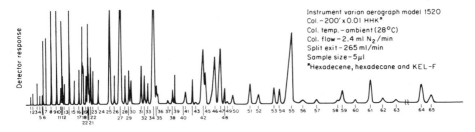

Instrument varian aerograph model 1520
Col.–200′ x 0.01 HHK*
Col. temp.–ambient (28°C)
Col. flow–2.4 ml $N_2$/min
Split exit–265 ml/min
Sample size–5 μl
*Hexadecene, hexadecane and KEL-F

**Fig. 24–12.** Separation of a crude oil distillation cut on a capillary column. [From H. M. McNair and E. J. Bonelli, "Basic Gas Chromatography," Varian Aerograph, Walnut Creek, California, 1969.]

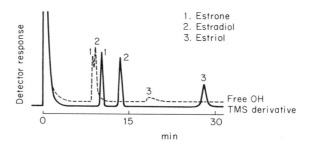

**Fig. 24-13.** Comparison of separation of sterols as parent estrogens and TMS derivatives. [From A. E. Pierce, "Silylation of Organic Compounds," Plenum, New York, 1968. Courtesy of Plenum Press.]

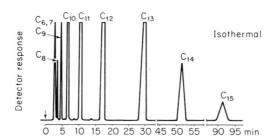

Conditions: 20′ by 1/16″ column, 3% Apiezon L on 100/120 mesh VarAport 30 at 150°C, 10 ml/min He.

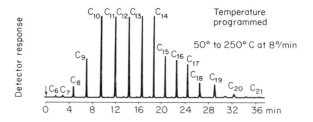

**Fig. 24-14.** Comparison of an isothermal and temperature-programmed gas chromatographic separation of a complex mixture of normal paraffins. [From H. M. McNair and E. J. Bonelli, "Basic Gas Chromatography," Varian Aerograph, Walnut Creek, California, 1969.]

introduced into the column at a low column temperature (often ambient temperature) and the temperature of the column is increased at a controlled, uniform rate up to a predetermined maximum temperature. Although non-linear programming is possible, most applications utilize a linear increase in temperature. In general, temperature programming is employed in the separation of complex mixtures where the components differ widely in their vapor pressures. For example, at a constant temperature, low boiling components appear early, and if the temperature of the column is too high, they emerge too rapidly as overlapped peaks. The higher boiling components will emerge too slowly if the temperature is too low and hence, appear as drawn out tailing peaks or they may not appear at all. This is illustrated in Fig. 24–14a, where a constant column temperature was used for the separation of a complex mixture of saturated hydrocarbons. By using a temperature program sharp, fully resolved peaks are obtained for the entire mixture. This is illustrated in Fig. 24–14b where the more complex mixture of saturated hydrocarbons is separated. It should be noted that the separation is completed in less than a third of time required as in Fig. 24–14a.

## Questions

1. How does gas chromatography differ from other forms of chromatography?
2. Describe the characteristics of the gas chromatograph and how it is used for qualitative and quantitative determinations.
3. What are the basic components of the gas chromatograph? Describe each in detail.
4. How is the detector in a gas chromatograph calibrated?
5. What types of compounds can be separated by gas chromatography?
6. How is gas chromatography applied to elemental analysis?
7. Explain how a thermal conductivity detector works.
8. Why does the temperature of the sample port have to be higher than the column temperature?
9. Explain why it is important for the internal standard peak to be close to the sample peak in the gas chromatogram.
10. Nitrogen is much more inexpensive than helium. Why isn't nitrogen used as the carrier gas when using a thermal conductivity detector?
11. Suggest a procedure that can be used to deposit the liquid phase uniformly over the inert solid support. Why is it necessary to have the liquid phase present in a uniform coating?
12. What problems would be encountered if a solid support had too much liquid phase coated over it? Consider problems before and after packing of the column.

## Problems

1. In Fig. 24–5, if the carrier gas flow rate is 60 ml/minute, calculate the retention volumes of each of the components.

2.* If the column used for the separation in Fig. 24–5 was 3.0 meters long, calculate the number of plates and the HETP for the column by using peak 3.

3.* A 4-$\mu$l mixture containing only A and B was separated. The height and peak width at $\frac{1}{2}$ $H$ were 75 mm and 9 mm and 64 mm and 17 mm for A and B, respectively. The height and peak width values for 1 $\mu$l of pure A were 58 mm and 6 mm and for 1 $\mu$l of pure B they were 38 mm and 14 mm. If the density of A and B is 1.32 g/ml and 1.60 g/ml, respectively, calculate the amount of A and B in the sample.

4. An unknown mixture containing the pesticide, 2,4-dichorophenoxyacetic acid (2,4-D), was separated by gas chromatography. The unknown weighed 10.0 mg and was dissolved in 5.00 ml of solvent. Calculate the % 2,4-D in the unknown if the following data were obtained.

| mg of 2,4-D per 5 ml | Sample size ($\mu$l) | Area of peak (mm²) |
|---|---|---|
| 2.0 | 5 | 12 |
| 2.8 | 5 | 17 |
| 4.1 | 5 | 25 |
| 6.4 | 5 | 39 |
| unknown | 5 | 20 |

5.* Benzene can be converted to ethylbenzene using ethylbromide and a Friedel–Crafts' catalyst. The amount of ethylbenzene in benzene, also used as the solvent, can be determined using toluene (A) as the internal standard. Calculate the amount of ethylbenzene (B) present in the unknown if the following data were obtained. A 5-$\mu$l sample size was used for each.

| wt A (mg) | wt B (mg) | Area A (mm²) | Area B (mm²) |
|---|---|---|---|
| 0.1 | 0.1 | 6.1 | 4 |
| 0.1 | 0.4 | 5.9 | 15 |
| 0.1 | 0.8 | 6.0 | 31 |
| 0.1 | 1.2 | 6.1 | 49 |
| 0.1 | unknown | 6.0 | 27 |

* Answers are listed at the end of the book for problems marked with an asterish.

6. A series of monosaccharides were isolated from a glycoprotein sample and subsequently silylated. Upon injection and separation the following results were obtained. Assuming that each component has the same response factor calculate the percent of each component.

| Compound | Peak area |
| --- | --- |
| Fructose | 30 |
| Xylose | 32 |
| Mannose | 42 |
| Galactose | 18 |
| Glucose | 16 |
| N-Acetylglucosamine | 57 |
| N-Acetylgalactosamine | 34 |

7.* A sample of meat (3.2045 g) was treated to extract lindane and the solution diluted to 1000.0 ml. Upon injection into the gas chromatograph a peak area (the peak corresponding to lindane) of 32.6 mm$^2$ was obtained. A standard (0.541 ng/10 ml) was injected and a peak area of 41.7 mm$^2$ obtained. If the amount of sample injected in each case was 5.0 $\mu$l, calculate the amount of lindane in the meat sample in ppb.

# Chapter
# Twenty-Five
# Sheet Methods

## INTRODUCTION

Sheet methods, such as paper and thin-layer chromatography, can be used to separate a wide variety of organic and inorganic mixtures. In principle, a sheet method can be considered to be an open, thin column. These techniques, in comparison to column techniques, have the advantages of being simple, rapid, inexpensive, and provide excellent resolving power. Although quantitative results are possible, the main application is in qualitative identification, since very small quantities of sample can be handled. Sheet methods can also be scaled up to handle purification on a preparative basis (up to 2 g). The latter application is particularly important in natural product chemistry, biological, and pharmaceutical investigations. In general, qualitative identification is possible by comparison of sample $R_f$ values to standards or by isolating the separated compound from the sheet and identifying it by chemical and/or instrumental means.

Since a sheet method is similar to a column method, it is not surprising that adsorption, partition, sieving, and ion exchange are readily carried out on a sheet. Paper chromatography is more suited to partition while all four are routine in thin-layer chromatography (adsorption and partition are used the most). Inorganic as well as organic mixtures can be separated by sheet methods. However, the applications are much more extensive in the latter class of samples.

For the most part the experimental techniques are similar for most sheet methods. Only in the preparation of the paper or thin layer is there a difference. For this reason each technique is briefly discussed from the standpoint of sheet preparation. Subsequently, the experimental procedures, which includes *pretreatment, sample application, developing chambers, development,* and *visualization,* are described. The last section briefly lists some typical applications.

# PAPER CHROMATOGRAPHY

Paper chromatography was introduced in 1944 and in its simplest form can be described as the passage of a mobile liquid through the porous structure of the paper which contains a stationary liquid phase. Development is terminated before the mobile phase reaches the edge of the paper so that the zones are distributed across the paper. Perhaps the major limitations of paper chromatography in comparison to thin-layer chromatography are that longer development times are needed, zones are not always sharply defined, accuracy in quantitative analysis is only fair, and development conditions are sometimes difficult to reproduce.

**Types of Paper.** A wide variety of papers, which are very uniform from lot to lot, are available commercially in different sizes, shapes, porosities, thicknesses, and chemical treatments (acid or base washed). In general, paper is composed of randomly directed cellulose fibers. The cellulose itself is a network of polymeric carbohydrate chains (molecular weight $> 100,000$) possessing hydrophilic character and crosslinked by a stable hydrogen-bonded system. Water or other very polar-type solvents are tightly held within the hydrophilic cellulose system and can be considered to be "different" from the bulk water or polar solvent.

The cellulose papers can be modified in several ways to alter its chromatographic behavior. For example, paper can be impregnated with diatomaceous earth, alumina, silica gel, and ion-exchange resins. These kinds of papers will exhibit properties of these adsorbents and consequently, influence the retention of the stationary liquid and the adsorption or partition sequence for a mixture. The ion exchange resin impregnated paper will have either cation or anion exchange properties.

If the paper is acetylated (hydroxy groups are converted to acetyl groups) the paper takes on a hydrophobic property. That is, it tends to retain a hydrophobic type solvent rather than a hydrophilic type solvent as a stationary phase. This type of application is referred to as reversed-phase chromatography. The paper can also be made hydrophobic by silicone treatment or by impregnating it with inert nonpolar-type organic polymers.

If very corrosive eluting conditions are required, a glass fiber type paper can be used. By special treatment, adsorption effects due to the glass can be minimized.

Paper chromatography is essentially partition chromatography and there are a wide variety of useful combinations of stationary and mobile phases. It is not necessary that the two systems be immiscible. The types of stationary phases that are used can be classified as aqueous, hydrophilic, and hydrophobic systems.

**Aqueous Stationary Phase.**     Water is readily held by paper. Therefore, water-equilibrated paper is attained by suspending the paper in a closed chamber whose atmosphere is saturated with water. If an aqueous buffered or salt phase is desired, the paper is drawn through the respective solution and then exposed to a water saturated atmosphere in a chamber. This type of system is particularly suited to the separation of moderately polar to very polar (ionic) mixtures.

**Hydrophilic Stationary Phase.**     An organic solvent can be used for the hydrophilic stationary phase. If the solvent is volatile enough, the paper can be equilibrated in a chamber whose atmosphere is saturated with the solvent. Alternatively, the stationary phase solvent is dissolved in a very volatile diluent and the paper is dipped into the solution. Subsequently, while drying in air, the volatile diluent evaporates leaving the stationary phase liquid uniformly distributed throughout the paper. Some of the more common hydrophilic solvents that are used are methanol, formamide, glycols, cellosolves, and glycerol.

**Hydrophobic Stationary Phase.**     The paper must be modified, as described previously, before it will exhibit a tendency to retain a hydrophobic stationary phase. Equilibration in the vapors of the solvent or the dipping technique in a solution of the solvent and a volatile diluent are used for introducing the hydrophobic solvent into the modified paper. Solvents such as dimethylformamide, kerosene, aromatic and aliphatic hydrocarbons, and certain oxygenated solvents are used.

**Table 25–1.  Ten Basic Mobile Phases for Paper Chromatography**[a]

---

a. Isopropanol–ammonia–water (9:1:2)
b. *n*-Butanol–acetic acid–water (4:1:5)
c. Water–phenol
d. Formamide–chloroform
e. Formamide–chloroform–benzene
f. Formamide–benzene
g. Formamide–benzene–cyclohexane
h. Dimethylformamide–cyclohexane
i. Kerosene–70% iospropanol
j. Paraffin oil–dimethylformamide–methanol–water

---

[a] Taken from K. Macek, "Chromatography," 2nd ed. (E. Heftmann, ed.), Reinhold, New York, 1967.

**Mobile Phase.**      Numerous combinations are possible for the mobile phase. Mixtures of two, three, or more solvents, solutions of salts, and solutions of buffers can be used. In some cases choosing the optimum eluting conditions is partially a trial-and-error process. However, certain guidelines can be used to predict eluting conditions. For example, the characteristics of the components in the mixture and the type of stationary phase being employed should be considered. Based on these considerations and the fact that close to thirty years of experience has been gained, specific eluting conditions can be listed. It is beyond the scope of this book to go into this type of detail and Table 25–1 lists a brief typical set of eluting conditions. Mixtures (a) to (c) are for the separation of hydrophilic substances, (d) to (g) are for intermediate hydrophilic substances, and (h) to (j) are for hydrophobic substances. It is also possible to use different ratios within each mixture to bring about a more gradual change in partitioning power.

# THIN-LAYER CHROMATOGRAPHY

Thin-layer chromatography is a simple, versatile, inexpensive chromatographic technique of major importance. Separations on the layer, which is a thin stationary phase coated on a glass, metal or plastic plate, are affected by adsorption, partition, exclusion, or ion exchange processes. In general, the first two are the most important and are the only ones that will be considered in this chapter. The thin-layer method has several advantages over paper chromatography. In addition to providing separations by processes other than partition, it provides sharper and faster separations, higher sensitivity, the separated compounds are more easily recovered, and it is more easily adapted to the separation of larger concentrations.

**Coating.**      In general, thin-layer chromatography is carried out experimentally like a sheet method but the sheet exhibits the properties of a column technique. Therefore, any of the stationary phases used in column experiments as stationary phases can be used in thin-layer chromatography, provided the support is available in small enough particle size and is held in some way in sheet form.

The layer material consists of a binder and the active support material. This mixture is made into an aqueous slurry, smeared on a glass, metal, or plastic plate, and the water is evaporated. This leaves a thin layer of the support held in a continuous layer by the binder (also gives the layer physical strength). The plates must be flat and clean, since this will affect layer uniformity and layer adhesion. Application of a sample to be separated and the development techniques are typical of sheet methods.

The layers must be uniform in thickness and in quantity of support per

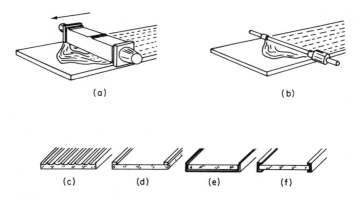

**Fig. 25-1.** Techniques for spreading thin layers. (a) Commercial spreader; (b) glass rod as spreader; (c) grooved glass plate for holding layer; (d) mounting strips or tape on the glass edge; (e) tray in which a glass plate lies; (f) attached guide rails. (Taken from E. Stahl, "Thin Layer Chromatography," 2nd ed., Springer-Verlag, New York, 1969).

unit area. Fortunately, commercial thin layers are excellent in satisfying these requirements. In addition, commercial thin layers are available bound to glass, metal (Al), and plastic, in various sizes, and impregnated in several different ways to aid development and visualization.

Silica gel and alumina are the two most useful supports in thin-layer chromatography. In general, acidic and neutral substances are separated on silica gel, while basic mixtures are resolved on alumina. Other supports that are useful are cellulose, diatomaceous earth, and a variety of organic polymers.

If a binder is not used, the adsorbents will not adhere very well to the glass or other plate material nor will they have mechanical strength. The binder that is used most often is calcium sulfate (plaster of Paris). Other binders that can be used are starch, plastic dispersions, and hydrated silicon dioxide. Whichever binder is considered, it is possible for the binder to exhibit adsorptive properties and thereby affect the chromatographic behavior of the thin layer.

Figure 25-1a represents a commercially available mechanical spreading device, while (b) to (f) represent simpler laboratory techniques for spreading the layer. Aside from commercial spreaders, technique (d) is the most common. The thickness of the layer is controlled by the number of tape layers that are used. For preparative scale separations, thick plates up to 2 mm are prepared, while for conventional separations plate thicknesses are about 0.1–0.5 mm. In (b) to (f) the slurry is poured on the plate and a glass rod is used to smear the slurry over the plate. Subsequently, the water is evaporated and the layer, depending on the support used, is activated by drying in the oven.

## EXPERIMENTAL TECHNIQUES

Many of the experimental techniques employed in paper chromatography and thin layer chromatography are the same. The exception is the preparation of the paper or layer. In this section the techniques common to the two methods are discussed.

Once the paper or layer is available, the experimental techniques can be divided into five basic steps. These are pretreatment, sample application, development, visualization, and interpretation of data. Each of these will be briefly discussed.

**Pretreatment.**    The extent of pretreatment will be determined by the type of chromatography (partition or adsorption) and the eventual application. Adsorbents have adsorption sites of varying activity and retain water most strongly. For the fullest impact of adsorption the adsorbent should be at a certain stage of dryness (activated). As a result, the temperature and length of time that layers are dried before use determine the number of active sites available, and consequently, the effectiveness of the plate. Therefore, the activating process (removal of water) must be controlled and the layers must be stored under conditions which maintain the degree of activity. Adsorbents should not be subjected to too high a drying temperature or for too long a heating period since the adsorbents can be chemically altered, which will lead to a modified adsorptive behavior.

Paper techniques, which deal primarily with partition, require introduction of the stationary liquid phase. If this phase is water, its introduction is straightforward since ordinary cellulose paper shows a high affinity for water. Techniques of preparing papers with other stationary phases have already been described. Usually, this is a problem only if a stationary phase other than water is to be used. If needed, the papers can be stored in an atmosphere saturated with the stationary liquid phase.

**Sample Application.**    Before applying the sample, the origin should be marked on the sheet by making a scratch or pencil mark off to the side. The origin, which is where the sample is to be applied, must be far enough from the edge of the sheet so that it does not become submerged in the solvent system used for development.

The type of spotting technique (sample application) is determined by the sample size, and the purpose for the separation. In general, amounts that can be safely separated by adsorption and partition are about 50 mg and 5 mg, respectively, for a 20 × 20 cm layer of 1-mm thickness. These are applied as solutions of very small volumes (1–100 $\mu$l). Since the zones can not be smaller than those originally applied, it is important to keep these as small and well defined as possible; otherwise, resolution is reduced. In some cases,

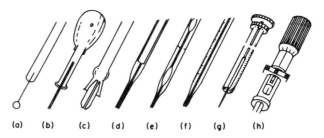

**Fig. 25–2.** Typical devices for sample application in sheet methods. (Taken from E. Stahl, "Thin Layer Chromatography," 2nd ed., Springer-Verlag, New York, 1969).

the sample spot can be increased in concentration by repeated addition of small aliquots of the sample. For this, the sample is dissolved in a volatile solvent and after application, the solvent is vaporized. This process is repeated until the desired amount of sample is applied. Also, each application must be made carefully so that a small well-defined zone is maintained.

Depending on the dimensions of the layer, more than one mixture can be applied. In general, the different samples should be applied at least 1–2 cm apart. Figure 25–2 illustrates several types of devices used for sample application. Design (a) is used primarily for qualitative purposes. If the loop in (a) is 0.4-mm diameter Pt wire and the cross-section diameter is 1.5 mm, a spot of about 10 $\mu$l is obtained. Designs (b) to (h) are used for quantitative as well as qualitative applications. These various micropipets, which come in many sizes, may or may not employ a bulb as shown in (b). Some are calibrated to deliver, while others have a graduated calibration. For the small-sized pipets, they are filled by capillary action and the bulb is used to expel the sample. The pipets in (b), known as "Microcaps" are intended for a single use. Precision capillaries, which are easily replaced and have capacities of 1–100 $\lambda$, are used. An accuracy of 1% is claimed for the Microcaps. Designs (g) and (h) are typical of microsyringes that can be used. (These are also used in gas chromatography.) The syringe must have a high precision and generally is calibrated so that different volumes can be expelled. Syringes with total capacities of 1–100 $\mu$l of suitable precision are readily available.

**Developing Chambers.** Successful separation by sheet methods requires that the chromatography take place in a chamber whose atmosphere is saturated with the mobile phase. This is conveniently carried out in a de-developing chamber. A typical design for a thin-layer and a paper technique is shown in Fig. 25–3a and b.

The actual design depends on whether ascending or descending chromatography is to be used. Ascending development is where the mobile phase passes from a reservoir up the sheet, while descending development is passage

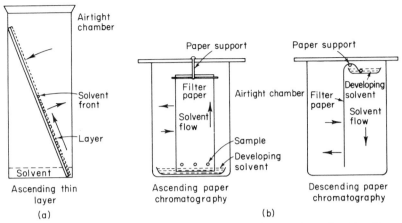

**Fig. 25–3.** Developing chambers for sheet methods. Arrows show the direction of the various solvent movements.

of the mobile phase from a reservoir down the sheet. Because thin layers have physical strength, ascending techniques are preferred for this type of chromatography (Fig. 25–3a). Descending techniques are usually used in paper chromatography (Fig. 25–3b) because of a lack of physical strength in the paper. However, by rolling the paper into a cylinder, an ascending technique can be used.

In ascending chromatography, the mobile phase travels upward in opposition to gravity by capillary action. This results in a slower development in comparison to the descending method in which the mobile phase travels downward assisted by gravity. Also, because of this difference, distance in the flow for the solvent front is controlled only by the length of the sheet in descending methods while in ascending methods the flow is limited by evaporation, and capillary characteristics of the sheet. Often it reaches a static point and will not rise any further.

Another common tank for thin-layer chromatography is known as the sandwich chamber (Fig. 25–4). Elution is still by the ascending technique. It is difficult to predict whether the sandwich or conventional type tank will yield the best results. Because saturation conditions are different, the chromatogram for the same sample will be different for the two types of tanks. Frequently, to ensure saturation within the chamber, filter paper is used to line the chamber walls.

**Development.** The key to reproducibility in sheet methods is the ability to control and reproduce development. It is not enough to just ensure that concentrations in the reservoir are the same. Time and temperature of

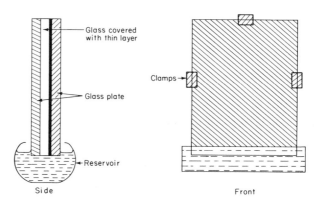

**Fig. 25–4.** A sandwich chamber for thin-layer chromatography. (Courtesy of Kontes Glass Co.)

development must be maintained since any change in the degree of saturation will influence the chromatographic results.

The choice of eluting conditions will depend on whether the separation is based on partition or adsorption. Once this decision has been made, one of several different development methods can be used.

After spotting the sample or series of samples, the separation is effected by ascending or descending techniques as previously outlined. Termination of the elution takes place when the sheet is removed from the tank. In some cases, the flow of the mobile phase is allowed to reach a specific distance. If this is not a predetermined distance, the travel of the solvent front must be marked.

After terminating the elution, the sheet is treated to evaporate the mobile phase. Mild heating in an oven or with a hair dryer are typical techniques for removal of the mobile phase. Subsequently, development can be repeated, followed by termination, mobile phase removal, development, and so on. This technique is known as multiple development. The same eluting condition can be used each time or it may be changed. For example, if the mixture contains solutes widely different in polarity, a weak eluting condition is used first. Those components of the sample which are easily eluted travel the furthest. The next development after drying is with an eluting agent of better eluting power and the elution termination is short of reaching the previous elution. This procedure can be repeated several times.

Two-dimensional development is a special form of multiple development. The sample is applied near one corner of the sheet and during development separation takes place along one side of the sheet. After termination and drying, the sheet is rotated 90° and development, usually with different eluting conditions, is carried out again. This is illustrated in Fig. 25–5.

Radial development is done by spotting the sample in the middle of a

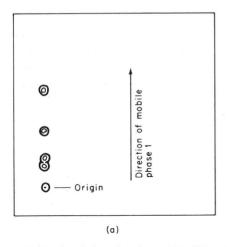

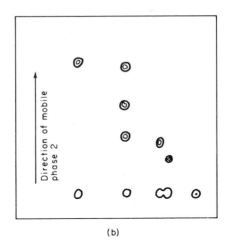

**Fig. 25–5.** A two-dimensional chromatogram for the separation of a six-component mixture. (a) Chromatogram in the first direction; (b) chromatogram in the second direction after rotating the sheet 90° and using a different mobile phase.

sheet. The mobile phase enters the sheet at the location of the spot by means of a wick and travels in a circular pattern from the center to the outer edges. Therefore, the separation appears as a series of concentric rings.

**Visualization.**     Visualization is the process of detecting the spots on the sheet after completion of development. The most direct detection can be based on the visible or fluorescent color of the spot itself. In the latter case an ultraviolet activating lamp is used to scan the sheet.

If neither of these techniques is possible, one of several other common chemical or instrumental techniques is used. It is very common to incorporate phosphors into the thin layer. Thus, wherever there is a spot, the fluorescence of the spot is quenched or enhanced with respect to the background upon scanning with a ultraviolet lamp. Another common technique is to spray the sheet with a fine mist of a solution which undergoes a reaction with the solutes on the sheet to produce colored or fluorescent derivatives. For example, amino acids are readily detected by reaction with ninhydrin (**I**).

Several other common spray solutions are listed in Table 25–2. Several of

**Table 25–2.   Several Typical Spray Reagents**

| Reagent | Types of compounds | Reagent | Types of compounds |
|---|---|---|---|
| Anisaldehyde in H$_2$SO$_4$ and HOAc | Carbohydrates | Dragendorff's reagent | Alkaloids and organic bases in general |
| Antimony trichloride in CHCl$_3$ | Steroids, steroid glycosides, aliphatic lipids, vitamin A | Ferric chloride | Phenols |
| | | Fluorescein–Br$_2$ | Unsaturated compounds |
| Bromcresol purple | a. Halogen ions, except F$^-$ | 8-Hydroxyquinoline–NH$_3$ | Inorganic cations |
| | b. Dicarboxylic acids | Ninhydrin | a. Amino acids |
| Bromcresol green | Carboxylic acids | | b. Aminophosphatides |
| 2-4-Dinitrophenylhydrazine (2,4-DNPH) | Aldehydes and ketones | | c. Amino sugars |
| | | Silver nitrate–NH$_4$OH–fluorescein | Halides |

the solutions (for example ninhydrin, acid–base indicators) are commercially available in aerosol spray cans.

If the solutes are radioactive or can be made radioactive, scanning with radioactive-sensitive instruments can be used to locate the zones. This is a common detection technique when working with trace quantities of biological compounds. Often it is necessary to tag the solutes with a radioactive atom.

After the spots are visualized, $R_f$ values are usually calculated immediately, since fading is possible. This is particularly true when chemical visualization is used. The zone is encircled by a pencil (scratched on a thin layer) and the

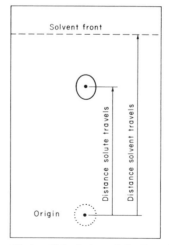

**Fig. 25–6.**   Calculation of the $R_f$ value.

sheet is photographed, or carefully copied into the research notebook. The distance the zone and solvent front travel are measured as shown in Fig. 25–6 and the $R_f$ value is calculated by Eq. (22-9).

Evaluation of $R_f$ values can be used for qualitative purposes by comparison to $R_f$'s for known compounds. Since $R_f$ is very sensitive to conditions of sheet treatment and development techniques and parameters, these variables must be carefully controlled. If at all possible the unknown and standard should be developed together on the same sheet. Alternatively, the zone can be separated from the sheet and identification can be made by conventional chemical and instrumental tests.

Quantitative results can be obtained by either of two common techniques. In one method the spot is removed from the sheet, the solute is separated from the sheet material, and then diluted to a known volume. Subsequently, a procedure for the determination can be chosen based on the concentration level, physical, and chemical properties of the solute. Unknowns and standards should be run in duplicate on the same sheet in order to minimize errors due to development variations.

The second method involves the measurement of spot area since it has been shown that the square root of the area is directly proportional to the logarithm of the amount of the substance. Spot areas are computed by a planimeter or by transferring the zone area to a piece of graph or photocopy paper. Squares are counted on the graph paper, while the areas are cut out of the photocopy and accurately weighed.

The areas on a visualized chromatogram can also be measured by densitometry (see Chapter 21). The visualized chromatogram is passed through a beam of a densitometer so that the relative size and density of the spots can be recorded. Absorption, fluorescence, and radioactivity scanning are also possible. A typical instrumental tracing for a sheet chromatogram is

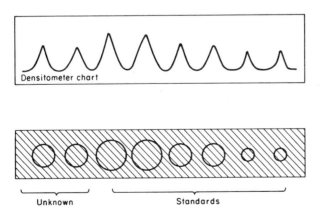

**Fig. 25–7.** An idealized densitometer tracing for the chromatogram. The area covered by the diagonal lines is scanned by the densitometer.

shown in Fig. 25–7. The areas under the curves on the chart are proportional to the amount of the solutes present.

The error is large in quantitative measurements and is determined by many factors including development variations, purity of reagents, and experimental manipulations. In general, the range of accuracy is 3–10%. The better accuracy is usually obtained for the determination when the spot is removed from the layer, while the poorer accuracy is obtained for direct measurements on the sheet.

## APPLICATIONS

Applications of paper and thin-layer techniques are numerous and appear to be endless. For example, a botanist can separate flavanoids that occur in plants. A technician is able to detect trace pesticides in water. A pharmaceutical or biochemical worker can test purity of a natural product or identify metabolites of a drug in animal excretion samples. In a forensic laboratory, the analyst can identify the presence of illicit drugs, compare inks, or identify poisons. The clinical worker provides qualitative and quantitative analyses of biological and metabolic samples which often enables the early detection of some diseases. Perhaps the two main reasons which have contributed to the growth of sheet methods are past success and the availability of reproducible sheets at modest prices from commercial suppliers.

## *Questions*

1. Justify or criticize the statment, "sheet methods are often referred to as open thin-column experiments."
2. List several hydrophobic solvents that are useful as eluting agents in sheet chromatography.
3. List several hydrophilic solvents that are useful as eluting agents in sheet chromatography.
4. Describe how the partitioning process occurs in paper chromatography.
5. Explain why paper is able to hold significant quantities of water.
6. List the factors which influence the $R_f$ value in paper-partition chromatography.
7. List the factors which influence the $R_f$ value in thin-layer partition and adsorption chromatography.
8. What is the purpose of the binder in thin-layer chromatography?
9. Describe a sampling technique that can be used for applying a large sample of a dilute solution.
10. Differentiate between ascending and descending chromatography.
11. Can longer chromatograms be obtained by ascending or descending chromatography? Explain.

12. If the two distances in Fig. 25–6 are 6.5 and 10 cm, calculate the $R_f$ value.
13. Compare the two processes, "development of a chromatogram" and "elution."
14. Explain how a densitometer is used for quantitative measurements in sheet methods.
15. List the factors which will influence the accuracy for quantitative analysis in sheet methods.

# Chapter Twenty-Six Ion Exchange

## INTRODUCTION

Ion exchange methods entail a reversible, stoichiometric exchange between the ions in the mobile liquid phase and the ions on the exchange sites on the stationary phase. It would appear that this method is limited only to the separation of ionic or partially ionic mixtures. Although this has been the major application, more recently, separation of nonionic organic molecules on ion exchangers has been investigated. In general, retention of these kinds of molecules is the result of the highly polar character of the ion exchanger and a careful selection of eluting conditions.

Ion exchangers are usually employed in a method for the separation of micro ($<$1-mg quantities) as well as macro ($>$1-g quantities) samples and are therefore, of practical use in quantitative analysis and purification. Because of the macro application, ion exchangers can also be used on a pilot plant and industrial scale. In general, the basic laboratory techniques common to adsorption and partition column methods are also applied in ion-exchange column methods.

A wide variety of very complex mixtures have been separated on ion exchangers. These include closely related elements such as Zr and Hf, rare earths, transuranium elements, halides, alkali metals, and transition metals. Complex mixtures of organic acids (carboxylic acids, phenols, etc.) and bases (amines, alkaloids, etc.), amino acids, peptides, and proteins have also been separated.

Ion exchangers can be utilized in the laboratory in many other ways: (1) Interfering ions can be removed. (2) Group separations are possible. (3) Samples can be concentrated. (4) Charges on ions can be established. (5) Samples can be purified. (6) Standard solutions can be prepared. (7) Formation constants of complexes can be determined. (8) The ion exchangers can be used as acid and base catalysts. Only a few of these many specialty applications will be considered.

If water is passed through two columns, the first containing a cation exchanger in the hydrogen form and the second an anion exchanger in the hydroxide form, deionized water can be prepared. The reactions, assuming all salts in the water are univalent cations ($M^+$) and anions ($X^-$), are

Cation exchanger: Resin—$SO_3^-H^+$ + MX $\rightleftharpoons$ Resin—$SO_3^-M^+$ + HX

Anion exchanger: Resin—$NR_3^+OH^-$ + HX $\rightleftharpoons$ Resin—$NR_3^+X^-$ + $H_2O$

Water purification for the laboratory using this procedure has two advantages in comparison to distillation. (1) The ion-exchange procedure is faster and cheaper, and (2) it easily satisfies the purity requirements of a scientific laboratory. Unfortunately, the economics of water purification on a large community scale is not presently feasible by the ion-exchange method.

Ion exchangers can be utilized in the preparation of solutions in several ways. For example, ionic salts can be removed from organic reaction mixtures. This often facilitates purification of the organic compounds through crystallization. A solution of a metal chloride or some other anion can be prepared by passing a metal nitrate solution through a column of an anion exchanger charged in the chloride form or anion form of interest. If a weighed amount of KCl is passed through a cation exchanger in the hydrogen form, and the effluent is collected and diluted to a known volume, a standard solution of HCl is prepared. This is possible since the exchange of ions is stoichiometric. Conversion of salts to an acid (or base if an anion exchanger–hydroxide form is used) can also be used for analysis (see Chapter 8). However, it should be apparent that total salt is being determined.

## PROPERTIES OF ION EXCHANGE

Ion exchangers of different origin, structure, and composition will often have different properties. Therefore, it is important to be able to characterize experimentally an ion exchange material. Fortunately, chemical suppliers are able to manufacture several different cation and anion exchangers whose properties are uniform from lot to lot.

Awareness of the properties of ion exchangers facilitates the selection of the correct ion exchanger for the particular problem. The more important properties are color; density; mechanical strength; particle size; capacity; selectivity; amount of crosslinking; swelling; porosity–surface area; and chemical resistance. The significance and methods for the establishment of these properties are described in detail in reference books dealing with ion exchange. Only the more important properties and reference to synthetic ion exchangers will be considered in this book.

Regardless of solvent the exchange of univalent ions may be represented by the following equation:

$$(A^+)_R + (B^+)_S \rightleftharpoons (A^+)_S + (B^+)_R \tag{26-1}$$

where R and S refer to the cation being in the resin phase and solution phase, respectively. Since the reaction reaches an equilibrium position, an equilibrium expression in terms of concentrations (activities would be more precise) can be written

$$K_A^B = \frac{[A^+]_S[B^+]_R}{[A^+]_R[B^+]_S} \tag{26-2}$$

If the experimental variables are held constant and the concentrations of A and B are low, $K_A^B$ is a constant and is an indication of the preference the ion exchanger shows for one ion over another. Upon choosing one ion as a reference, to which the others are compared, a selectivity scale is obtained. A typical scale is shown in Table 26–1.

In using ion exchangers, a successful application requires that the rate of exchange be rapid. Several properties of the ion exchanger will influence the exchange rates. Therefore, the optimum ion exchanger conditions must re-

Table 26–1.  Selectivities for Several Cations and Anions on a Typical Strongly Acidic and Basic Exchangers, Respectively[a]

| Cation | Selectivity[b] | Anion | Selectivity[c] |
|--------|------------|-------|------------|
| $H^+$ | 1.0 | $OH^-$ | 1.0 |
| $Li^+$ | 0.9 | Benzene sulfonate | 500 |
| $Na^+$ | 1.3 | Salicylate | 450 |
| $NH_4^+$ | 1.6 | Citrate | 220 |
| $K^+$ | 1.75 | $I^-$ | 175 |
| $Rb^+$ | 1.9 | Phenolate | 110 |
| $Cs^+$ | 2.0 | $HSO_4^-$ | 85 |
| $Ag^+$ | 6.0 | $NO_3^-$ | 65 |
| $Mn^{2+}$ | 2.2 | $Br^-$ | 50 |
| $Mg^{2+}$, $Fe^{2+}$ | 2.4 | $CN^-$ | 28 |
| $Zn^{2+}$, $Co^{2+}$, $Cu^{2+}$, $Cd^{2+}$, $Ni^{2+}$ | 2.6–2.9 | $Cl^-$ | 22 |
| $Ca^{2+}$ | 3.4 | $HCO_3^-$ | 6.0 |
| $Sr^{2+}$ | 3.85 | $IO_3^-$ | 5.5 |
| $Hg^{2+}$, $Pb^{2+}$ | 5.1–5.4 | Formate | 4.6 |
| $Ba^{2+}$ | 6.15 | $F^-$ | 1.6 |

[a] Taken from J. Inczedy, "Analytical Applications of Ion Exchange," Pergamon, New York, 1966.

[b] Values are relative to hydrogen.

[c] Values are relative to hydroxide.

flect a compromise between these properties which include particle size, mechanical strength, crosslinking, swelling, and porosity–surface area.

The common synthetic ion exchangers are based on a polystyrene type polymer. Polystyrene beads (a linear polymer) would have little mechanical strength and upon adding functional groups to the polymer (sulfonic acid or quaternary ammonium group) solubility is greatly enhanced. If the polymer is crosslinked by the incorporation of divinylbenzene, mechanical strength is imparted to the copolymer and also, it is insoluble in all solvents, even when the highly polar exchange groups are added to the polymer.

The exchange sites are located primarily inside the copolymer bead (see Fig. 22–3) and are available only if the copolymer swells. When the copolymer swells, the polymer chains spread apart which forms narrow passageways throughout the copolymer bead. When an ion exchanger is placed in a polar solvent, swelling occurs, while if it is in a nonpolar solvent, contraction occurs. The electrolyte concentration will also influence the degree of swelling.

Crosslinking provides mechanical strength and retards ion exchanger swelling. However, if the internal portion of the ion exchanger is not available (poor swelling), the exchange (amount and rate) is greatly reduced. An optimum condition for analytical work is to incorporate 8% crosslinking into the copolymer. This represents a compromise in the mechanical strength and swelling properties.

Particle size and porosity–surface area contribute directly to rate of exchange. Small particles and large surface area will increase the rate. The particle size that is used will depend on the column length, diameter, and whether gravity or pumping is used to control the flow of the mobile phase. In general, most laboratory applications are successful with 100–200 mesh (0.147–0.074 mm) ion-exchange particles.

Although the term porosity is used with the kinds of ion exchangers described here, the ion exchangers are not truly porous supports like alumina and zeolites. The small openings are a function of swelling and are not permanent channels or cavities.

Total capacity of an ion exchanger is a measure of the total amount of exchangeable ions expressed per unit weight of dry ion exchanger (mmole/g) or water swollen ion exchanger (mmole/ml). It is determined by taking a weighed sample of the ion exchanger, placing it in a column and passing a solution of KCl through the column in large excess. For the cation exchanger in the H form, the effluent will be acidic which can be titrated with standard base. For the anion exchanger, the solution will be basic, if the exchanger is in the OH form, and the effluent is titrated with standard acid. If the exchanger is in the Cl form, a $KNO_3$ solution is used and the exchanged $Cl^-$ is determined by a $Ag^+$ titration. From the titration data and weight of the exchanger, the capacity can be calculated. Capacities for the more common cation and anion exchanger are about 5.0 and 3.5 mmole/g, respectively. These capacities are considerably larger than capacities for naturally occurring ion exchangers and is one of the main reasons why synthetic ion exchangers are preferred for most applications.

In general, ion exchangers are stable toward strong acids at all concentrations, strong bases, and almost all organic solvents. Depending on the form of the exchanger, cation exchangers (strong acid) are stable to about 150°C, while anion exchangers (strong base) are stable to about 70°C. Anion exchangers are never stored in the hydroxide form because they rapidly decompose, particularly if they are dry; storage is usually in the Cl form.

Because of the differences in rates of exchange and the fact that they may vary significantly for different kinds of separations, flow rates in column experiments must be controlled. In general, flow rates are in the range of 0.5 ml/minute to 5 ml/minute.

## BASIS FOR SEPARATIONS

Separations of inorganic mixtures are based on one of three different principles.

1. At low concentration the extent of exchange increases with increasing valency of the exchanging ion.

2. At low concentration and constant valence the extent of exchange increases with increasing atomic number.

3. The extent of exchange is strongly influenced by the formation of complexes.

Although the first two are useful, the major approach to the separation of inorganic mixtures, particularly for metal ions, is through the formation of complexes. Furthermore, separation due to charge and size are greatly influenced by concentration and temperature. As these parameters increase, the observed exchanges are often different than the expected trends.

Figure 26–1 illustrates several separations based on size. It should be noticed that the order of appearance of the components of the mixture in the effluent is what would be predicted from selectivity experiments (see Table 26–1).

**Use of Complexing Agents.**   Although many different inorganic and organic complexing agents are useful as eluting agents for the separation of metal ions, the one ligand that has been extensively studied and has wide utility is chloride ion. The reaction between a metal ion and chloride ion is a stepwise formation of a series of complexes and can be represented by

$$M^{n+} + Cl^- \rightleftharpoons MCl^{n-1}, MCl_2^{n-2}, \text{etc.} \tag{26-3}$$

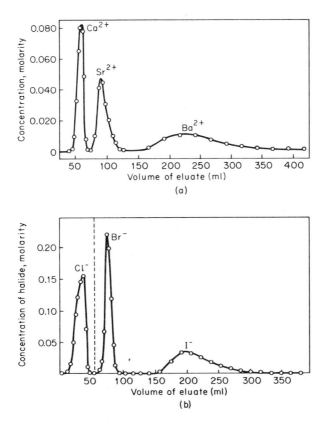

**Fig. 26–1.** Separation based on size. (a) Colloidal Dowex-50, 19 cm × 2.5 cm²; 1.2 $M$ ammonium lactate, 0.56 cm/minute. (b) Dowex 1-× 10,100–200 mesh, 6.7 cm × 3.4 cm²; 55 ml 0.50 $M$ NaNO₃, then 2.0 $M$ NaNO₂; 1.0 cm/minute. (Taken from R. Kunin, "Symposium Ion Exchange and Chromatography," No. 195, American Society for Testing Materials, Philadelphia, 1958.)

The complexes that are present are determined by the concentration of chloride ion in the solution where hydrochloric acid is used as the source of chloride ion. In general, it can be considered that the cation (metal ion) is converted to an anion (metal–chloro complex). Therefore, the separation is one of separation of anions from cations since at different chloride ion concentration the metal ions are there as cations or complexes.

It is possible to determine the appropriate eluting conditions through a trial and error process whereby each metal ion is placed on a column of the ion exchanger and different HCl concentrations are used as the eluting agent. Thus, the HCl concentration required to keep the metal ion on the ion exchanger and that required to take it off can be established.

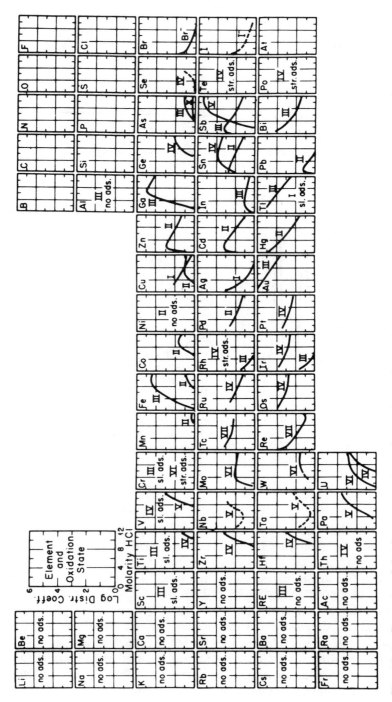

**Fig. 26–2.** Retention of the elements on a strong base anion exchanger from hydrochloric acid. no ads., no adsorption for $0.1 < M$ HCl $< 12$; sl. ads., slight adsorption in $12\ M$ HCl ($0.3 \le D_v \le 1$); str. ads., strong adsorption $D_v \gg 1$. (Taken from K. A. Kraus and F. Nelson, "Symposium on Ion Exchange in Analytical Chemistry," No. 195, American Society for Testing Materials, Philadelphia, 1958.)

However, a more fruitful approach is to determine distribution coefficients ($K_D$) for the metal ions as a function of HCl concentration by one of several batch or column procedures. The most widely used method is a batch method and involves equilibrating a known weight of ion exchanger with a known volume of a standard solution of the metal ion. Subsequently, the amount of metal ion remaining in the solution or on the ion exchanger is determined. From these data the $K_D$ can be calculated [see Eq. (22-8)].

One system which has been extensively studied is the separation of metal ions on strong base anion exchangers in the chloride form using hydrochloric acid. Distribution coefficients were measured for all metal ions and several

**Table 26–2. Approximation Elution Order for Metals Using Hydrochloric Acid[a,b]**

| $M$ HCl | Element and oxidation state |
|---|---|
| 12 | (A) "Nonadsorbables": |
| | Alkali metals |
| | Alkaline earths |
| | Rare earths(III), Al(III), Ni(II), Y(III), Tl(I), Ac(III), Th(IV) |
| | (B) Slightly adsorbed: |
| | $D_v \approx 1$: Sc(III), Ti(III), V(IV), Cr(III), As(V) |
| | $D_v = 1$ to 3: Mn(II), In(III) |
| | (C) Slightly or not absorbed, readsorbable at low-$M$ HCl: Cu(I), Rh(III), Ag(I), Ir(III), Pb(II) |
| 9.5 | Ti(IV) |
| 8 | Hf(IV) (Tends to tail) |
| 7.5 | Zr(IV) (Tends to tail) |
| 7 | Fe(II), V(V) |
| 6 | U(IV) |
| 5.5 | Ge(IV) |
| 4.5 | Co(II), As(III) |
| 3 | Cu(II) (May elute early) |
| 1.5 | Ga(III) |
| | Sb(V) (Slow to remove) |
| 1 | Pa(V) (Tends to tail) |
| | U(VI) |
| | Mo(VI) (Tends to tail, may elute in more than one band) |
| | Fe(III) (Tends to tail) |
| 0.25 | In(III) (May elute early) |
| 0.02 | Zn(II) |
| 0.001 | Cd(II) |

[a] Taken from K. A. Kraus and F. Nelson, "Symposium on Ion Exchange in Analytical Chemistry," No. 195, American Society for Testing Materials, Philadelphia, 1958.

[b] *Note.* The following ions are difficult to remove in chloride media:

(a) Strong chloride complexes: Ru(IV), Rh(IV), Pd(II), Pd(IV)(?), Sn(II), Sb(III), Os(IV), Ir(IV), Pt(II)(?), Au(III), Hg(II), Tl(III), Bi(III).

(b) Oxyanions: Cr(VI), Tc(VII), Re(VII).

(c) Hydrolyzable ions: Nb(V), Ta(V), W(VI), Sn(IV).

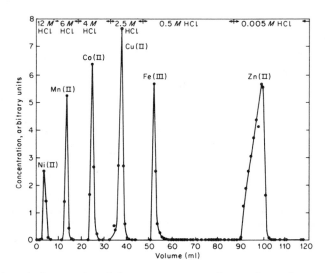

**Fig. 26–3.** A typical separation of elements on a strong base anion exchanger using hydrochloric acid. (Taken from K. A. Kraus and F. Nelson, "Symposium on Ion Exchange in Analytical Chemistry," No. 195, American Society for Testing Materials, Philadelphia, 1958.)

anions as a function of HCl concentration (0.01–12 $F$). These data are reproduced in form of a periodic table in Fig. 26–2.

Prediction of whether a separation for a given pair of metal ions is possible is made by comparing their $K_D$ values. An acid concentration for the sample is chosen such that both metal ions are retained (high $K_D$). The first HCl eluting mixture that is used is one in which one metal has a large $K_D$ while the other is very small. An ideal situation would be where one $K_D$ is greater than 10 while the other approaches zero. The second metal ion is eluted by changing the HCl concentration to a level that produces a very low $K_D$ for that metal ion. If several metal ions are present, more care must be exercised in choosing the HCl concentration and for very complex metal mixtures, group separations are often made.

The data in Fig. 26–2 can be summarized as shown in Table 26–2 where the HCl concentrations listed are the minimum levels of acid needed to elute metal ions. Below this value the metal ions are retained. Several typical chromatograms are shown in Fig. 26–3.

Quantitative separations are easily attained for many mixtures. It is also possible to separate a trace quantity of a metal from a macro amount of another metal ion. Once the metal ions have been separated and collected in separate vessels, any procedure for their determination is suitable providing it is sensitive enough. Therefore, in choosing the method for the determination, little attention is given toward interferences (there should be none except

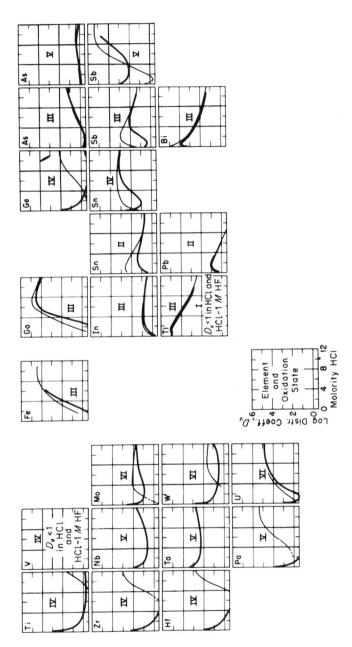

**Fig. 26–4.** Retention of several elements on a strong base anion exchanger from hydrochloric–hydrofluoric acid mixtures. (- - -)Distribution coefficients in absence of HF; (—) distribution coefficients in HCl–HF mixtures [usually 1 $M$ HF except zirconium (IV), hafnium (IV), niobium (V), tantalum (V), and protactinium (V) where $M$ HF = 0.5). (Taken from K. A. Kraus and F. Nelson, "Symposium on Ion Exchange in Analytical Chemistry," No. 195, American Society for Testing Materials, Philadelphia, 1958.)

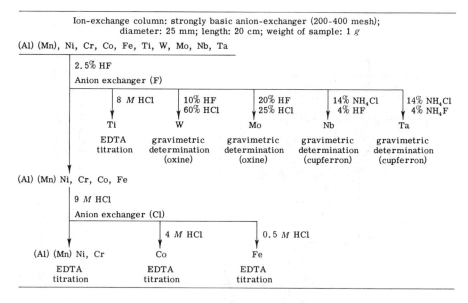

**Fig. 26–5.** A flow diagram for the separation and analysis of a high temperature alloy. (Taken from J. Inczedy, "Analytical Applications of Ion Exchange," Pergamon, New York, 1966.)

chloride ion). The more common quantitative techniques which compliment this separation procedure are spectrophotometry, atomic absorption, emission methods, and EDTA titrations. The latter is particularly useful because of the simplicity, speed, and versatility of the method.

Many other inorganic ligands can be used in a similar way. These include HF, HBr, HI, $HNO_3$, $H_2SO_4$, $Na_2CO_3$, and $H_3PO_4$. Most of these (except the halogen acids) are more selective than HCl because they form complexes with only a few metal ions. Figure 26–4 shows that $K_D$ values for several metal ions using HF–HCl mixtures and an anion exchanger. Figure 26–5 shows a flow diagram and recommended procedure for the analysis and separation of a complex metal mixture utilizing a combination of these eluting conditions.

In the presence of large HCl concentrations, many metal ions are retained by anion exchangers as chloro complexes. If the HCl concentration is reduced, the complexes break up and the metal ions come off the exchanger. The exact opposite behavior should be expected if a cation exchanger (strong acid) is used. That is, high distribution values should be obtained where HCl concentration is low (complexes are not or only partially formed) and low values at large HCl concentration (complexes are formed). This expected result has been experimentally verified and provides an alternate separation scheme for metal ions.

The changes in $K_D$ as a function of HCl concentration are not as large in comparison to values observed with the anion exchanger. The principal reason for this is that the binding of a cation to the anionic exchange site on the cation exchanger is much stronger than the binding of its complex to the anion exchanger. Even the build-up of large concentrations of $H^+$ will not cause a sharp drop in the $K_D$ for multiple charged cations. However, many practical separations are possible.

**Separation of Organic Mixtures.** Organic compounds that have ionic character (acidic or basic) can be separated on anion and cation ion exchangers. Both weak and strong acidic or basic type exchangers are used. However, the separations are not restricted only to organic molecules possessing ionic character. Many nonionic organic compounds are retained by ion exchangers through adsorption, salting out, size, and other exchanger–solute interactions. By proper choice of eluting conditions separations are possible.

In separating organic molecules several other properties must be considered. Organic molecules change in size by a large factor relative to size variations in inorganic ions. Therefore, the porous-like property of resins becomes a more important and determining property for the separation of organic compounds. In fact, size of the organic molecule can be the principle factor which determines the separation.

Since many organic compounds have limited or no solubility in water,

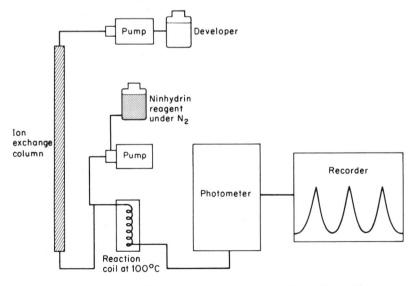

**Fig. 26–6.** A general design for an automatic amino acid analyzer.

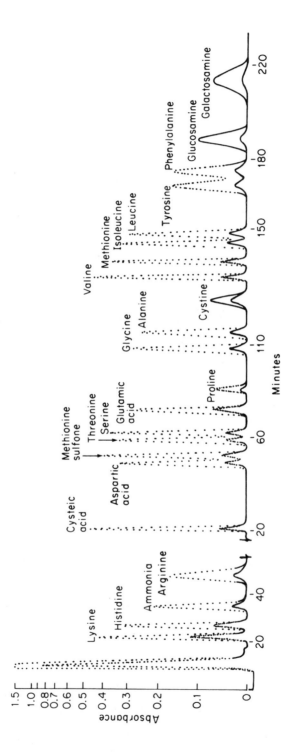

**Fig. 26–7.** A typical chromatogram for the separation of an amino acid mixture on a strongly acidic cation exchanger with an amino acid analyzer. Sample load, 0.25 μmole. Conditions: column temperature, 55°C; flow rate, 6.8 ml/hour. Basic amino acids were separated on a 0.9 × 5 cm column and the acidic and neutral amino acids were separated on a 0.9 × 57 cm column. [Taken from M. M. Benson and M. M. Patterson, *Anal. Chem.* **37**, 1108 (1965).]

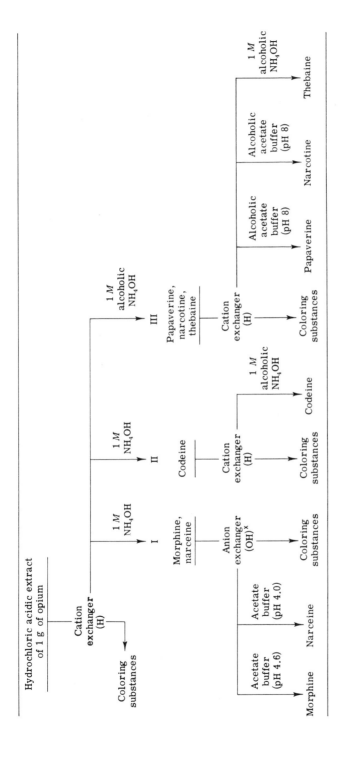

**Fig. 26-8.** A flow diagram for the separation of a mixture of opium alkaloids on cation and anion exchangers.

organic solvents must be used in the eluting mixture. In general, depending on the solvent, rates of exchange (or retention) are slower, degree of resin swelling is less, dissociation in the exchanger is less, and polymer degradation increases. Even the total capacity of the exchanger toward an organic solute may be quite different than found for the exchange capacity for the exchanger. This is possible since retention of the organic molecule may be due to adsorption or some other exchanger–organic molecule interaction as well as to exchange.

One area of considerable interest is the separation of amino acids, peptides, and proteins. These molecules contain ionizable groups which are strongly influenced by pH. Thus,

$$H_3N^+CHCO_2H \underset{low\ pH}{\overset{}{\rightleftharpoons}} \underset{R}{H_3^+NCHCO_2^-} \overset{high\ pH}{\underset{}{\rightleftharpoons}} \underset{R}{H_2NCHCO_2^-}$$

By changing pH of the solution the dipolar ionic character of amino acids, etc. is altered and this principle can be used for their separation on cation exchangers. In fact, ion-exchange chromatography has played an important role in the investigation of amino acids, peptides, and proteins.

In general, the variables in the separation of amino acids are choice of exchanger, pH, concentration of ionic species in the eluting agent, and the type of side chain on the amino acid. Currently, amino acid analysis is done automatically. A typical design is shown in Fig. 26–6. The amino acid mixture is quantitatively applied to the cation exchanger in the column and the eluting agent (citrate buffers) is pumped through the column (this is often a gradient in which pH is gradually increased). The effluent is mixed with

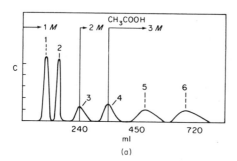
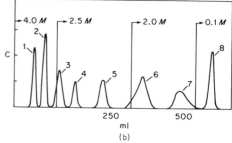

**Fig. 26–9.** Examples of separations of organic molecules on ion-exchangers. (a) Separation of alcohols on cation resin by the technique known as solubilization chromatography. (1) *tert*-butyl alcohol; (2) *n*-amyl alcohol; (3) *n*-hexyl alcohol; (4) *n*-heptyl alcohol; (5) *n*-octyl alcohol; (6) *n*-nonyl alcohol. (b) Separation of alcohols on anion resin by the technique known as salting-out chromatography. (1) Glycerine; (2) methyl alcohol; (3) propylene glycol; (4) ethyl alcohol; (5) isopropyl alcohol; (6) *tert*-butyl alcohol; (7) *sec*-butyl alcohol; (8) *n*-butyl alcohol. (Taken from J. Inczedy, "Analytical Applications of Ion Exchangers," Pergamon, New York, 1966.)

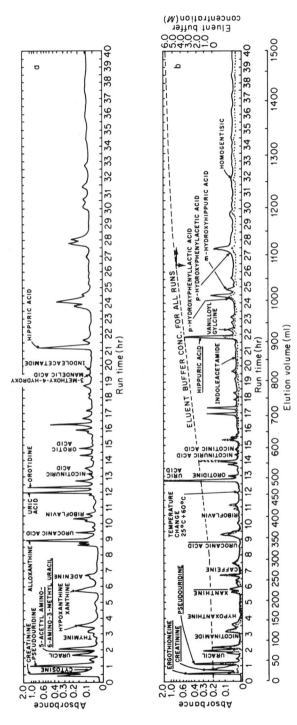

**Fig. 26–10.** Chromatograms for the separation of the components of urine with a high pressure column of anion exchange resin. (a) 2.0 ml urine of 20-year-old youth of family with hereditary nephritis; absorbance at 260 nm. (b) 0.5 cm³ urine, a composite from eight normal subjects; (- -) absorbance at 260 nm; ( . . . ) absorbance at 280 nm. Some identified peaks are labeled. [Taken from R. L. Jolley and C. D. Scott, *Clin. Chem.* **16,** 687 (1970).]

ninhydrin in a reaction chamber and passed through a spectrophotometer. When an amino acid emerges from the column, the ninhydrin reaction takes place and absorption is observed. A recording trace of absorbance vs ml of eluting agent provides the chromatogram.

An actual chromatogram for the separation of amino acids is shown in Fig. 26–7. Flow rates are usually about 0.5 ml/minute, column temperature is maintained at 50°C, and complete separation and analysis is possible in as little as 4 hours. By careful control of the elution conditions and selection of the spectrophotometer and its electronics as little as $10^{-9}$ mole (1 nanomole) of an amino acid can be quantitatively determined. In fact the amount of amino acids left by a single fingerprint on a beaker have been estimated.

For analysis of peptides or proteins, they are first hydrolyzed and the amino acids are separated, identified, and determined. These data do not, however, allow the prediction of the actual sequence of the amino acids in the peptide or protein. Other studies must be used for this purpose. In general, the technique used in the separation of amino acids can be used for the separation of peptide and protein mixtures.

Many different organic mixtures can be separated on ion exchange resins. Several examples are illustrated in Figs. 26–8 and 26–9. Columns of ion exchangers can also be used in a high pressure technique. One application combining the power of ion exchange in combination with high pressure elution is shown in Fig. 26–10 where a chromatogram for the separation of an urine sample of a normal patient is compared to one having nephritis. At the time of this investigation over 100 different peaks were found in the urine sample. Many of these were not known to be present before this separation. Subsequently, many have been identified and the possibility exists that the chromatogram can be used for medical diagnostic purposes.

## Questions

1. What property of the common strong acid and strong base ion exchangers accounts for their insolubility in water?
2. Why are ion-exchange beads always put in a column as a slurry rather than by dry packing?
3. Suggest an experimental procedure which can be used to determine which ion, $H^+$ or $Ag^+$, is preferred by a strong acid cation exchanger.
4. What species will be present in the effluent if the following solutions are passed through a strong acid ion exchanger in the K-form.
   a. A solution of NaCl
   b. A solution of $NaH_2PO_4$
   c. A solution containing HCl and $H_2SO_4$

5. Can ion exchangers be used in a batch operation in analytical chemistry? Give examples.

6. Forty (40) ml of a 0.1500 $F$ KCl solution was passed through a column of a strong acid exchanger in the H form and the effluent collected and diluted to volume in a 250 ml volumetric flask. Write the reaction(s) that take place and calculate the concentration of the species in the volumetric flask.

7. Thirty (30) ml of a 0.2000 $F$ $Na_3PO_4$ solution is passed through a column of strong base exchanger in the Cl form and the effluent is collected and diluted to volume in a 200-ml volumetric flask. Write the reaction(s) that take place and calculate the concentration of the species in the flask.

8. If a strong acid ion exchanger has an exchange capacity of 4.70 mmole/g of resin, calculate the number of milligrams of $Na^+$, of $Mg^{2+}$ that can be taken up by 2.250 g of resin.

9. Alkali metal ions can be separated on a cation resin using 0.1 $F$ HCl as eluting agent. If the mixture contains the group from $Li^+$ to $Cs^+$, predict the order of elution.

10. Suggest the HCl eluting agents that would be used to separate the following mixtures of Cl-form anion exchanger.

    a. Fe(III)–Co(II)

    b. Co(II)–Ni(II)

    c. Mg(II)–Cd(II)–Zn(II)–Co(II)–Cu(II)

    d. $UO_2$(II)–Th(IV)

11. If a peptide is hydrolyzed and each amino acid is identified and determined, why is it that this information can not be used to predict the amino acid sequence in the peptide?

12. Suggest an experimental procedure for the determination of the distribution coefficient for Cu (II) on an anion exchanger (Cl form) at 6.0, 1.0, and 0.1 $F$ HCl.

# Chapter
# Twenty-Seven
# Solvent Extraction

## INTRODUCTION

Solvent extraction or liquid–liquid extraction is a partitioning process where a solute distributes itself between two immiscible phases. In general, the basic process is the same as in partition chromatography; the difference lies in the experimental technique of performing the operation in the laboratory. Historically, solvent extraction came first and awareness of this kind of system certainly contributed to the early understanding of partition chromatography.

Solvent extraction is a standard method of purification for organic molecules and can be used for quantitative separations for both inorganic and organic systems. Only inorganic systems will be considered in this chapter.

Although some of the newer chromatographic techniques provide an alternate method for a particular separation, solvent extraction has still retained much of its popularity as a versatile separation method. Probably the main reason for this is that solvent extraction is easily performed, simple in methodology, reproducible, fast, and versatile. For example, the required apparatus can be as simple as a separatory funnel. In addition, the method is applicable to trace as well as macro levels and often provides very clean and complete separations.

## DISTRIBUTION LAW

The distribution law states that a solute will distribute itself between two immiscible liquid phases so that at equilibrium the ratio of the concentrations of the solute in the two phases at a particular temperature will be a constant, provided the solute has the same molecular weight in each phase. The equilibrium across the interface and through the bulk of the two phases is il-

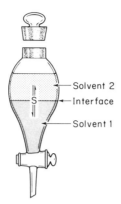

**Fig. 27–1.**   Distribution of solute S across an interface between two solvent phases.

lustrated in Fig. 27–1. For a solute S distributing between solvents 1 and 2, the equation for the distribution is

$$K = \frac{[S_2]}{[S_1]} \tag{27-1}$$

where $K$ is the partition coefficient (distribution coefficient) for a given temperature, independent of concentration, and expressed as equilibrium concentrations in the two phases. [The similarity between Eq. (27-1) and the chromatographic partition coefficient (Chapter 22) should be noted.]

The distribution law is very useful and frequently describes the actual experimental observation. However, it is not exact because it does not account for any equilibrium steps that might occur in either of the two phases. For this reason the distribution law is redefined in terms of a distribution ratio, $D$, where concentrations in the two phases are expressed as total concentrations rather than equilibrium concentrations.

$$D = \frac{\text{total concentration in solvent 2}}{\text{total concentration in solvent 1}} \tag{27-2}$$

This means that all species containing the one of interest are included (formal concentration) in solvent 2 and solvent 1.

The percent extracted, $\%E$, can be related to $D$ through the equation

$$\%E = \frac{100D}{D + (v_1/v_2)} \tag{27-3a}$$

where $v_1$ and $v_2$ are volumes of solvents 1 and 2. If the volumes are equal, Eq. (27-3a) simplifies to

$$\%E = \frac{100D}{D + 1} \tag{27-3b}$$

Percent extraction is a useful term for representing experimental data. For example, it is customary to plot $\%E$ vs the experimental parameters (pH, extractant concentration, etc.). Representation of the data in this way facilitates selection of the optimum conditions for the separation.

In quantitative extractions one solvent is usually water and the other is organic. Therefore, in the general equations, Eqs. (27-1) to (27-3), solvent 1 is water (w) and solvent 2 is the organic phase (o) and these equations become

$$K = \frac{[S_o]}{[S_w]}; \quad D = \frac{\text{total concentration}_o}{\text{total concentration}_w}; \quad \%E = \frac{100D}{D + (v_w/v_o)}$$

where the point of interest is extraction of the solute into the organic layer.

From Eq. (27-3), it can be seen that as the $\%E$ approaches 100%, $D$ approaches infinity as a limit. Furthermore, if the $\%E$ is in the range 99 to 100%, $D$ must be in the range 99 to infinity. Consequently, as $D$ approaches 100 and using a single extraction of equal volumes, the extraction can be considered to be quantitative.

Quantitative extraction is not restricted only to systems where $D \rightarrow 100$ or larger. A convenient technique is to perform the extraction several individual times or to do it continuously. (In principle, continuous extraction is similar to column partition chromatography.) Only the batch technique will be considered mathematically in this chapter.

Assume that $v_w$ ml of an aqueous solution containing $x$ g of solute is extracted with $v_o$ ml of an immiscible organic solvent. After equilibrium is established in the first extraction, $x_w^1$ g are in the aqueous phase and $x_o$ g are in the organic phase, or

$$\text{Concentration in aqueous phase} = \frac{x_w^1}{v_w}$$

$$\text{Concentration in organic phase} = \frac{(x - x_w^1)}{v_o} = \frac{x_o}{v_o}$$

Therefore,

$$D = \frac{c_o}{c_w} = \frac{x - x_w^1}{v_o} \bigg/ \frac{x_w^1}{v_w}$$

and

$$x_w^1 = x \left( \frac{v_w}{Dv_o + v_w} \right)$$

If another extraction of the aqueous layer is made with $v_o$ of solvent, $x_w^2$ will be the weight of solute remaining in the aqueous layer. Therefore, after the second extraction

$$x_w^2 = x_w^1 \left( \frac{v_w}{Dv_o + v_w} \right)$$

However, substitution for $x_w^1$ gives

$$x_w^2 = x \left( \frac{v_w}{Dv_o + v_w} \right)^2$$

If the same volume for the two immiscible phases are used in $n$ successive extractions, the weight of the solute after $n$ extractions, $x_n$, remaining in the aqueous layer is given by

$$x_n = x \left( \frac{v_w}{Dv_o + v_w} \right)^n \tag{27-4}$$

Favorable extraction is accomplished, therefore, by keeping $v_o$ small and performing repeated extractions. In other words, many extractions with small volumes of the extracting phase provide more complete extraction than one extraction with a large volume of the extracting phase.

# TYPES OF INORGANIC EXTRACTION SYSTEMS

For convenience, it is possible to classify extraction systems on the basis of the extractable species. These are chelate systems and ion-association systems. Furthermore, the classifications are for systems in which the extractable species is distributed between an organic solvent and water phase.

**Chelate System.** Extractable species in the chelate group are chelates formed between ligands and metal ions that are neutral in charge. These are often very soluble in organic solvents such as hydrocarbons and chlorinated hydrocarbons. The solubility in the organic solvent is usually not a limiting factor but rather, it is the solubility of the metal–chelate species in water. Because of this low solubility, the extraction is often limited to low concentrations of the metal ion.

The selectivity of the extraction varies widely and often parallels the selectivity of the chelating agent. For example, dimethylglyoxime is highly selective for the precipitation of Ni(II). Similarly, the Ni–DMG complex is selectively extracted from a slighly basic citrate solution with chloroform. Almost all other cations are not extracted.

In contrast, 8-hydroxyquinoline and diethyldithiocarbamate sodium salt will precipitate over 25 different metals. These precipitates can be extracted into $CHCl_3$. By control of pH and introduction of masking agents, some

Table 27–1.   Metals Extracted by 8-Hydroxyquinoline
and Diethyldithiocarbamate[a,b]

8-Hydroxyquinoline

| 1 | H | | | | | | | | | | | | | | | | | He |
|---|---|---|---|---|---|---|---|---|---|---|---|---|---|---|---|---|---|---|
| 2 | Li | Be | | | | | | | | | | | B | C | N | O | F | Ne |
| 3 | Na | Mg | | | | | | | | | | | Al | Si | P | S | Cl | A |
| 4 | K | Ca | Sc | Ti | V | Cr | Mn | Fe | Co | Ni | Cu | Zn | Ga | Ge | As | Se | Br | Kr |
| 5 | Rb | Sr | Y | Zr | Nb | Mo | Tc | Ru | Rh | Pd | Ag | Cd | In | Sn | Sb | Te | I | Xe |
| 6 | Cs | Ba | La* | Hf | Ta | W | Re | Os | Ir | Pt | Au | Hg | Tl | Pb | Bi | Po | At | Rn |
| 7 | Fr | Ra | Ac** | | | | | | | | | | | | | | | |

| | Ce* | Pr | Nd | Pm | Sm | Eu | Gd | Tb | Dy | Ho | Er | Tm | Yb | Lu |
|---|---|---|---|---|---|---|---|---|---|---|---|---|---|---|
| | Th** | Pa | U | Np | Pu | Am | Cm | Bk | Cf | Es | Fm | Mv | No | Lw |

Diethyldithiocarbamate

| 1 | H | | | | | | | | | | | | | | | | | He |
|---|---|---|---|---|---|---|---|---|---|---|---|---|---|---|---|---|---|---|
| 2 | Li | Be | | | | | | | | | | | B | C | N | O | F | Ne |
| 3 | Na | Mg | | | | | | | | | | | Al | Si | P | S | Cl | A |
| 4 | K | Ca | Sc | Ti | V | Cr | Mn | Fe | Co | Ni | Cu | Zn | Ga | Ge | As | Se | Br | Kr |
| 5 | Rb | Sr | Y | Zr | Nb | Mo | Tc | Ru | Rh | Pd | Ag | Cd | In | Sn | Sb | Te | I | Xe |
| 6 | Cs | Ba | La* | Hf | Ta | W | Re | Os | Ir | Pt | Au | Hg | Tl | Pb | Bi | Po | At | Rn |
| 7 | Fr | Ra | Ac** | | | | | | | | | | | | | | | |

| | Ce* | Pr | Nd | Pm | Sm | Eu | Gd | Tb | Dy | Ho | Er | Tm | Yb | Lu |
|---|---|---|---|---|---|---|---|---|---|---|---|---|---|---|
| | Th** | Pa | U | Np | Pu | Am | Cm | Bk | Cf | Es | Fm | Mv | No | Lw |

[a] Taken from R. Belcher and C. L. Wilson, "New Methods of Analytical Chemistry," 2nd. ed., © 1964 by R. Belcher and C. L. Wilson. Reprinted by permission of Van Nostrand Reinhold Company.

[b] Shaded area, complete extraction.

**Table 27–2.   Examples of Chelates Useful in Extraction**

| | |
|---|---|
| Acetylacetone | Sodiumdiethyldithiocarbamate |
| 8-Hydroxyquinoline | Potassium xanthate |
| Dimethylglyoxime | Quinalizarin |
| *N*-Nitrosophenylhydroxylamine ammonium salt (cup- | Salicylaldoxime |
|    ferron) | 1-Nitroso-2-naphthol |
| Diphenylthiocarbazone | |
| Toluene-3,4-dithiol | |

selectivity can be incorporated into the extraction. Table 27–1 lists the metals completely extracted by these two chelates, while Table 27–2 lists several other common chelates useful in extraction.

**Ion-Association Systems.**     The principal characteristic of this group is that extractable species are formed by the association of ions. Three general types can be suggested. First, the metal ion may be associated with a large sized counter ion or be part of a charged chelate containing large bulky organic groups associated with a counter ion. Examples of these two are compound **I**, which is extractable into xylene, and compound **II**, which is extractable into chloroform, respectively.

$$[(C_6H_5CH_2)_3NH^+]_2(ZnCl_4^{2-})$$

**I**

**II**

Compound **I** can be used for the extraction of zinc ion, while **II** is useful for the extraction of nitrate or perchlorate ion.

The second type of ion-association system is a coordination type compound formed between metal ions and anions such as halide, thiocyanate, and nitrate ions in which the extracting solvent plays a vital role in the coordination sphere of the metal being extracted. Usually, the organic solvent used is an oxygen containing solvent (ether, ketone, alcohol, or ester) which is immiscible with water. Since the solvent is part of the extracted species, it is often called an *oxonium* extraction system. A typical example is the extraction of Fe(III) from HCl solution into diethyl ether. The extracted species has the formula

$$\{[(C_2H_5)_2OH]^+\}\{(FeCl_4)[(C_2H_5)_2O]_2^-\}$$

**Table 27–3.  Examples of Ligands Useful in Ion Association Extraction**

| | |
|---|---|
| Fluoride | Carboxylic acids |
| Chloride | High molecular weight amines |
| Bromide | Tetraphenylarsonium salts |
| Iodide | Tetraphenylphosphonium salts |
| Nitrate | Heterocyclic polyamines |
| Thiocyanate | |
| Organo phosphorous acids, esters, and oxides | |

The third type of ion-association system includes extractable species that are salts formed between the ion of interest and a counterion of large molecular weight. They dissolve in organic solvents because of the formation of collodial aggregates or micelles. A typical example would be a metal ion salt of a fatty acid which can be extracted into $CHCl_3$.

Additional examples of these three general types of ion association extraction systems are shown in Table 27–3.

## EXTRACTION TECHNIQUES AND VARIABLES

The simplest type of extraction system for quantitative purposes is a batch extraction. Continuous extraction techniques are also available; however, they will not be discussed in this book.

In a batch extraction, a given volume of a solution containing the ion of interest is brought in contact with a given volume of the organic solvent in a pear-shaped separatory funnel. Adjustment of the experimental conditions in the aqueous layer will determine the extent of extraction. After equilibrium is reached, which is usually hastened by vigorous agitation, the two layers are separated by draining off one layer at a time.

There are many variations of the basic pear-type separatory funnel. For example, micro as well as macro funnels are available. Some are designed to drain off the top layer rather than the bottom layer. This type is useful for systems where the organic layer is less dense than the aqueous layer and therefore, appears at the top. The particular design that is selected will depend on the application and personal preference. However, in general, the ordinary pear-shaped separatory funnel is used most often.

**Extraction Variables.**    Since extraction involves first the conversion of the species into a chelate or ion-association system, experimental variables in the aqueous layer dealing with this are of vital importance. These variables have been discussed briefly in Chapter 15. In solvent extraction applications, the most important ones are control of pH and concentration of the ligand.

Of equal importance is the choice of the extracting solvent. For quantitative purposes, the solute must have a large distribution value in the particular solvent, while the distribution values for the other components in the mixture must be very low. In choosing a solvent, the generalization of "like dissolves like" is used. Therefore, the properties of the ion-association system or chelate being extracted are considered (see Chapter 15).

Several other properties of the solvent must be considered. These are:

1. Ease of recovery of the solute from the organic solvent. Usual procedures are boiling or stripping. The latter refers to a back extraction of the extracted species from the organic layer into a new aqueous layer.
2. Degree of miscibility of the two phases.
3. Specific gravities (densities) of the two phases.
4. Tendency to form emulsions.
5. Toxicity and flammability.
6. The possibility of using a mixed organic solvent system.

## SUMMARY

It is beyond the scope of this chapter to suggest the many types of useful, quantitative extractions that are possible. Some specific examples have already been cited (Tables 27–1 to 27–3).

In the extraction procedure the species of interest is extracted from the remaining components in the aqueous solution into an organic phase. Alternatively, the interferences are extracted into the organic phase and the species of interest is retained in the aqueous layer. Therefore, the entire approach should be one to use experimental conditions in which a favorable separation factor is possible. That is, the ratio of the individual partition coefficients (see Chapter 22) should be very large or the species of interest should have a large $D$ while the interfering species should have a very small $D$ value.

The general practical value of the extraction method, like any other separation method, is that once the interferences are removed the eventual method chosen to complete the analysis can be based on the concentration level of the species present rather than on the types of interferences present. For example, the extraction technique in combination with the EDTA titration is a selective, fast, powerful, accurate general method for the analysis of metal ions.

In addition to the selectivity provided by the isolation of a metal ion as a chelate or ion-association species into an organic layer, the presence of the organic solvent can enhance the sensitivity of the determination of the metal ion. For example, many of the ligands useful for extraction are also useful as spectrophotometric reagents for the metal ions. Having the metal ion as the chelate or ion-associated species in the organic layer often results in a large molar absorptivity and therefore an increased sensitivity. In emission methods, the organic layer containing the metal–ligand species can be directly introduced into the source. Often, the presence of the organic solvent, in comparison to water, leads to an increase in the sensitivity by a factor of 10 to 100 times.

## Questions

1. Compare the differences and similarities between partition chromatography and solvent extraction.
2. Why are metal chelates more likely to be extracted than inorganic salts by organic solvents?
3. Explain why it is preferred to have the extracted species in the more dense phase in a solvent extraction.
4. Suggest experimental and chemical reasons that will affect the quantitative aspects of an extraction.
5. Many metal ions are extracted into organic solvents as the 8-hydroxyquinoline complex. If the general reaction is

$$2 \quad \text{(quinolin-8-ol)} \; + \; M^{2+} \; \rightleftharpoons \; \text{(metal complex)} \; M \; + \; 2\,H^+$$

   explain why the extraction is dependent on pH.
6. Suggest a procedure that might be used to carry out an extraction continuously.
7. A frequent procedure is to extract a trace metal ion from a complex mixture into an organic phase and introduce the organic phase directly into the flame in a flame photometer. Often, the sensitivity is greater for the metal ion in the organic solvent than if it were in water. Explain this observation.
8. Calculate the distribution coefficient for iodine between $CCl_4$ and water, if 98.5% of the iodine in 50 ml of water is extracted into 50 ml of $CCl_4$.
9. A metal complex has a distribution coefficient of 3.94 for its extraction from water into methyl isobutyl ketone. If the aqueous and ketone solvents are used in 50-ml quantities, calculate the number of extractions that are required to remove 99.9% of the metal from the aqueous solution.

# Chapter
# Twenty-Eight
# Introduction to
# Electrochemistry

## INTRODUCTION

Electroanalytical techniques can be grouped into two main categories. These are voltammetric methods at zero current (potentiometry) and voltammetric methods at finite current (often referred to as voltammetry). In potentiometry, which was discussed in detail in Chapters 10, 11, and 13, potentials are measured while no significant amount of current is allowed to flow. In voltammetry, which is briefly surveyed in this chapter, current is permitted to flow and electrolysis takes place in the electrochemical cell. Electrolysis can be described as a process whereby solution components are converted from one oxidation state to another at an electrode–solution surface by means of a current flow.

In an electrolysis system the electrode that receives electrons from the external driving force and transmits these electrons to the reactant in solution is the cathode; its surface is where reduction occurs. The other electrode receives electrons from the solution and is the site of oxidation. This electrode is the anode.

## THE ELECTROLYTIC CELL

Potentiometry and the application of the galvanic cell were described in Chapters 10 and 11. As described in these chapters, this cell is able to release energy spontaneously in the form of a potential. In comparison, an electrolysis cell is one whereby an external potential must be applied to provide energy to force an electrochemical reaction to take place.

For example, for a zinc–silver cell

$$Zn \mid ZnCl_2(1\ F), AgCl_{(s)} \mid Ag$$

the spontaneous cell reaction would be

$$Zn + 2AgCl_{(s)} \rightarrow Zn^{2+} + 2Cl^- + 2Ag$$

and a voltage of 0.985 V would be produced.

If, however, a potential which is greater than and opposing the spontaneous voltage is applied, the reverse reaction

$$2Ag + 2Cl^- + Zn^{2+} \rightarrow 2AgCl + Zn \qquad (28-1)$$

is forced to take place. In this case the zinc electrode becomes the cathode and the reaction corresponds to the "charging" of the galvanic cell.

The measurements described in Chapters 10, 11, and 13 were made such that the current flow between the electrodes was infinitesimally small. Thus, the concentration of the electroactive species in solution was not altered. In contrast, the methods discussed in this chapter employ a significant current flow which alters the concentration of the sample (electrolysis of the electro-active species).

**Concentration Polarization.**    When a potential is applied across a cell, the unstirred solution should act as a metallic conductor and follow Ohm's law

$$E = iR \qquad (28-2)$$

where $E$ is the potential in volts, $i$ is the current in amperes, and $R$ is the resistance in ohms. The potential–current relationship should be a linear

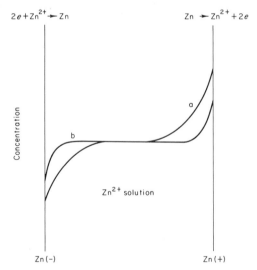

**Fig. 28–1.**  The concentration gradient of zinc (II) at the zinc anode and cathode. (a) Gradient for an unstirred solution; (b) gradient for a stirred solution.

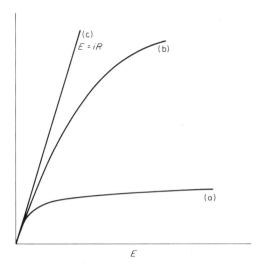

**Fig. 28–2.** Deviation of solution from Ohm's law. (a) Deviation for an unstirred solution; (b) deviation for a stirred solution; (c) ideal metallic conductor.

function; unfortunately, this is not the case with the electrochemical cell. For the cell

$$\text{Zn} \mid \text{ZnCl}_2 \ (1 \ M) \mid \text{Zn}$$

if the potential between the two zinc electrodes is increased, the concentration gradient at the two electrodes changes as shown in Fig. 28–1. The zinc anode is oxidized:

$$\text{Zn} \rightarrow \text{Zn}^{2+} + 2e$$

while the zinc ion is reduced at the cathode:

$$\text{Zn}^{2+} + 2e \rightarrow \text{Zn}$$

The overall effect is that the concentration of zinc ion at the two electrodes is markedly different. As the potential is increased across the electrode pair, the concentration of zinc ion at each electrode becomes even more dissimilar and the potential at each electrode changes according to the Nernst equation:

$$E_{\text{Zn}^{2+},\text{Zn}} = E^{\circ}_{\text{Zn}^{2+},\text{Zn}} - \frac{0.0592}{2} \log \frac{1}{[\text{Zn}^{2+}]} \tag{28-3}$$

This results in a back emf and a current–voltage curve similar to curve a in Fig. 28–2 is obtained. This effect is called concentration polarization. The curve ultimately levels off because diffusion to and from each electrode reaches a steady state, thus, a limiting current is observed. At this point the system is defined as a *diffusion controlled system*.

If, on the other hand, the solution is stirred, the concentration gradient is less than that observed in an unstirred solution. This results in a smaller back emf such that for a given potential the current will be greater in the stirred solution (curve b, Fig. 28–2) in comparison to an unstirred solution.

Under both circumstances, where the solution is diffusion controlled or stirred, a deviation from the ideal metallic conductor occurs (Fig. 28–2c). Thus, an excess voltage must be applied to maintain a current comparable to a metallic conductor.

**Equilibrium and Nonequilibrium Electrode Systems.**     Consider the galvanic cell

$$Zn \mid ZnCl_2 \ (1 \ F), \ AgCl_{(s)} \mid Ag$$

where the reaction is

$$2AgCl_{(s)} + Zn = 2Ag + Zn^{2+} + 2Cl^-$$

The equilibrium potential, $E_{eq}$, produced by this cell at a stationary condition will be 0.985 V. The overall current, $i$, in the cell is given by

$$i = i_c + i_a$$

where $i_c$ is the current $(+)$ due to the reduction reaction and $i_a$ is the current $(-)$ due to the oxidation reaction. If the equilibrium potential is applied, $i_c = i_a$, and no net current flows. This is illustrated in Fig. 28–3, where both cathodic and anodic reactions are shown.

When a voltage is applied that is different from the equilibrium potential, electrolysis takes place and the composition of the electrode–solution interface approaches a new equilibrium condition. Either the cathodic or anodic direction can be followed.

If the cell

$$Pt \mid ZnCl_2 \ (1 \ F), \ AgCl_{(s)} \mid Ag$$

is considered, the cell will not provide a spontaneous (galvanic) potential because the platinum electrode is not in equilibrium with the system. Because of this situation, cathodic current is observed only when an external potential is applied. As the voltage is increased only a very small increase in current (residual current) is noted (see Fig. 28–3) until the value of approximately 0.985 V is obtained. At this voltage the zinc is reduced and deposited on the platinum electrode (working electrode). This process is accompanied by an increase in current. If the solution is unstirred (diffusion controlled), the current will reach a limiting value and an increase in potential will not increase the current flow. The current at this point is referred to as the diffusion current and the point of initial current flow is the decomposition potential.

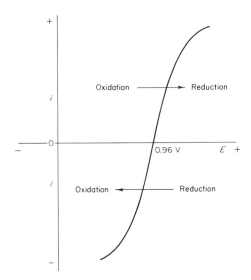

**Fig. 28-3.** Change in current as a potential is applied to a galvanic cell.

In order to obtain the characteristics of the working electrode it is necessary to measure the potential vs an internal reference. A discussion of reference electrodes is presented in Chapter 10.

The total potential, $E_t$, across the electrochemical cell is the difference between the anodic potential, $E_a$, and the cathodic potential, $E_c$, plus the potential due to the resistance of the solution ($iR$ drop). By placing a standard reference electrode in the solution the potential at the cathode may be measured independent of the anode since the reference electrode maintains a constant voltage. Thus, the measurement of the potential of the working electrode will be with respect to a standard (reference electrode).

## ELECTRON TRANSFER

The overall electrode reaction proceeds at a rate governed by (1) the rate at which electrons are transferred between the electrode and electroactive species and (2) the rate of movement of the electroactive material (mass transport) to the electrode surface. The overall rate constant, $k$, for the electron transfer is composed of a cathodic rate constant, $k_c$, and an anodic rate constant, $k_a$:

$$\text{Ox} + n e \underset{\text{Anodic }(k_a)}{\overset{\substack{k \\ \hline \\ \text{Cathodic }(k_c)}}{\rightleftharpoons}} \text{Red}$$

For very fast electron transfer rates, $k$ is large and the electrode process is said to be reversible. If $k$ is small, the process is said to be irreversible.


## MASS TRANSPORT

During electrolysis, the decrease of reactant concentrations at the electrode surface affects both electrode potential and electrolysis current. Therefore, the mass transport rate, which is the rate at which reactants move from the bulk of the solution to the electrode surface, is an important factor in electrolysis. Mass transfer is achieved by three basic processes: migration, diffusion, and convection.

**Migration.** Migration is an ionic movement process due to electrical gradients where ions are attracted by the electrode of the opposite charge. This is illustrated in Fig. 28–4.

Migration is usually undesirable in voltammetry. It is eliminated by adding a supporting electrolyte which is (1) electrochemically inert and (2) is at a much higher concentration than the electroactive species (100 times or larger).

**Diffusion.** Diffusion is the movement of the electroactive species due to a concentration gradient. Each electroactive species has a characteristic diffusion rate through a diffusion layer on a stationary electrode in an unstirred solution. It is dependent upon the species properties and those of the solvent and is nearly independent of the surrounding electrolytes. The limiting current, $i_{\lim}$, flowing during electrolysis is given by

$$i_{\lim} = \frac{nFDCA}{d}$$

where $n$ is the number of electrons, $F$ the Faraday, $D$ the diffusion coefficient, $C$ the concentration, $A$ the area of the electrode, and $d$ is the diffusion layer thickness at the electrode surface.

When some other means of mass transfer is operative, such as reproducible stirring, the diffusion layer reaches a study-limiting value. Diffusion is relatively slow when compared to mechanical stirring.

**Convection.** Convection is the movement caused by mechanical or thermal agitation. Mechanical stirring increases the mass transport rate and reduces the diffusion layer, which leads to an increase in the limiting current. Hence, the rate of stirring must be controlled in order to obtain a constant current. Control may be achieved at zero stirring (unstirred solution) or at

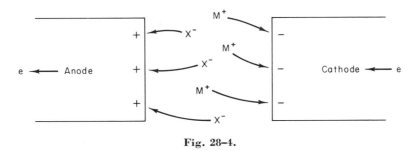

**Fig. 28–4.**

some finite rate. If an unstirred solution is employed over long periods of time, temperature control is necessary since thermal convection will occur.


## FARADAY'S LAWS

The quantity of matter converted during electrolysis is related to the quantity of electricity passed through the solution. This relationship was formulated as two laws of electrolysis by Michael Faraday in 1834 after studying electrolytic cells in detail.

1. The amount of chemical change produced by application of an electric current is proportional to the quantity of electricity passed.

2. The amounts of different substances reacted or deposited by an equivalent quantity of electric charge is proportional to their chemical equivalent weights.

The quantity of electricity passed through a cell is measured by the product of the current flow and the time. If one equivalent weight of a substance is oxidized or reduced, 1 Faraday $(F)$ of electricity must have passed through the solution. A Faraday is defined as 96,500 coulombs, where a coulomb is that quantity of electricity passing through a conductor at 1 ampere current for 1 second.

If Faraday's two laws are combined, a quantitative relationship between the amount of a substance oxidized or reduced and the quantity of electricity passed through the solution can be formulated

$$w = \frac{i \times t \times \text{equiv wt}}{F} \tag{28-4}$$

where $w$ is the weight of a substance oxidized or reduced, $i$ is the current in amperes, $t$ is the time in seconds, and $F$ is the Faraday (96,500 coulombs). The equivalent weight of the substance is the formula weight or atomic weight divided by the number of electrons lost or gained per molecule or atom during the electrochemical process. This occurs only if the electrochemical process is 100% current efficient for reduction of the copper and silver solutions, where the current efficiency of an electrochemical process is the ratio of the current consumed for a particular reaction and the total current flow. If only one electrochemical reaction occurs in the cell, the process is 100% current efficient.

Consider two electrolytic cells connected in series, one containing a solution of silver(I) and the other a solution of copper(II). For every Faraday of electricity that passes through the cells one gram atomic weight of silver (or 107.87 g) and 0.5-g atomic weight of copper (or 31.77 g) is plated.

*Example 28-1.*    Calculate the weight of silver and copper plated out when 0.1 Faraday is passed through two cells connected in series containing the respective solutions assuming 100% current efficiency for each solution.

$$Ag^+ + e \rightarrow Ag$$

$$Cu^{2+} + 2e \rightarrow Cu^{2+}$$

The equivalent weight for silver is 107.9/1. Hence,

$$i \times t = 9{,}650 \text{ coulombs} = 0.1 \text{ Faraday}$$

$$w = \frac{9{,}650 \times 107.9/1}{96{,}500}$$

$$w = 10.79 \text{ g of silver}$$

The equivalent weight of copper is 63.54/2. Hence,

$$w = \frac{9{,}650 \times 63.54/2}{96{,}500}$$

$$w = 3.177 \text{ g}$$

*Example 28-2.*    A solution of zinc is electrolyzed for 30 seconds using a current of 1.0 mA. Calculate the weight of zinc plated on the electrode (assume 100% current efficiency) for the electrochemical reaction.

$$Zn^{2+} + 2e \rightarrow Zn$$

$$\text{equivalent weight of zinc} = 65.38/2$$

$$\text{number of coulombs passed} = i \times t = 1.0 \times 10^{-3} \text{ A} \times 30 \text{ second}$$

$$w = \frac{1 \times 10^{-3} \times 30 \times 65.38/2}{96{,}500}$$

$$w = 1.02 \times 10^{-5} \text{ g}$$

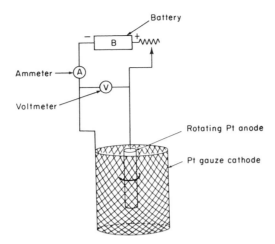

Battery

B

Ammeter → A

Voltmeter

Rotating Pt anode

Pt gauze cathode

**Fig. 28–5.**   Configuration of platinum electrodes for electrodeposition.

**Electrogravimetry.**     Electrogravimetry is a method in which a metal is quantitatively electroplated on to an electrode, usually platinum. The amount of metal plated is determined by the difference in the weight of the electrode before and after electroplating.

In the electroplating procedure a current is forced to flow between a platinum gauze cathode and platinum wire anode using the circuit shown in Fig. 28–5. After complete deposition of the metal the cathode is removed, dried, and weighed. Two other objectives in electrolysis are (1) electrolytic deposition to achieve a separation of the deposited species from the rest of the solution (electroseparation) and (2) electrolytic deposition for preparative purposes (electroplating).

**Constant  Current  Electrodeposition.**     In  this  technique  a  fixed cathodic current (where a cathodic deposition is of interest) is applied to the electrolysis cell and the species that is most easily reduced is deposited on the cathode. If only one electroactive species is present, it is completely deposited. However, if several electroactive species are present, selective deposition is often not possible since the potential changes at the electrode. The electrolysis setup in Fig. 28–5 is used for constant-current methods. Because a large cathodic current can be applied, the plating process is rapid.

**Controlled Potential Electrodeposition.**     In this technique the potential at the cathode (where a cathodic reaction is of interest) is not allowed to change to a potential at which a second species is electroactive. Thus, a potential large enough to bring about electrolysis of the first but not so large to deposit the next electroactive species is applied to the cell. The electrolysis setup shown in Fig. 28–5 must be modified before it can be used for controlled potential electrodeposition. The modification includes insertion of a reference electrode to measure the potential at the cathode and circuitry to prevent this electrode from departing from the predetermined potential value. A difference of about 0.2 V is required to selectively deposit one electroactive species in the presence of the other. By changing the potential, sequential deposition is possible.

Even though controlled-potential electrodeposition is more selective than constant-current methods, the latter technique is used more often because this method of electrolysis is much faster. In constant-current methods the current is held constant, while in controlled-potential methods the current is high as the electrolysis is initiated, but drops to zero as the deposition approaches completion.

Perhaps the most used electrogravimetric procedure is one in which copper is determined. In this procedure, the reactions at the platinum cathode and anode are

$$\text{Cathode:} \quad Cu^{2+} + 2e \rightarrow Cu$$

$$\text{Anode:} \quad 2H_2O \rightarrow O_2 + 4H^+ + 4e$$

Electrolysis is carried out in sulfuric or nitric acid solution. The acid concentration should not be too high or else hydrogen gas is formed at the cathode due to reduction of hydronium ion. This reaction interferes with the copper plate causing it to crumble and fall off the electrode. To prevent reduction of hydronium ion, nitrate salts can be added to the solution which is reduced to ammonia according to the reaction

$$NO_3^- + 10H^+ + 8e \rightarrow NH_4^+ + 3H_2O$$

In this way a uniform coating of copper, which adheres to the electrode surface, is obtained. Also, in electrolysis, the electrodes must be clean and free of materials, such as grease, which prevent the metal from adhering to the electrode.

Some ions tend to interfere with the electrodeposition of copper. Chloride ion is probably the most significant for the following reasons:

1. The anion is oxidized to chlorine gas at the anode. Once formed it can react with the copper metal deposited on the cathode and also with the platinum electrode. Ultimately, low weights are obtained.

2. Chloride interferes with the reduction of copper(II) to copper metal, because the anion stabilizes the copper(I) ion by complexation.

## COULOMETRY

Coulometry applies to those techniques which are based on measurement of the coulomb ($i \times t$). Consider an electrolytic cell in which a metal ion is to be determined by measurement of the product of the current and time. Either a constant current or a controlled potential can be used. If a constant current is applied to the cell, the potential must vary as governed by the Nernst equation. As the electrolysis proceeds, the potential difference between the cathode and anode increases as the concentration of the oxidized species decreases. Thus, if another reducible component (another metal ion or solvent) is present, it too will eventually enter into the reaction and the current will be used for its reduction as well as for the sample species. On the other hand, if a constant potential is used, the current decreases as a function of time due to the decrease in current carriers in solution. This, however, eliminates the interference from competing reactions.

Both techniques are commonly used in analysis. Constant-current coulometry is used primarily for coulometric titrations, whereas constant-potential coulometry provides the selectivity of potential control. It is important to note that quantitative coulometric measurements do not need standard solutions for calibration, since the coulometer is internally calibrated by measuring the current–time product.

**Constant-Current Coulometry.**    A diagram of an instrument used for constant-current coulometry is presented in Fig. 28–6. The basic components are a battery or power supply, a large resistor to maintain a constant current, and the electrolysis cell. The cell is separated into anode and cathode compartments to limit the diffusion of reduced or oxidized species to the opposite electrode.

Constant-current coulometry is used most often for coulometric titrations. Before the titration is started, four things must be accomplished. First, an appropriate solvent or compound must be found which will provide the components necessary to allow the electrochemical generation of the titrant.

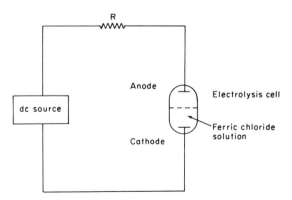

**Fig. 28–6.**  Diagram for a simple constant-current coulometer. (From G. W. C. Miller and G. Phillips, "Coulometry in Analytical Chemistry," Pergamon Press, Oxford, 1967.)

Second, the current level must be chosen so that the current density, $i/A$ where $A$ is the area of the electrode surface, at the electrode is sufficiently low enough to ensure 100% current efficiency. Third, the appropriate current polarity must be chosen in order to generate the titrant at the electrode. Finally, a method for end-point detection must be selected. Once these steps are outlined, the coulometric titration is almost identical to a classic titration except that the buret is replaced by the working electrode.

For example, bromine is a useful titrant in organic reactions, except that it is very difficult to prepare as a standard solution. However, it can be generated and standardized simultaneously by coulometry. By placing a bromide salt in solution and electrochemically oxidizing it to bromine, the standard titrant is generated *in situ*. The amount of bromine generated is determined by measuring the current and time involved. For example, olefins can be titrated by this technique. The sample and a large excess of soluble bromide salt is placed in a titration cell in a spectrophotometer. The current is initiated and the reactions that occur are

$$2Br^- \rightarrow Br_2 + 2e \qquad \text{(electrochemical)} \qquad (28\text{-}5)$$

$$\underset{\text{(olefin)}}{\overset{}{\Large \backslash\!\!C=C\!\!/}} \longrightarrow \underset{\overset{|}{Br}\ \overset{|}{Br}}{-\overset{|}{C}-\overset{|}{C}-} \qquad \text{(chemical)} \qquad (28\text{-}6)$$

After all the olefin has reacted, excess bromine builds up which is detected by an increase in absorbance. The time required to reach this point is carefully measured. The amount of olefin is then calculated from these data.

One of the more practical electrogenerated titrants is silver(I). By placing a silver wire anode into a sample solution, a measured amount of silver(I) can be generated by controlling the current flow* and monitoring the time of electrolysis. For the determination of chloride, the reactions are

$$Ag \rightarrow Ag^+ + 1e \qquad \text{anodic reaction}$$

$$Ag^+ + Cl^- \rightarrow AgCl \qquad \text{titration reaction}$$

The Cotlove chloridometer, which is used for the determination of chloride in urine, operates on this principle. A spiral of wire is immersed in the test solution containing chloride and is used as the anode. Upon electrolysis, the generated silver(I) removes the chloride by precipitation and the end point is detected potentiometrically. This chloridometer can be used also for the titration of mercaptans, other halogens, and sulfhydryl groups.

A list of other examples of coulometric titrations is presented in Table 28–1. Most of these can be determined on a micro to ultramicro level since time and current can be measured with a high degree of accuracy.

---

* The current efficiency must be 100% since the number of coulombs is measured by the product of current and time.

**Table 28–1. Examples of Coulometric Analysis**

| Substance | Titrant | Substance | Titrant |
|-----------|---------|-----------|---------|
| Ag(I) | $Br^-$, $Cl^-$, $I^-$ | Acetic acid | $OH^-$ |
| Ce(IV) | Fe(II) | Ascorbic acid | $I_2$ |
| $CN^-$ | Hg(II) | Cyclohexene | $Br_2$ |
| $S^{2-}$ | $I_2$ | Cysteine | Hg(II) |
| Sb(III) | $Br_2$ | Hydrazine | $Br_2$ |
| U(VI) | Ti(III) | Pyridine | $H_3O^+$ |

*Example 28–3.* One-hundred milliliters of a solution of chloride are coulometrically titrated with silver ion using a current of 1.00 mA. Calculate the concentration of the chloride if the end point is detected after 102 seconds. The silver ion titrant is produced at the anode (silver) in a cell shown in Fig. 28–7. The end point is detected potentiometrically using a Ag–SCE electrode pair.

Using Faraday's laws, the weight of silver can be calculated.

$$wt_{Ag^+} = \frac{i \times t \times at\ wt_{Ag^+}/n}{F}$$

$$Moles_{Ag^+} = moles_{Cl^-} = \frac{wt_{Ag^+}}{at\ wt_{Ag^+}} = \frac{i \times t}{nF}$$

$$Moles_{Cl^-} = \frac{(1.0 \times 10^{-3})(1.02 \times 10^2)}{(1 \times 96,500)} = 1.1 \times 10^{-6}\ mole/100\ ml$$

$$C_{Cl^-} = 1.1 \times 10^{-5}\ F\ Cl^-$$

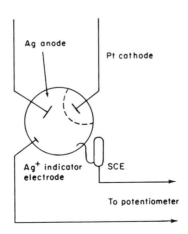

**Fig. 28–7.** Coulometric cell for titration of chloride with silver.

End points can be detected by many instrumental and chemical procedures. For example, if Ce(IV) is to be titrated with generated Fe(II)

$$Fe^{3+} + e \rightarrow Fe^{2+} \qquad \text{(electrochemical)}$$

$$Fe^{2+} + Ce^{4+} \rightarrow Fe^{3+} + Ce^{3+} \qquad \text{(chemical)}$$

the titration can be monitored potentiometrically.

If a color change occurs at the end point of the reaction, such as an acid–base indicator, the end point can be detected visually or spectrophotometrically. Other methods of end-point detection include monitoring the current of one of the reaction components at its decomposition potential (amperometry, see Chapter 29) or by following the change in conductance of the solution.

**Controlled-Potential Coulometry.**    In controlled-potential coulometry the quantity of electricity passed through a solution is measured while the potential is held constant. The technique is generally used for the determination of micro amounts of metal ions in solution.

As stated earlier, controlled-potential coulometry adds a certain amount of specificity in comparison to controlled current electrolysis. For example, assume two electroactive species, which have different decomposition potentials, are present in the same solution. The current–voltage characteristics of each ion in a stirred solution are shown in Fig. 28–8. If species I is to be analyzed, the potential must be controlled such that species II does not react in the electrochemical cell. This voltage must be above the decomposition potential

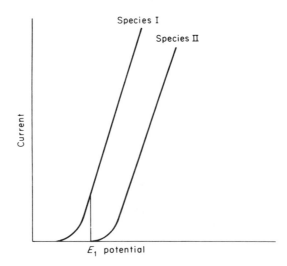

**Fig. 28–8.**   Potential-current scans for two species can be separated by controlled-potential coulometry.

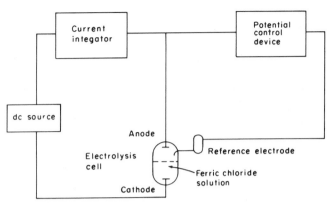

**Fig. 28–9.** Block diagram of a controlled-potential coulometry instrument. (From G. W. C. Milner and G. Phillips, "Coulometry in Analytical Chemistry," Pergamon, Oxford, 1967.)

of species I but below that of species II. By electrolyzing the solution at this potential ($E_1$), the amount of species I may be determined in presence of species II without prior separation. If, however, the concentration of species II is desired, then species I must be removed prior to electrolysis.

Although a certain amount of selectivity is gained, there are added complications. Since the characteristics of the sample solution change during the electrolysis, it is difficult to control the potential over long periods of time. Thus, the instrument which is used is more complex than that used for controlling the current. A block diagram of an instrument is shown in Fig. 28–9.

A second complication is the need of measuring the quantity of electricity passed during the electrolysis. Since the potential remains constant, the current decreases as the concentration of the electroactive species is depleted. This decrease in current is defined by

$$\log \frac{i_t}{i_{t=0}} = -Kt \qquad (28\text{-}7)$$

where $i_t$ is the current at any time $t$, $i_{t=0}$ is the current at $t = 0$, and $K$ is a constant which is dependent upon electrode design and configuration. A plot of $\log i_t/(i_{t=0})$ vs $c$ results in a straight line of slope $-K$.

If the initial current can be measured, the concentration of the bulk solution can be determined by comparison with a calibration curve. As the concentration of the electroactive species in solution increases, the initial current should also increase. Unfortunately, this is not a very accurate method of quantitatively using controlled-potential coulometry. Thus, it becomes necessary, as in the case of controlled-current electrolysis, to measure the product of current and time. This measurement may be made by many methods including mechanical, electronic, and electrochemical devices.

The classical method of determining the amount of electricity passed through a solution is by utilizing an electrochemical coulometer. If an electrochemical cell such as

$$Ag \mid AgNO_3 \mid Ag$$

is placed in series with the test cell, the amount of electricity which passes through the test solution can be monitored by weighing the cathode in the cell before and after electrolysis. From the electrochemical reaction

$$Ag^+ + 1e \rightarrow Ag$$

the amount of silver plated on the electrode is directly related to $i \times t$ (number of coulombs). The number of coulombs passing through the silver coulometer will therefore be the same as the number passing through the test cell.

## Questions

1. Describe the difference between an electrolytic and a galvanic cell.
2. Describe concentration polarization.
3. What are the differences between equilibrium and nonequilibrium electrode processes?
4. Describe the significance of each step in the electrogravimetric determination of copper.
5. What are the differences between constant-current coulometry and constant-potential electrodeposition? coulometry?
6. Explain why the addition of electrolyte minimizes the migration current in electrolysis.
7. Mercaptans, RSH, can be determined by coulometric titration with $Ag^+$. Explain how this is accomplished. List all reactions and describe the cell that would be used.
8. Why is a controlled potential not used in coulometric titrations?
9. Suggest how a strong acidic and basic titrant can be generated coulometrically?
10. Explain why standard solutions are not required in coulometric titrations.
11. Compare the accuracy and precision of the following methods of determining chloride ion in solution:
    a. Gravimetric determination by precipitation with silver ion
    b. Potentiometric titration with silver ion
    c. Coulometric titration with the Cotlove chloridometer
    d. Volhard titration

# Problems

1.* What are the weights of each of the following ions deposited by 4.251 coulombs?
a. Cu from Cu(II); b. Cd from Cd(II); c. Ag from Ag(I); d. iodide on a silver anode.

2.* How many coulombs are required to reduce 56.783 g of silver(I) to the metal?

3. What weight of copper(II) would be deposited from solution if the solution were electrolyzed for 12.35 minutes at 2.674 mA current?

4.* The amount of chloride in a cerebral spinal fluid sample is to be determined by titration with electrogenerated silver(I). If 0.2 ml of sample were titrated and 2.705 mg of silver(I) were required to reach the end point, what is the concentration (mEq/liter) of the chloride in the spinal fluid?

5. The amount of chloride in a serum sample is to be determined by the method described in Problem 4. If a 0.20-ml sample was titrated to the end point with 2.16 mg of silver, calculate the concentration of chloride in the sample in mEq/liter.

6. A normal person excretes 75–200 mEq of Cl⁻ in urine per 24-hour period. If the excretion volume over that period of time was 1.564 liters and a 0.20-ml sample was titrated, calculate the maximum and minimum amounts of silver (in grams) a Cotlove chloridometer would use.

---

* Answers are listed at the end of the book for problems marked with an asterisk.

# Chapter
# Twenty-Nine
# Polarography

## INTRODUCTION

Polarography is an electrolysis technique in which microelectrolysis is performed at a dropping mercury electrode (DME) in an unstirred solution. The data are obtained as a current–voltage curve that is characteristic of the electroactive species.

Both reduction and oxidation can be studied at the DME. In the former, the working electrode (DME) is made the cathode while in the latter, it is the anode. For this reason reduction processes are often referred to as cathodic polarography, cathodic waves, or cathodic processes. Similarly, oxidation can be described as anodic polarography, anodic waves, or anodic processes. Because of the similarity of the two, most of the discussion will deal with the reduction process.

In general, polarography can be used in qualitative and quantitative analysis. Under favorable circumstances as little as $10^{-6}$ mole/liter of electroactive species can be detected and determined. The measured currents will range from 1 to 100 $\mu$A.

## INSTRUMENTATION

The current–voltage curve for an electroactive species is obtained by the polarographic instrument shown in Fig. 29–1. The polarograph is composed of a voltage supply (battery), a slide wire to vary the potential, an ammeter to measure the current passing through the cell, and the polarographic cell which contains the DME and a reference electrode. The potential on the dropping mercury electrode is varied by changing the contact point on the slide wire. At each applied potential the current is measured with the ammeter. Although not shown in Fig. 29–1, a switch, which changes the DME and SCE

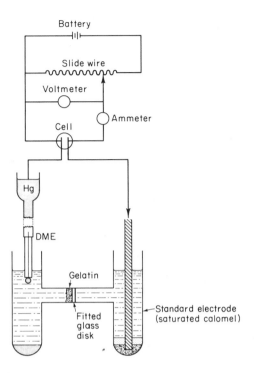

**Fig. 29–1.** The polarographic instrument and cell. (See text for explanation.)

from cathode and anode to anode and cathode, respectively, is included in the circuit. In modern polarographic instruments the potential is scanned and current monitored automatically.

**The Cell.** A typical cell (Fig. 29–1) consists of a reference electrode (see Chapter 10) and a working electrode (DME) separated by a sintered glass disc or other porous media. If an electroactive species is placed in the solution and the voltage of the DME is continuously made more negative (with respect to the reference electrode), a curve called a polarogram will be obtained. A typical polarogram is shown in Fig. 29–2.

The polarogram in Fig. 29–2 can be divided into four main parts. In part (a), little current flows because the potential is not negative enough to cause reduction of the electroactive species in the solution. When the decomposition potential is reached, reduction takes place and the current rises as the applied potential is made more negative (b). Eventually, the current levels off (limiting current) and rises very gradually (c) until the potential is sufficiently negative to cause reduction of the next electroactive species (d). The order of reduction is such that the more easily reduced species is reduced first.

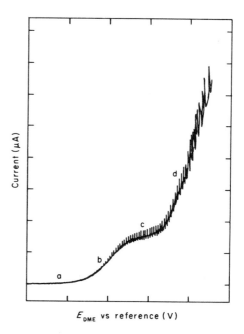

Fig. 29–2.   The polarogram. (See text for explanation.)

The instrumental response appears as an oscillation. This is the result of the growth and termination of each mercury drop. As the drop grows, the current increases and when the drop falls the electrical contact is terminated. Therefore, the current starts to go to zero. However, a new drop immediately forms, completing the circuit, and the process repeats itself. In general, it is customary to use the tops of the oscillation to represent the average current at each applied voltage.

**The DME.**    The dropping mercury electrode is made by attaching a mercury reservoir to a fine capillary tube. With this system fine drops of mercury emerge from the capillary tip. The capillary bore (about 0.03–0.05 mm), the capillary length (about 5–12 cm), and the height of the reservoir above the capillary will determine the mercury flow (drop time).

The DME is the working electrode in the sample solution and it is at this electrode where the diffusion controlled electrolysis takes place. A second electrode, a reference electrode such as the saturated calomel electrode, is also placed in the sample solution and the current–voltage characteristics of this system are determined.

There are several reasons for using the DME.

1. There is a continuous renewal of the electrode surface.
2. Surface area of the drop is reproducible for any given capillary.

3. There is a high hydrogen overvoltage on Hg. Therefore, reduction of alkali metal ions can be observed.

4. Amalgam formation, which tends to make many metal ions more easily reducible, is possible.

5. Currents reach a steady value rapidly with the DME.

Depending on what ions are in the solution and the type of solvent used, the potential range for the DME is about $+0.2$ to $-3.0$ V vs SCE. If more positive potential values are required, a stationary micro platinum electrode (approximately $+0.7$ to $-1.3$ V) can be used. Several other solid electrodes are also available (Au, C, SiC, $B_4C_3$, and W). Most of these have their greatest application in anodic polarography.

## THE POLAROGRAPHIC WAVE

Several different kinds of currents contribute to the polarographic wave. A polarogram for the reduction of $Cd^{2+}$ in 0.1 $F$ KCl is shown in Fig. 29–3. The three main currents are the residual current, diffusion current, and the supporting electrolyte current.

The electroactive species should not act as a current carrier except after passage of the decomposition potential. If they are current carriers prior to this potential, the resulting current, is called the migration current. In order to minimize migration current, a supporting electrolyte is added to the solution in large concentrations relative to the electroactive species. Further-

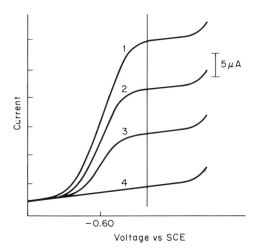

**Fig. 29–3.** Change in cadmium reduction wave with change in concentration. The supporting electrolyte is 0.1 $M$ potassium chloride. Scan 1, $5.40 \times 10^{-4}$ $M$; Scan 2, $3.70 \times 10^{-4}$ $M$; Scan 3, $2.30 \times 10^{-4}$ $M$; Scan 4, supporting electrolyte only.

more, the added electrolyte has to be electrochemically inactive in the potential range of interest and it must be free of trace levels of active interfering ions. The current due to the supporting electrolyte is called the residual current. In Fig. 29–3 the residual current would be measured in the absence of Cd(II).

After the decomposition potential is passed a limiting current is reached. The limiting current ($i_l$) is the sum of the diffusion current ($i_d$) and the residual current ($i_r$). Therefore,

$$i_d = i_l - i_r \tag{29-1}$$

Addition of the electrolyte has insured that the ions, which reach the electrode, do so only by diffusion (migration current is negligible).

The diffusion current can be described in the following way considering that a reduction process takes place. After the applied voltage reaches the decomposition potential, reduction takes place at the electrode surface.

$$Cd^{2+} + 2e = Cd(Hg) \tag{29-2}$$

The concentration of the oxidized species is depleted and a concentration gradient is produced around the DME. As $Cd^{2+}$ is reduced, more $Cd^{2+}$ reaches the stationary electrode from the bulk solution by diffusion. Therefore, the limiting current is dependent upon the rate of diffusion of the ions to the electrode and the current levels off. If the diffusion is increased, the limiting current is increased.

At more negative potentials, the current raises rapidly because of reduction of the supporting electrolyte (supporting electrolyte current in Fig. 29–3). The actual potential at which this reduction takes place will depend on the electrolyte used.

The diffusion current is proportional to the concentration at the electrode through

$$i_d \propto c \tag{29-3}$$

This is illustrated in Fig. 29–3, where a series of polarograms for different concentrations of Cd(II) in 0.1 $F$ KCl are shown.

The diffusion current is related to the concentration, $c$, through the Ilkovic equation

$$i_d = 607 n c D^{1/2} m^{2/3} t^{1/6} * \tag{29-4}$$

where $i_d$ is the diffusion current as an average in microamp, $n$ is the number of electrons in the reaction per mole, $c$ is the concentration in mmole/liter,

---

* $i_d = 706 n c D^{1/2} m^{2/3} t^{1/6}$ where 706 is used because current is measured at its maximum value; i.e., the top of the oscillation in the polarogram.

$D$ is the diffusion coefficient in cm$^2$/sec, $m$ is the rate of flow of Hg in mg/sec, and $t$ is the drop time in seconds. If the mass and drop time are held constant, Eq. (29-4) reduces to

$$i_d = K'c \tag{29-5}$$

where $K'$ includes the parameters which are held constant.

In Figs. 29-2 and 29-3, it is assumed that dissolved oxygen has been removed. This is required since oxygen is polarographically active and is reduced in waves at $-0.05$ V and $-0.94$ V. Therefore, before measuring a polarogram, nitrogen gas is bubbled through the solution in the polarographic cell. If a very fine stream of nitrogen bubbles are used, the oxygen will be removed in about 3–5 minutes. In most applications, the polarogram is run with the solution under a blanket of nitrogen which prevents oxygen from being redissolved in the sample solution.

**Half-Wave Potential.**    In an electrochemically reversible reaction the wave is governed by the Nernst equation. Thus, the midpoint of the polarographic wave should be independent of the concentration. This point is defined as the half-wave potential ($E_{1/2}$) and is characteristic of the species undergoing reduction (or oxidation) for a specified set of experimental conditions. As an approximation, $E_{1/2}$ values correspond to the standard reduction potential for the reaction being investigated.

The $E_{1/2}$ is related to the potential at the DME by the Heyrovsky–Ilkovic equation

$$E_{DME} = E_{1/2} + \frac{0.0592}{n} \log \frac{i_d - i}{i} \tag{29-6}$$

where $i$ is the current for the potential at the $E_{DME}$. This equation, which applies only to reversible systems, is useful for several reasons. For example, a plot of $E_{DME}$ (choose potentials along the rising portion of the wave) vs $\log (i_d - i)/i$ gives a straight line whose slope is $0.059/n$ and intercept is the $E_{1/2}$ (when $i = i_d/2$, $\log (i_d - i)/i = 0$, and $E_{DME} = E_{1/2}$). Thus, the half-wave potential and the number of electrons participating in the electrochemical reaction can be determined.

The half-wave potential can also be determined by a geometrical treatment of the polarogram. This is shown in Fig. 29-4.

The half-wave potentials for many inorganic ions have been determined under a variety of experimental conditions. Table 29-1 contains a partial listing of half-wave potentials. In general, if the half-wave potentials differ by at least 0.3 V a well-defined polarographic wave for each species is observed. This is illustrated in Fig. 29-5.

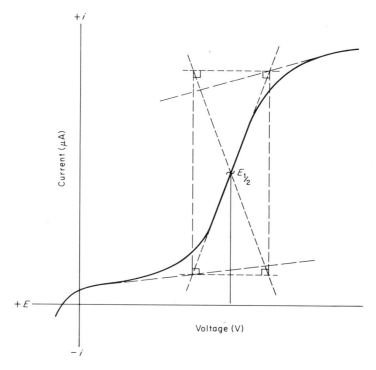

**Fig. 29–4.** Geometric treatment of the polarographic wave.

**Table 29–1. Examples of $E_{1/2}$ Values for Inorganic Substances**

| Substance | Supporting electroyte | $E_{1/2}$ vs SCE (V) |
|---|---|---|
| Al(III) | 0.1 $F$ KCl | −1.70 |
| As(III) | 2 $F$ Acetic acid, 2 $F$ NH$_4$C$_2$H$_3$O$_2$ | −0.92 |
| Br$^-$ | 0.1 $F$ KNO$_3$ | +0.12 |
| BrO$_3{}^-$ | 0.1 $F$ H$_2$SO$_4$ | −0.41 |
| Cd(II) | 0.1 $F$ KCl | −0.60 |
| Cr(II) | 0.1 $F$ KCl | −0.34 |
| Cu(II) | 1 $F$ NaOH | −0.41 |
| Fe(II) | 0.1 $F$ KCl | −1.3 |
| In (III) | 0.1 $F$ KCl | −0.561 |
| Mn(II) | 1.0 $F$ KCl | −1.364 |
| Ni(II) | 0.1 $F$ KCl | −1.1 |
| Pb(II) | 0.1 $F$ KCl | −0.40 |
| Sn(II) | 0.1 $F$ HCl | −0.83 |
| U(VI) | 0.1 $F$ KCl | −0.185 |
| Zn(II) | 0.1 $F$ KCl | −0.995 |

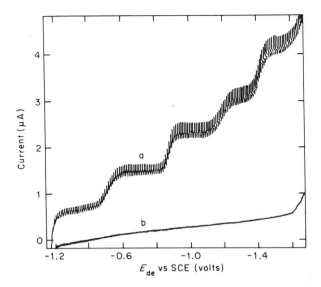

**Fig. 29–5.** Polarograms of (a) approximately 0.1 m$M$ each of silver(I), thallium(I), cadmium(II), nickel(II), and zinc(II) listed in the order in which their waves appear, in 1 $F$ ammonia–1 $F$ ammonium chloride containing 0.002% Triton X-100; (b) the supporting electrolyte alone. (From L. Meites, "Polarographic Techniques," Interscience, New York, 1965.)

## PRECAUTIONS IN POLAROGRAPHY

Obtaining precise, accurate, reproducible polarograms is dependent upon successfully handling several experimental, instrumental, and chemical problems. These will be briefly discussed in this section.

The purity of the supporting electrolyte is important since it is used in high concentration (0.1–1.0 $F$). Trace levels of impurities in the electrolyte can given rise to an interfering polarographic wave. Impurities in the mercury can also result in interfering polarographic waves. Therefore, it is necessary in some cases to purify the supporting electrolyte salt and the Hg. Since nitrogen gas can contain reducible impurities, it must be specially purified tank nitrogen or purified in the laboratory before using it for deaeration.

If the electrochemical reaction at the DME involves hydrogen ion, the reduction (or oxidation) will be affected by pH. Thus, pH must be controlled. Often the buffer that is used for controlling the pH also serves as the supporting electrolyte.

Many polarographic waves suffer from polarographic maxima, which is a distortion of the polarographic wave. Instead of the current leveling off at the limiting current, wild oscillations are observed and are the result of absorption phenomena. Adding surface-active agents, such as gelatin, dye ions, and others, will eliminate the maxima. However, the concentration of the maximum suppressor must be carefully controlled since it can affect the diffusion current.

It would appear that it would be very difficult to reproduce a polarographic wave from capillary to capillary. Although the capillary determines several of the parameters in the Ilkovic equation, it is not necessary to exactly reproduce the capillary length, diameter, or Hg reservoir height. If the product $m^{2/3}t^{1/6}$ is reproduced, data from capillary to capillary or for capillaries at different mercury pressures are comparable since $i_d \propto m^{2/3}t^{1/6}$.

## APPLICATIONS OF POLAROGRAPHY

Half-wave potentials are characteristic of a particular electroactive species. Therefore, if solvent, pH, and supporting electrolyte are carefully controlled, the $E_{1/2}$ can be used for qualitative applications.

Since $i_d$ is proportional to concentration, quantitative determinations are possible to detect as little as $10^{-6}$ mole/liter in favorable circumstances, the usual range of analysis is about $10^{-2}$ to $10^{-5}$ mole/liter. Furthermore, reversibility of the electrochemical reaction is not a requirement for successful application to analysis. However, for an irreversible reaction the experimental and instrumental conditions must be carefully regulated. Polarographic qualitative and quantitative analysis can be applied to inorganic and organic systems.

Two general techniques are utilized in quantitative polarography. Since $i_d$ is proportional to concentration [Eq. (29-5)], a calibration curve can be prepared by measuring $i_d$ for a series of standards of known concentration at a potential on the limiting current part of the wave. In using this calibration curve, $m^{2/3}t^{1/6}$ must be reproduced.

The second method is based on the internal standard principle. A known concentration of an electroactive reference ion is added to the unknown solution and the $i_d$'s for the two separated waves are measured. The concentration of the unknown is calculated from

$$c_{unk} = c_{std} \frac{i_{d\ unk}}{i_{d\ std}} \tag{29-7}$$

Modest changes in temperature or capillary characteristics are compensated for by this method.

Procedures for the determination of the ions shown in Table 29–1 and for many others in a variety of different industrial, mineral, biological, and environmental samples are available. In general, the accuracy of polarographic determinations is about $\pm 1$–2%.

A typical example is the determination of titanium in aluminum alloys.

**Table 29–2.  Comparison of Polarography and Certified Analysis for the Determination of Titanium**[a]

| Sample | Certified value of titanium (%) | Polarographic values of titanium content | |
| --- | --- | --- | --- |
| | | Individual determinations (%) | Mean (%) |
| National Bureau of Standards | | | |
| Aluminum-silicon alloy no. 87 | 0.16 | 0.167, 0.163, 0.166, 0.160, 0.154 | 0.162 ± 0.008 |
| Aluminium Laboratories, Ltd. | | | |
| Alcan 123 CAC | 0.12 | 0.116, 0.121, 0.113, 0.107, 0.110 | 0.113 ± 0.008 |
| Alcan 123 CAE | 0.15 | 0.148 | 0.148 |
| Alcan 125 CAE | 0.17 | 0.182, 0.162, 0.179 | 0.174 ± 0.012 |
| Alcan 135 CAD | 0.14 | 0.137, 0.121, 0.143, 0.136 | 0.134 ± 0.013 |
| Misc. standard sample | 0.19 | 0.185, 0.179, 0.190, 0.176 | 0.183 ± 0.007 |

[a] From R. P. Graham and A. Hitchen, *Analyst* **77,** 533 (1952).

A 2-g sample of the alloy is weighed and dissolved in 30 ml of 6.5 $F$ sodium hydroxide. Water (170 ml) is added and the suspension filtered. The precipitate is redissolved in 26 ml of 4.5 $F$ sulfuric acid, neutralized ($NH_4OH$), supporting electrolyte added (25 ml of 2 $F$ sulfuric acid and 15 g of tartaric acid) and the solution diluted to a known volume (100 ml). Standards of known Ti concentration are prepared in the same way. Typical data which were obtained for this determination are shown in Table 29–2. The purpose of the tartaric acid is to complex many of the metal ions in the solution and effectively remove them as interferences (masking).

Because of the effect of complexation on polarographic behavior, the polarographic technique can be used to study the properties of certain complexes. If the two reversible reduction processes are

$$M^{n+} + ne \rightarrow M$$

$$MX_p{}^{n+} + ne \rightarrow M + pX$$

it can be shown that

$$\Delta E_{1/2} = (E_{1/2})_{\text{complex}} - (E_{1/2})_{\text{uncomplexed}} = \frac{0.0592}{n} \log K - p \frac{0.0592}{n} \log [X]$$

$$(29\text{-}8)$$

where $K$ is the formation constant for the complex, $p$ is the coordination number, and $[X]$ is the molar concentration of the complexing agent.

As the concentration of the complex is varied (by holding the metal ion concentration constant and varying the ligand concentration), the $E_{1/2}$ of the complex changes. Therefore, plotting $E_{1/2}$ vs log $[X]$ yields a line of

**Table 29–3.  Table of Organic Functional Groups Which Are Polarographically Active**

| Type of compound | Substance | Supporting electrolyte | $E_{1/2}(V)$ |
|---|---|---|---|
| Aldehyde | Acetaldehyde | 0. 1 $F$ LiOH | $-1.73$ |
| | Benzaldehyde | McIlvaine buffer (pH = 2.2) | $-0.96, -1.32$ |
| | Formaldehyde | 0.2 $F$ KOH | $-1.59$ |
| Ketone | Benzophenone | McIlvaine buffer (pH = 1.3) | $-0.90$ |
| | Terramycin | Phosphate buffer (pH = 8.8) | $-1.16$ |
| Vitamin | Thiamine | Phosphate buffer (pH = 7.2) | $-1.26$ |
| | Riboflavin | Phosphate buffer (pH = 7.2) | $-0.40$ |
| | Ascorbic acid | Phosphate buffer (pH = 7.0) | $-0.2$ |
| Aromatic nitro | o-Dinitrobenzene | Phthalate buffer (pH = 2.5) | $-0.12$ |
| | Nitrobenzene | Phthalate buffer (pH = 3.5) | $-0.30$ |
| Heterocyclic | Acridine | Citrate buffer (pH = 4.0) | $-0.32$ |
| | | Phosphate buffer (pH = 7.3) | $-0.51$ |
| | | Borate buffer (pH = 11.8) | $-0.93$ |
| Hormone | Testosterone | Britton–Ribonson buffer (pH = 7) 50% ethyl alcohol | $-1.20$ |

slope $-p(0.059/n)$ and an intercept of $0.059/n \log K$. Since $n$ is known, $p$ and $K$ can be calculated.

**Organic Polarography.**    Many organic functional groups are polarographically active. Table 29–3 contains a partial list of these groups. The exact location of the $E_{1/2}$ is dependent on the position of the functional group in the molecule. Thus, steric, resonance, and conjugative effects strongly influence the polarographic behavior. In addition, the reduction (or oxidation) in aqueous solutions usually involves hydrogen ion and the reaction can be represented as

$$\text{Organic}_{ox} + ne + n\text{H}^+ \rightarrow \text{H}_n\text{organic}_{red} \qquad (29\text{-}9)$$

This means that the pH of the solution must be carefully controlled. If nonaqueous solutions (such as acetonitrile or dimethylformamide) are used, radical formation is often part of the reactions occurring at the electrode. Figure 29–6 illustrates the effect of pH on the reduction of pyruvic acid.

Most organic systems, particularly in aqueous solution, do not yield reversible polarographic waves. This does not prevent the method from being useful in quantitative analysis of organic functional groups. In general, the techniques as described in inorganic applications are also used in organic applications. Deaeration, supporting electrolyte, adjustment of $m^{2/3}t^{1/6}$, and careful measurement of $i_d$ are still required. However, because of irreversibility, the experimental conditions must be rigidly controlled.

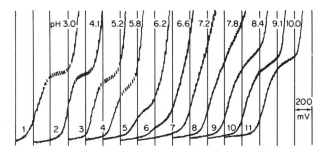

**Fig. 29–6.** The effect of pH on the reduction of pyruvic acid. Curves (1) and (2) start at −0.6 V, (3)–(7) at −0.8 V, and (8)–(11) at −1.0 V vs SCE. (From J. Heyrovsky and P. Zuman, "Practical Polarography," Academic Press, New York, 1968.)

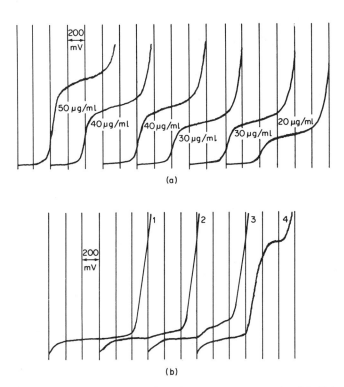

**Fig. 29–7.** Typical polarograms for selected organic compounds. (a) Polarograms of morphine with change in concentration. Curves start at −0.2 V vs a Hg pool. (b) Polarograms of aldehydes in commercially available spirits. Polarograms are taken in 0.1 $M$ lithium hydroxide mixed with 5 ml of distillate: (1) extra fine spirit; (2) and (3) fine spirit; (4) crude spirit. Curves start at −0.8 V vs Hg pool. (From J. Heyrovsky and P. Zuman, "Practical Polarography," Academic Press, New York, 1968.)

Frequently, in organic polarographic analysis the compound is converted to a derivative which is more easily examined polarographically. For example, molecules containing aromatic groups can be nitrated or nitrosated; the nitro group or nitroso group is introduced into the molecule, respectively.

Another useful polarographic wave in organic analysis is the H-catalytic wave. This wave is the result of hydrogen evolution and, briefly, it comes about because the molecule contains a labile hydrogen ion. Catalytic currents as low as $10^{-7}$ $M$ can be detected and compounds, such as cystine (other —SH compounds), proteins, alkaloids and some dyes, can be determined by observing the catalytic wave. Perhaps the most widely used procedure employs the "Brdicka reaction." The sample, for example, blood serum, is dissolved and to the solution is added an $NH_3$–$NH_4Cl$ buffer and a hexammine cobalt(III) chloride complex. In the presence of protein, this mixture yields a catalytic wave whose $i_d$ can be correlated to protein concentration in the blood.

The polarographic method can be used for the determinations of organic functional groups in organic industry, environment, pharmacy, biochemistry, food industry, medicine, and for microanalysis. Figure 29–7 illustrates some typical polarographic waves for organic substances.

## AMPEROMETRIC TITRATIONS

In an amperometric titration, the change in the limiting current is measured at a fixed potential as a function of the addition of a titrant. For example, Pb(II) can be titrated with a standard sulfate solution.

$$Pb^{2+} + SO_4^{2-} \rightarrow PbSO_{4(s)}$$

Electrolyte, alcohol (to suppress the solubility of $PbSO_4$), and a working electrode are placed in the $Pb^{2+}$ sample solution. The sample is deaerated, the potential set at a value in which a limiting current is observed, and titrant is added from a buret. Since $i_d$ is proportional to $Pb^{2+}$ concentration and its concentration decreases with precipitation, $i_d$ must decrease. The titration curve, therefore, appears as two straight lines with their intersection corresponding to the end point. This is illustrated in Fig. 29–8.

Elimination of charging current oscillations and improved sensitivity are possible by using a rotating solid electrode such as Pt. In general, the sample being titrated should be in the order of $10^{-3}$ to $10^{-5}$ $M$ and under favorable conditions an accuracy of 0.1% is possible. The shape of the titration curve will depend on the reaction, whether the reactant, titrant, or product are electroactive, and the applied potential that is selected.

**The Oxygen Cathode.** One of the useful applications of amperometry is the oxygen monitor or the oxygen cathode. The cathode, which is a platinum

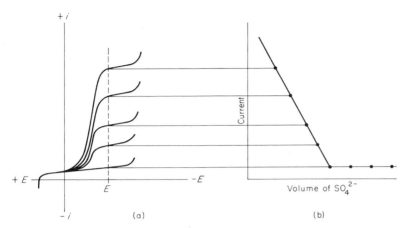

**Fig. 29–8.** Origin of the amperometric titration curve for the titration of $Pb^{2+}$ with $SO_4^{2-}$. (a) Polarograms of $Pb^{2+}$ as $SO_4^{2-}$ is added; (b) amperometric titration curve.

or gold electrode, is used for the measurement of dissolved oxygen and operates on the principle of monitoring the rate of the cathodic reduction of oxygen above its reduction potential. The amount of dissolved oxygen is determined, after calibration, by the amount of current that flows between the indicator (working) electrode and the reference (generally Ag–AgCl) electrode.

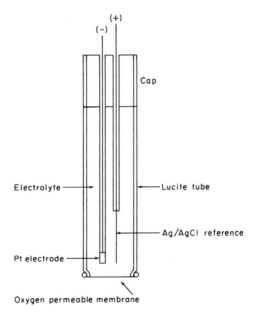

**Fig. 29–9.** The Clark electrode.

A diagram of a typical oxygen electrode is shown in Fig. 29–9. This electrode, which is known as a Clark electrode, is routinely used for monitoring dissolved oxygen in biological systems. The working (Pt) electrode and the Ag–AgCl electrode are incased in a lucite tube and the end is covered with an oxygen permeable membrane. Hence, the measurement is based only on the diffusion of oxygen across the membrane. The membranes are prepared from cellophane, polyethylene, or nylon.

In a typical measurement, the electrode is inserted into an airtight container which holds the sample. The atmosphere above the solution is expelled by pushing the electrode into the solution. Subsequently, the potential across the electrode pair is initiated and the amount of oxygen is registered on a meter.

## Questions

1. How does polarography differ from other electroanalytical techniques?
2. Define the components of a simple polarograph.
3. How is a polarographic cell designed?
4. What are the advantages of a DME?
5. Describe the polarographic wave.
6. What is the decomposition potential?
7. How is the $E_{1/2}$ determined?
8. What precautions should be taken in polarography?
9. How is a quantitative determination accomplished by polarography?
10. Describe an amperometric titration.

## Problems

1.* The data below were obtained for a series of standard solutions of zinc ion. What is the concentration of an unknown zinc solution if its diffusion current is 10.2 $\mu A$?

| $[Zn^{2+}]$ ($M \times 10^3$) | $i_d$ ($\mu A$) | $[Zn^{2+}]$ ($M \times 10^3$) | $i_d$ ($\mu A$) |
|---|---|---|---|
| 0.00 | 2.6 | 0.80 | 16.8 |
| 0.20 | 6.7 | 1.00 | 20.5 |
| 0.40 | 12.9 | 1.20 | 24.6 |

---

* Answers are listed at the end of the book for problems marked with an asterisk.

2.* The polarogram of a chloramphenicol solution gives an $i_d$ of 40.3 $\mu$A. To 10.0 ml of the original solution, 5.00 ml of 2.5 mg/ml chloramphenicol standard is added. This polarogram gives an $i_d$ of 65.2 $\mu$A. What is the concentration of the unknown solution?

3.* A solution of thiamine hydrochloride (100.0 ml) is buffered and diluted to 1000 ml with pH 11.5 buffer. From the data below determine the total amount of thiamine (in grams) in the sample.

| Conc. thiamine HCl (mg/ml) | $i_d$ ($\mu$A) | Conc. thiamine HCl (mg/ml) | $i_d$ ($\mu$A) |
|---|---|---|---|
| 0.40 | 32.0 | 1.20 | 95.8 |
| 0.60 | 49.2 | 1.60 | 130.0 |
| 0.90 | 75.0 | Unknown (after dilution) | 63.2 |

4.* A solution of streptomycin sulfate gives an $i_d$ of 24.3 $\mu$A. A standard solution containing 600 $\mu$g/ml gives an $i_d$ of 40.3 $\mu$A. What is the concentration of the unknown solution?

# Chapter
# Thirty
# Experimental
# Techniques

## INTRODUCTION

The main goals of the laboratory experiments in this book are to introduce the techniques required to successfully perform quantitative measurements and to demonstrate the principles discussed in the chapters. The experiments require the use of laboratory techniques that are basic to many areas of science and technology. Therefore, practice, insight, and confidence should be gained through the laboratory experience. These will, in turn, be valuable in other science courses and in future scientific or technological professions.

It is important to attempt to relate the experiment to the principle. Not only is this valuable for understanding the experiment, but also it provides an intellectual challenge. By thinking and reasoning logically, the importance of each part of the experimental procedure and how each of these contributes to the overall measurement can be realized. The laboratory should not be thought of as following a cookbook or as a form of automation.

Mistakes will be made in the laboratory. Whether they are due to poor technique or to lack of understanding of the chemistry associated with the experiment, it is important to learn from the mistakes so that they will not be repeated.

The development of *good laboratory habits* will prove beneficial:

1. Keep the working area clean and uncluttered. Clean up all spilled chemicals and dispose of chemicals according to the directions of the instructor.

2. Distilled water should be used for all solutions. The final rinsing of the equipment should be with distilled water. Glassware can be cleaned by using a

soap solution. Other cleaning solutions, such as alcoholic KOH or sulfuric acid–potassium dichromate solution should be used *only under the direction of the instructor.*

3. Use the laboratory time efficiently. The experimental procedure and the principles upon which the experiment are based should be read before entering the laboratory. Organize what has to be done before and not during the laboratory period. Determine beforehand where interruption in the procedure is possible and not possible. Not all experiments can be completed in one laboratory period.

4. Cooperation in the laboratory is required. Practical jokes and horseplay have no place in the laboratory.

5. Be patient and diligent with the manipulations in the laboratory. Speed will come with experience and careful planning. Usually, attempting to work excessively fast in relation to the amount of experience already gained will lead to carelessness. This can result in poor data as well as accidents.

6. Become accustomed to thinking in terms of the concepts, accuracy, precision, and error. This laboratory will probably have higher goals in these areas than were required previously in other laboratories.

7. Use common sense. This applies to the interpretation of the experimental procedure, use of the chemicals, and relationship to the other individuals in the laboratory.

8. The laboratory notebook is very important since it is the record of the experimental results. If the notebook is adequately kept, it should be possible to find the original data taken and the results for the analysis with a minimum of effort. Enough detail should be present so that it is possible to trace the course of the analysis. This becomes valuable if the presence of an error is suspected. The notebook, therefore, should contain all data, observations, and conclusions recorded in ink on bound pages. (Loose sheets of paper should never be used.) All data should be recorded while performing the experiment, not at a later time. Do not commit the observation or other data to memory for later recording. Erroneous work should be crossed out, not erased, since these data may actually turn out to be useful. Figure 30–1 illustrates a typical style for a laboratory notebook.

9. Be prepared in case of emergency. Although all the experiments in this text are safe if done properly, accidents can happen. Do not cause accidents through your own carelessness. Locate the fire extinguishers and other emergency equipment in the laboratory. Be prepared to treat burns from hot objects or corrosive acids. Usually, this involves a thorough washing with water under a faucet or shower. Fast treatment is particularly important if the eyes are involved. Figure 30–2 illustrates the many safety items that you should know how to use. The following are safety rules that you should follow.

*Left Page Contains Calculations*

**Standardization**

1  $mg\ KHP = ml_{NaOH} \times F_{NaOH} \times \dfrac{reaction}{ratio} \times KHP$

$445.0 = 21.58 \times F \times 1 \times 204.2$

$F_{NaOH} = 0.1010\ F$

2  etc. + average
3

**Unknown**

1  $\%KHP = \dfrac{ml_{NaOH} \times F_{NaOH} \times \dfrac{reaction}{ratio} \times KHP \times 100}{wt\ sample,\ mg}$

$\%KHP = \dfrac{19.81 \times 0.1011 \times 1 \times 204.2 \times 100}{849.7}$

$\%KHP = 48.12\ \%$

2  etc. + average
3

*Statistical calculations should also be given here*

---

*Right Page Contains Experimental Data*

Experiment 1

*Determination of Potassium Acid Phthalate (KHP) in an Unknown*

Date begun: May 1, 1973                      Unknown number: 99

Date completed: May 3, 1973

Titrant: NaOH

Reaction: [phthalic structure with $CO_2H$ / $CO_2K$] + NaOH → [phthalate structure with $CO_2Na$ / $CO_2K$] + $H_2O$

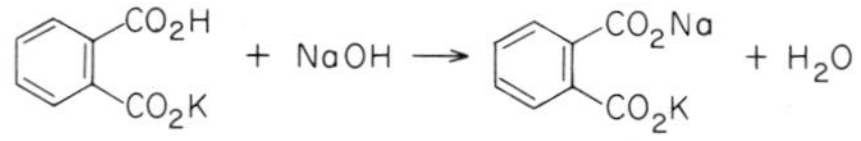

**Standardization of Titrant**

| Determination | 1 | 2 | 3 |
|---|---|---|---|
| Weight bottle + KHP | 18.1462 | 17.7012 | 17.2411 |
| Weight bottle + KHP – sample | 17.7012 | 17.2411 | 16.8141 |
| Weight pure KHP | 0.4450 g | 0.4601 g | 0.4270 g |
| Volume titrant | 21.58 ml | 22.22 ml | 20.74 ml |
| $F_{NaOH}$ | 0.1010 F | 0.1014 F | 0.1008 F |
| | average | 0.1011 F | *(Statistical treatment included if required)* |

**Unknown**

| Determination | 1 | 2 | 3 |
|---|---|---|---|
| Weight bottle + Unk | 19.4612 | 18.6115 | 17.7911 |
| Weight bottle + Unk – sample | 18.6115 | 17.7911 | 16.9244 |
| Weight pure KHP | 0.8497 g | 0.8204 g | 0.8667 g |
| Volume titrant | 19.81 ml | 19.08 ml | 20.25 ml |
| % KHP | 48.12 % | 48.01 % | 48.22 % |
| | average | 48.12 % | *(Statistical treatment included if required)* |

**Fig. 30–1.** A typical handling of experimental data in a laboratory notebook.

# SAFETY RULES

1. ALWAYS WEAR SAFETY GLASSES.
2. "Horseplay" is strictly forbidden. Enjoy chemistry, but be mature.
3. No smoking, eating, or drinking in the lab.
4. Always add acids to water, never water to acids.
5. Return caps and lids to all reagent bottles immediately.
6. Never return reagents to stock bottles.
7. Dispose of unused or contaminated reagents properly. Throw solids in waste jars. Flush water-soluble liquids down the sink with a large excess of water. Water-insoluble liquids should be placed in a waste can or jar.
8. Always lubricate glass tubing, thermometers, or funnels before inserting them into a stopper. Always wrap toweling around them while inserting. Keep hands together.
9. Be very cautious when testing for odors.
10. Never aim the opening of a test tube or flask at yourself or at anyone else.
11. Use a pipet bulb for all solutions other than water or dilute salt solutions.
12. Use hoods whenever poisonous or irritating fumes are evolved.
13. Never leave an experiment unattended while it is being heated or is reacting rapidly.
14. Makeshift equipment and setups are the first step toward an accident.
15. Under no conditions are unauthorized or unsupervised experiments to be performed.
16. Never wear loose clothing which might drag into an experiment or open flame.
17. Do not wear open shoes (sandals) or bare feet in the lab.
18. Never wear contact lenses to lab, even under safety glasses. Chemical irritants may infuse under the lens and cause irreparable eye damage.
19. If you have long hair, tie it back to keep it out of flames.
20. Report any accident, however minor, to your instructor or advisor at once. Fill out an accident report and return it to the main office.
21. AT ALL TIMES THINK ABOUT WHAT YOU ARE DOING!

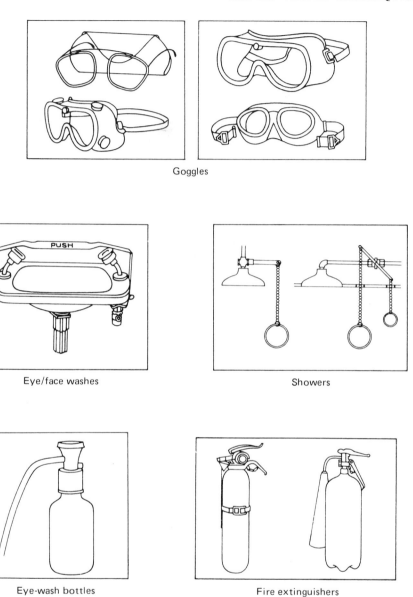

Goggles

Eye/face washes                              Showers

Eye-wash bottles                          Fire extinguishers

**Fig. 30-2.**   Safety equipment that you should know how to use and where it is stored.

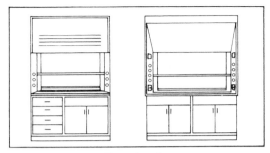

Fume hoods

Fire blanket

**Fig. 30-2.** (*Cont.*)

# MASS, WEIGHT, AND WEIGHING TECHNIQUES

Mass is an intrinsic property of matter and is constant throughout the universe. Weight, in contrast, is the force with which a body is attracted by the gravitation between it and a larger body, in our case the earth. The weight of an object may vary slightly in different locations on the earth.

One of the most important tools in quantitative analysis is the analytical balance. With it objects of unknown mass are compared with objects of known mass until the difference between the two is less than the limits of detection for the balance.

Since the analytical balance compares an unknown mass to a known mass under the same gravitational force, a change in the geographical location or gravitational attraction will not effect the value determined. Therefore, it is customary to use the terms "mass" and "weight" interchangeably when dealing with the analytical balance.

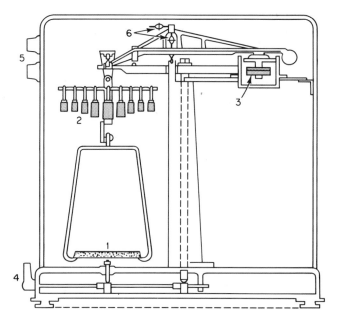

**Fig. 30–3.** A typical single-pan balance: (1) balance pan; (2) balance weights; (3) damping weight; (4) pan release; (5) weight settings; (6) sensitivity adjustments.

**Analytical Balance.**    For most analytical purposes a balance with a maximum load of 100–200 g and the ability to weigh an object to ±0.1 mg is all that is required. Most analytical balances encountered by analytical chemists today are of the single-pan variety and are designed for weighing by the substitution method. A typical balance of this type is illustrated in Fig. 30–3.

The single-pan balance has weights in the range 0.1 g to a total of 200 g suspended on the same side of the fulcrum as the pan. A counterweight is on the opposite side of the beam. When an object is placed on the pan, the instrument registers an out-of-balance position. To restore a balanced position, weights on the pan side are removed from the beam until the sum of the weights is within 0.1 g of the object weight but still on the low side. The beam is then allowed to find its rest point. This occurs rapidly due to a strong damping of the swinging motion. It should be noted that the balance is always under a constant load, and, therefore, the balance operates at a constant sensitivity.

The weights that are removed are identified by a series of dials on the front of the balance. The beam deflection, which reads up to 100 mg, also includes a vernier that allows measurement down to 0.1 mg. Reading of the beam deflection is simplified by projecting its scale, which is magnified, on a small screen. Thus, the dials give the weight in the range 0.1–200 g, the beam deflection in the range 0.001–0.100 g, and the vernier in the range 0.0001–0.001 g.

Protection of the knife edge and beam is provided by a three-position beam-arrest knob. One position arrests the pan, one is a partial release, and the third is a complete release. The partial release position is used while finding the weight range of the object.

The *single-pan balance* has three major *advantages* over the equal-arm two-pan balance :

1. The weight can be obtained faster.
2. Sensitivity is constant over the entire weight range.
3. A separate set of weights is not needed since the weights in the balance are never touched.

The *sensitivity of an analytical balance* is defined as the magnitude of deflection produced by one unit of weight. Several factors in the construction contribute to the sensitivity of the balance:

1. Sensitivity is directly proportional to balance arm length; an increase in the arm length increases the sensitivity.
2. Sensitivity is inversely proportional to the mass of the system (beam, pans, and load); an increase in this mass reduces the sensitivity. For this reason lightweight materials (Mg, Mg–Al) are used for the construction.
3. Sensitivity is inversely proportional to the distance between the support point and the center of gravity of the oscillating system; an increase in this distance decreases the sensitivity.
4. All moving parts must be as close to frictionless as possible; an increase in friction results in a decrease in sensitivity.

A wide variety of balances, which are designed to make compromises among these competing effects, are commercially available. In addition to the single-pan balance described above, more sensitive balances, balances which are useful down to 0.01 mg, and balances capable of accurately weighing very large samples are also available.

**Weighing Errors.** There are five main sources of error in the use of the analytical balance.

DEFECTIVE BALANCE. Errors due to balance construction or operation of the weights are possible. In a single-pan balance, in which the working parts are enclosed, these errors are difficult to discover unless they happen to be large. As a matter of practice the calibration and sensitivity should be checked periodically. Defects may also occur as the result of corrosion, chipped knife edges, dust, and magnetic damping errors.

STATIC EFFECTS.    A static charge can be imparted to the balance from the operator or from the objects placed on the pan. If a balance becomes charged, the net effect is a force being applied to the pan, and, consequently, the weight will be in error. Semimicro and micro balances of the equal-arm type are particularly susceptible to this type of error.

TEMPERATURE EFFECTS.    If there is a temperature gradient in the balance case, convection currents will be present. These will cause drafts against the pan and lead to erroneous weights. If the gradient is large, beam lengths (from fulcrum to each end) may actually change due to expansion and lead to an error. For these reasons hot objects should never be weighed. The balance should also be protected from sources of heat or drafts and never be positioned so that sunlight falls directly on it.

BUOYANCY EFFECTS.    When an object is placed on the balance pan, the net downward force on the pan is due to the mass of the object minus the force due to the buoyancy of air on the object. The same can be stated for weights being placed on the pan. At balance

$$m_o = m_w + (\text{buoyancy}_o - \text{buoyancy}_w) \tag{30-1}$$

where the subscripts o and w refer to object and weight, respectively, and $m$ is mass. Therefore, the true weight of an object is not equal to its weight in air unless the buoyancy effects in Eq. (30-1) happen to be equal. The difference in buoyancy is known as the *buoyancy correction*, and accounting for it will provide the true weight of the object in vacuum. Buoyancy corrections are very small and are needed only where *exact* weights are required. In most cases the correction is negligible unless the densities of the object and the weights are significantly different.

OPERATIVE ERRORS.    Operative errors are the worst and most frequent type of problem but yet are one of the easiest to correct or control. In general, they occur as the result of carelessness and improper handling of the balance. The worst offenses are those which can lead to permanent damage to the balance. For example, spillage of chemicals can lead to etching of the pan or other parts of the balance. Sudden jarring, or adding or removing weights or the sample from the pan while it is suspended can damage the knife edges in the balance. These kinds of errors will permanently affect the accuracy and sensitivity of the balance. Other operative errors, such as misreading the weight scale, spilling of weighed samples, and poor handling of the object during the weighing process are perhaps more obvious and need not be discussed further.

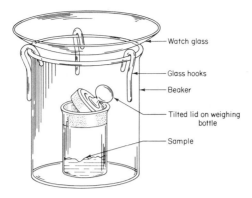

Watch glass

Glass hooks

Beaker

Tilted lid on weighing bottle

Sample

**Fig. 30–4.**   A technique for protecting a sample during drying in an oven.

VOLATILE CONTAMINANTS.   Often, objects will change weight during the weighing process. The most common reason for this is that $H_2O$ or $CO_2$ is being taken up from the atmosphere. Even the container holding the sample is capable of this action. Volatile contaminants are usually minimized by weighing techniques and location of the balance. For example, the balance can be kept in a low-humidity, controlled temperature room or in a dry box. Often, a small portion of desiccant is placed in the balance.

Frequently a sample contains small amounts of water which must be removed if a dry sample weight is required. Hence, the sample is dried in an oven at a temperature that is high enough to remove the water but not high enough to cause sample decomposition. The arrangement for oven drying illustrated in Fig. 30–4 protects the sample from dust and dirt that might accidently fall into the weighing bottle. Also, the beaker is more easily handled with tongs than the weighing bottle itself. Before weighing, the sample and its container are cooled to room temperature in a desiccator.

**Weighing the Sample.**    Replicate weighed samples are almost always required in routine quantitative analysis. Two general techniques are used for obtaining the weighed samples. The first technique is referred to as weighing by difference. The entire sample, which is usually in a weighing bottle, is weighed. Subsequently, the weighing bottle is removed from the balance pan and a sample is taken with a spatula and transferred to another container. The weighing bottle is reweighed and the sample weight is found by difference. This procedure can then be repeated for additional samples. None of the sample particles should remain on the spatula. Also, if $n$ number of weighed portions are needed, only $n + 1$ weightings will be required.

The second method involves weighing the sample directly on the balance pan. Weighing paper or a watch glass is used to hold the sample to prevent

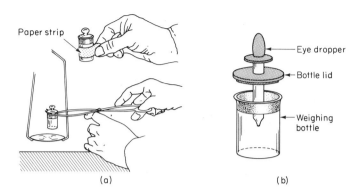

**Fig. 30–5.** Common weighing techniques: (a) two methods for transferring weighing bottles; (b) a modified weighing bottle for liquid samples.

chemical corrosion of the balance pan. Once the sample is weighed, it is transferred to another vessel. In direct weighing $2n$ weighings are required for $n$ number of weighed portions.

Figure 30–5a shows a common technique of transferring weighed weighing bottles. Another technique is to use a chamois. The paper-strip method shown in Fig. 30–5a has the principal advantage of being simple without introducing handling errors. The object or the sample container *should not be* manipulated barehanded during the weighing process when using an analytical balance. Weighing a liquid is a little more troublesome. One useful technique is to use a small bottle as shown in Fig. 30–5b. After its weight is recorded, the vial is removed from the balance, the eye dropper is used to transfer a few drops of the sample, and the entire system reweighed.

It is unwise to keep removing sample until a desired weight is transferred. With practice, portions can be taken which are approximately near the desired sample weight. Remember, as the number of transfers increase, the chance of error increases. This is particularly important if weighing by difference is used.

## VOLUME (VOLUMETRIC TECHNIQUES)

Volumetric glassware is used for the measurement of volumes. However, due to calibration, some kinds of glassware provide a more accurate measurement of volume than others. Those with low accuracy are beakers, flasks, and graduated cylinders. Those with high accuracy are volumetric flasks, pipets, and burets.

Volumetric flasks are calibrated *to contain* a stated volume at a specified temperature from 1 ml to several liters and are used to prepare solutions of known concentration. For example, if 1.000 g of a substance is placed in a 1.000-liter volumetric flask and dissolved, the final concentration of the solution will be 1.000 g/liter (or 1.000 mg/ml). The concentration can be expressed in several ways that are useful in quantitative analysis (see Chapter 3).

Pipets and burets are calibrated *to deliver* (TD) a certain volume of liquid from one mark to the next or to the orifice through gravity flow at a specified temperature. Pipets are made to deliver one fixed volume (volumetric or transfer pipet) or variable volumes (graduated or measuring pipet). Generally, volumetric pipets are available in many sizes from 0.100 to 100.0 ml. Pipets (micropipets) are also available in smaller sizes, but they are of a different design than the macropipets.

The main purpose of a buret is to deliver fractional volumes of a solution of known concentration or to deliver a volume of a solution of known concentration which reacts stoichiometrically with another reactant. Burets are graduated with capacities from 1.00 to 100.0 ml and the type of graduation will vary according to its volume. For example, a 5.000-ml microburet can be read to ±0.002 ml, while a 100.0-ml buret can be read only to ±0.02 ml. The calibration of volumetric glassware is for water unless otherwise stated, for dilute aqueous solutions, and for a specified temperature (usually 20°C).

Several other useful volumetric devices are also available. Syringes can be used at the micro level (capacity as low as 0.10 $\mu$l) and macro level to deliver gases, liquids, or solutions. Variations of the syringe are also available. The solution is forced from the reservoir by turning a piston and the volume delivered is read from a dial which records the piston turning. These are also available in the micro and macro range. Automatic burets are of several styles. The important part of these devices is the motor, for it must push the piston at an accurate, known, reproducible rate. Since burets are very basic devices in quantitative analysis, it is not surprising that considerable talent has gone into developing their design, calibration, accuracy, convenience, and versatility for applications to ordinary as well as special chemical reactions.

## COMMON LABORATORY OPERATIONS

A complete procedure for a quantitative analysis consists of several separate operations and techniques. Many of these are encountered frequently and can therefore be considered to be common in the laboratory. Often success in the laboratory will hinge on being able to perform these steps efficiently and carefully. The necessary skills will normally come with practice. However, reaching this level can be hastened by doing them correctly from the beginning. The remaining part of this section discusses the common laboratory operations and apparatus used in quantitative analysis.

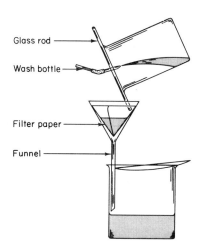

Glass rod

Wash bottle

Filter paper

Funnel

**Fig. 30-6.** Transfer of a solution and precipitate to a filtering setup.

**Filtering.**    The goal in filtering is to separate a solid phase from its contact solution (mother liquor). The filtering technique that is selected will be determined by the size of the particles being filtered, the chemical reactivity of the particles, and the purpose of the filtration (to be quantitative or not). The most common technique is to collect the solid on filter paper. It may also be collected on fritted glass plates, asbestos or filter pulp mats, and membranes.

Filter papers are available in different porosities and a paper is chosen according to the size of the particles being filtered. Fortunately, the manufacturers are able to prepare filter paper reproducibly and their recommendations as to where a particular paper should be used are very reliable.

Figure 30-6 illustrates the transfer of a solution and precipitate to a setup for filtering through filter paper.

If the weight of the filtered precipitate is desired (in a gravimetric precipitation method of analysis), the filter paper weight must be eliminated. For this purpose an ashless paper is used. This means that when the paper is combusted no paper residue will remain to be detected by the balance. The usual procedure is to carefully wrap up the moist filter paper containing the precipitate (after completing the filtration) and place it into a crucible which has been brought to constant weight. A porcelain crucible is most often used; other crucibles are made of silica or platinum. Initial drying is at low temperature ($<100°C$) until the paper is dry. The temperature is then raised so that the paper chars. Subsequently, a higher temperature is used for complete combustion of the carbon residue, leaving only the precipitate behind. This entire process can be done with burners (Bunsen or Meeker) or muffle furnaces; the combustion method chosen will be determined by the temperatures to which

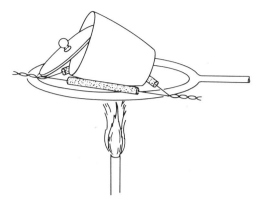

**Fig. 30–7.**   Charring of a filtered precipitate in a porcelain crucible over a Meeker burner.

the precipitate can be exposed. Figure 30–7 illustrates the charring of filter paper over a Meeker burner.

Filter paper cannot always be used. For example, the precipitate might slowly react with the cellulose or the precipitate might not be stable at the temperature needed for charring the paper. In these cases filtering crucibles such as those shown in Fig. 30–8 are used. In the crucible in (a) the filtering disk is a porous glass frit. Like filter paper, several different porosities are available. The disk in the crucible in (b), called a Gooch crucible, is perforated with large holes and an asbestos mat or other filtering substance is placed over these holes. Both types of filtering crucibles can be brought to constant weight, although it is more difficult for the Gooch type. This is one of the reasons why the Gooch crucible is not frequently used in modern quantitative analytical procedures (see Chapter 7).

Synthetic membrane filters have been developed in recent years. Their main application is not in gravimetry, but as filtering devices in separations. For example, dust in air, high molecular weight proteins, or other large molecules can be filtered. The membranes are remarkably uniform, microporous screens with millions of pores per square centimeter of surface area.

(a)

(b)

**Fig. 30–8.**   Filtering crucibles: (a) porous glass fritted disk crucible; (b) porcelain perforated disk crucible, a Gooch crucible.

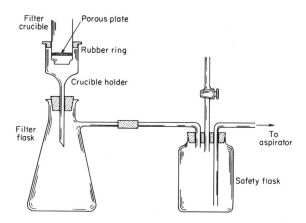

**Fig. 30–9.** Filtration under suction.

Furthermore, the manufacturers can produce synthetic membranes in many distinct pore sizes. They are also able to guarantee that all pores are of exactly the same size. For example, a readily available membrane has a pore size of 0.025 micrometer ($\mu$m) with a tolerance of $\pm0.003$ $\mu$m.

Filtration by the force of gravity (Fig. 30–6) is often a slow procedure. To hasten the filtration, the liquid may be drawn through the filter by suction. The most common source of suction is a water aspirator and the setup for such a filtration is shown in Fig. 30–9. For very difficult filtrations a vacuum pump can be used as the source of suction.

**Drying.** A desiccator provides a water-free atmosphere for storing a sample for extended periods. Frequently, this storage occurs during the cooling of the sample prior to weighing. The desiccator is usually made of glass or aluminum, has a space to fill with desiccant, and may be provided with an outlet for evacuation. The *desiccant* is a compound which has a high affinity and capacity for water and, therefore, keeps the atmosphere within the desiccator dry. Obviously, repeated opening of the desiccator will introduce atmosphere containing water. Therefore, the desiccant must take up water rapidly and have a much larger affinity for the water than the sample being stored. Figure 30–10 shows a typical desiccator containing a sample in a crucible.

Table 30–1 lists several common desiccants in the order of their effectiveness. Because of their chemical reactivity, it should be apparent that some of these desiccants should not be used in an aluminum desiccator.

Care should be exercised when placing hot samples into the desiccator. If a very hot object (for example, a sample in an open bottle or crucible) is put in

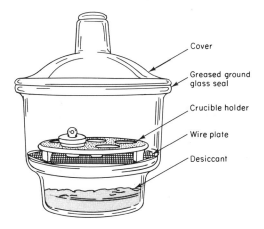

**Fig. 30-10.** A desiccator.

the desiccator and the lid tightly closed, a partial vacuum will develop as the object cools. Subsequently, upon opening the desiccator air will suddenly rush into the desiccator and spill the sample by blowing it out of its container. In general, the hot object is allowed to cool in air for 30–60 seconds before placing it in the desiccator.

**Measuring Volume.** Proper usage of all volumetric glassware requires a correct interpretation of the level of the liquid with respect to the graduation on the glassware. The bottom of the meniscus (lowest point of solution level) should be read if at all possible. This becomes difficult only when the solution

**Table 30-1. List of Common Desiccants**

| Preparation | Formula | Residue of water in mg per liter of air after drying at 25°C |
|---|---|---|
| Calcium chloride (granular) | $CaCl_2$ | 0.14–0.25 |
| Calcium oxide | $CaO$ | 0.2 |
| Sodium hydroxide (fused) | $NaOH$ | 0.16 |
| Magnesium oxide | $MgO$ | 0.008 |
| Calcium sulfate (anhydrous) | $CaSO_4$ | 0.005 |
| Sulfuric acid (concentrated) | $H_2SO_4$ (95–100%) | 0.003–0.3 |
| Silica gel | $(SiO_2)_x$ | (~0.001) |
| Magnesium perchlorate (anhydrous) | $Mg(ClO_4)_2$ | 0.0005 |
| Phosphorus pentoxide | $P_2O_5$ | <0.000025 |

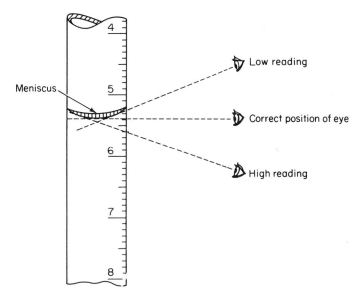

**Fig. 30–11.** Technique for reading a meniscus.

is highly colored. Often, holding a small piece of white paper or cardboard behind the meniscus aids in its detection. Also, the eye should be at the same height as the meniscus. The general technique of reading the meniscus is shown in Fig. 30–11.

*Volumetric flasks* are used for preparing solutions of known concentration. Therefore, it is important to transfer a known weight of a solute into the flask without losing any of it. Two techniques can be used. In the first, the weighed sample is transferred to the flask directly through a powder funnel (a very short, wide-stemmed funnel). Solvent is added and the solution carefully swirled until the solute dissolves. Subsequently, more solvent is added up to the mark on the volumetric flask, with the last few milliliters added very carefully.

Alternatively, the weighed solute is transferred to a beaker and dissolved with solvent. The solution is then carefully and completely transferred to the volumetric flask. More solvent is added to the flask to complete the dilution to known volume.

Both methods require that the solute be fully dissolved before the final dilution is made. When dilution is complete, the stopper is placed in the flask and complete mixing is ensured by placing the thumb over the stopper and inverting the flask back and forth at least five times.

The procedure that is adopted for filling and emptying *pipets* should be one that ensures a complete, quantitative, transfer of an amount of a solution or solvent that corresponds to the calibration on the pipet. Usually, the pipet is rinsed with a small portion of the sample that is to be pipetted. If the pipet is wet with water, several rinsings should be performed. Thus, the entire inner wall of the pipet is wetted with the sample solution. Also, by observing the drainage of the solution from the pipet it is possible to tell if the pipet is clean. If the liquid tends to form beads, the pipet is dirty and should be cleaned.

In obtaining an aliquot, the solution is drawn up to a level of about 2–3 centimeters (cm) above the calibration mark. The pipet is removed from the solution and the stem is wiped dry with tissue or filter paper. Subsequently, the solution is allowed to slowly run out the pipet, which is held vertically, until the meniscus reaches the calibration mark.

Delivery of the aliquot should be rapid and free with the pipet held in a vertical position (see Fig. 30–12). When the liquid level reaches the pipet tip, the tip is touched to the side of the container for about 4–5 seconds. The pipet should not be blown dry unless it is calibrated for this procedure.

Suction of the solution is done either orally or by a suction bulb. Bulbs should always be used when the solution is corrosive, highly volatile, or toxic; in fact, as a general rule, suction bulbs should always be used.

Delivery of a solution from a *buret* is often determined by the experimental procedure. Usually, the stopcock is opened in such a way that its fastest delivery is obtained. However, the delivery is stopped a few milliliters short of the complete transfer of the required volume and this remaining amount is added slowly and carefully. In a titration procedure one hand is used to swirl

**Fig. 30-12.** Handling of a pipet in a vertical position.

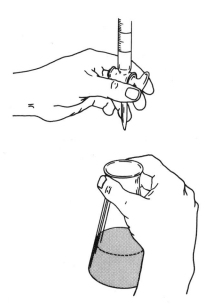

**Fig. 30-13.** Technique for manipulating the stopcock of a buret.

the titration vessel and the other is used to control the stopcock (see Fig. 30–13). The buret tip should be above the solution in the titration vessel and a wash bottle can be used to rinse titrant that collects at the buret tip into the solution.

Prior to using the buret it should be rinsed with the titrant. The buret should be in a vertical position and drainage should not leave beads of titrant on the inner glass walls of the buret. If the latter happens, the buret is dirty and should be cleaned before it is used.

## GRAPHS

Two types of graphs commonly used to handle experimental data in analytical chemistry are calibration curves and titration curves.

A *calibration curve* (or analytical working curve) is usually a graphical correlation of instrument response to concentration. Figure 30–14 illustrates a typical calibration curve. Frequently, the graph is nonlinear, and it can cover a wide concentration range or response range. In these cases the data can be plotted in log or even −log form.

The actual experimental points are shown in Fig. 30–14. If the correlation among the data is linear, a line is drawn which provides the best fit for all the

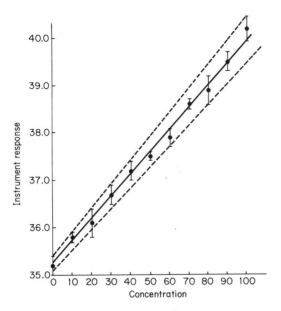

**Fig. 30–14.** A typical analytical calibration curve.

data. Usually, in drawing the line an estimate of the error associated with each point is also considered in establishing the best fit. A convenient statistical treatment which can be done manually or very easily by computer is to fit the data by a "least-squares" method.

The concentration of an unknown is established by treating the unknown in the same way as the standards that were used for preparing the calibration curve. When the instrumental response for the unknown has been determined, its concentration can be read directly from the calibration curve (Fig. 30–14).

Graphical linear relationships often provide other useful information about the system. The equation describing the line in Fig. 30–14 takes the general form

$$y = b + mx$$

where $b$ is the intercept, $m$ is the slope, and $y$ and $x$ are the ordinate and abscissa, respectively. Depending on the exact form of the equation, values for $b$ and $m$ can be the information that is being sought. For example, the $b$ term might be proportional to the equilibrium constant and the $m$ term might be proportional to the stoichiometry of the reaction.

Determination of $b$ is obvious (see Fig. 30–14). For $m$ the following equation is used:

$$m = \text{slope} = \frac{y_2 - y_1}{x_2 - x_1}$$

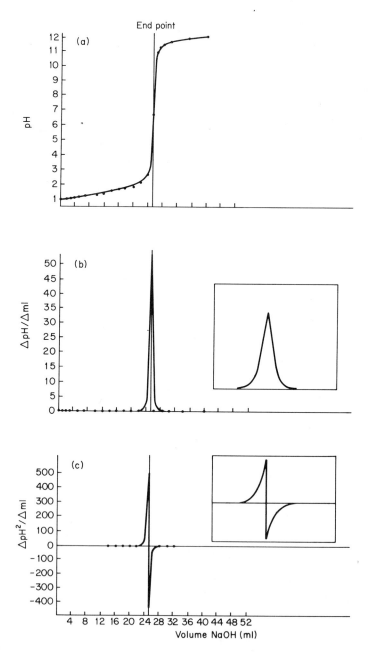

**Fig. 30-15.** Treatment of titration data: (a) titration curve; (b) first differential titration curve; (c) second differential titration curve.

The values for $y$ and $x$ should be read from the line rather than using the actual experimental points. (The experimental points are not necessarily on the line; see Fig. 30–14.)

Plotting of titration curves is a common procedure in volumetric analytical methods. Usually the titration curve, which is a plot of a change in a property of the solution against the amount of added titrant, is S-shaped. A typical titration curve is shown in Fig. 30–15a.

The titration curve represents a chemical reaction.

$$A + B \rightarrow \text{products}$$

Assume B is the titrant and the curve in Fig. 30–15a is a plot of some property of A as a function of the concentration of B. When the amount of added B is exactly stoichiometric to A (stoichiometric point), the slope changes on the rising portion of the curve. The determination of this point (end point) is necessary in analysis, since this corresponds to the amount of B required to react with A.

An end point can be found by one of several ways. The fastest is its location by inspection as shown in Fig. 30–15a. A second method is to determine the rate of change in the property plotted on the $y$ axis with respect to a small constant increment of titrant. These values, $\Delta Y/\Delta V$, are plotted vs volume added and the resulting graph is the first derivative of the data (see Fig. 30–15b).

The second derivative can also be taken (see Fig. 30–15c). Although the calculation of the second derivative is time-consuming, it provides a more exact measurement of the end point since the point of inflection appears as zero on the plot.

The time needed for computing the first and second derivatives can be eliminated by appropriate electronic circuits. With instruments so equipped the experimental data are produced on a chart paper or on an oscilloscope as either the first or second derivative.

## Questions

1. What is the difference between mass and weight?
2. Does the analytical balance measure mass or weight?
3. In past years the equal-arm balance was routinely used in the analytical laboratory. Discuss the limitations of this balance.
4. Compare the equal-arm to the single-pan analytical balance in terms of advantages and limitations.
5. List the factors that influence the sensitivity of a single-pan balance.

6. How would one prevent a balance from attaining a static charge and if it does how is it removed?
7. Describe a suitable technique for weighing a solid sample, a liquid sample, and a gaseous sample to a 0.1-mg accuracy.
8. Describe a technique for calibrating a 50-ml volumetric flask.
9. How is the volume of a 25-ml pipet affected if the solution being pipetted is at 70°C?
10. What is ashless filter paper?
11. Explain why certain filter paper is used for large crystals while other types are used for small crystals.
12. Why isn't $H_2SO_4$ or $P_2O_5$ used as a desiccant in a aluminum desiccator?
13. Graph the following experimental data:

| Mass (g) | Volume (cm³) |
|---|---|
| 15.0 | 2.0 |
| 40.0 | 4.6 |
| 61.0 | 8.0 |
| 71.0 | 9.3 |
| 88.0 | 10.9 |

a. What is the significance of the slope?
b. If the mass were determined as ±0.3 g and the volume as ±0.3 cm³, show how this error range would affect each point on the graph.
14. Graph the following experimental data:

| Temperature (°C) | Solubility (g/100 ml) |
|---|---|
| 10 | $0.3 \times 10^{-4}$ |
| 15 | $0.6 \times 10^{-4}$ |
| 20 | $1.1 \times 10^{-4}$ |
| 30 | $2.8 \times 10^{-4}$ |
| 40 | $5.5 \times 10^{-4}$ |
| 50 | $8.6 \times 10^{-4}$ |

Predict the solubility at 5°C and 70°C.
15. Graph the following experimental data using pH as the ordinate.

| pH | Volume of NaOH added (ml) | pH | Volume of NaOH added (ml) |
|---|---|---|---|
| 2.88 | 0.00 | 6.35 | 39.00 |
| 3.48 | 2.00 | 8.73 | 40.00 |
| 3.91 | 5.00 | 11.09 | 41.00 |
| 4.28 | 10.00 | 11.77 | 45.00 |
| 4.75 | 20.00 | 12.05 | 50.00 |
| 5.23 | 30.00 | 12.30 | 60.00 |
| 5.71 | 36.00 | 12.52 | 80.00 |

a. Determine the pH at the point of inflection (point of greatest slope).
b. Determine the pH at halfway to the conditions in part (a).
c. Plot the first and second derivative of the graph. Use the line drawn in the graph as the source of additional data.

# Experiments

## 1. Calibration of a Buret: Practice with the Balance

REFERENCE: Chapter 3.

INTRODUCTION: Two basic types of equipment in an analytical laboratory are balances and volume-measuring devices. Often the beginner will assume that these devices are accurate and that they know how to use them. This experiment is designed to provide practice in the use of the analytical balance and to check the calibration on one of the volume devices used in the laboratory. Burets or pipets can be used.

PROCEDURE: Clean the buret using a mild detergent and buret brush. Thoroughly rinse and test the buret for dirt by allowing water to drain from the buret. If the buret is clean, an even film of water will develop on the surface. If the buret is still dirty, water droplets will form as the liquid is allowed to drain from the buret. Sometimes it is necessary to clean the buret with dichromate cleaning solution. However, care must be taken when using this cleaning agent since it can cause serious burns.

After the buret has been cleaned and rinsed, fill it with distilled water so that the meniscus is above the zero mark. (Be sure there are no air bubbles in the buret tip.) Record the temperature of the water. Weigh a glass-stoppered weighing bottle to the nearest milligram.

Zero the meniscus exactly to the mark on the buret. Remove the water droplet at the buret tip by touching it to a piece of glass. Slowly, allow a little less than 5 ml of water to drain into the weighing bottle, wait 30 seconds for drainage from the wall of the buret, and adjust the meniscus to read exactly 5.00 ml of water drained. Touch the tip of the buret with the weighing bottle, seal, and weigh. Repeat the procedure for the 5.00 to 10.00, 10.00 to 15.00, and 15.00 to 20.00-ml mark in individual weighing bottles. Calculate the volume of water contained in each 5-ml portion from the data obtained and table of densities provided by the instructor.

Determine the volume for a pipet and a water sample provided by the instructor.

## 2.   The Importance of Sampling and Statistical Handling of Data

REFERENCE:   Chapters 3, 4, and 30.

INTRODUCTION:   An important part of any analysis is to obtain a representative sample. If the sample is homogeneous, any portion of it can be taken. If the sample is not homogeneous, a statistical sampling is required. This experiment illustrates the importance of sampling, practice with the buret, and statistical handling of data.

PROCEDURE:   Prepare a solution containing 20 ml of mineral oil and 170 ml of 0.1 $F$ HCl in a 250-ml Erlenmeyer flask. Stopper the flask and shake vigorously until the oil droplets appear to be suspended in the aqueous layer. Pipet 20 ml of the solution into a 250-ml Erlenmeyer flask and add about 25 ml of distilled water. Titrate the sample with 0.1 $F$ NaOH using phenolphthalein as indicator. Repeat the procedure *seven* (7) times. (Salad dressing containing acetic acid can be used in place of the HCl–mineral oil mixture.)

1. Compare the results and comment about the method of sampling.
2. For the seven results calculate the following:

| | |
|---|---|
| a. Average | e. Apply the $Q$ test |
| b. Median | f. Apply the $4\bar{d}$ test |
| c. Average deviation | g. Apply the $3s$ rule |
| d. Standard deviation | |

## 3.   Gravimetric Determination of Sulfate

REFERENCE:   Chapters 3, 7, and 30.

INTRODUCTION:   If a solution containing sulfate ion is mixed with one containing barium ion, a barium sulfate precipitate forms.

$$SO_4^{2-} + Ba^{2+} \rightarrow BaSO_{4(s)}$$

This reaction is readily applied to the gravimetric determination of sulfate and less often applied to the determination of barium. The method is generally for macro amounts of sulfate but if certain precautions are taken micro amounts of sulfate can be quantitatively precipitated. Furthermore, sulfur in other oxidation states in inorganic and organic samples can be determined, provided the sulfur is first oxidized to an oxidation state of $6 + (SO_4^{2-})$.

This particular gravimetric method is typical of most gravimetric procedures. The principal laboratory techniques that are illustrated in this experiment are: (1) formation of a precipitate; (2) digestion, which results in the formation of crystals that are easily filtered; (3) filtration; (4) weighing objects at constant weight; (5) calculations involving gravimetric factors. The chemistry of the system illustrates the formation, isolation, and weighing of a pure, stoichiometric precipitate.

PROCEDURE: Three porcelain crucibles should be cleaned and brought to constant weight over a Meeker burner (see Chapter 30, Laboratory Techniques). The sulfate sample is dried for 1 hour at 100°C and after cooling 0.4-g samples (3 samples) are weighed into 400-ml beakers. Dissolve the sample with 100 ml of water and 3 ml of conc. HCl. Place in separate beakers 50 ml of the $BaCl_2$ solution (50 ml/0.4 g of sample) and bring each sample solution and each $BaCl_2$ solution to a boil. Rapidly add the $BaCl_2$ solution to the sulfate sample and stir for 1 minute. Allow the precipitate to settle and add 1–2 ml of the $BaCl_2$ solution to check for complete precipitation. Heat each solution—precipitate for 1 hour without boiling. After this digestion period, cool the solution and filter through Whatman No. 42 ashless filter paper. All of the precipitate must be quantitatively transferred to the filter paper; use the rubber policeman and the wash bottle. Wash the precipitate with a minimum amount of water or until the filtrate shows an absence of chloride ion. (Check with dilute $AgNO_3$ solution.) Carefully fold the filter paper and transfer to a weighed porcelain crucible. Dry the paper in the oven at 100°C and then carefully char the paper over a Meeker burner (do not burn the paper and use the crucible lid to prevent losses from the crucible). Bring the crucible and precipitate to constant weight. Find the weight of barium sulfate and calculate the percent $SO_3$ in the sample.

## 4. Gravimetric Determination of Chloride

REFERENCE: Chapters 3, 7, and 30.

INTRODUCTION: The gravimetric determination of chloride is a typical and widely used gravimetric method. The stoichiometry of the reaction is given by

$$Ag^+ + Cl^- \rightarrow AgCl_{(s)}$$

The procedure includes (1) formation of a precipitate, (2) digestion at room temperature, (3) filtration, (4) weighing at constant weight. Filter paper is not used to collect the AgCl precipitate because cellulose can act as a reducing agent and reduce the $Ag^+$ in AgCl to silver metal. AgCl can also be decomposed photochemically.

$$2AgCl \xrightarrow{h\nu} 2Ag_{(s)} + Cl_{2(s)}$$

If the precipitate is exposed excessively to daylight, it will be grayish rather than white due to the formation of silver metal. Usually, this transformation is not easily detected by the analytical balance unless the amount of precipitate is very small.

PROCEDURE: Three sintered glass crucibles of medium porosity are brought to constant weight (see Chapter 30) by heating in an oven at 120°C for at least 1 hour.

The unknown chloride is dried for 1 hour at 120°C. After cooling three samples in the range 0.4–0.6 g are carefully weighed and transferred to 400-ml beakers. About 100–150 ml of water and 1 ml of conc. $HNO_3$ are added to dissolve the sample. The solution is brought to boiling and 0.2 $M$ $AgNO_3$ is added slowly, while stirring the hot solution. The AgCl precipitate should form immediately. Complete precipitation is determined by allowing the precipitate to settle and adding small amounts of $AgNO_3$ to the aqueous layer. If no cloudiness appears precipitation is complete. Avoid adding a large excess of $AgNO_3$. Allow the hot solution to sit for 1 hour. The precipitate is filtered through the sintered glass crucibles, using the techniques shown in Figs. 30–6 and 30–9. A rubber policeman can be used to aid in the transfer of the precipitate. The precipitate should be washed with small amounts of dilute nitric acid (a couple of drops of $HNO_3$ per 100 ml). The filtrate should be checked to be sure that precipitation was complete by adding one drop of the $AgNO_3$ solution to it. The sintered glass crucibles are then brought to constant weight using the procedure outlined previously.

Calculate the %Cl in the sample.

### 5. Homogeneous Precipitation of Nickel: Determination of Nickel in a Nickel Ore

REFERENCE: Chapters 3, 7, and 30.

INTRODUCTION. Nickel can be homogeneously precipitated as the insoluble dimethylglyoxime (DMG) complex by an *in situ* preparation of the complexing agent from hydroxylamine and biacetyl. The properties of DMG and its nickel complex were discussed in Chapter 7 (see also Chapter 15) and will not be repeated here. In the procedure, it is necessary to control the pH, and tartaric acid is added to mask other metal ions common to a nickel ore. The reactions are the following:

$$2H_2NOH + \begin{array}{c} CH_3-C{=}O \\ | \\ CH_3-C{=}O \end{array} \rightarrow \begin{array}{c} CH_3-C{=}NOH \\ | \\ CH_3-C{=}NOH \\ (DMG) \end{array} + H_2O$$

$$2DMG + Ni^{2+} \rightarrow Ni(DMG)_2 + 2H^+$$

PROCEDURE: Dry an unknown nickel sample to constant weight in an oven at 100°C. Weigh accurately three samples of about 0.5 g into clean porcelain crucibles. Add 3 g of potassium pyrosulfate to each container and mix. Fuse the mixture over a burner into a homogeneous melt, cool, and place in a 600 ml beaker containing approximately 200 ml of hot distilled water. After the sample has dissolved, remove and rinse the crucible.

To the sample solution add 0.5 g each of tartaric acid and hydroxylamine hydrochloride. Adjust the pH to about 7.2 with dilute ammonium hydroxide and subsequently add 10 ml of a biacetyl solution (1 ml biacetyl/100 ml $H_2O$) and stir. Let the solution stand for 2 hours and then heat to just below the boiling point of water for 2 more hours. Allow the solution to cool and filter the precipitate through a medium porosity sintered glass crucible which have been dried to constant weight. Wash the precipitate with ethanol, dry to constant weight (110°C), and weigh. Calculate the percent nickel in the original sample.

If the unknown is received as a nickel solution, dilute to volume, take an aliquot, and start the procedure at the point where the tartartic acid is added.

### 6.   Determination of Acetic Acid in Vinegar

REFERENCE:   Chapters 3, 8, and 30.

INTRODUCTION:   Acetic acid is a weak acid having a $K_a$ of $1.76 \times 10^{-5}$. It is widely used in industrial chemistry as glacial acetic acid (it has a specific gravity of 1.053 and is 99.8 wt/wt%) or in solutions of varying concentration. In the food industry it is used as vinegar, a dilute solution of glacial acetic acid.

The stoichiometry of the titration is given by

$$HC_2H_3O_2 + NaOH \rightarrow NaC_2H_3O_2 + H_2O$$

PROCEDURE:   The procedure requires the preparation and standardization of a 0.1 $F$ NaOH solution. This is described in Experiment 7.

A 10-ml aliquot of vinegar is carefully pipetted into a 100-ml volumetric flask and diluted to volume with water. A 20-ml aliquot is removed from the flask by a pipet and transferred to a 250-ml Erlenmeyer flask. Approximately 40 ml of water and a few drops of phenolphthalein are added. The mixture is carefully titrated with the standard NaOH solution until a faint pink color of the indicator persists. Calculate the formality of the vinegar and the g of $HC_2H_3O_2$ per ml.

### 7.   Determination of a KHP Unknown

REFERENCE:   Chapters 3, 8, and 30.

INTRODUCTION:   This analysis is a typical acid–base type of determination. A sodium hydroxide titrant is prepared, standardized and used to determine the KHP (potassium acid phthalate) content of a homogeneous sample (the unknown sample is a mixture of KHP and inert material). The reaction taking place during the titration is

TITRANT PREPARATION:   Heat 1 liter of distilled water in an appropriate container and boil for 2–3 minutes. Cover and cool to room temperature. Obtain 5.5–6 ml of 50% NaOH from the instructor (keep the stock bottle closed when not in use) and transfer the solution to a clean plastic bottle. Add the 1 liter of distilled water, mix thoroughly, and keep the bottle closed. The boiling procedure and closing of the NaOH containers is necessary to exclude carbon dioxide.

STANDARDIZATION:   Either dry primary standard KHP at 110°C in a weighing bottle or obtain already dried KHP from the instructor. Carefully weigh the weighing bottle and its contents. Transfer a KHP sample, 0.3–0.45 g, to a 250-ml conical flask. Weigh the weighing bottle again and find the weight of KHP by difference. Three samples should be weighed. Dissolve the sample in 40–50 ml of water, add three drops of phenolphthalein indicator, and titrate to the pink color with NaOH solution. Repeat the titration for each sample and calculate the formality of the NaOH solution.

TITRATION OF THE UNKNOWN: The titration of the unknown is performed in the same way as the standardization. Dry the sample in a weighing bottle at 110°C and cool in a desiccator before weighing. A sample weight of 0.5–0.9 g is probably suitable. However, since it is desirable for best accuracy to have the titration fall in the 17–24-ml range often, one sample is weighed to obtain an approximate titration value. From this, a weight range for a 17 to 24-ml titration can be predicted. Calculate the percent KHP in the sample.

## 8.  Determination of Soda Ash ($Na_2CO_3$)

REFERENCE:  Chapters 3, 8, and 30.

INTRODUCTION:  Carbonate ion is a weak difunctional base and is readily determined by titration with a strong acid. The titration can be conveniently carried out to the formation of $HCO_3^-$ or to $H_2CO_3$.

$$CO_3^{2-} + H_3O^+ \rightarrow HCO_3^- + H_2O$$

$$HCO_3^- + H_3O^+ \rightarrow H_2CO_3 + H_2O$$

A sharper end point is obtained if the titration is carried out to the formation of $H_2CO_3$ and therefore, better accuracy in the analysis is possible. Since $H_2CO_3$ is really carbon dioxide dissolved in water, the end point is further improved by heating the solution at the end point. This expels the $CO_2$.

The titration of soda ash ($Na_2CO_3$) is an important industrial determination. Closely related to this procedure is the determination of $NaOH/Na_2CO_3$ mixtures, $Na_2CO_3/NaHCO_3$ mixtures, and $NaHCO_3$. It is also possible to determine the $HCO_3^-/CO_2$ content of blood by titration.

PROCEDURE:  The procedure involves the preparation of a 0.1 $F$ HCl solution, its standardization, and determination of an unknown. Methyl red is used as the indicator for the titrations.*

*Preparation of the HCl Titrant.*  Approximately 9 ml of concentrated HCl are diluted to 1 liter of distilled water. The acid solution should be stored in a clean plastic bottle and mixed thoroughly before using it.

---

* Methyl orange (yellow to orange), Bromcresol green (blue to green), or a mixed indicator of methyl orange and a blue dye (green $\rightarrow$ grey $\rightarrow$ purple) can be used. If these indicators are used, heating should be omitted.

*Standardization of the HCl Titrant.*   Dry primary standard $Na_2CO_3$ for 2 hours at 110°C in a clean weighing bottle. After cooling carefully weigh at least three samples (0.09–0.11 g) into Erlenmeyer flasks. Dissolve the $Na_2CO_3$ sample with approximately 25–40 ml of distilled water, add 3 to 5 drops of methyl red indicator, and titrate with the HCl solution until the indicator just begins to turn from yellow to red. Heat the solution to boiling (boil for 1 minute) and cool before proceeding with the addition of more HCl titrant to the methyl red end point. If the indicator color has faded, add more indicator.

*Titration of Unknown.*   The unknown sample is dried at 110°C for 2 hours, cooled, and 3 portions are carefully weighed into Erlenmeyer flasks. The remaining procedure is the same as the procedure used for the standardization. However, sample sizes of the unknown should be taken so that an end point at 15 to 23 ml is obtained. Report the % $Na_2CO_3$ in the sample.

## 9.   Determination of Total Salt

REFERENCE:   Chapters 3, 8, and 30.

INTRODUCTION:   It is important to know the quantity of salts (ionic substances) present in many industrial products. One simple, straightforward, and accurate method for this type of analysis is to convert the salts into acids (or bases). The acids (or bases) are then titrated with standard strong base (acid) solution.

The technique for this conversion is to pass the sample through a column of strongly acidic cation resin or a strongly basic anion resin.

$$Resin-SO_3^-H^+ + M^+X^- \rightarrow Resin-SO_3^-M^+ + H^+X^-$$

$$Resin-NR_3^+OH^- + M^+X^- \rightarrow Resin-NR_3^+X^- + M^+OH^-$$

The effluent from the column is collected and titrated.

It should be noted that the resin is insoluble in water, the exchange is stoichiometric according to charge, and exchange is reversible. Since cation exchange resin in the H-form is more stable, this kind of resin, rather than the OH-form anion resin, is used.

This procedure is widely used in the pharmaceutical industry for the titration of drugs that are used as salts. Typical examples are tetraalkyl ammonium salts, amine hydrochlorides, amine sulfates, and amine perchlorates.

PROCEDURE:   A standard 0.1 *F* NaOH solution is required in this procedure. If the NaOH solution was standardized for a previous experiment, its concentration should be rechecked. (See Experiment 5.)

A column of the ion exchange resin is prepared by making a slurry of the resin in water and transferring the slurry to a glass column. *The resin column should never be allowed to run dry.* About 50 ml of distilled water are passed through the column to wash and settle the resin. After this, a 20-ml aliquot of the unknown solution (obtain unknown from the instructor in a 100-ml volumetric flask) is added to the funnel and allowed to pass through the column at a flow rate of about 1 ml/minute. The effluent should be collected in a Erlenmeyer flask or a beaker. A few ml of distilled water are used to wash the funnel when the sample is completely into the column. An additional 75–100 ml is passed through the column. The collected sample and 75–100 ml wash solution is then titrated with standard NaOH using phenolphthalein as indicator. Calculate the mmoles of salt/ml for the unknown solution.

## 10.  Determination of Ammonia by the Kjeldahl Method

REFERENCE:  Chapters 3, 8, 13, and 30.

INTRODUCTION:  The Kjeldahl Method for the determination of nitrogen is widely used in research, industrial, biological, and food laboratories. The method involves the conversion of nitrogen in the nitrogen-containing sample into $NH_3$ by boiling with $H_2SO_4$. After cooling, the solution is made basic by adding NaOH and the $NH_3$ is steam distilled into a trapping solution (see Chapter 8). The analysis is completed by an acid–base titration. Alternatively the $NH_3$ can be determined by an ammonia ion-selective electrode (see Chapter 13).

In the procedure provided here the oxidation step is omitted since the sample is supplied as an ammonium salt. If nitrogen samples are supplied, the instructor will provide the details for the oxidation.

PROCEDURE:  A standard 0.1 $F$ NaOH and 0.1 $F$ HCl solution are required. The procedures for preparing these are given in Experiments 7 and 8.

An unknown solution of an ammonium salt is provided in a 100-ml volumetric flask. This is diluted to volume with water and a 20-ml aliquot is carefully pipetted into the distillation flask in the equipment shown in Fig. 8–9. A 50-ml aliquot of the standard HCl solution is carefully transferred to the collection flask. Approximately 125 ml of $H_2O$ and 25 ml of 10% NaOH are carefully added to the flask. All connections must fit tightly. The condenser is turned on, the solution is heated, and the distillation carried out for approximately 30 minutes or until all the ammonia is transferred. Remove the collection flask from the distillation, turn off the heat and condenser water, and disconnect the condenser. Wash the inside of the condenser with a few ml of water into the collection flask. Add a few drops of methyl red indicator and titrate the remaining acid with the standard NaOH. Calculate the mg of $NH_3$/ml in the 100-ml volumetric flask.

## 11.  Nonaqueous Titration of an Amine in Glacial Acetic Acid

REFERENCE:  Chapters 3, 8, 9, and 30.

INTRODUCTION:  The titration of amines in nonaqueous solvents is frequently used as a control method in the organic chemical industry and for the assay of many pharmaceutical products. For example, amines can be determined in raw petroleum and petroleum products. These determinations are important since the amount of amine present will seriously affect the conversion of the raw petroleum into useful products. In pharmaceutical preparations, alkaloids, sulfonamides, antihistamines, narcotics, and amino acids are some of the more frequently titrated types of compounds.

The nonaqueous titration of organic acids is also a useful general analytical technique. In this experiment a common amine is titrated with a strong acid titrant in glacial acetic acid. In general, nonaqueous titrations are affected by the introduction of water and carbon dioxide into the system and therefore, the laboratory techniques used in the experiment should prevent these errors. If possible, the titration should be carried out under a blanket of nitrogen.

PROCEDURE: The same procedure is used for standardization and for the titration of the unknown. For standardization tris(hydroxymethyl)aminomethane (mol wt 121.1) is used as the primary standard.

Weigh accurately three samples of the amine unknown (0.14–0.19 g). Dissolve one of the samples in 30 ml of glacial acetic acid in a 180-ml tall form beaker and add 2 to 3 drops of crystal violet indicator (0.1% in acetic acid). The solution is then titrated with 0.1 $F$ $HClO_4$ in $HC_2H_3O_2$ or 0.1 $F$ 2,4-dinitrobenzene sulfonic acid in $HC_2H_3O_2$ until a pure blue color is obtained. Calculate the percent $NH_2$ in the sample.

PERCHLORIC ACID IN ORGANIC SOLVENTS AND AMINE PERCHLORATES CAN BE DANGEROUS IF NOT PROPERLY HANDLED: The solution should not be allowed to go to dryness. Immediately, after completing the titration discard the titrated sample according to the directions of the instructor.

## 12. Determination of Arsenic by Titration with Iodine

REFERENCE: Chapters 3, 12, and 30.

INTRODUCTION: Iodine is widely used as a titrant in inorganic and organic analysis. This experiment illustrates a typical iodine titration procedure and provides a procedure for standardizing an iodine titrant for other applications.

PROCEDURE

*Preparation of a 0.05 F Iodine Solution.* Dissolve 20 g of KI in 50 ml of water. Carefully add about 12.8 g of $I_2$ and stir until dissolved. Dilute the mixture to 1 liter and store in a glass container in a cool, dark place.

*Standardization of the Iodine Solution.* Weigh accurately three 0.1-g samples of $As_2O_3$ of known purity into 400-ml beakers. Add 50 ml of 2% NaOH solution. Cool and carefully add 10 ml of 20% $H_2SO_4$. Cool again and add slowly $NaHCO_3$ (solution will effervesce) until in excess. Dilute to about 150–200 ml, add 2 ml of starch indicator solution, and titrate with the iodine solution to the first blue color.

The starch indicator solution is prepared by taking 1.0 g of starch, adding a few milliliters of water to make a starch paste, and pouring the paste into 100 ml of boiling water while stirring. Boil for an additional minute, cool, and add 2–3 g of KI. The starch indicator solution should be kept in a stoppered glass bottle.

*Determination of a Arsenic Unknown.*   Three samples of the arsenic unknown are carefully weighed (about 0.5 g assuming 20% $As_2O_3$) and treated as in the standardization procedure.

Calculate the %$As_2O_3$ in the unknown.

## 13. Determination of Oxalate by Titration with Potassium Permanganate

REFERENCE:   Chapters 3, 12, and 30.

INTRODUCTION:   Potassium permanganate is widely used as a titrant in inorganic and organic analysis. In this experiment, oxalate ion is determined by titration with potassium permanganate titrant. This procedure is also the one that can be routinely used for standardizing $KMnO_4$ for other applications.

Since many transition metal ions can be precipitated as oxalate salts this procedure can be modified to allow the determination of these metal ions. In general, excess oxalate is added, the metal oxalate precipitate filtered, and then the precipitate is titrated with $KMnO_4$.

PROCEDURE

*Preparation of 0.02 F Potassium Permanganate Solution.*   About 3.2 g of $KMnO_4$ is dissolved per liter of solution. This solution is brought to boil and kept hot for 1 hour, allowed to cool, preferably overnight, and the solution is filtered (remove solid $MnO_2$) through a sintered glass crucible of fine porosity. Store the titrant in a clean glass-stoppered bottle and keep in the dark.

*Standardization of the Potassium Permanganate Solution.*   Primary standard $Na_2C_2O_4$ is dried for 1 hour at 100°C. After cooling, three samples (0.13 g) are accurately weighed and transferred to 400-ml beakers. Approximately 200 ml of $H_2O$ containing 10 ml of 1:1 $H_2SO_4$:$H_2O$ are added and the solution is brought to boiling. Once dissolution is complete, the hot solution is titrated with the $KMnO_4$ titrant at a rate not to exceed 10 ml/min  with vigorous stirring. The end point is indicated by a persistant purple coloration due to excess titrant.

*Determination of the $Na_2C_2O_4$ Unknown.*   The unknown sample is dried at 100°C for 1 hour. After cooling, three samples are carefully weighed (about 0.6 g assuming 20% $Na_2C_2O_4$) and titrated by the same procedure used for standardization.

Calculate the %$Na_2C_2O_4$ in the unknown.

## 14.   Iodometric Determination of Copper in Brass

REFERENCE:   Chapters 3, 12, and 30.

INTRODUCTION:   Copper(II) and iodide undergo the reaction

$$2Cu^{2+} + 4I^- \rightarrow 2CuI_{(s)} + I_2$$

where copper(II) is reduced in a one-electron step and each iodide is oxidized in a one-electron step. Even though copper(I) precipitates as a near white CuI precipitate it does not interfere in the titration of the liberated $I_2$ with thiosulfate titrant.

$$I_2 + 2S_2O_3{}^{2-} \rightarrow 2I^- + S_4O_6{}^{2-}$$

Iodine forms a complex with starch that is intensely blue. The disappearance of this color can be used for detecting the end point of the titration. Copper(I) iodide adsorbs iodine. Consequently, in the titration the starch indicator will turn colorless for a few seconds and revert back to a blue color as the adsorbed iodine goes into solution. Addition of potassium thiocyanate eliminates this end-point problem.

PROCEDURE: Prepare 500 ml of 0.1 $F$ sodium thiosulfate solution. The water should be boiled and 0.1 g of $Na_2CO_3$ should be added per 500 ml of solution.

The procedure for the unknown and standardization of the sodium thiosulfate solution is the same. Pure copper wire is used for the primary standard.

Weigh three 0.15-g (0.13 g for the wire) portions of the brass sample into conical flasks. Add 10–15 ml of 1:3 nitric acid and carefully heat on a hot plate until dissolved *following the instructor's instructions.* Cool before titration.

Take each sample and add 35 ml of water. Heat to boiling, add 0.5 g of urea, and boil for an additional minute. Cool and neutralize the excess acid with 6 $F$ $NH_3$ (solution should turn dark blue). Add 5 ml of acetic acid, 3 g of KI, and allow the solution to set for several minutes. Titrate the solution with the sodium thiosulfate solution until the reaction mixture has a faint brown color. Add 3 ml of the starch indicator solution and 10 ml of 5% potassium thiocyanate solution. Stir thoroughly and continue the titration to the disappearance of the blue color. Report the %Cu in the sample.

### 15. Determination of Commercial Hydrogen Peroxide

REFERENCE: Chapters 3, 12, and 30.

INTRODUCTION: Commercial hydrogen peroxide can be titrated with $KMnO_4$ or by an iodometric method. The $KMnO_4$ titration is less preferred because of interferences due to preservatives, such as boric acid, salicyclic acid, and glycerol, and to stabilizers, which are often added to commercial hydrogen peroxide solutions.

Hydrogen peroxide will react with iodide in acid solution to produce iodine by the reaction

$$H_2O_2 + 2H^+ + 2I^- \rightarrow I_2 + 2H_2O$$

The liberated iodine is then titrated with the standard thiosulfate solution.

The reaction rate for $H_2O_2$–$I^-$ is slow but increases with increasing acid concentration. Addition of ammonium molybdate will also catalyze the reaction. If this is used, the titration should be carried out under a $CO_2$ atmosphere since it also catalyzes the air oxidation of HI.

Hydrogen peroxide is commercially available as percent by weight or as 10-, 20-, 40-, or 100-volume aqueous solutions. The volume numbers correspond to the amount of available $H_2O_2$. For example, a 10-volume solution will yield 10 times its volume of $O_2$ measured at 760 mm and 0°C. A 10-volume solution is 3% $H_2O_2$ by weight.

STANDARDIZATION OF 0.1 $F$ SODIUM THIOSULFATE SOLUTION:    Approximately 25 g of [*25*] $Na_2S_2O_3 \cdot 5H_2O$ are dissolved per liter of solution. [500 ml] If the solution is to be kept for over a week, boiled distilled water should be used.

Potassium iodate will react with KI in acid solution according to the reaction and the stoichiometrically liberated iodine can be titrated with the sodium thiosulfate.

$$IO_3^- + I^- + 6H^+ \rightarrow 3I_2 + 3H_2O$$

Potassium iodate is dried at 100°C for 1 hour. After cooling 1.4 g of the $KIO_3$ is carefully weighed and dissolved in water and diluted to 500 ml in a volumetric flask. An aliquot (25.00 ml) of the iodate solution is transferred to a flask, 10 ml of 10% KI solution, 3 ml of 1 $F$ $H_2SO_4$, and 100 ml of water are added. The liberated iodine is then titrated with the thiosulfate until the solution reaches a pale yellow color. At this point, 2 ml of a starch solution (see Experiment 10) are added and the titration continued until the blue color disappears. Three aliquots should be titrated and the standardization should be repeated if the sodium thiosulfate is to be used for more than a week. The $KIO_3$ solution is stable for over a month if kept in a stoppered bottle.

DETERMINATION OF THE HYDROGEN PEROXIDE:    The $H_2O_2$ solution should be diluted to about 2-volume strength. For a 20-volume commercial $H_2O_2$ solution, exactly 10 ml of it is diluted to 250 ml with water in a volumetric flask.

A 25.0-ml aliquot is taken and carefully added to 100 ml of 1 $F$ $H_2SO_4$ containing 1 g of KI in a Erlenmeyer flask. Mix thoroughly, stopper, and allow the mixture to stand for 15 minutes. Subsequently, the liberated iodine is titrated with standard sodium thiosulfate. When most of the iodine has reacted, 2 ml of starch solution is added and the titration continued to the disappearance of the blue color.

An alternate procedure, which provides better accuracy, is to add the 25.0-ml aliquot to 100 ml of 1 $F$ $H_2SO_4$. Several chips of dry ice, 10 ml of 10% KI, and 3 drops of 3% ammonium molybdate solution are added. The liberated iodine is immediately titrated with standard thiosulfate. More dry ice is added if needed.

Calculate the weight of $H_2O_2$ per 1000 ml of the original solution.


## 16. Determination of Ferric Iron and Total Iron in Pharmaceutical Preparations

REFERENCE:    Chapters 3, 12, and 30.

INTRODUCTION:    A wide variety of anemia related disorders (iron deficiency) can be corrected by taking "iron" pills. In general, the pills or solutions contain ferrous salts plus other ingredients. Two typical commercial preparations are Fergon (Sterling Drug, Inc.) and Feosol (Smith, Kline, & French Pharmaceutical Co.). Both of these, which are in pill form, are readily available at the drug store. Fergon is a ferrous gluconate based tablet and Feosol is a ferrous sulfate based tablet. In this experiment, the amount of ferric iron per tablet (which may be very small) and total amount of iron per tablet will be determined through oxidation–reduction titrations. Additional details for these quality control analyses are available in "The National Formulary."

PROCEDURE: A standard solution of potassium dichromate (0.05 $F$) (see Experiment 17) and sodium thiosulfate (0.1 $F$) (see Experiment 13) are required. Approximately 30 Fergon or Feosol tablets are weighed and are powdered and kept in a closed weighing bottle until used.

FERRIC IRON: Dissolve 5 g of the accurately weighed sample in 100 ml water, 10 ml of 12 $F$ HCl, and add 3 g of KI. Stir thoroughly and allow the solution to stand for 5–10 minutes in the dark. Titrate any liberated iodine with standard sodium thiosulfate using starch as the indicator.

Calculate the milligrams of Fe(III) per gram and milligrams of Fe(III) per tablet.

TOTAL IRON: Weigh a 1–1.5-g sample accurately and dissolve it in a mixture of 75 ml of water and 15 ml of 12 $F$ HCl in an Erlenmeyer flask. Heat the solution almost to boiling and add 1 ml of the $SnCl_2$ solution. Avoid large excesses of $SnCl_2$. Immediately, cool the solution and add 10 ml of the $HgCl_2$ solution. A very small quantity of white precipitate should appear. If it is gray, start again and use less $SnCl_2$ solution. (The amount of $SnCl_2$ required will be determined by the amount of $Fe^{3+}$ in the tablet.)

Add carefully 10 ml of concentrated $H_2SO_4$ and 15 ml of concentrated $H_3PO_4$ to the solution. Cool, add 5–9 drops of diphenylamine sulfonate indicator, and titrate with the standard potassium dichromate solution.

Calculate the milligrams of Fe per gram and milligrams of Fe per tablet.

The $SnCl_2$ solution is prepared by dissolving 30 g of $SnCl_2 \cdot 2H_2O$ in 50 ml of 12 $F$ HCl and diluting to 1 liter with water. The solution is stable for several weeks if kept in a tightly closed bottle.

The $HgCl_2$ solution is prepared by dissolving 5 g of $HgCl_2$ in 100 ml of water.

A 0.2% solution of sodium diphenylamine sulfonate is prepared in water.

### 17. Potentiometric Titration of Iron(II) with Potassium Dichromate

REFERENCE: Chapters 3, 10, 11, 12, and 30.

INTRODUCTION: In general, the same basic laboratory techniques are required in carrying out all redox potentiometric titrations. However, the experimental conditions such as acidity, protection from oxygen, presence of a catalyst, and adjustment of oxidation state of the sample being titrated may vary widely. In this experiment familiarity with potentiometry is the main goal. This is illustrated by titrating an iron (II) solution with a standard solution of $K_2Cr_2O_7$ using a Pt–SCE electrode pair.

PROCEDURE: A 0.05 $F$ solution of $K_2Cr_2O_7$ is prepared by accurately weighing about 2.9 g of primary standard $K_2Cr_2O_7$ (dry at 110°C for 30 minutes) and diluting to exactly 200 ml with 1 $F$ H_2SO_4 (calculate the exact formality for the weight taken).

A carefully measured aliquot of the $Fe^{2+}$ solution (0.11 $F$) which is dissolved in 1 $F$ $H_2SO_4$ is taken, the Pt–SCE electrode pair are inserted, and the potential is recorded according to the directions of the instructor. Titrant is added from a buret at 2–3-ml increments at the beginning and 0.1-ml increments at the equivalence point allowing equilibrium to be reached before the potential is recorded.

Plot the data in the following way and select the end point from each graph.

    a. Potential (vs SCE) vs ml of titrant.
    b. Potential (vs NHE) vs ml of titrant.
    c. First derivative of potential (vs SCE) vs milliliter of titrant (see Chapter 30).
    d. Second derivative of potential (vs SCE) vs milliliter of titrant (see Chapter 30).

If the $Fe^{2+}$ solution is provided as an unknown, report the milligrams of Fe in the iron unknown solution. Alternatively, the iron content of an iron ore can be determined. The instructor will provide directions for dissolving the iron ore and for adjusting the oxidation state of the iron to 2+ (see Chapter 13).

### 18. Potentiometric Titration of Sulfanilamide

REFERENCE: Chapters 3, 10, 11, 12, and 30.

INTRODUCTION: Potentiometry is used for end-point detection in a wide variety of applications. Suitable color indicators are not available for all oxidation–reduction titrations and potentiometric titrations are ideally suited to automation.

The determination of sulfanilamide in commercial tablets is a typical routine pharmaecutical assay. Sulfanilamide will undergo a reaction with sodium nitrite to form a diazonium salt [see reaction (11–7)] and the course of this reaction can be followed potentiometrically with a Pt–SCE electrode pair.

PROCEDURE: Prepare a 0.1 $F$ $NaNO_2$ solution which will be standardized against pure sulfanilamide. The basic procedure for the standardization and the titration of the unknown are the same. The sulfanilamide sample is dissolved in 50 ml of water and 10 ml of concentrated HCl in a 250 ml beaker. A calomel and Pt electrode are inserted into the solution. The operation of the potentiometer will be explained by the instructor. The titrant is placed in a buret and a few milliliters of titrant are added while the solution is stirred magnetically. (Do not stir vigorously or allow the stir bar to hit the electrodes.) After the system reaches equilibrium record the potential and add more titrant and repeat the process. At the beginning of the titration 3–5 ml additions are sufficient, while in the vicinity of the stoichiometric point 0.1-ml increments should be used. Repeat the titration, plot the data, and from the graph determine the volume of titrant required for the titration.

For standardization weigh accurately about 0.3 g of the pure sulfanilamide. For assay of the sulfanilamide tablets weigh accurately 20 tablets and carefully powder them in a mortar avoiding any loss. Weigh accurately about 0.3 g of this powder for each sample to be titrated. (The instructor may provide single tablets or a solution for assay instead.) Add the water, acid, and stir to dissolve the sample. Because of filler materials in the tablet, it may not all dissolve.

From the data calculate the total amount of sulfanilamide in the original 20 tablets. Express the assays as the average weight of sulfanilamide per tablet and percent of labeled strength.

### 19. Potentiometric Titration of an Organic Acid; Determination of p$K_a$ and Formula Weight

REFERENCE: Chapters 3, 7, 13, and 30.

INTRODUCTION: The formula weight of an acid in a given titration can be calculated if the sample weight of the pure acid, the formality of the titrant solution, the volume of titrant needed, and the reaction ratio are known. Also, the p$K_a$ value for a monoprotic acid can be shown to be approximately equal to the pH value at the point corresponding to the midpoint of the titration (50% neutralized).

It should be emphasized that p$K_a$ values obtained this way may differ from the true value unless a constant temperature is maintained, ionic strength is maintained, and approximations at the midpoint pH are not made.

In this experiment, the entire titration curve is obtained by means of the measurement of the pH after the addition of each increment of the titrant solution. A glass electrode and a saturated calomel electrode are used and the pH is measured with a pH meter.

PROCEDURE: A 0.1 $F$ standardized NaOH solution is required. If the NaOH solution is old, its concentration should be rechecked vs KHP. (See Experiment 5.)

Standardize the pH meter with the standard buffers. The electrodes should be carefully rinsed with water before placing them in the buffer solution or in the sample solution.

Carefully weigh out the organic acid and transfer to a 250-ml beaker. The approximate sample weight should be calculated assuming a molecular weight of 150, 15 ml titration, and 0.1 $F$ NaOH titrant. Dissolve the sample in 100 ml of distilled water. Insert the electrodes (be sure that they are washed free of buffer solution), and titrate with the NaOH solution. Collect points at 0.5 to 1-ml increments. As the stoichiometric point is approached (pH will increase), take points at smaller increments (0.1 ml). Continue the titration until the pH reaches a value of 10 and increases only slightly with increased titrant. Two titrations are required. The meter should always be at standby when not in use or when the electrodes are removed from the solution.

A titration curve for each titration should be plotted. The p$K_a$, stoichiometric point, and molecular weight × reaction ratio should be determined from the graphs according to the directions of your instructor.

### 20. Measurements with the Fluoride Electrode

REFERENCE: Chapters 13 and 30.

INTRODUCTION: The fluoride electrode developes a potential proportional to the logarithm of the activity of the fluoride ion in the sample. For each tenfold change in fluoride ion activity, the electrode exhibits (at 25°C) a 59.16 mV change. If the concentration of fluoride ion equals fluoride activity, a calibration curve can be prepared by plotting potential vs fluoride molarity or vs ppm $F^-$. (See Fig. 13–11 for the activity-concentration relationship.) The potential for the unknown solution is measured and its concentration deduced from the calibration curve.

PROCEDURE: A series of NaF solutions are prepared in the concentration range of $10^{-1}$ to $10^{-5}$ $M$ by dilution of a standard NaF solution. The potential of each of these solutions is measured according to the directions of the instructor and a calibration curve is constructed. The potential for the unknown solution is measured and its concentration is determined from the calibration curve.

Report the fluoride ion concentration of the unknown solution in $M$ and ppm units, Fluoride ion concentration in several other samples can also be determined. However, these should be treated so that they contain a suitable pH and ionic strength. Hence, the dilution should be such that 9 parts of 15% $NaC_2H_3O_2$ and 1 part unknown are mixed. The standards must also be prepared in the same 9:1 ratio.

Several samples that can be tested are:

a. Carbonated beverages of all kinds (remove $CO_2$ by heating)

b. Coffee and tea

c. Toothpaste

d. Drinking and natural water

e. Detergents (add some sodium citrate to the solution)

f. Saliva

g. Teeth (dissolve in $HClO_4$, care must be used in handling the $HClO_4$ or an explosion may occur, add sodium citrate to the final solution)

## 21. Fajan's Method: Titration of Chloride with Silver

REFERENCE: Chapters 7, 14, and 30.

INTRODUCTION. There are several methods that are routinely used for the determination of chloride. Although the method in this experiment and the Volhard method will provide accurate analysis of chloride in biological samples, the procedures are not used for this purpose because it is felt that they are not suitable for multiple analysis in the clinical laboratory. However, these two methods are routinely used to calibrate and check the accepted clinical procedures.

This experiment illustrates a typical precipitation titration and a chemical way for detecting its end point. There are also several excellent instrumental ways of detecting the end point of this titration. In this procedure, the indicator is dichlorofluorescein, and dextrin is added to retard coagulation of the precipitate and promote surface adsorption on the precipitate.

PROCEDURE: Prepare and standardize a solution of silver nitrate as follows. Dissolve about 4.3 g of silver nitrate in 250 ml of distilled water and store in a dark bottle. Weigh accurately three 0.2-g portions of dried NaCl into 250-ml Erlenmeyer flasks and dissolve in 50 ml of distilled water. Add 10 drops of dichlorofluorescein solution and 0.1 g of solid dextrin to each flask. The solution should be neutral to slightly acidic. Titrate each of the three samples to the indicator end point (solid, white → pink) and calculate the concentration of the standard solution.

Obtain an unknown chloride sample from your laboratory instructor. Dry the sample for one hour at 110°C. After cooling, weigh three 0.4-g samples accurately and dissolve each in 50 ml of distilled water. Add the indicator and dextrin in the amounts described previously and titrate with the silver nitrate solution. Calculate the percent Cl in the original sample.

## 22. Determination of Magnesium and Calcium in Hard Water and Egg Shells

REFERENCE: Chapters 3, 15, 16, and 30.

INTRODUCTION: In this experiment water hardness (Mg–Ca) is determined by titration with EDTA. The water sample will not contain $Fe^{3+}$ and other interfering metal ions. Hence, the masking effect of $CN^-$ will not be required.

PROCEDURE: *Preparation of standard 0.05 F EDTA.* EDTA is available as the disodium dihydrate salt. This salt is dried at 70°C for 1 hour, cooled, and 9.3 g are carefully weighed into a 500-ml volumetric flask and diluted to volume with distilled water. (The exact concentration should be calculated according to the weight taken.)

*Determination of the Ca–Mg Unknown.* The unknown in the volumetric flask is diluted to volume. An accurately measured aliquot (25.00 ml) is taken, 5 ml of pH 10 $NH_3/NH_4Cl$ buffer, and enough Eriochrome Black T solid (the indicator and $Na_2SO_4$ are mixed at a 1:200 ratio) to impart a color to the solution are added. Titrate with the standard EDTA solution to a color change of red to blue. If available, a pH meter should be used to insure that the pH remains at pH 10 during the titration.

Report the results as mg of Mg/ml of unknown solution.

*Ca–Mg in Egg Shells.* The eggs are broken, the albumin and yolk are discarded, and the shells are carefully washed with distilled water. The shells are air-dried for several days and then ground into a fine powder with a mortar and pestle. Three samples of 0.100 g each are accurately weighed into porcelein crucibles and ashed at 700°C in a muffle furnace for at least 16 hours. Subsequently, the crucibles are cooled, and 2 ml of water and 1 ml 12 F HCl are carefully added. When the residue has dissolved, the solution is transferred to a flask and titrated with the standard EDTA titrant using the previously outlined procedure.

Calculate the mg of Ca per gram of egg shell. Depending on the type of egg the Mg–Ca composition of the shell is about 99% Ca. Also note that interferences of other metal ions that may be present at the trace level are not accounted for.

## 23.   EDTA Titration of Zinc and Other Transition Metals

REFERENCE:  Chapters 3, 15, 16, and 30.

INTRODUCTION:  Naphthyl azoxine S (NAS) is a useful indicator for the EDTA titration of 25 different metals. This is summarized in Table 15–8.

The color change for the direct titration is from a pale yellow to a pale pink color. For some metals the color change is not too sharp and for this reason a small measured amount of Zn or Cu is added to the sample. A blank which corresponds to the ml of EDTA equal to the Zn or Cu must then be subtracted from the total volume of EDTA added.

PROCEDURE:  A sample containing about 0.1–0.2 mmole of metal ion to be titrated is dissolved and diluted to approximately 100 ml. Add several drops of pyridine buffer and adjust the pH to about 6. Add 1 or 2 drops of 0.5% by weight NAS indicator solution (aqueous) and titrate with standard 0.01 $F$ EDTA. The standard EDTA solution is prepared as described in Experiment 22.

Calculate the mg of Zn in the unknown.

## 24.   Spectrophotometric Determination of Manganese

REFERENCE:  Chapters 3, 12, 17, 19, and 30.

INTRODUCTION:  Manganese(II) ion and its common salts are colorless. However, oxidation of $Mn^{2+}$ to $Mn^{7+}$ produces the purple $MnO_4^-$ species. This color is very intense and consequently, small amounts of manganese can be detected. Not many of the common, simple inorganic ions have this property. For example, $Co^{2+}$, $Cu^{2+}$, and $Ni^{2+}$ solutions are pink, blue, and green, respectively; however, their colors are not very intense (molar absorptivities are low). This method is widely used for the determination of manganese and is applied to many steel, alloy, and ore samples where Mn is a minor constituent.

Several different oxidation procedures are available for the oxidation of manganese. Perhaps the best is with potassium metaperiodate. This oxidizing agent has the advantages of not requiring a catalyst, has long stability, and produces soluble products. The reaction is

$$2Mn^{2+} + 5IO_4^- + 3H_2O \rightarrow 2MnO_4^- + 5IO_3^- + 6H^+$$

Chloride ion must be absent since it will be oxidized to chlorine. Also, it is important to maintain high acidity to ensure complete oxidation to $MnO_4^-$; at low acidity $MnO_2$ is formed.

PROCEDURE:  You should turn in a clean 100-ml volumetric flask (labeled with your name) to your instructor. Upon its return dilute carefully to volume with water. In two separate beakers, pipet carefully 5- and 10-ml aliquots of the unknown. Add 15 ml of concentrated $HNO_3$, 20 ml of water, 0.5 g of $KIO_4$, and boil for 3–5 minutes.

Cool to room temperature and transfer to 100-ml volumetric flasks and dilute to volume. Prepare a series of standards (4) in the same way. Measure the absorbance of the unknown and standards at 530 nm using water as a blank. Report the concentration of Mn in mg Mn/ml of unknown solution.

If the percent Mn of a steel sample is to be determined, a special dissolving technique must be included in the procedure. These directions will be provided by the instructor.

## 25. Spectrophotometric Determination of Iron with 1,10-Phenanthroline

REFERENCE: Chapters 3, 15, 17, 19, and 30.

INTRODUCTION: Many different reagents are available for the spectrophotometric determination of iron. Perhaps the most useful of these are 1,10-phenanthroline(I) and its derivatives. These reagents are very sensitive and, in general, are applied to the determination of trace levels of iron. Trace level iron in alloys, ores, minerals, commercial products of all kinds, pharmaceuticals, and biological samples can be determined with 1,10-phenanthroline. However, in a biological sample, iron must first be freed from its strong binding with the complex organic chemicals in the biological sample.

In the procedure the iron must be converted quantitatively to the $Fe^{2+}$ oxidation state. 1,10-Phenanthroline is added and an intensely colored complex between $Fe^{2+}$ and 1,10-phenanthroline forms. The reactions for these steps are

$$Fe^{3+} + \text{reducing agent} \longrightarrow Fe^{2+}$$

Adjustment of the oxidation state is usually done with hydroxylamine hydrochloride. Although there are many other ways of reducing $Fe^{3+}$ to $Fe^{2+}$, the hydroxylamine hydrochloride reagent is the best since it does not interfere in the absorption measurement. The color intensity of the complex is very sensitive to pH. Therefore, the pH of the solution must be carefully controlled.

PROCEDURE: A clean 100-ml volumetric flask (labeled with your name) should be turned in to your instructor. After receiving your unknown sample dilute the solution to volume with water. Carefully pipet 10-, 15-, and/or 20-ml aliquots of the unknown into separate 100-ml volumetric flasks. Add 10 ml of saturated sodium acetate solution,

10 ml of the 10% hydroxylamine hydrochloride solution, wait for 5 minutes, and add 10 ml of the 0.1% 1.10-phenanthroline solution. Allow 10 minutes, dilute to volume, and read the absorbance at 510 nm using water as a blank. Prepare the standards in the same way using exactly 1-, 2-, 3-, 4-, and 5-ml aliquots (in 100-ml volumetric flasks) of iron solution that is 100 ppm Fe. Report the results as ppm Fe in the unknown solution.

## 26. Spectrophotometric Determination of Lead with Dithizone

REFERENCE: Chapters 3, 17, 19, 27, and 30.

INTRODUCTION: Dithizone (diphenylthiocarbazone) exists in solution as a tautomeric mixture

$$HS-C \overset{N-NHC_6H_5}{\underset{N=NC_6H_5}{\Big\langle}} \rightleftarrows S=C \overset{NH-NHC_6H_5}{\underset{N=NC_6H_5}{\Big\langle}}$$

It forms complexes with many metal ions and can be used for their spectrophotometric determination at the microgram level. The main application is the determination of bismuth, copper, lead, and mercury. Selectivity is achieved by pH control and the use of masking agents. Since the metal complexes are insoluble, the complex is extracted into an organic solvent.

The two main problems that are encountered in this determination are: (1) The reagent is very sensitive and extra care must be employed to avoid introducing metal-ion impurities. (2) The reagent is easily oxidized to diphenylthiocarbadiazone, $S=C(N=NC_6H_5)_2$, which does not form metal complexes.

PROCEDURE: An unknown in a 100-ml volumetric flask is obtained from the instructor and diluted to volume. A 10.0-ml aliquot of this solution is taken and to this is added 75 ml of a $NH_3$–$Na_2SO_3$ solution. (This solution is prepared by diluting 35 ml of conc. $NH_3$ and 0.15 g $Na_2SO_3$ to 100 ml.) The pH is adjusted to pH = 9.4 with dilute HCl or $NH_3$, transferred to a separatory funnel, and 8.0 ml of 0.005% dithizone in $CHCl_3$ is added. An additional 15 ml of $CHCl_3$ is added, the mixture is shaken, the phases allowed to separate, and the chloroform layer is drawn off into a 25-ml volumetric flask. After diluting to volume with $CHCl_3$ the absorbance is read at 510 nm. A blank should be prepared minus the lead sample.

A calibration curve is prepared from a standard solution of pure lead chloride in which 0.0060 g of $PbCl_2$ is dissolved in 1 liter of water. Aliquots of this are treated as described in the procedure. Calculate the lead concentration of the unknown in $\mu g$ of Pb/ml and the molar absorptivity for the Pb–dithizine complex.

### 27.  Spectrophotometric Determination of Phosphorus in Egg Shells

REFERENCE:  Chapters 3, 17, 19, and 30.

INTRODUCTION:  The determination of phosphorus in biological fluids, fertilizers, organic molecules, and industrial products is a very important analysis. In this experiment a trace level of inorganic phosphorus as phosphate is determined in egg shells by a spectrophotometric procedure. This procedure with certain modifications is routinely used in the clinical laboratory for the routine analysis of inorganic phosphate in biological fluids.

Phosphate ion and molybdate ion ($MoO_3{}^{2-}$) react to form the product phosphomolybdate ($[PO_4 \cdot 12MoO_3]^{3-}$). This product is reduced to a material of unknown structure called molybdenum blue and imparts a blue color to the solution. Several different reducing agents can be used. The more common ones are hydrazine hydrosulfate, $SnCl_2$, 1,2,4-aminonaphtholsulfonic acid, and ascorbic acid.

PROCEDURE*:  The eggs are broken, the albumin and yolk are discarded, and the shells are carefully washed with distilled water. The shells are air dried for several days and then ground into fine particles with a mortar and pestle. Three samples of 0.250 g each are accurately weighed into porcelain crucibles and ashed at 700°C in a muffle for at least 16 hours. Subsequently, the crucibles are cooled, and 2 ml of water and 1 ml 12 $F$ HCl are carefully added. When the residue has dissolved, the solution is transferred to a 100-ml volumetric flask and diluted to volume.

A 5.00-ml aliquot is taken and transferred to a 50-ml volumetric flask. Subsequently, 5 ml of the sodium molybdate solution and 3 ml of the reducing solution are added with careful stirring. The mixture is diluted to volume, and after waiting for at least 6 minutes for color development (time for development for standards and unknown should be the same) the absorbance is determined at 660 nm. A blank using all reagents minus the phosphorus standard should be used.

The molybdate solution is made by dissolving 12.5 g of ammonium molybdate in about 100 ml of water and added to 150 ml of 5 $F$ $H_2SO_4$. Dilute to 500 ml total volume and mix thoroughly. The solution is stable if kept in a closed container.

The reducing solution is made by dissolving ascorbic acid in water at a ratio of 1 g per 100 ml or a 1% solution.

A calibration curve is prepared by dilutions of a standard solution containing 200 mg P/liter. (The solution should also contain 8 ml conc. HCl/liter.) Analytical reagent $NaH_2PO_4$, $NaH_2PO_4 \cdot H_2O$, $K_2HPO_4$, or $KH_2PO_4$ are suitable sources of phosphorus.

Calculate the mg P/g of egg shell.

### 28.  The Simultaneous Spectrophotometric Determination of a Permanganate and Dichromate Mixture

REFERENCE:  Chapters 3, 12, 17, 19, and 30.

---

* The instructor may also supply a P unknown solution in a volumetric flask in addition to the egg shells.

INTRODUCTION: If two or more components in a mixture absorb differently, their individual concentrations can be determined spectrophotometrically even though their absorption spectra overlap. The reason this is possible is because absorbancies are additive. Measurements must be made at two different wavelengths if the mixture contains two components, three wavelengths if three components are in the mixture, etc. The appropriate Beers law equations are written for each absorption measurement and providing all absorptivities are known the equations are solved simultaneously to give the concentrations of each component.

In this experiment Cr and Mn as $Cr_2O_7^{2-}$ and $MnO_4^-$ are spectrophotometrically determined in a mixture. The equations needed for this are (the student should derive these—see Chapter 19):

$$A_{\lambda_1} = \epsilon_{\lambda_1}^{Cr_2O_7^{2-}} bc^{Cr_2O_7^{2-}} + \epsilon_{\lambda_1}^{MnO_4^-} bc^{MnO_4^-}$$

$$A_{\lambda_2} = \epsilon_{\lambda_2}^{Cr_2O_7^{2-}} bc^{Cr_2O_7^{2-}} + \epsilon_{\lambda_2}^{MnO_4^-} bc^{MnO_4^-}$$

Although the measurements are made in the visible region in this experiment, the general procedure applies to measurements in the ultraviolet as well as the infrared.

PROCEDURE: A standard solution of $K_2Cr_2O_7$ (0.003 $F$) and $KMnO_4$ (0.002 $F$) in 0.25 $F$ $H_3PO_4$ are provided. Carefully prepare dilute solutions in 0.25 $F$ $H_2SO_4$ of each standard solution and record the absorption of each as a function of wavelength. If this is done manually, the main number of points should be collected in the 440 and 545-nm region. Calculate the absorptivities of each at 440 and 545 nm.

The unknown is obtained from the instructor in a 100-ml volumetric flask and diluted to volume with 0.25 $F$ $H_2SO_4$. The absorbance is recorded at 440 and 545 nm. A blank of 0.25 $F$ $H_2SO_4$ should be used in all measurements. Calculate the mg of Cr and Mn per ml in the 100 ml unknown.

This procedure can be modified to determine Mn and Cr in a steel sample. The instructor will provide the directions for dissolving the steel and oxidizing the Mn and Cr to $MnO_4^-$ and $Cr_2O_7^{2-}$; see Experiment 24.

## 29. Determination of Glucose in Blood: A Study of the Variables

REFERENCE: Chapters 3, 17, 19, and 30.

INTRODUCTION: The one experimental technique which is encountered most often in clinical analysis is the conversion of the unknown into a colored species. Subsequently, its absorbance is measured and its concentration is ascertained from a calibration curve. Frequently, the chemistry associated with the color formation is not understood and for this reason the experimental parameters must be carefully controlled and reproduced in order to obtain precise and accurate results. The procedure outlined in this experiment is the clinical procedure for the determination of glucose in blood and completion of the experiment should demonstrate the kind of experimental control required in clinical analysis.

PROCEDURE: With a micropipet, transfer 10 $\mu$l of blood into a micro test tube and add 100 $\mu$l of 3% trichloroacetic acid. Mix with a vibrator and let the solution stand for a few minutes. Centrifuge, extract 40 $\mu$l of the supernatant liquid with a micropipet and transfer to a micro test tube. Add 200 $\mu$l of o-toluidine reagent,* seal with parafilm and place in boiling water for exactly 8 minutes. Cool rapidly with tap water, wait exactly 10 minutes and measure the absorbance in a microcuvet at 630 nm.

Prepare a standard by dissolving 0.100 g of dried glucose in 100 ml saturated benzoic acid solution. Extract 10 $\mu$l of standard and treat exactly as the unknown. Measure the absorbance after treatment and calculate the concentration (mg glucose/ 100 ml blood) of the unknown in the original sample.

VARIABLES ASSOCIATED WITH THE METHOD. A number of variables which feasibly could affect this determination become readily apparent. Design a series of experiments which will demonstrate the reasons for the following requirements in the procedure.

1. After the color developing reagent is added, why is the mixture placed in boiling water for exactly 8 minutes?
2. After the sample is cooled with tap water, why is the spectrophotometric measurement delayed for 10 minutes?
3. Why is the absorbance measured at 630 nm?
4. What concentration level of o-toluidine is required?

Before initiating the experiments, outline and discuss your procedures with the laboratory instructor. The experiment can be scaled up to the milliliter level if microequipment is not available. A glucose unknown sample can be substituted for the blood sample.

### 30.  Spectrophotometric Determination of Aldehydes and Ketones

REFERENCE: Chapters 3, 17, 19, and 30.

INTRODUCTION: Many reactions are available which can be used to convert an organic molecule into a derivative which absorbs in the visible region. These reactions in almost all cases are not specific for a single organic compound but rather are for all compounds containing the same reaction site. Rates of reaction, however, are different and often these are affected by other structural features in the molecule.

This experiment is typical of derivative formation and subsequent determination by absorption measurements. The color development is based on the reaction between 2,4-dinitrophenylhydrazine (**I**) and a carbonyl compound in acid solution. In this experiment cinnamaldehyde (**II**) is being determined. The stoichiometry is given by

---

\* The reagent is prepared by dissolving 0.270 g of thiourea and 10 ml of o-toluidine in 90 ml of glacial acetic acid.

The reaction is routinely used for the determination of other aldehydes and ketones including many steroids.

PROCEDURE: The 2,4-nitrophenylhydrazine (2,4-DNP) solution is prepared by adding 1 g of 2,4-DNP and a few drops of conc. HCl to 100 ml of MeOH and refluxing for approximately 2 hours or until dissolved. If tightly stoppered the solution is stable for several months; if not, it should be used only over a one- to two-week period.

The instructor will provide the cinnamaldehyde unknown in a 25-ml volumetric flask which is diluted to volume with MeOH. A 5-ml aliquot is carefully transferred to a large test tube and to this is added 1 ml of the 2,4-DNP reagent and 2 drops of conc. HCl. The test tube is placed in boiling water and carefully heated for about 10 minutes. After cooling the solution, which should be very dark, it is transferred to a 25-ml volumetric flask. The test tube is washed with MeOH, 5 ml of 10% NaOH in 80% MeOH is added, and the entire solution diluted to volume with MeOH. The solution should be a wine red color and the absorbance is read at 480 nm. A calibration curve is prepared in the same manner by dilution of the standard cinnamaldehyde solution $(1 \times 10^{-4} F)$ provided by the instructor. The measurements should be made against a blank containing all reagents except the aldehyde.

Calculate the mg of cinnamaldehyde per ml in the 25 ml unknown. Also, report the molar absorptivity for the cinnamaldehyde–2,4-DNP derivative.


### 31. Determination of Sodium in Several Different Samples by Flame Photometry

REFERENCE: Chapters 3, 17, 20, and 30.

INTRODUCTION: Prior to flame photometry, the determination of sodium (potassium and other alkali metals) involved tedious, time-consuming methods. In general, useful volumetric titration and gravimetric procedures are limited to only a few reagents. In contrast, the flame method provides rapid, accurate results and can be applied to trace levels as well as to semimacro levels. The technique is routinely used in industry and clinical laboratories for the analysis of sodium, potassium, and many other metals. In this experiment, a calibration curve of flame emission vs sodium concentration is prepared and the sodium content of several unknowns are determined. (In clinical analysis, an internal standard method employing lithium is usually used for sodium and/or potassium in blood, urine, and other biological fluids.)

PROCEDURE: A standard stock solution of sodium ion is prepared by dissolving an accurately weighed sample of primary standard sodium carbonate (about 0.23 g) in a small amount of dilute HCl and diluting to 1 liter in a volumetric flask (0.1 g Na/liter or 100 ppm Na). (Combine the acid and sodium carbonate carefully and without loss since effervescence will occur.) Prepare a series of solutions containing 4, 7, 10, 15, 25, and 40 ppm by dilution of the standard sodium solution. Measure the flame emission at 589 nm for each standard solution six times according to the directions provided by the instructor. Evaluate the data for each standard statistically and plot the calibration curve.

The instructor will provide an unknown solution and it should be diluted until its emission falls in the range of the calibration curve. The sodium content of tap water should also be determined. Measure each sample six times and statistically treat the data.

The sodium content for biological samples can be determined by the same procedure. For example, normal serum will contain 138–146 mmole of sodium/liter, while the sodium content in urine will vary widely (a normal adult will excrete approximately 75–200 mmole of sodium per 24 hours). There is no special sample treatment required for urine or serum. Therefore, an aliquot is taken of such a size and diluted with water so that through convenient dilutions the flame emission will fall within the range of the calibration curve.

Results for the solution unknowns should be reported as ppm sodium present while for the biological samples the results should be reported as mmole/liter of sample.

VARIABLES TO CONSIDER: The following parameters will change the emission intensity of the sodium in the flame.

1. Determine the optimum slit width by observing the intensity of emission as a function of slit width.

2. Determine the optimum fuel to oxygen ratio by keeping the fuel constant and varying the oxidant flow while a 10-ppm sodium solution is nebulized through the flame. Then, while maintaining the optimum oxidant flow vary the flow of the fuel. Plot the emission of sodium vs both the fuel and oxidant flow and determine the optimum ratio.

### 32. Fluorimetric Determination of Quinine

REFERENCE: Chapters 3, 17, 21, and 30.

INTRODUCTION: In inorganic chemistry the most frequent applications of fluorescence are for the determination of metal ions as fluorescent organic complexes. In organic chemistry many individual types of organic molecules will fluoresce or can be converted to fluorescent molecules. Three such molecules, which are routinely determined in the pharmaceutical industry, are riboflavin (vitamin B₂), thiamine (vitamin B₁),

I

and quinine (**I**). Detailed procedures for these analyses are available in the "United States Pharmacopeia" and the "National Formulary."

PROCEDURE:   The procedure for the unknown and the preparation of the calibration curve are the same.

The standard quinine solution is prepared by dissolving 0.100 g of quinine in 0.05 $F$ $H_2SO_4$ and diluting to volume in a 1-liter volumetric flask with 0.05 $F$ $H_2SO_4$.

An accurately measured 10-ml aliquot of the standard solution is diluted to 1 liter in a volumetric flask with 0.05 $F$ $H_2SO_4$. Aliquots of this solution ranging from 10 to 60 ml are accurately measured and transferred to 100-ml volumetric flasks and diluted to volume with 0.05 $F$ $H_2SO_4$. The fluorescence of these solutions are measured at a fluorescent wavelength of 480 nm. Directions for using the fluorometer will be provided by the instructor.

The unknown solution is provided in a 100-ml volumetric flask and should be diluted to volume with 0.05 $F$ $H_2SO_4$. Determine its concentration by measuring its fluorescence and using the calibration curve.

Calculate the milligrams of quinine per milliliter of unknown solution.

### 33.   Interpretation of Infrared Spectra

REFERENCES:   a. "Sadtler Standard Spectra," Sadtler Research Laboratories, Philadelphia, Pa.

b. C. E. Meloan, "Elementary Infrared Spectroscopy," Macmillan, New York, 1963.

c. R. Silverstein and G. Bassler, "Spectrometric Identification of Organic Compounds," Wiley, New York, 1967.

d. Chapter 18.

PROCEDURE:   Obtain a series of infrared spectra from your instructor. Using this text and the references given below, identify the functional groups present in each spectra. Suggest possible structures for each of the molecules.

**34. Separation of Methylene Blue and Fluorescein by Column Chromatography**

REFERENCE: Chapters 22, 23, and 30.

INTRODUCTION: Column chromatography is a technique which can be applied to the separation of many complex mixtures. In this experiment methylene blue and fluorescein will be separated and the effect of eluting agent will be observed.

PROCEDURE: Prepare a slurry of 25 g of alumina (neutral 100–200 mesh) in ethyl alcohol. After placing a glass wool plug in the end of a chromatography column (2 cm diam. × 25 cm), pour in the slurry, and tap to remove air bubbles. At the alumina settles, allow the ethanol to drain to within 1 inch of the top of the column. Do not allow the column to go dry at any time.

Prepare a dye mixture by dissolving 5 mg each of methylene blue and fluorescein (sodium salt) in 5 ml of ethyl alcohol. Allow the eluent in the column to drain to within 1 mm of the top of the alumina. Subsequently stop the eluent flow and carefully add the dye solution to the top of the column by means of a dropper. Allow the material to pass on the column by draining such that the solution is 1 mm from the alumina. Add 2 ml of ethyl alcohol and drain to the appropriate height again. Fill the tube with ethyl alcohol and elute the methylene blue from the column. Collect the eluent in 5-ml aliquots in vials.

After all the methylene blue has been collected, rate the relative concentration of each vial (from the depth of blue) from 1 to 10. Plot the relative concentration of each vial vs the tube number to obtain the chromatogram.

Repeat the experiment using 50% water/50% ethyl alcohol and 100% water eluents. Compare the data for the three eluting agents.

**35. Parameters Affecting the Separation of a Hydrocarbon Mixture**

REFERENCE: Chapters 22, 24, and 30.

INTRODUCTION: Gas chromatography is probably the most valuable separation technique developed to date. The technique, however, involves the optimization of certain parameters which directly affect the separation of mixtures. In this experiment, the effects of column temperature and carrier gas flow on the separation of a series of hydrocarbons will be determined.

PROCEDURE: Prepare 5 ml of a mixture of pentane, hexane, and heptane (1:1:1) and seal in a vial equipped with a rubber septum. Using an appropriate column (6 ft, 10% SE 30, 10% OV-1, or 10% tri-m-cresyl phosphate) set the following parameters on the gas chromatograph.

1. Column temperature, 60°C
2. Injection part temperature, at least 150°C
3. Detector temperature, at least 150°C
4. Gas flow, 60 cm³/min

After the instrument has come to equilibrium, inject 2 $\mu$l of the mixed hydrocarbon sample. Mark the injection on the recorder trace and allow the three components to elute. Repeat the same experiment with the column temperature at 90°C, 110°C, 125°C, and 150°C while keeping the flow rate constant at 60 cm$^3$/min. Determine which temperature is optimum (separation in the minimum time period) and adjust the column to that temperature. Obtain chromatograms of the hydrocarbon mixture for carrier gas flow to that value.

Obtain from your instructor an unknown mixture containing any or all of the hydrocarbons. Inject 2 $\mu$l of sample and determine qualitatively what constitutes the sample. From integration of the peak areas, determine the percent of each component. (The detector response for each component can be determined by use of the 1:1:1 mixture of the hydrocarbons.)

### 36. Separation of Amino Acids by Paper Chromatography

REFERENCE:   Chapters 22, 25, and 30.

INTRODUCTION:   Paper chromatography is a technique used to separate complex mixtures of drugs, metal ions, amino acids, and dyes. Its application is limited only by the number of papers and eluting agents available.

One of the classic applications of paper chromatography is the separation of amino acids. The sample is spotted on the paper, eluted, and subsequently visualized by reaction with ninhydrin to form a purple or brown color. Qualitative identification is made by calculation of the $R_f$ values for each spot. Quantitative determination can be made by comparison of the intensity of the color with standards.

PROCEDURE:   Prepare solutions of 0.1 g/100 ml each of valine, threonine, glycine, cystine, and isoleucine. Turn on a hot plate and allow it to come to temperature. Obtain an 8 in. $\times$ 10 in. sheet of chromatographic paper from your instructor and place six equally spaced small circles with a lead pencil (on the 10-in. edge) $\frac{1}{2}$ in. from the bottom. Do not handle the paper with your bare hands since amino acids will be transferred to the paper. (Forceps or surgical gloves are invaluable in this experiment.) Label the paper at the top with the name of each of the five amino acids and label the sixth unknown.

Place three small drops of an amino acid on the circle and dry carefully over the hot plate between each drop application.

Repeat the application for each amino acid (3 drops/circle). Obtain an unknown mixture of amino acids from your instructor and spot the last circle (3 times).

Roll the sheet into a cylinder (all spots should be on the bottom) and staple the top and bottom. Into an 800-ml beaker place a 95% ethyl alcohol–5% water mixture to a depth of about $\frac{1}{4}$ in. and cover with a watch glass. Allow the system to come to equilibrium (about 10 minutes) and place the paper cylinder into the beaker. Cover the beaker and allow the eluting agent to rise about 6 in. Remove the paper and immediately mark the position of the solvent front with a lead pencil.

After the chromatogram has dried, spray the paper with 0.2% ninhydrin solution in water saturated with butyl alcohol. Dry the chromatogram over the hot plate until colored spots develop.

Calculate the $R_f$ values of the amino acids and the spots for the unknown. Identify qualitatively the composition of the unknown.

## 37.  Separation of Vitamins by Thin-Layer Chromatography

REFERENCE:   Chapters 22, 25, and 30.

INTRODUCTION:   Experimentally, thin-layer chromatography is an efficient method of separating complex mixtures. Specifically, the method has been applied to the detection of impurities in pharmacential dosage formalations, detection and isolation of illicit drugs, and the separation of synthetic organic mixtures.

Thin-layer chromatography is similar to paper chromatography in that the sample is spotted on a coated plate. The sample is subsequently separated by elution and detected by a color forming chemical reaction or fluorescence. The mixture is qualitatively identified by calculation of the $R_f$ values at the respective spots. Quantitatively the spots are determined by the intensity of fluorescence or the intensity of the color of the compound on the plate. Each method requires that standards be prepared.

PROCEDURE:   Obtain from your instructor six $20 \times 3$ cm silica gel $^{254}$F precoated TLC plates, 1-ml solutions (0.01 g/ml) each of vitamin E, ascorbic acid, vitamin $D_3$, nicotinamide, $dl$-$\alpha$-tocopherol, vitamin A acetate, and an unknown solution. Spot each plate with one of the solutions using a capillary tube, allow the solvent to evaporate, and place in a developing chamber containing an 80/20 mixture of cyclohexane/diethyl ether.

Allow the solvent to move about 15 cm, remove, and mark the solvent front with a pencil. After the plates have been allowed to dry, place the chromatograms under a short-wave UV light (254 nm) and observe the vitamins as dark spots on a green fluorescing background.* Circle each of the spots, calculate the $R_f$ values, and determine the composition of the unknown solution.

## 38.  Separation of a Minor Constituent in Presence of a Major Component

REFERENCE:   Chapters 15, 22, 26, and 30.

---

* An alternate method of visualizing the separated compounds is to spray with a 20 g/100 ml antimony(III) chloride solution in chloroform. The chloroform, however, should be treated for 3 hours with aluminum oxide activated basic (100 g/liter of chloroform) before the solution is prepared. After spraying, vitamin A will turn blue immediately with rapid fading, vitamin D gradually turns yellow-orange, and vitamin E is only visible after heating the plate at 100°C for 5 minutes.

INTRODUCTION: A frequent goal in metal analysis is the determination of the minor constituent metals present in the sample. This kind of analysis is necessary, since these minor metals can influence the bulk physical and chemical properties of the metal. Often, by introducing minor concentration of other metals, favorable, as well as undesirable metallurgical properties are obtained.

A 100:1 Co–Zn mixture is separated in this experiment on a column of anion resin and the amount of Zn present in the sample is determined by EDTA titration. The conditions for the EDTA titration (see Experiment 22) do not permit the titration of zinc in the presence of cobalt or vice versa. Therefore, if the separation is not complete, poor results will be obtained.

PROCEDURE: The sample is obtained from the instructor in a 100-ml volumetric flask and diluted to volume with 2 $F$ HCl.

The column is prepared by mixing some 100–200 mesh, Dowex 1 $\times$ 8, Cl-form resin with 2 $F$ HCl and pouring this slurry into the glass column ($\sim$1-cm diameter). (*The resin should not be dry packed.*) The column arrangement is illustrated in Fig. 22–5a. Resin is added as a slurry until a column height of 5–6 in. is obtained. Excess resin should not be thrown away but returned to the instructor. Passage of 2 $F$ HCl is continued through the column until the resin bed is completely settled. Subsequently, the level of the 2 $F$ HCl is adjusted to be about $\frac{1}{2}$ in. above the top of the resin. An accurately measured 10-ml aliquot of the sample is passed into the column at a flow rate of 0.5 ml/minute. The sample is carefully washed into the column with 2 $F$ HCl and this eluting mixture is continued until all the Co is removed. The eluting agent is then switched to 0.001 $F$ HCl. (Do not pour the 0.001 $F$ HCl directly into 2 $F$ HCl in the reservoir above the column.) Approximately 100 ml of 0.001 $F$ HCl, or enough to elute the zinc, should be passed through the column and the effluent collected in a beaker. The flow rate throughout this procedure should be about 0.5 ml/minute. This effluent should then be carefully evaporated to a few milliliters to remove excess HCl, diluted, and pH adjusted to 6.0. The Zn in this solution is then titrated with EDTA according to the procedure in Experiment 22.

Calculate the milligrams of Zn per milliliter and total milligrams of Zn in the unknown.

## 39. Electrogravimetric Determination of Copper

REFERENCE: Chapters 3, 10, 11, 28, and 30.

INTRODUCTION: Electrolysis is one of the most accurate methods for the determination of copper. Basically, the procedure involves the reduction of Cu(II) to copper metal at a platinum electrode. After the metal is quantitatively plated from solution the cathode is weighed to determine the amount of copper.

PROCEDURE: Obtain 100 ml of sample from your laboratory instructor along with a set of platinum electrodes. Clean the electrodes* by dipping them into concentrated

---

* The platinum electrodes must be handled with extreme care because they are fragile. Also, handle only the wire portion so grease from the hands does not contaminate the electrodes.

nitric acid for a few minutes. Rinse with distilled water, then with 95% ethyl alcohol, dry at 110°C for 15 minutes, and weigh the cylindrical cathode after cooling. Attach the electrodes to the electrolysis apparatus. Pipet exactly 25 ml of the unknown solution into a tall form beaker, add 3 ml of concentrated sulfuric acid, and 1 ml of concentrated clear nitric acid. Center the electrodes in the beaker and add distilled water such that only 1 cm of the cathode is exposed above the solution. Turn on the stirrer and adjust the current level to approximately 2 A according to the directions of the instructor. The current will decrease as the concentration of the copper in solution decreases. After 45 minutes, add enough distilled water to completely cover the electrode. If copper is deposited on this fresh area, continue the electrolysis for another 15 minutes.

After deposition is complete, remove the beaker from the apparatus (while the current is still flowing). Shut down the electrolysis equipment and remove the cathode. Rinse the electrode with ethyl alcohol, dry, and weigh. Calculate the concentration of copper in the original solution in moles/liter.

A brass sample or copper ore can be analyzed for copper by this technique. However, it is necessary to dissolve the sample and adjust the experimental conditions before proceeding with the electrolysis. If these kinds of samples are to be analyzed, the instructor will provide the details for dissolution of the sample.

### 40. Polarographic Determination of Cadmium

REFERENCE: Chapters 10, 11, 28, and 29.

INTRODUCTION: Polarography can be used for the determination of trace amounts of many different metal ions. One of these is cadmium ion. Cadmium ion also serves as an excellent test ion for evaluating the experimental properties of the polarographic method because its reduction is reversible and well defined. Hence, this experiment illustrates polarographic analysis by a calibration curve and the experimental parameters in a polarographic measurement.

PROCEDURE: A series of standard solutions containing 0, $10^{-3}$, $10^{-4}$, $10^{-5}$, and $10^{-6}$ $F$ cadmium ion are prepared from a standard solution provided by the instructor (prepared by dissolving pure cadmium metal in acid solution and diluted to volume). Each solution should also be 0.1 $F$ in $KNO_3$ and 0.01% in gelatin.

Each solution is placed in the polarographic test cell and the oxygen is removed by bubbling nitrogen gas through the solution for 10–15 minutes. The height of the mercury column is adjusted so the flow rate is approximately 1 drop per 5 seconds. The polarogram is then recorded from 0 to $-2.0$ V according to the directions provided by the instructor. Select the appropriate potential and plot a calibration curve of $i_d$ vs concentration.

An unknown cadmium solution is obtained from the instructor and diluted such that its $i_d$ will lie on the calibration curve. The supporting electrolyte and gelatin concentrations should be the same as those used for the calibration curve.

Determine the concentration of the unknown cadmium solution, the $n$ value, and the $E_{1/2}$. In addition, record the polarogram for a fixed cadmium concentration at different drop times, in the presence of 0.1 $F$ KCl rather than KNO$_3$, and in the absence of gelatin. Explain your observations.

## 41. Polarographic Determination of Riboflavin

REFERENCE: Chapters 10, 11, 28, and 29.

INTRODUCTION: Riboflavin (vitamin B$_2$) has a half-wave potential at 0.47 V (vs SCE) in a phosphate buffer of pH 7.5. This polarographic reduction wave can be used routinely for the assay of riboflavin in a variety of pharmaceutical preparations. In general, the method is specific, useful for trace levels, and free from interferences of filler and excipient materials usually found in the pharmaceutical preparations. In this procedure, the determination is based on the preparation of a calibration curve for pure riboflavin.

PROCEDURE. The solutions for the calibration curve and the unknown sample are prepared similarly. Weigh accurately 25 mg of the pure reference riboflavin plus 2.5 g of pure sodium salicylate into a 250-ml volumetric flask, dissolve, and dilute to volume with distilled water. *Protect the solution from light.* Pipet 10-, 15-, 20-, and 25-ml aliquots of the standard into 50-ml volumetric flasks, add 10 ml of the pH 7.5 buffer (0.5 $M$ or 62.5 g Na$_2$HPO$_4$·2H$_2$O–10.2 g KH$_2$PO$_4$/500 ml), stir, and dilute to volume. Record the polarogram of each of these solutions after removal of oxygen according to the directions of the instructor. Select an appropriate potential and plot the diffusion current, $i_d$, versus concentration ($\mu$g/ml) of riboflavin.

If a solution unknown is provided, make proper dilutions such that the final solution is 1 % in salicylate, 0.1 $M$ in pH 7.5 buffer, and contains about 20 to 50 $\mu$g riboflavin/ml. For a tablet containing riboflavin, carefully record its weight, pulverize it, and transfer a weight of the powder quantitatively to a volumetric flask so that the previous concentrations can be prepared. Record the polarogram for the unknown, determine its $i_d$, and ascertain its concentration from the calibration curve. For a solution unknown report the results in mg/ml, while for the tablet assay report the results in mg/tablet and percent of the label claim.

## Appendix I
# MATHEMATICAL OPERATIONS

### Exponents

The use of exponents enables one to write many numbers in a simplified form. Examples:

$$0.0001 = 1.0 \times 10^{-4}$$
$$0.00033 = 3.3 \times 10^{-4}$$
$$2000000 = 2.0 \times 10^{6}$$
$$1.55 = 1.55 \times 10^{0}$$
$$0.37 = 3.7 \times 10^{-1}$$
$$0.00446 = 4.46 \times 10^{-3}$$

Note that in each case the exponent is the number of places the decimal point has been moved. For numbers less than 1, the sign of the exponent is negative, and for numbers greater than 1, the sign is positive.

It is most convenient to convert numbers to the standard form used above, $A \times 10^n$, where

$A$ is the number greater than 1 with the decimal point after the first digit.

$n$ is a whole number.

In the first example listed above, $A = 1.0$ and $n = -4$. Remember, that when powers of 10 are multiplied, exponents are added; when divided, exponents are subtracted.

### Logarithms

*1. Methods of Writing Numbers*

Numbers can be written as fractions, as decimals, or in exponential form. For many mathematical operations the use of decimals instead of fractions has obvious advantages. Another convenient method for writing a number (especially those which are much less than 1) is in semiexponential form with the decimal point after the first significant figure, and the exponent an integer. A number can be converted to exponential form by finding the log of the number. The log of the number is the exponent of the power to which 10 must be raised to give that number.

$$N = 10^a$$

$$\log_{10} N = a$$

(Logs can be based on numbers other than 10, but the base 10 is the most common and convenient.) Below are some examples of a number written in several ways.

| Fraction | Decimal | Semiexponential | Exponential |
|----------|---------|-----------------|-------------|
| 1/50 | 0.02 | $2 \times 10^{-2}$ | $10^{-1.7}$ |
| — | 31.5 | $3.15 \times 10^{1}$ | $10^{1.5}$ |
| — | 315 | $3.15 \times 10^{2}$ | $10^{2.5}$ |
| — | 0.000315 | $3.15 \times 10^{-4}$ | $10^{-3.5}$ |

In chemistry it is frequently necessary to convert a number from decimal or semiexponential form to exponential form (i.e., to find the log of the number.) It is also important to be able to convert a number from exponential form back to decimal or semiexponential form.

2.  *How to Find the Log of a Number*

   (a) Write the number, $N$, in the form

$$N' \times 10^c$$

where $N'$ is the number written with the decimal point after the first significant figure, and $c$ is a whole number equal to the number of places the decimal point must be moved. When the decimal point is moved to the right, $c$ is positive; when the decimal point is moved to the left, $c$ is negative. Examples:

| Number, $N$ | Number in form, $N' \times 10^c$ |
|---|---|
| 14.3 | $1.43 \times 10^1$ |
| 143 | $1.43 \times 10^2$ |
| 0.295 | $2.95 \times 10^{-1}$ |
| 0.00295 | $2.95 \times 10^{-3}$ |
| 2.95 | $2.95 \times 10^0$ |

   (b) Find the log of $N'$ (the mantissa of log $N$) in a log table or slide rule by finding the log of $N$ in the table, disregarding any decimals in $N$. The value found in the log table with a decimal point before the first figure (zero included) is the log of $N'$. The log of the original number, $N$, is given by

$$N = c + \log N'$$

Examples:

| Number, $N$ | Log $N$ |
|---|---|
| 14.3 | $1 + \log 1.43 = 1.16$ |
| 143 | $2 + \log 1.43 = 2.16$ |
| 0.295 | $-1 + \log 2.95 = -0.53$ |
| 0.00295 | $-3 + \log 2.95 = -2.53$ |
| 2.95 | $0 + \log 2.95 = 0.47$ |

3.  *How to Find a Number When Its Log Is Known*

   (a) *Relationships.* Suppose we have a whole number or number in decimal form, $N$, which is equal to $10^a$ when written in exponential form.

$$N = 10^a$$
$$\log_{10} N = a$$
$$N = \text{antilog } a$$

The problem is to convert the exponential form of a number to its decimal form (or semiexponential form). To do this, we must find the antilog of the exponent, $a$.

   (b) *To find the antilog of an exponent, $a$.* Write $a$ in the form $m + c$ (mantissa + characteristic). When this is done

$$N = \text{antilog } m \times 10^c$$

If $a$ is negative, write $m + c$ so that $m$ is positive and $c$ is negative.

Examples:

| $a$ | $m + c$ | $N$ |
|---|---|---|
| 2.95 | $0.95 + 2$ | $8.9 \times 10^2$ |
| 4.30 | $0.30 + 4$ | $2.0 \times 10^4$ |
| 0.50 | $0.50 + 0$ | $3.15 \times 1 = 3.15$ |
| $-0.50$ | $0.50 - 1$ | $3.15 \times 10^{-1}$ |
| $-4.30$ | $0.70 - 5$ | $5.0 \times 10^{-5}$ |
| $-10.19$ | $0.81 - 11$ | $6.5 \times 10^{-11}$ |

(Antilog $m$ is found using a log table to find the number that has a log equal to $m$. If available a table of antilogs may be used to find the antilog of $m$ directly.)

*4. Arithmetic Operations With Exponential Numbers*

(a) *Multiplication and division.* To multiply two or more exponential numbers, add the exponents. To divide one exponential number by another, subtract the exponents. Examples:

$$10^2 \times 10 = 10^3$$
$$10^3 \times 10^2 = 10^5$$
$$10^{-3} \times 10^2 = 10^{-1}$$
$$10^{0.5} \times 10^{4.1} \times 10^{0.3} = 10^{4.9}$$

$$10^2/10 = 10 \qquad 10^5/10^{-3} = 10^8$$
$$10^5/10^3 = 10^2 \qquad 10^{-5}/10^{0.5} = 10^{-5.5}$$

(b) *Taking the root of an exponential number, or raising an exponential number to a power.* To raise an exponential number to a power, multiply the exponent by the power to which it is to be raised. Examples:

$$(10^2)^2 = 10^4 \qquad \sqrt{10^6} = (10^6)^{1/2} = 10^3$$
$$(10^{-3})^2 = 10^{-6} \qquad (10^6)^3 = (10^6)^{1/3} = 10^2$$
$$(10^{-3})^3 = 10^{-9} \qquad \sqrt{10^{-5.2}} = (10^{-5.2})^{1/2} = 10^{2.6}$$

## The Quadratic Formula

A quadratic equation can be solved algebraically. For the general equation

$$ax^2 + bx + c = 0$$

it can be shown that

$$x = \frac{-b \pm \sqrt{b^2 - 4ac}}{2a}$$

# Appendix II
## SOLUBILITY PRODUCT CONSTANTS

| Compound | $K_{sp}$ | Compound | $K_{sp}$ |
|---|---|---|---|
| AgBr | $5.2 \times 10^{-13}$ | FeS | $1.5 \times 10^{-19}$ |
| AgCl | $1.8 \times 10^{-10}$ | $Hg_2Cl_2$ | $1.1 \times 10^{-18}$ |
| $Ag_2CrO_4$ | $1.1 \times 10^{-12}$ | $Hg_2Br_2$ | $1.4 \times 10^{-21}$ |
| $Ag_2Cr_2O_7$ | $2.0 \times 10^{-7}$ | $Hg_2I_2$ | $1.2 \times 10^{-28}$ |
| AgI | $8.3 \times 10^{-17}$ | $MgF_2$ | $6.4 \times 10^{-9}$ |
| $Ag_2O$ | $2.6 \times 10^{-8}$ | $Mg(NH_4)PO_4$ | $3.0 \times 10^{-13}$ |
| $Ag_3PO_4$ | $1.2 \times 10^{-10}$ | $MgCO_3$ | $2.6 \times 10^{-5}$ |
| AgSCN | $1.0 \times 10^{-12}$ | $Mg(OH)_2$ | $3.4 \times 10^{-11}$ |
| $Ag_2SO_4$ | $1.7 \times 10^{-5}$ | NiS | $1.4 \times 10^{-24}$ |
| $Al(OH)_3$ | $5.0 \times 10^{-33}$ | $PbCl_2$ | $1.6 \times 10^{-5}$ |
| $BaCO_3$ | $8.1 \times 10^{-9}$ | $PbCO_3$ | $5.6 \times 10^{-14}$ |
| $BaCrO_4$ | $3.0 \times 10^{-10}$ | $PbCrO_4$ | $1.8 \times 10^{-14}$ |
| $BaSO_4$ | $1.3 \times 10^{-10}$ | $PbF_2$ | $3.7 \times 10^{-8}$ |
| $BaC_2O_4$ | $1.7 \times 10^{-7}$ | $PbI_2$ | $1.1 \times 10^{-9}$ |
| $Bi_2S_3$ | $1.6 \times 10^{-72}$ | $Pb(IO_3)_2$ | $9.8 \times 10^{-14}$ |
| $CaF_2$ | $4.9 \times 10^{-11}$ | $Pb_3(PO_4)_2$ | $1.5 \times 10^{-32}$ |
| $Ca(OH)_2$ | $3.7 \times 10^{-6}$ | $PbSO_4$ | $1.1 \times 10^{-8}$ |
| $Ca_3(PO_4)_2$ | $1.0 \times 10^{-26}$ | PbS | $4.2 \times 10^{-28}$ |
| $CaSO_4$ | $1.2 \times 10^{-6}$ | $SrCO_3$ | $1.6 \times 10^{-9}$ |
| CdS | $3.6 \times 10^{-29}$ | $SrF_2$ | $2.8 \times 10^{-9}$ |
| $Cu(OH)_2$ | $3.0 \times 10^{-20}$ | $SrSO_4$ | $2.8 \times 10^{-7}$ |
| $Cu_2S$ | $1.0 \times 10^{-45}$ | TlCl | $1.7 \times 10^{-4}$ |
| $Fe(OH)_3$ | $1.1 \times 10^{-36}$ | $ZnCO_3$ | $3.0 \times 10^{-6}$ |

## Appendix III

# IONIZATION CONSTANTS OF WEAK ACIDS AND BASES

| Acid | $K_a$ | | $pK_a$ | Acid | $K_a$ | | $pK_a$ |
|------|-------|---|--------|------|-------|---|--------|
| Acetic | $1.76 \times 10^{-5}$ | | 4.75 | Hydro-fluoric | $6.76 \times 10^{-4}$ | | 3.17 |
| Benzoic | $6.46 \times 10^{-5}$ | | 4.19 | | | | |
| Carbonic | $4.47 \times 10^{-7}$ | $(K_1)$ | 6.35 | Hydrogen sulfide | $1.00 \times 10^{-7}$ | $(K_1)$ | 7.00 |
| | $4.68 \times 10^{-11}$ | $(K_2)$ | 10.33 | | $1.2 \times 10^{-13}$ | $(K_2)$ | 12.92 |
| Chloro-acetic | $1.40 \times 10^{-3}$ | | 2.85 | Lactic | $1.59 \times 10^{-4}$ | | 3.80 |
| Dichloro-acetic | $3.32 \times 10^{-2}$ | | 1.48 | Nitrous | $4.6 \times 10^{-4}$ | | 3.34 |
| | | | | Oxalic | $5.90 \times 10^{-2}$ | $(K_1)$ | 1.23 |
| EDTA-(H$_4$Y) | $1.00 \times 10^{-2}$ | $(K_1)$ | 2.00 | | $6.40 \times 10^{-5}$ | $(K_2)$ | 4.19 |
| | $2.16 \times 10^{-3}$ | $(K_2)$ | 2.66 | Phenol | $1.28 \times 10^{-10}$ | $(K_1)$ | 9.89 |
| | $6.92 \times 10^{-7}$ | $(K_3)$ | 6.17 | Phosphoric | $7.15 \times 10^{-3}$ | $(K_1)$ | 2.15 |
| | $5.50 \times 10^{-11}$ | $(K_4)$ | 10.26 | | $6.2 \times 10^{-8}$ | $(K_2)$ | 7.21 |
| Formic | $1.77 \times 10^{-4}$ | | 3.75 | | $4.8 \times 10^{-13}$ | $(K_3)$ | 12.32 |
| Glycine | $4.5 \times 10^{-3}$ | $(K_1)$ | 2.35 | Sulfurous | $1.54 \times 10^{-2}$ | $(K_1)$ | 1.81 |
| | $1.66 \times 10^{-10}$ | $(K_2)$ | 9.78 | | $1.02 \times 10^{-7}$ | $(K_2)$ | 6.99 |
| Hydro-cyanic | $4.93 \times 10^{-10}$ | | 9.31 | Trichloro-acetic | $2.0 \times 10^{-1}$ | | 0.30 |

| Base | $K_b$ | | $pK_b$ | Base | $K_b$ | | $pK_b$ |
|------|-------|---|--------|------|-------|---|--------|
| Ammonia | $1.79 \times 10^{-5}$ | | 5.75 | Piperidine | $1.6 \times 10^{-3}$ | | 2.80 |
| Aniline | $4.27 \times 10^{-10}$ | | 9.37 | Pyridine | $1.78 \times 10^{-9}$ | | 8.75 |
| Diethyl-amine | $1.29 \times 10^{-4}$ | | 3.89 | Triethyl-amine | $1.02 \times 10^{-3}$ | | 2.99 |
| Ethanol-amine | $4.0 \times 10^{-5}$ | | 4.40 | Trimethyl-amine | $6.31 \times 10^{-5}$ | | 4.20 |
| Ethylamine | $4.7 \times 10^{-4}$ | | 3.33 | Tris-(hydroxy-methyl) amino-methane | $1.26 \times 10^{-6}$ | | 5.90 |
| Ethylene-diamine | $5.15 \times 10^{-4}$ | $(K_1)$ | 3.29 | | | | |
| | $3.66 \times 10^{-7}$ | $(K_2)$ | 6.44 | | | | |
| Hydroxyl-amine | $2.26 \times 10^{-9}$ | | 8.02 | | | | |
| Imidazole | $1.23 \times 10^{-7}$ | | 6.91 | Urea | $1.2 \times 10^{-14}$ | | 13.82 |

# Appendix IV
# REDUCTION POTENTIALS

## Partial List of Standard Reduction Potentials

| Half-reaction | $E°_{Ox, Red}$ (V) |
|---|---|
| $S_2O_8^{2-} + 2e \rightleftharpoons 2SO_4^{2-}$ | $+2.0^a$ |
| $Co^{3+} + 1e \rightleftharpoons Co^{2+}$ | $+1.84$ |
| $H_2O_2 + 2H^+ + 2e \rightleftharpoons 2H_2O$ | $+1.77$ |
| $MnO_4^- + 4H^+ + 3e \rightleftharpoons MnO_2 + 2H_2O$ | $+1.70$ |
| $Ce^{4+} + 1e \ (1 \ F \ HClO_4) \rightleftharpoons Ce^{3+}$ | $+1.70$ |
| $2BrO_3^- + 12H^+ + 10e \rightleftharpoons Br_2 + 6H_2O$ | $+1.52$ |
| $MnO_4^- + 8H^+ + 5e \rightleftharpoons Mn^{2+} + 4H_2O$ | $+1.51$ |
| $Cl_2 + 2e \rightleftharpoons 2Cl^-$ | $+1.36$ |
| $Cr_2O_7^{2-} + 14H^+ + 6e \rightleftharpoons 2Cr^{3+} + 7H_2O$ | $+1.33$ |
| $Tl^{3+} + 2e \rightleftharpoons Tl^+$ | $+1.28$ |
| $O_2 + 4H^+ + 4e \rightleftharpoons 2H_2O$ | $+1.23^a$ |
| $2IO_3^- + 12H^+ + 10e \rightleftharpoons I_2 + 6H_2O$ | $+1.19$ |
| $Br_2 + 2e \rightleftharpoons 2Br^-$ | $+1.09$ |
| $2Hg^{2+} + 2e \rightleftharpoons Hg_2^{2+}$ | $+0.907$ |
| $Ag^+ + 1e \rightleftharpoons Ag$ | $+0.799$ |
| $Fe^{3+} + 1e \rightleftharpoons Fe^{2+}$ | $+0.771$ |
| $O_2 + 2H^+ + 2e \rightleftharpoons H_2O_2$ | $+0.682$ |
| $Hg_2SO_{4(s)} + 2e \rightleftharpoons 2Hg + SO_4^{2-}$ | $+0.615$ |
| $H_3AsO_4 + 2H^+ + 2e \rightleftharpoons H_3AsO_3 + H_2O$ | $+0.559$ |
| $I_2 + 2e \rightleftharpoons 2I^-$ | $+0.536$ |
| $H_2SO_3 + 4H^+ + 4e \rightleftharpoons S + 3H_2O$ | $+0.45$ |
| $Cu^{2+} + 2e \rightleftharpoons Cu$ | $+0.337$ |
| $Hg_2Cl_{2(s)} + 2e \rightleftharpoons 2Hg + 2Cl^-$ | $+0.268$ |
| $AgCl_{(s)} + 1e \rightleftharpoons Ag + Cl^-$ | $+0.222$ |
| $Sn^{4+} + 2e \rightleftharpoons Sn^{2+}$ | $+0.154$ |
| $Cu^{2+} + 1e \rightleftharpoons Cu^+$ | $+0.153^a$ |
| $S_4O_6^{2-} + 2e \rightleftharpoons 2S_2O_3^{2-}$ | $+0.09^a$ |
| $2H^+ + 2e \rightleftharpoons H_2$ | $0.000$ |
| $Pb^{2+} + 2e \rightleftharpoons Pb$ | $-0.126$ |
| $Ni^{2+} + 2e \rightleftharpoons Ni$ | $-0.23$ |
| $Co^{2+} + 2e \rightleftharpoons Co$ | $-0.28$ |
| $PbSO_{4(s)} + 2e \rightleftharpoons Pb + SO_4^{2-}$ | $-0.356$ |
| $Cd^{2+} + 2e \rightleftharpoons Cd$ | $-0.402$ |
| $Cr^{3+} + 1e \rightleftharpoons Cr^{2+}$ | $-0.41$ |
| $2CO_2 + 2H^+ + 2e \rightleftharpoons H_2C_2O_4$ | $-0.49$ |
| $Zn^{2+} + 2e \rightleftharpoons Zn$ | $-0.763$ |
| $Al^{3+} + 3e \rightleftharpoons Al$ | $-1.66^a$ |
| $Na^+ + 1e \rightleftharpoons Na$ | $-2.71$ |
| $K^+ + 1e \rightleftharpoons K$ | $-2.93$ |
| $Li^+ + 1e \rightleftharpoons Li$ | $-3.05$ |

$^a$ Calculated values.

## Partial List of Formal Reduction Potentials

| Half-reaction | Conditions | $E^f_{Ox, Red}$ (V) |
|---|---|---|
| $Ce^{4+} + 1e \rightleftharpoons Ce^{3+}$ | 1 F $HClO_4$ | +1.70 |
| | 1 F $HNO_3$ | +1.60 |
| | 1 F $H_2SO_4$ | +1.44 |
| | 1 F HCl | +1.28 |
| $Cr_2O_7^{2-} + 14H^+ + 6e \rightleftharpoons 2Cr^{3+} + 7H_2O$ | 0.1 F HCl | +0.93 |
| | 1.0 F HCl | +1.00 |
| | 3.0 F HCl | +1.08 |
| | 0.1 F $HClO_4$ | +0.84 |
| | 1.0 F $HClO_4$ | +1.03 |
| $Fe^{3+} + 1e \rightleftharpoons Fe^{2+}$ | 0.5 F HCl | +0.71 |
| | 1.0 F HCl | +0.70 |
| | 5 F HCl | +0.64 |
| | 10 F HCl | +0.53 |
| | 2 F $H_3PO_4$ | +0.46 |
| | 1 F $H_2SO_4$ | +0.68 |
| $H_3AsO_4 + 2H^+ + 2e \rightleftharpoons H_3AsO_3 + H_2O$ | 1 F HCl | +0.577 |
| | 1 F $HClO_4$ | +0.577 |
| $2H^+ + 2e \rightleftharpoons H_2$ | 1 F HCl | +0.005 |
| | 1 F $HClO_4$ | +0.005 |
| $Sn^{4+} + 2e \rightleftharpoons Sn^{2+}$ | 1 F HCl | +0.14 |
| | 2 F HCl | +0.13 |
| $Ti^{4+} + 1e \rightleftharpoons Ti^{3+}$ | 1 F $H_3PO_4$ | -0.05 |
| | 5 F $H_3PO_4$ | -0.15 |
| | 0.2 F $H_2SO_4$ | -0.01 |
| | 1 F $H_2SO_4$ | +0.06 |
| | 2 F $H_2SO_4$ | +0.12 |
| | 4 F $H_2SO_4$ | +0.20 |

## Reduction Potentials for Reference Electrodes

| Half-reaction | Conditions | $E^f_{Ox, Red}$ (V) |
|---|---|---|
| $Hg_2SO_4 + 2e \rightleftharpoons 2Hg + SO_4^{2-}$ | $K_2SO_4$(saturated) | +0.640 |
| $Hg_2Cl_{2(s)} + 2e \rightleftharpoons 2Hg + 2Cl^-$ | KCl(saturated) | +0.241 |
| $AgCl_{(s)} + 1e \rightleftharpoons Ag + Cl^-$ | KCl(saturated) | +0.199 |

## Appendix V

# LOG FORMATION CONSTANTS FOR METAL COMPLEXES

### Log Formation Constants for Metal–NTA and EDTA Complexes

| Metal ion | NTA | EDTA | Metal ion | NTA | EDTA |
|-----------|-----|------|-----------|-----|------|
| $Ag^+$ | | 7.32 | $Lu^{3+}$ | | 19.06 |
| $Al^{3+}$ | | 16.13 | $Mg^{2+}$ | 5.41 | 8.69 |
| $Ba^{2+}$ | 4.82 | 7.76 | $Mn^{2+}$ | 7.44 | 13.58 |
| $Bi^{3+}$ | | $\sim$23 | $Nd^{3+}$ | 11.09 | 16.75 |
| $Ca^{2+}$ | 6.41 | 10.70 | $Ni^{2+}$ | 11.26 | 18.56 |
| $Cd^{2+}$ | 9.54 | 16.59 | $Pb^{2+}$ | 11.8 | 18.3 |
| $Ce^{3+}$ | 10.7 | 16.05 | $Pd^{2+}$ | | 18.5 |
| $Co^{2+}$ | 10.6 | 16.21 | $Pr^{3+}$ | 10.89 | 16.55 |
| $Co^{3+}$ | | $\sim$36 | $Sc^{3+}$ | | 23.1 |
| $Cr^{3+}$ | | $\sim$23 | $Sm^{3+}$ | 11.39 | 17.2 |
| $Cu^{2+}$ | 12.68 | 18.79 | $Sn^{2+}$ | | $\sim$22 |
| $Dy^{3+}$ | 11.62 | 17.75 | $Sr^{2+}$ | 4.98 | 8.63 |
| $Er^{3+}$ | | 18.15 | $Tb^{3+}$ | | 17.6 |
| $Eu^{2+}$ | | 7.7 | $Th^{4+}$ | | 23.2 |
| $Eu^{3+}$ | | 17.35 | $Ti^{3+}$ | | 17.7 |
| $Fe^{2+}$ | 8.84 | 14.33 | $TiO^{2+}$ | | 17.3 |
| $Fe^{3+}$ | 15.87 | 25.1 | $Tm^{3+}$ | | 18.59 |
| $Ga^{3+}$ | | 20.27 | $V^{2+}$ | | 12.70 |
| $Gd^{3+}$ | 11.43 | 17.2 | $V^{3+}$ | | 25.9 |
| $Hf^{4+}$ | | 19.2 | $VO^{2+}$ | | 18.77 |
| $Hg^{2+}$ | | 21.8 | $Y^{3+}$ | 11.41 | 18.0 |
| $Ho^{3+}$ | | 18.1 | $Yb^{3+}$ | 12.09 | 18.7 |
| $In^{3+}$ | | 24.95 | $Zn^{2+}$ | 10.45 | 16.5 |
| $La^{3+}$ | 10.37 | 15.30 | $Zr^{4+}$ | | 19.9 |

## Log Formation Constants for Other Metal Complexes

| Metal ions | $NH_3$ | $Cl^-$ | $CN^-$ | $I^-$ | $OH^-$ |
|---|---|---|---|---|---|
| $Ag^+$ $K_1$ | 3.4 | 2.7 | 21.1 | 13.9 | 2.3 |
| $K_2$ | 4.0 | 1.8 | 0.7 | −0.2 | 1.3 |
| $K_3$ | | 0.3 | −1.1 | | 1.2 |
| $Cd^{2+}$ $K_1$ | 2.6 | 1.6 | 5.5 | 2.4 | |
| $K_2$ | 2.1 | 0.5 | 5.1 | 1.0 | |
| $K_3$ | 1.4 | −0.6 | 4.7 | 1.6 | |
| $K_4$ | 0.9 | | 3.6 | 1.2 | |
| $K_5$ | −0.3 | | | | |
| $K_6$ | −1.7 | | | | |
| $Cu^{2+}$ $K_1$ | 4.1 | | | | 6.0 |
| $K_2$ | 3.5 | | | | |
| $K_3$ | 2.9 | | | | |
| $K_4$ | 2.1 | | | | |
| $K_5$ | −0.5 | | | | |
| $K_6$ | | | | | |
| $Fe^{3+}$ $K_1$ | | 0.6 | | | 11.0 |
| $K_2$ | | 0.1 | | | |
| $K_3$ | | −1.0 | | | |
| $Ni^{2+}$ $K_1$ | 2.8 | | | | |
| $K_2$ | 2.2 | | | | |
| $K_3$ | 1.7 | | | | |
| $K_4$ | 1.2 | | | | |
| $K_5$ | 0.7 | | | | |
| $K_6$ | | | | | |
| $Zn^{2+}$ $K_1$ | 2.3 | | | | 4.4 |
| $K_2$ | 2.3 | | | | |
| $K_3$ | 2.4 | | | | |
| $K_4$ | 2.1 | | | | |

## Appendix VI

# DISTRIBUTION DIAGRAMS FOR
# SELECTED COMPOUNDS

### Acetic Acid

$\alpha_0$ = degree of dissociation
$\alpha_1$ = degree of formation

Fraction of $HC_2H_3O_2$ is given to the left of $\alpha_1$ and the fraction of $C_2H_3O_2^-$ is given to the right of $\alpha_1$.

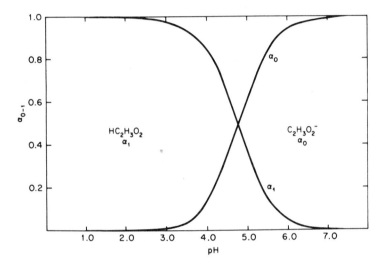

**Fraction of Phosphate Present as the Different Protonated Species**

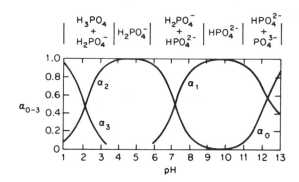

**Fraction of Triethylenetetraamine ($N_4$) Present as the Different Protonated Species**

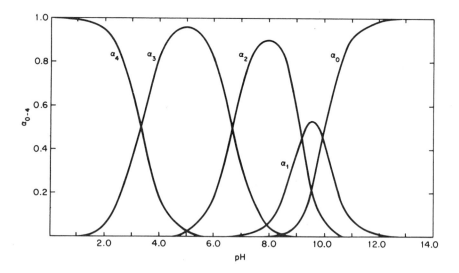

## Appendix VII
# FOUR-PLACE LOGARITHMS

| $n$ | 0 | 1 | 2 | 3 | 4 | 5 | 6 | 7 | 8 | 9 |
|---|---|---|---|---|---|---|---|---|---|---|
| 10 | 0000 | 0043 | 0086 | 0128 | 0170 | 0212 | 0253 | 0294 | 0334 | 0374 |
| 11 | 0414 | 0453 | 0492 | 0531 | 0569 | 0607 | 0645 | 0682 | 0719 | 0755 |
| 12 | 0792 | 0828 | 0864 | 0899 | 0934 | 0969 | 1004 | 1038 | 1072 | 1106 |
| 13 | 1139 | 1173 | 1206 | 1239 | 1271 | 1303 | 1335 | 1367 | 1399 | 1430 |
| 14 | 1461 | 1492 | 1523 | 1553 | 1584 | 1614 | 1644 | 1673 | 1703 | 1732 |
| 15 | 1761 | 1790 | 1818 | 1847 | 1875 | 1903 | 1931 | 1959 | 1987 | 2014 |
| 16 | 2041 | 2068 | 2095 | 2122 | 2148 | 2175 | 2201 | 2227 | 2253 | 2279 |
| 17 | 2304 | 2330 | 2355 | 2380 | 2405 | 2430 | 2455 | 2480 | 2504 | 2529 |
| 18 | 2553 | 2577 | 2601 | 2625 | 2648 | 2672 | 2695 | 2718 | 2742 | 2765 |
| 19 | 2788 | 2810 | 2833 | 2856 | 2878 | 2900 | 2923 | 2945 | 2967 | 2989 |
| 20 | 3010 | 3032 | 3054 | 3075 | 3096 | 3118 | 3139 | 3160 | 3181 | 3201 |
| 21 | 3222 | 3243 | 3263 | 3284 | 3304 | 3324 | 3345 | 3365 | 3385 | 3404 |
| 22 | 3424 | 3444 | 3464 | 3483 | 3502 | 3522 | 3541 | 3560 | 3579 | 3598 |
| 23 | 3617 | 3636 | 3655 | 3674 | 3692 | 3711 | 3729 | 3747 | 3766 | 3784 |
| 24 | 3802 | 3820 | 3838 | 3856 | 3874 | 3892 | 3909 | 3927 | 3945 | 3962 |
| 25 | 3979 | 3997 | 4014 | 4031 | 4048 | 4065 | 4082 | 4099 | 4116 | 4133 |
| 26 | 4150 | 4166 | 4183 | 4200 | 4216 | 4232 | 4249 | 4265 | 4281 | 4298 |
| 27 | 4314 | 4330 | 4346 | 4362 | 4378 | 4393 | 4409 | 4425 | 4440 | 4456 |
| 28 | 4472 | 4487 | 4502 | 4518 | 4533 | 4548 | 4564 | 4579 | 4594 | 4609 |
| 29 | 4624 | 4639 | 4654 | 4669 | 4683 | 4698 | 4713 | 4728 | 4742 | 4757 |
| 30 | 4771 | 4786 | 4800 | 4814 | 4829 | 4843 | 4857 | 4871 | 4886 | 4900 |
| 31 | 4914 | 4928 | 4942 | 4955 | 4969 | 4983 | 4997 | 5011 | 5024 | 5038 |
| 32 | 5051 | 5065 | 5079 | 5092 | 5105 | 5119 | 5132 | 5145 | 5159 | 5172 |
| 33 | 5185 | 5198 | 5211 | 5224 | 5237 | 5250 | 5263 | 5276 | 5289 | 5302 |
| 34 | 5315 | 5328 | 5340 | 5353 | 5366 | 5378 | 5391 | 5403 | 5416 | 5428 |
| 35 | 5441 | 5453 | 5465 | 5478 | 5490 | 5502 | 5514 | 5527 | 5539 | 5551 |

| $n$ | 0 | 1 | 2 | 3 | 4 | 5 | 6 | 7 | 8 | 9 |
|------|------|------|------|------|------|------|------|------|------|------|
| 36 | 5563 | 5575 | 5587 | 5599 | 5611 | 5623 | 5635 | 5647 | 5658 | 5670 |
| 37 | 5682 | 5694 | 5705 | 5717 | 5729 | 5740 | 5752 | 5763 | 5775 | 5786 |
| 38 | 5798 | 5809 | 5821 | 5832 | 5843 | 5855 | 5866 | 5877 | 5888 | 5899 |
| 39 | 5911 | 5922 | 5933 | 5944 | 5955 | 5966 | 5977 | 5988 | 5999 | 6010 |
| 40 | 6021 | 6031 | 6042 | 6053 | 6064 | 6075 | 6085 | 6096 | 6107 | 6117 |
| 41 | 6128 | 6138 | 6149 | 6160 | 6170 | 6180 | 6191 | 6201 | 6212 | 6222 |
| 42 | 6232 | 6243 | 6253 | 6263 | 6274 | 6284 | 6294 | 6304 | 6314 | 6325 |
| 43 | 6335 | 6345 | 6355 | 6365 | 6375 | 6385 | 6395 | 6405 | 6415 | 6425 |
| 44 | 6435 | 6444 | 6454 | 6464 | 6474 | 6484 | 6493 | 6503 | 6513 | 6522 |
| 45 | 6532 | 6542 | 6551 | 6561 | 6571 | 6580 | 6590 | 6599 | 6609 | 6618 |
| 46 | 6628 | 6637 | 6646 | 6656 | 6665 | 6675 | 6684 | 6693 | 6702 | 6712 |
| 47 | 6721 | 6730 | 6739 | 6749 | 6758 | 6767 | 6776 | 6785 | 6794 | 6803 |
| 48 | 6812 | 6821 | 6830 | 6839 | 6848 | 6857 | 6866 | 6875 | 6884 | 6893 |
| 49 | 6902 | 6911 | 6920 | 6928 | 6937 | 6946 | 6955 | 6964 | 6972 | 6981 |
| 50 | 6990 | 6998 | 7007 | 7016 | 7024 | 7033 | 7042 | 7050 | 7059 | 7067 |
| 51 | 7076 | 7084 | 7093 | 7101 | 7110 | 7118 | 7126 | 7135 | 7143 | 7152 |
| 52 | 7160 | 7168 | 7177 | 7185 | 7193 | 7202 | 7210 | 7218 | 7226 | 7235 |
| 53 | 7243 | 7251 | 7259 | 7267 | 7275 | 7284 | 7292 | 7300 | 7308 | 7316 |
| 54 | 7324 | 7332 | 7340 | 7348 | 7356 | 7364 | 7372 | 7380 | 7388 | 7396 |
| 55 | 7404 | 7412 | 7419 | 7427 | 7435 | 7443 | 7451 | 7459 | 7466 | 7474 |
| 56 | 7482 | 7490 | 7497 | 7505 | 7513 | 7520 | 7528 | 7536 | 7543 | 7551 |
| 57 | 7559 | 7566 | 7574 | 7582 | 7589 | 7597 | 7604 | 7612 | 7619 | 7627 |
| 58 | 7634 | 7642 | 7649 | 7657 | 7664 | 7672 | 7679 | 7686 | 7694 | 7701 |
| 59 | 7709 | 7716 | 7723 | 7731 | 7738 | 7745 | 7752 | 7760 | 7767 | 7774 |
| 60 | 7782 | 7789 | 7796 | 7803 | 7810 | 7818 | 7825 | 7832 | 7839 | 7846 |
| 61 | 7853 | 7860 | 7868 | 7875 | 7882 | 7889 | 7896 | 7903 | 7910 | 7917 |
| 62 | 7924 | 7931 | 7938 | 7945 | 7952 | 7959 | 7966 | 7973 | 7980 | 7987 |
| 63 | 7993 | 8000 | 8007 | 8014 | 8021 | 8028 | 8035 | 8041 | 8048 | 8055 |
| 64 | 8062 | 8069 | 8075 | 8082 | 8089 | 8096 | 8102 | 8109 | 8116 | 8122 |
| 65 | 8129 | 8136 | 8142 | 8149 | 8156 | 8162 | 8169 | 8176 | 8182 | 8189 |
| 66 | 8195 | 8202 | 8209 | 8215 | 8222 | 8228 | 8235 | 8241 | 8248 | 8254 |
| 67 | 8261 | 8267 | 8274 | 8280 | 8287 | 8293 | 8299 | 8306 | 8312 | 8319 |

| $n$ | 0 | 1 | 2 | 3 | 4 | 5 | 6 | 7 | 8 | 9 |
|---|---|---|---|---|---|---|---|---|---|---|
| **68** | 8325 | 8331 | 8338 | 8344 | 8351 | 8357 | 8363 | 8370 | 8376 | 8382 |
| **69** | 8388 | 8395 | 8401 | 8407 | 8414 | 8420 | 8426 | 8432 | 8439 | 8445 |
| **70** | 8451 | 8457 | 8463 | 8470 | 8476 | 8482 | 8488 | 8494 | 8500 | 8506 |
| **71** | 8513 | 8519 | 8525 | 8531 | 8537 | 8543 | 8549 | 8555 | 8561 | 8567 |
| **72** | 8573 | 8579 | 8585 | 8591 | 8597 | 8603 | 8609 | 8615 | 8621 | 8627 |
| **73** | 8633 | 8639 | 8645 | 8651 | 8657 | 8663 | 8669 | 8675 | 8681 | 8686 |
| **74** | 8692 | 8698 | 8704 | 8710 | 8716 | 8722 | 8727 | 8733 | 8739 | 8745 |
| **75** | 8751 | 8756 | 8762 | 8768 | 8774 | 8779 | 8785 | 8791 | 8797 | 8802 |
| **76** | 8808 | 8814 | 8820 | 8825 | 8831 | 8837 | 8842 | 8848 | 8854 | 8859 |
| **77** | 8865 | 8871 | 8876 | 8882 | 8887 | 8893 | 8899 | 8904 | 8710 | 8915 |
| **78** | 8921 | 8927 | 8932 | 8938 | 8943 | 8949 | 8954 | 8960 | 8965 | 8971 |
| **79** | 8976 | 8982 | 8987 | 8993 | 8998 | 9004 | 9009 | 9015 | 9020 | 9025 |
| **80** | 9031 | 9036 | 9042 | 9047 | 9053 | 9058 | 9063 | 9069 | 9074 | 9079 |
| **81** | 9085 | 9090 | 9096 | 9101 | 9106 | 9112 | 9117 | 9122 | 9128 | 9133 |
| **82** | 9138 | 9143 | 9149 | 9154 | 9159 | 9165 | 9170 | 9175 | 9180 | 9186 |
| **83** | 9191 | 9196 | 9201 | 9206 | 9212 | 9217 | 9222 | 9227 | 9232 | 9238 |
| **84** | 9243 | 9248 | 9253 | 9258 | 9263 | 9269 | 9274 | 9279 | 9284 | 9289 |
| **85** | 9294 | 9299 | 9304 | 9309 | 9315 | 9320 | 9325 | 9330 | 9335 | 9340 |
| **86** | 9345 | 9350 | 9355 | 9360 | 9365 | 9370 | 9375 | 9380 | 9385 | 9390 |
| **87** | 9395 | 9400 | 9405 | 9410 | 9415 | 9420 | 9425 | 9430 | 9435 | 9440 |
| **88** | 9445 | 9450 | 9455 | 9460 | 9465 | 9469 | 9474 | 9479 | 9484 | 9489 |
| **89** | 9494 | 9499 | 9504 | 9509 | 9513 | 9518 | 9523 | 9528 | 9533 | 9538 |
| **90** | 9542 | 9547 | 9552 | 9557 | 9562 | 9566 | 9571 | 9576 | 9581 | 9586 |
| **91** | 9590 | 9595 | 9600 | 9605 | 9609 | 9614 | 9619 | 9624 | 9628 | 9633 |
| **92** | 9638 | 9643 | 9647 | 9652 | 9657 | 9661 | 9666 | 9671 | 9675 | 9680 |
| **93** | 9685 | 9689 | 9694 | 9699 | 9703 | 9708 | 9713 | 9717 | 9722 | 9727 |
| **94** | 9731 | 9736 | 9741 | 9745 | 9750 | 9754 | 9759 | 9763 | 9768 | 9773 |
| **95** | 9777 | 9782 | 9786 | 9791 | 9795 | 9800 | 9805 | 9809 | 9814 | 9818 |
| **96** | 9823 | 9827 | 9832 | 9836 | 9841 | 9845 | 9850 | 9854 | 9859 | 9863 |
| **97** | 9868 | 9872 | 9877 | 9881 | 9886 | 9890 | 9894 | 9899 | 9903 | 9908 |
| **98** | 9912 | 9917 | 9921 | 9926 | 9930 | 9934 | 9939 | 9943 | 9948 | 9952 |
| **99** | 9956 | 9961 | 9965 | 9969 | 9974 | 9978 | 9983 | 9987 | 9991 | 9996 |

# PROBLEM ANSWERS

## Chapter 3

1a.  0.2039 $F$;  2a.  127.8 ml;  3a.  5.00 g;  4a.  5.00 g;  5a.  10.5 mmoles;
6a.  11.97 $F$;  7a.  4.178 ml;  8a.  25.45%;  9a.  127 ppm;  10a.  3.33 ml;
11a.  102.0 mmoles;  12a.  1.962 g

## Chapter 4

1a.  51.45;  2.  mean of 3.6056, median of 3.6053, average deviation of 0.0012, standard deviation of 0.00187, no elimination by $4\bar{d}$ or $3\sigma$ test, $Q$ test not possible with Table 4–1, in Table 4–2 using 1.83 at 90% data not in range 3.6066 to 3.6046 can be eliminated, using 1.83 at 90% gives 3.6056 ± 0.00099;  4.  mean of 0.3395, average deviation of 0.0035, standard deviation of 0.00146, cannot eliminate any data by mean, $\sigma$, $Q$, or $t$ test;  6. mean of 78.0, average deviation of 10.9, standard deviation of 14.5, cannot eliminate any data by mean, $\sigma$, $Q$, or $t$ text

## Chapter 7

1a.  $9.11 \times 10^{-9} M$;  1b.  $8.0 \times 10^{-5} M$;  2a.  $1.79 \times 10^{-4}$ (mole/liter)$^2$;  2b. $1.24 \times 10^{-7}$ (mole/liter)$^3$;  3b.  $4.9 \times 10^{-9}$ mole/liter;  4a.  2.53 g/500 ml;  4b. 0.265 g/500 ml;  5a.  $3.60 \times 10^{-10}$ ppm;  7.  0.138 mg/ml;  10.  28.37%;  13. 29.3 mg;  14.  3.110 mg/ml;  16.  12.31%;  17.  14.66%;  20.  93.59%; 23.  20.35 mg $CO_2$, 5.059 mg $H_2O$;  24.  24.58%

## Chapter 8

1c.  2.70;  2a.  2.79;  2b.  1.93 (approx.), 1.95 (exact);  3b.  1.37;  3d. 8.89;  4a.  8.95;  4b.  5.28;  5.  $2.5 \times 10^{-9}$;  7.  4.68;  8.  4.62;  10. 0.253;  11.  9.30;  12.  0.725 g;  15a.  1.69;  15c.  12.83;  16.  pH = 1.30 at 0%, pH = 1.37 at 10%, pH = 1.51 at 25%, pH = 1.75 at 50%, pH = 2.08 at 75%, pH = 3.51 at 90%, pH = 7.00 at 100%, pH = 11.83 at 125%, pH = 12.33 at 200%; 17.  pH = 2.41 at 0%, pH = 3.05 at 10%, pH = 3.52 at 25%, pH = 4.00 at 50%, pH = 4.48 at 75%, pH = 4.95 at 90%, pH = 8.47 at 100%, pH = 12.29 at 125%, pH = 12.78 at 200%;  20.  1.71 if $H_3PO_4$ is treated as a monoprotic acid and ionization of $H_3PO_4$ is negligible, if ionization of $H_3PO_4$ is accounted for pH = 1.81;  21.  3.98 g;  24. 10.23;  27.  $2.14 \times 10^{-3} M$;  29.  10;  31.  10.61;  33.  $3.72 \times 10^{-6}$ (mole/ liter)$^3$;  34.  0.08637 $F$;  36.  0.05720 $F$;  39.  40.41%;  40.  92.22% $Na_2CO_3$, 6.73% $NaHCO_3$;  45.  1.03%;  47.  98.91%;  49.  88.09%

## Chapter 9

1. 0.08427 $F$; 4. 45.17% sulfathiazole, 28.68% sulfapyridine; 5. 55.58%; 6. 182 mg; 8. 13.30%; 9. 97.73%; 10. 508.9 mg/tablet

## Chapter 10

1a. $Ni^{2+} + Zn = Ni + Zn^{2+}$, +0.533 V; 1e. $2Br^- + 2Fe^{3+} = 2Fe^{2+} + Br_2$, −0.319 V; 2a. +0.842 V; 2c. +0.330 V; 2e. +1.48 V; 3b. −0.063 V; 4a. $I_2 + Pb = 2I^- + Pb^{2+}$, $E_{cell} = +0.899$ V, $K = 2.32 \times 10^{22}$, spontaneous; 4c. $2MnO_4^- + 5Tl^+ + 16H^+ = 2Mn^{2+} + 5Tl^{3+} + 8H_2O$, $E_{cell} = +0.09$ V, $K = 7.10 \times 10^{38}$, spontaneous; 7a. $1.02 \times 10^{18}$; 7e. $1.67 \times 10^{-11}$; 8b. +0.570 V; 9. $1.34 \times 10^{-23}$ (mole/liter)$^2$; 12. $1.79 \times 10^{-7}$; 14. 0.104 $M$; 16. −12.18 kcal

## Chapter 11

1a. +1.422 V; 2a. 1.181 V; 3a. +0.652 V at 25%, +0.68 V at 50%, +1.06 V at 100%, +1.404 V at 125%, +1.44 V at 200%; 4a. 27.33 ml;

5b. $$E_{sp} = \frac{E^\circ_{Cr_2O_7^-, Cr^{3+}} + E^\circ_{Fe^{3+}, Fe^{2+}}}{7} - \frac{0.0592}{7} \log \frac{2[Cr^{3+}]}{[H^+]^{14}} ;$$

6. $6.29 \times 10^{-9}$ mg; 8a. $Fe(C_{12}H_8N_2)_3^{2+}/Fe(C_{12}H_8N_2)_3^{3+} = 4.89$; 8b. $Fe(C_{12}H_8N_2)_3^{2+}/Fe(C_{12}H_8N_2)_3^{3+} = 2.05 \times 10^{-2}$

## Chapter 12

1. 0.03692 $F$; 4. 105.5 mg; 5. 0.1768 $F$; 7. 2.22 mg $H_2O$/ml titrant; 9. 12.94%; 11. 24.51; 14. ~~8.84~~%; 4.42 15. 0.1646 liter; 17. 0.09317 $F$; 19. ~~91.77%~~ 45.99%

## Chapter 13

1a. 0.513 V; 2a. 1.25; 3. $7.94 \times 10^{-7}$; 6a. +5 mV; 8. 0.374%, 3.737 mg/g

## Chapter 14

1a. 10.54; 2. $pPb^{2+} = 1.60$, $pIO_3^- = 5.71$; 3. $pBa^{2+}$ = undefined and $pSO_4^{2+} = 1$ at 0 ml, $pBa^{2+} = 8.71$ and $pSO_4^{2+} = 1.18$ at 10 ml, $pBa^{2+} = 8.52$ and $pSO_4^{2-} = 1.37$ at 20 ml, $pBa^{2+} = 7.94$ and $pSO_4^{2-} = 1.95$ at 40 ml, $pBa^{2+} = pSO_4^{-2} = 4.95$ at 50 ml, $pBa^{2+} = 2.04$ and $pSO_4^{2-} = 7.85$ at 60 ml, $pBa^{2+} = 1.78$ and $pSO_4^{2-} = 8.11$ at 70 ml; 5. $5.51 \times 10^{-5}$ mg; 6. 0.08061 mole/liter; 8. 16.11%; 11. 131.6 g/liter; 13. 83.48 mg

## Chapter 15

1. 9.61%; 2. 13.97 g; 3. 0.07974 $F$; 4. 128.6 mg/tablet; 8. 225.7 mg/24 hours

## Chapter 16

1. $1.54 \times 10^8$ at pH = 3, $2.16 \times 10^{12}$ at pH = 5, $2.96 \times 10^{15}$ at pH = 7, $3.21 \times 10^{17}$ at pH = 9, $5.24 \times 10^{18}$ at pH = 11; 2. $4.86 \times 10^9$; 3. $K_{MY'}(Cd) = 7.78 \times$

$10^{10}$, $K_{MY'}(Mg) = 9.80 \times 10^2$;      5.   7.15 at pH = 2, 8.14 at pH = 4, 7.92 at pH = 6, 6.32 at pH = 8

## Chapter 17

1a.   600 nm;      1h.   $1.0 \times 10^{13}$ nm;      2a.   $1.82 \times 10^5$ cal/mole;      2d.   700 cal/mole; 4a.   $5 \times 10^{14}$ Hz;      4h.   $3.0 \times 10^4$ Hz;      5a.   61.7%;      5e.   6.17%;      6a.   0.495; 6b.   1.27;      7a.   0.742;      8a.   $7.1 \times 10^3$ liter mole$^{-1}$ cm$^{-1}$;      9a.   0.398;   10. $1.05 \times 10^{-4}$ $M$

## Chapter 19

1.   $1.75 \times 10^4$ liter mole$^{-1}$ cm$^{-1}$;      3.   $3.11 \times 10^{-5}$ mole/liter;      4.   0.0498 g/liter; 7a.   0.187;      7b.   0.324, 47.4%;      7c.   54.9 ppm;      12.   3.52 ppm;      14.   $1.98 \times$ $10^6$;      15.   5.61 mg/100 ml;      18.   1.23;      20.   $1.368 \times 10^{-3}$ $F$;      22.   $1.773 \times 10^4$ liter mole$^{-1}$ cm$^{-1}$

## Chapter 20

1.   0.020%;      4.   263 ppm

## Chapter 21

1.   $1.12 \times 10^{-3}$ mole/liter;      4.   9.33 mg;      6.   0.128 mg/liter

## Chapter 24

2.   1670 plates, 1.80 mm;      3.   0.640 $\mu$g A/$\mu$l, 0.818 $\mu$g B/$\mu$l;      5.   0.7 mg/ 5 $\mu$l;      7. 13.2 ppb

## Chapter 28

1a.   140   mg;      1b.   248   mg;      1c.   238   mg;      1d.   280   mg;      2.   50,780   coulombs;      4.   125.5 mEq/liter

## Chapter 29

1.   $4.8 \times 10^{-4}$ $M$;      2.   0.876 mg/ml;      3.   0.78 g;      4.   362 $\mu$g/ml

# Index

# ATOMIC WEIGHTS
## Based on Carbon-12

| Element | Symbol | Atomic Number | Atomic Weight | Element | Symbol | Atomic Number | Atomic Weight |
|---|---|---|---|---|---|---|---|
| Actinium | Ac | 89 | | Erbium | Er | 68 | 167.26 |
| Aluminum | Al | 13 | 26.9815 | Europium | Eu | 63 | 151.96 |
| Americium | Am | 95 | | Fermium | Fm | 100 | |
| Antimony | Sb | 51 | 121.75 | Fluorine | F | 9 | 18.9984 |
| Argon | Ar | 18 | 39.948 | Francium | Fr | 87 | |
| Arsenic | As | 33 | 74.9216 | Gadolinium | Gd | 64 | 157.25 |
| Astatine | At | 85 | | Gallium | Ga | 31 | 69.72 |
| Barium | Ba | 56 | 137.34 | Germanium | Ge | 32 | 72.59 |
| Berkelium | Bk | 97 | | Gold | Au | 79 | 196.967 |
| Beryllium | Be | 4 | 9.0122 | Hafnium | Hf | 72 | 178.49 |
| Bismuth | Bi | 83 | 208.980 | Helium | He | 2 | 4.0026 |
| Boron | B | 5 | 10.811 | Holmium | Ho | 67 | 164.930 |
| Bromine | Br | 35 | 79.909 | Hydrogen | H | 1 | 1.00797 |
| Cadmium | Cd | 48 | 112.40 | Indium | In | 49 | 114.82 |
| Calcium | Ca | 20 | 40.08 | Iodine | I | 53 | 126.9044 |
| Californium | Cf | 98 | | Iridium | Ir | 77 | 192.2 |
| Carbon | C | 6 | 12.01115 | Iron | Fe | 26 | 55.847 |
| Cerium | Ce | 58 | 140.12 | Krypton | Kr | 36 | 83.80 |
| Cesium | Cs | 55 | 132.905 | Lanthanum | La | 57 | 138.91 |
| Chlorine | Cl | 17 | 35.453 | Lead | Pb | 82 | 207.19 |
| Chromium | Cr | 24 | 51.996 | Lithium | Li | 3 | 6.939 |
| Cobalt | Co | 27 | 58.9332 | Lutetium | Lu | 71 | 174.97 |
| Copper | Cu | 29 | 63.54 | Magnesium | Mg | 12 | 24.312 |
| Curium | Cm | 96 | | Manganese | Mn | 25 | 54.9381 |
| Dysprosium | Dy | 66 | 162.50 | Mendelevium | Md | 101 | |
| Einsteinium | Es | 99 | | Mercury | Hg | 80 | 200.59 |